ICSU Short Reports Volume 3

Third European Bioenergetics Conference

Proceedings of the 3rd EBEC Meeting
Hannover, F.R.G.
September 2–7, 1984

Sponsored by the IUB-IUPAB Bioenergetics Group

Edited by

Günter Schäfer

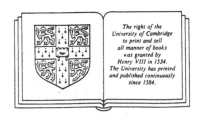

The right of the
University of Cambridge
to print and sell
all manner of books
was granted by
Henry VIII in 1534.
The University has printed
and published continuously
since 1584.

CAMBRIDGE UNIVERSITY PRESS
Cambridge
London   New York   New Rochelle
Melbourne   Sydney

CAMBRIDGE UNIVERSITY PRESS
Cambridge, New York, Melbourne, Madrid, Cape Town, Singapore, São Paulo, Delhi

Cambridge University Press
The Edinburgh Building, Cambridge CB2 8RU, UK

Published in the United States of America by Cambridge University Press, New York

www.cambridge.org
Information on this title: www.cambridge.org/9780521106825

First published 1985
This digitally printed version 2009

A catalogue record for this publication is available from the British Library

Library of Congress Catalogue Card Number: 85-47623

ISBN 978-0-521-30813-7 hardback
ISBN 978-0-521-10682-5 paperback

ORGANIZING COMMITTEE

G. SCHÄFER

E. BÄUERLEIN, B. HESS, M. KLINGENBERG, A KRÖGER, W. NEUPERT,
D. OSTERHELT, W. SEBALD, H. SIES, A. TREBST

PROGRAMME COMMITTEE

E. BÄUERLEIN, B. BEECHEY, B. CANNON, Z. DRAHOTA, D. GAUTHERON
A. GOFFEAU, J.B. JACKSON, M. KLINGENBERG, G. SCHÄFER
G.L. SOTTOCASA, K. van DAM, A. VINOGRADOV

ACKNOWLEDGEMENTS

The organizers of the Third EBEC express their thanks to the following institutions for generous support of the conference:

COMMISSION OF THE EUROPEAN COMMUNITIES - DIRECTORATE FOR SCIENCE, RESEARCH AND DEVELOPMENT

DEUTSCHE FORSCHUNGSGEMEINSCHAFT

THE GOVERNMENT OF LAND NIEDERSACHSEN

INTERNATIONAL UNION OF BIOCHEMISTRY (IUB)

INTERNATIONAL UNION OF PURE AND APPLIED BIOPHYSICS (IUPAB)

DEUTSCHER AKADEMISCHER AUSTAUSCHDIENST (DAAD)

GESELLSCHAFT FÜR BIOLOGISCHE CHEMIE

BOEHRINGER, MANNHEIM

Contents

SYMPOSIA AND COLLOQUIA LECTURES

POSTER PRESENTATIONS

DETAILED LISTING OF REPORTS

SYMPOSIA AND COLLOQUIA LECTURES

POSTER PRESENTATIONS

I. MECHANISM AND COMPONENTS OF ELECTRON TRANSPORT

II. GENERATION AND UTILIZATION OF PROTON MOTIVE FORCE

III. H^+-ATPases: STRUCTURE, FUNCTION AND REGULATION

IV. ATP-DRIVEN CATION PUMPS

VI. BIOENERGETICS OF INTEGRATED CELLULAR SYSTEMS

VII. BIOENERGETIC ASPECTS OF METABOLIC REGULATION

VIII. BIOSYNTHESIS AND MOLECULAR GENETICS
OF MEMBRANE COMPONENTS

IX. LIGHT-DRIVEN BIOENERGETIC SYSTEMS

X. SELECTED SPECIAL TOPICS

FOREWORD

The scientific significance and impact of this Third European
Bioenergetics Conference (EBEC), held at Hannover, Germany, can be
anticipated from the approximately 500 participants, compared with
250 attending the First EBEC in Urbino and almost 400 at the
Second in Lyon in 1982. The applicants numbered about 700, but
due to physical limitations, it was impossible for the Organizers
to accommodate all of them.

Thus the old adage that "there isn't two without three" is
verified. The EBEC series goes on satisfactorily and judging from
the attendance, the highly promising topics dealt with, the papers
listed in the programme and the qualifications of the chairmen and
speakers, the high standard of these meetings was surely
maintained. The structure of the conference allowed for
comprehensive and informative talks while stimulating discussion
groups provided a constructive opportunity for informal and useful
exchange of ideas, suggestions, details for our joint efforts, in
a cordial atmosphere warmed by a flow of the superb German beer.

The EBEC meetings represent the concerted expression of all
the people actively working in the field. The timing, the place,
the programme committee, the organizers and the general scientific
content are agreed upon at the general assemblies of the national
delegates of bioenergetics groups from all the world -the missing
ones are warmly invited to join the group - which are held during
the EBEC sessions and are thus directly determined by what is
really going on in the field. Then, for the next two years, all
the burden of the detailed scientific organization, fund raising
and practicalities falls on the shoulders of the Organizing
Committee, whose members are never thanked enough for the
tremendous effort they have to put in these things. But the
experience always proves successful and rewarding.

Bioenergetics is a clearly expanding field of biochemistry,
biophysics and biomedical sciences in general, Evolution has taken
millions of years to develop the sophisticated systems performing
energy conversion in biomembranes. When one considers the
complexity of these systems, which have masses of 10^5-10^6 million
daltons, it is understandable why elucidation of the detailed
mechanism of energy transfer in cells has eluded and is still
eluding us. However, as we appreciate these days, our knowledge
of molecular genetics, protein structure, kinetics and reaction

mechanisms of these systems is becoming so advanced and integrated that a speedy solution of these issues may be expected.

Paradoxically, however, as more problems are solved, more arise. We have reached the convinction that the major device for energy transfer in biomembranes is provided by proton circulation or "proticity", a basic concept of biology that is essentially due to Mitchell's far-reaching intuition, but as soon as this was realized we immediately started asking ourselves which are the molecular mechanisms by which protonic forces are generated, transmitted from one system to the other and utilized to drive endergonic processes like pyrophosphate bond formation and uphill solute transport? And we can see how easily detailed anticipations can be falsified and replaced by new ones. But this apparently is the way in which scientific knowledge seems to progress.

To what extent can vectorial catalysis account for energy transfer, which are the attributes of cooperativity in membrane proteins and their contribution to energy transfer? How has Nature learned to evolve and conserve the essential structural and functional features that we are disclosing in these intricate systems responsible for biological conversion of energy? It is highly rewarding and stimulating to see that we can ask and attempt with confidence to answer questions like these in which some of the essence of life is hidden. At this meeting we debated these problems in depth and it is really satisfactory to have reached such a stage if we consider the controversies and disputes nourished for years and the trepidation with which we started to organize these meetings.

On behalf of all of us I would like to express our gratitude to the Officers of the International Union of Biochemistry and the Union of Pure and Applied Biophysics, who have paid credit to us and have encouraged and supported with generosity our conferences. Thanks are also due to the other bodies supporting us: The Commission of the European Community; Department of Science Research and Development; the Deutsche Forschungsgemeinschaft; the Land Niedersachsen and the local authorities and firms.

..At last but not least, thanks to our German hosts, Günter Schäfer and his colleagues, for working so hard for us!

Sergio Papa
Chairman of IUB-IUPAB Bioenergetics Group

FROM THE ORGANIZERS

The successful 1st and 2nd EBEC meetings held in Urbino (1980) and Lyon (1982) have made this joint venture of the European Bioenergetics Study Groups one of the most important scientific events in the field of bioenergeties.

Under the auspices of the IUB-IUPAB Bioenergetics Group it was agreed that the 3rd EBEC meeting would take place in 1984 in Germany. The city of Hannover, centrally located, was chosen as a place with excellent facilities. There the meeting was organized by the German Bioenergetics Study Group as an executive body of the Gesellschaft fur Biologische Chemie, which took the risk and responsibility of the enterprise.

Although it has become increasingly difficult to find sponsorship and travel grants for large international conferences, the interest in the 3rd meeting was overwhelming. Despite the increase in registration fees we had about 500 active conference members participating. Unfortunately technical reasons like accommodations, prevented us from accepting more applicants.

Due to the continuous expansion of our field, it has become impossible to cover bioenergetics in a single meeting. Rather each EBEC offers the advantage to the organizers to stress specific focal points and to give an individual character to a particular event. The 3rd EBEC tried to present symposia, colloquia and discussion groups - replacing the former "Round Tables" - from unifying points of view, avoiding classical separation of fields like photosynthesis and respiration etc.

Microbial bioenergetics forms a substantial topic of this EBEC because these systems have provided most successful experimental models during the past several years. A further topic, the relation of protein structure and function, was selected for the symposia sessions. The selection of speakers strictly followed the rule suggested by the IUB-IUPAB program committee, that of giving preference to those who did not present lectures at the preceding EBEC. Whereas the colloquia of communicated papers are closely linked to symposia topics, the poster topics are more closely related to the various discussion groups.

As an innovation the "EBEC-Reports" are to be officially published shortly after the meeting in the ICSU Short Reports series. This makes the communications a quotable publication, following an obvious demand (many quotations of EBEC Reports are already found in the literature) and underlines the importance of the meeting.

The chairman of the board thanks all members of the committees for their valuable assistance and all colleagues who contributed actively to make the 3rd EBEC a successful scientific event

Greetings from Hannover!

Günter Schäfer
for the organizers of the 3rd EBEC

ICSU SHORT REPORTS:
A MESSAGE FROM THE ICSU PRESS

The ICSU Press was established in 1983 as the publishing house of the International Council of Scientific Unions. Our aims include innovations in publication and we are pleased with the reception accorded to the Short Reports, of which this is Volume 3, produced in association with Cambridge University Press.

The ICSU Short Reports address a problem associated with the widespread use of posters to present short communications at scientific meetings. These posters are extremely informative. Their presenters go to a great deal of pains to put their message across in a compact, easily understandable fashion, but the poster is displayed for no more than half a day and in a large meeting often is not seen even by participants in whose area of research it falls. All that remains for posterity is the abstract that was prepared many months before the meeting, is out of date at the time of the meeting and contains but a fraction of the information imparted by the poster.

The reporting format used here is an attempt to allow the presenter to commit the poster to paper. The two-page spread can be used to convey a great deal of information by way of introduction, methodology, results, conclusions, bibliography, diagrams, figures and tables. Produced to a very tight deadline, the short report acts as an up-to-date, permanent record of the poster, and the immediate availability of the volume of short reports to libraries and persons who could not attend the meeting makes this a very timely and we hope effective method of publication. In the same way the short report format is used, as in this volume, to allow the invited speakers also to present a digest of their contribution in a timely and informative manner and which, through the bibliography, directs the person who wishes to explore the subject in more depth to the original seminal papers on which the speaker's contribution is based.

It is a pleasure to acknowledge that this reporting concept was brought to the ICSU Press by a prominent member of the Board of Management, Professor Lars Ernster, who at the same time is the Secretary General of ICSU.

We thank all the contributors to this third volume of the ICSU Short Reports. When referring to any paper from this volume you should use the citation:

ICSU Short Reports, 3 (1985) first page-last page.

Symposia and Colloquia Lectures

ANAEROBIC ELECTRON TRANSPORT IN BACTERIAL ENERGY CONVERSION

A.H. Stouthamer
Biological Laboratory, Vrije Universiteit, Postbus 7161,
1007 MC Amsterdam, The Netherlands

Bacteria can utilize a large variety of compounds as terminal electron acceptor instead of oxygen, e.g. oxidized nitrogen compounds. The first step in nitrate dissimilation is a reduction to nitrite. Two ways of dissimilatory nitrite reduction can be distinguished. Nitrite can be converted to gaseous products (nitrous oxide or nitrogen) or to ammonia. The ATP gain during these reductions is dependent on: 1. the respiratory chain for electron transport towards these nitrogenous oxides. 2. the side of the membrane at which the reduction takes place. 3. the number of proton pumps present in the respiratory chain of the organism. Generally cytochrome b is the electron donor for the reduction of nitrate and cytochrome c the electron donor for the reduction of nitrite (to nitrous oxide or ammonia) and of nitrous oxide. The reduction of nitrate always occurs at the cytoplasmic side of the membrane. However nitrite and nitrous oxide are reduced at the periplasmic side of the membrane. By a reduction at the periplasmic side of the membrane the charge separation is reduced. Oxidation of some substrates may occur at the periplasmic side of the membrane, by which the charge separation is increased. Electron transfer and proton translocation are shown in Fig. 1 for Paracoccus denitrificans and Campylobacter sputorum. The site 2 region in the electron transport chain in P. denitrificans has been split in two proton-translocating segments. During electron transport to nitrate site I and IIa and during electron transport to oxygen, nitrite and nitrous oxide site I, IIa and IIB are passed. After aerobic growth sometimes also site III is present. In Escherichia coli the $\rightarrow H^+/O$ and $\rightarrow H^+/NO_3^-$ ratios are 4 and measurements of molar growth yields indicate that the stoichiometry of ATP formation is the same with oxygen and nitrate. E. coli contains site I and IIa which function in electron transport to oxygen and nitrate. In C. sputorum only site IIa is present. The charge separation during electron transport in these organisms is shown in Table 1. The results are in accordance with calculations of the stoichiometric ATP gain from molar growth yields using a $\rightarrow H^+/ATP = 3$. These findings can not be explained easily by a Q cycle model for electron transfer and energy generation. For this purpose a b cycle model seems more appropriate.

References
1. Boogerd, F.C., van Verseveld, H.W. and Stouthamer, A.H. (1982). Biochim. Biophys. Acta. 723, 415-427.
2. De Vries, W., van Berchum, H. and Stouthamer, A.H. (1984). Ant. van Leeuwenhoek 50, 63-73.

3. Stouthamer, A.H., van 't Riet, J. and Oltmann, L.F. (1980) in Diversity of bacterial respiratory systems (Knowles, C.J., ed) Vol. 2, pp. 19-49. CRC Press, Boca Raton, Florida, U.S.A.

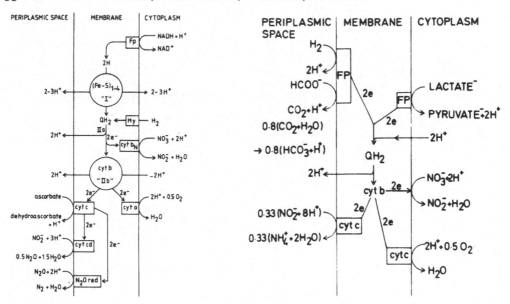

Fig. 1. Simplified scheme for proton translocation and electron transport in P. denitrificans (left) and C. sputorum (right). Fp, flavoprotein, QH_2, ubiquinol; Q, quinone; Hy, hydrogenase; Fe-S, iron-sulphur center.

Table 1. Number of charges translocated across the cytoplasmic membrane during flow of $2e^-$ ($\rightarrow q^+/2e^-$ ratio) from various substrates to an electron acceptor in P. denitrificans, E. coli and C. sputorum.

			$\rightarrow q^+/2e^-$			
	P. denitrificans		E. coli		C. sputorum	
Electron acceptor reaction	NADH	H_2 or succinate	NADH or	lactate formate	lactate	H_2 or formate
$O_2 \rightarrow H_2O$	7	4	4	2	2	4
$O_2 \rightarrow H_2O$(cyt aa$_3$)	9	6	-	-	-	-
$NO_3^- \rightarrow NO_2^-$	5	2	4	2	2	4
$NO_2^- \rightarrow \frac{1}{2}N_2O$	5	2	-	-	-	-
$N_2O \rightarrow N_2$	5	2	-	-	-	-
$NO_2^- \rightarrow NH_4^+$	-	-	-	-	0	2

METHANOGENESIS

R. Thauer
Mikrobiologie, FB Biologie, Philipps-Universität
Marburg, D-3550 Marburg/Lahn (FRG)

Methanogenic bacteria ferment H_2 and CO_2, formate, methanol, methylamines and/or acetate to methane. Per mol methane 0.5 to 1.0 mol ATP is generated. Coupling is propably via the chemiosmotic mechanism (1). Sodium ions are somehow involved (2). The exergonic step of methanogenesis is catalyzed by methyl-CoM reductase which contains a nickel porphinoid as prosthetic group (3-6).

References:
1. Blaut, M. and Gottschalk, G. (1984) Eur. J. Biochem., in press.
2. Schönheit, P. and Perski, H.J. (1983) FEMS Letters 20, 263-267
3. Ellefson, W.L., Whitman W.B. and Wolfe, R.S. (1982) Proc. Natl. Acad. Sci. USA, 79, 3707-3710.
4. Livingston, D.A., Pfaltz, A., Schreiber, J., Eschenmoser, A., Ankel-Fuchs, D., Moll, J., Jaenchen, R. and Thauer,R.K. (1984). Helv. Chim. Acta, 67, 334-351.
5. Ankel-Fuchs, D., Jaenchen, R., Gebhardt, N.A. and Thauer,R.K. (1984) Arch. Microbiol., in press.
6. Thauer, R.K., Brandis-Heep, A., Diekert, G., Gilles, H.-H., Graf, E.-G., Jaenchen, R. and Schönheit, P. (1983) Naturwissenschaften, 70, 60-64.

BIOENERGETIC PROBLEMS OF ALKALOPHILIC BACTERIA

T.A. Krulwich and A.A. Guffanti
Department of Biochemistry, Mount Sinai School of Medicine
of the City University of New York,
New York, New York   10029 USA

Alkalophiles live at external pH values of 10.5 and higher, but maintain a cytoplasmic pH at $\leq$ pH 9.5. While the $\Delta\psi$, positive out, is substantial, the total $\Delta\bar{\mu}_H+$ is low at optimal pH values because of the large "reversed" ΔpH [1].

Several lines of evidence indicate a crucial role for Na^+ in pH homeostasis [2]. Na^+ may function via an electrogenic Na^+/H^+ antiporter in respiring cells. The antiporter is inhibited by $[H^+]_{in}$ and is, accordingly, relatively inactive at neutral pH. Indeed, pH homeostasis in the neutral pH range may be a problem for most alkalophiles. In highly Na^+-dependent alkalophiles, e.g., B. firmus RAB, Na^+/H^+ antiport activity may have an exclusive role in "acidifying" the interior. Na^+ circulation is completed by a plethora of $Na^+/$solute symporters. Na^+ is the major coupling ion for transport. The K_m for Na^+-dependent symporters correlates with that of the Na^+/H^+ antiporter in a given species. In less Na^+-dependent species, e.g., B. alcalophilus, there may be an auxiliary role for a K^+/H^+ antiporter.

The membranes of alkalophilic bacilli contain enormous quantities of respiratory chain components, especially large amounts of b- and c-type cytochromes. There are several b-species, a single c-cytochrome, and cytochromes aa_3, as determined by midpoint potentials. A Rieske Fe-S protein, at the same potential as the cytochrome c, is present in B. alcalophilus [3,4]. Cytochrome oxidase purified from B. firmus RAB contains subunits of 56K and 40K as well as a mole of bound cytochrome c. Cells of B. firmus RAB, using endogenous substrates exhibit high H^+/O ratios at pH 9.0. The H^+/O ratios are much lower at neutral pH [5]. Observations of structural/functional features of the alkalophile respiratory chain suggest that this chain is especially adapted to function well at high pH.

ATP synthesis via oxidative phosphorylation occurs at low $\Delta\bar{\mu}_H+$ values in cells and in ADP + P_i-loaded vesicles. No requirement for or stimulation by Na^+ is found [1]. Rather, an H^+-translocating F_1F_0-ATPase appears to function at low $\Delta\bar{\mu}_H+$ values, as long as that $\Delta\bar{\mu}_H+$ is derived from respiration. Artificial $\Delta\psi$ values of comparable magnitudes to those formed by respiration do not energize ATP synthesis at pH 9.0. Even at pH 7.0, respiration is more efficacious than a comparable, valinomycin-induced K^+ diffusion $\Delta\psi$ in energizing ATP synthesis [6]. A localized pathway for H^+ movements between pumps and sinks may be particularly crucial for alkalophiles and, hence, particularly amenable to study (e.g., genetically) in these organisms.

Current studies are directed at genetic as well as biochemical approaches to the interesting questions of alkalophile bioenergetics. It has been possible, thus far, to transform B. firmus RAB to antibiotic resistance using plasmids from Bacillus subtilis.

References

1. Krulwich, T.A. and Guffanti, A.A. (1983) Adv.Microbial.Physiol. 24, 173-214.
2. Krulwich, T.A. (1983) Biochim.Biophys.Acta 726, 245-264.
3. Lewis, R.J., Prince, R., Dutton, P.L., Knaff, D. and Krulwich, T.A. (1981) J.Biol.Chem. 256, 10543-10549.
4. Kitada, M., Lewis, R.J. and Krulwich, T.A. (1983) J.Bacteriol. 154, 330-335.
5. Lewis, R.J., Krulwich, T.A., Reynafarje, B. and Lehninger, A.L. (1983) J.Biol.Chem. 258, 2109-2111.
6. Guffanti, A.A., Fuchs, R.T., Schneier, M., Chiu, E. and Krulwich, T.A. (1984) J.Biol.Chem. 259, 2971-2975.

SODIUM ION TRANSPORT AS PRIMARY ENERGY SOURCE FOR ATP
SYNTHESIS

P. Dimroth and W. Hilpert
Institut für Physiologische Chemie der TU München
Biedersteiner Straße 29, 8000 München 40, FRG.

In certain anaerobic bacteria, Na^+ transport is driven
by decarboxylation reactions [1]. These decarboxylations
are indispensible steps of the respective fermentation
pathways and proceed by a large negative free energy change
(6-7 kcal/mol) which provides the driving force for the
transport. In this way, part of the decarboxylation energy
is conserved and can be used by the cell to drive endergonic
membrane reactions. Three different enzymes have been recognized to function as sodium transport decarboxylases.
These are oxaloacetate decarboxylase from Klebsiella aerogenes [2], methylmalonyl-CoA decarboxylase from Veillonella
alcalescens [3] and glutaconyl-CoA decarboxylase from Acid-
aminococcus fermentans [4].

Growing cells of V. alcalescens maintain a Na^+ concentration gradient of about 1:10 from inside to outside,
probably due to the action of methylmalonyl-CoA decarboxylase. The usefulness of this Na^+ gradient as a driving force
for endergonic chemical synthesis has been demonstrated by
the reversion of malonyl-CoA decarboxylation with a reconstituted proteoliposomal system. Malonyl-CoA was formed from
acetyl-CoA and Co_2 when a Na^+ concentration gradient inside
> outside was applied but not after dissipating the Na^+
gradient with monensin. The Na^+ gradient could either be
imposed by diluting Na^+-loaded vesicles into a Na^+-free
buffer or by the Na^+ pump oxaloacetate decarboxylase with
proteoliposomes containing oxaloacetate decarboxylase in
addition to methylmalonyl-CoA decarboxylase. The latter
system is infact a transcarboxylase in catalyzing the
formation of pyruvate and malonyl-CoA from oxaloacetate and
acetyl-CoA and vice versa. This transcarboxylation strictly
depends upon the circulation of Na^+ ions and is thus completely different from the classical transcarboxylase from
Propionibacterium shermanii. With the reconstituted proteoliposomal systems decarboxylation and Na^+ uptake were
tightly coupled, so that per decarboxylation of 1 mol oxaloacetate or malonyl-CoA 2 mol Na^+ ions were translocated
through the membrane.

In certain organisms, a decarboxylation derived Na^+ gradient is the only source of energy to support ATP synthesis. As an example, Propionigenium modestum, a strictly anaerobic bacterium, grows from the fermentation of succinate to propionate and Co_2 [5]. Intermediates of the fermentation pathway are succinyl-CoA and methylmalonyl-CoA. Decarboxylation of methylmalonyl-CoA to propionyl-CoA proceeds by generation of a Na^+ gradient which is the only biologically useful energy gained in this fermentation process. This Na^+ gradient in turn drives ATP synthesis via a Na^+ stimulated ATPase which is present in high amounts in the bacterial membrane. These decarboxylation-dependent energy conservations are distinct from all other membrane-linked energy conservation mechanisms since no electron transport chains are involved and since Na^+ and not H^+ is functioning as the coupling ion.

1. Dimroth, P. (1982) Biosci. Rep. 2, 849-860.

2. Dimroth, P. (1980) FEBS Lett. 122, 234-236.

3. Hilpert, W. & Dimroth, P. (1982) Nature (Lond.) 296, 584-585.

4. Buckel, W. & Semmler, R. (1982) FEBS Lett. 148, 35-38.

5. Schink, B. & Pfennig, N. (1982) Arch. Microbiol. 133, 209-216.

THE REGULATION OF SOLUTE TRANSPORT IN BACTERIA

W.N. Konings, M.G.L. Elferink, K.J. Hellingwerf, B. Poolman,
J.M. van Dijl
Dept. of Microbiology, University of Groningen, Kerklaan 30,
9751 NN HAREN, The Netherlands

Solute transport across the cytoplasmic membrane of bacteria can
occur via several mechanisms: the primary transport systems, which
convert chemical or light energy into electrochemical energy; the
secondary transport systems which convert one form of electrochemical energy into another form and the group translocation systems
which modify the solute chemically during the translocation process
(1). The primary transport systems comprise the systems which generate an electrochemical gradient of protons (or sodium) such as the
membrane-bound proton translocating ATPase and the cytochrome-linked
electron transfer systems. The secondary transport systems are
driven by the electrochemical gradients.

The proton motive force or its components, the membrane potential
and the pH-gradient across the cytoplasmic membrane has been demonstrated to be the driving force for the uptake of many solutes in
bacteria (1). Several solutes are translocated by proton-solute symport systems but the number of protons and charges translocated with
the solute can vary (2). This variable proton-solute stoichiometry
has been studied extensively for lactate transport in Escherichia
coli (3,4) and Streptococcus cremoris (5,6). These results have
supplied experimental support for the Energy Recycling Model which
postulates that the efflux of endproducts in fermentative bacteria
leads to the generation of a proton motive force and supplies metabolic energy to the organisms (7).

Recent studies in Rhodopseudomonas sphaeroides (8,9) and E. coli
(10) have shown that the rate of secondary transport processes is
determined by the proton motive force but in addition independently
of the proton motive force by the rate of electron transfer in the
electron transfer systems: the cyclic electron transfer system in
Rps. sphaeroides grown under anaerobic conditions in the light and
the respiratory chain in E. coli and Rps. sphaeroides grown under
aerobic conditions in the dark. Kinetic analysis of solute uptake
indicates that the activity of the electron transfer systems determines the fraction of active transport carrier molecules in the
cytoplasmic membrane (9). A similar regulation by the rate of electron transfer has been reported for photophosphorylation in Rps.
capsulata (11) and Rps. sphaeroides (12).

The regulation by electron transfer is exerted on proton solute-

and on sodium-solute symport systems which argues against the involvement of localized chemiosmotic phenomena in this process.

Studies in E. coli have shown that the membrane-bound protein Enzyme II of the PEP-dependent sugar group translocation system (13,14) and the carrier proteins of lactose and proline (15,16) contain redox-active disulphide groups. The redox state of these groups is determined by the redox potential of the environment and by the proton motive force. It is very likely that the activity of these proteins is also controlled by a redox interaction with (a) components(s) of the electron transfer system and that this direct interaction can explain the regulation of the transport activity by the activity of the electron transfer systems.

REFERENCES

1. Konings, W.N. and Michels, P.A.M. (1980) in Diversity of Bacterial Respiratory Systems (Knowles, C.J., ed.), pp. 33-86, CRC Press, Boca Raton
2. Konings, W.N. (1980) Trends in Biochem.Sci. 2, 257-262
3. Brink, B. ten and Konings, W.N. (1980) Eur.J.Biochem. 111, 59-66
4. Brink, B. ten, Lolkema, J.S., Hellingwerf, K.J. and Konings, W.N. (1981) FEMS Microb.Lett. 12, 237-240.
5. Otto, R., Sonnenberg, A.S.M., Veldkamp, H. and Konings, W.N. (1980) Proc.Natl.Acad.Sci. USA 77, 5502-5506
6. Brink, B. ten and Konings, W.N. (1982) J.Bacteriol. 152, 682-686
7. Michels, P.A.M., Michels, J.P.J., Boonstra, J. and Konings, W.N. (1979) FEMS Microb.Lett. 5, 357-364
8. Elferink, M.G.L., Friedberg, I., Hellingwerf, K.J. and Konings, W.N. (1983) Eur.J.Biochem. 129, 583-587
9. Elferink, M.G.L., Hellingwerf, K.J., Nano, F.E., Kaplan, S. and Konings, W.N. (1983) FEBS Lett. 164, 185-189
10. Elferink, M.G.L., Hellingwerf, K.J., Belkum, M.J. van, Poolman, B. and Konings, W.N. (1984) FEMS Microb.Lett. 21, 293-298
11. Baccarini-Melandri, A., Casadio, R. and Melandri, B.A. (1977) Eur.J.Biochem. 78, 389-402
12. Venturoli, G. and Melandri, B.A. (1982) Biochim.Biophys.Acta 680, 8-16
13. Robillard, G.T. and Konings, W.N. (1981) Biochemistry 20, 5025-5032
14. Robillard, G.T. and Konings, W.N. (1982) Eur.J.Biochem. 127, 597-604
15. Konings, W.N. and Robillard G.T. (1982) Proc.Natl.Acad.Sci. USA 79, 5480-5484
16. Poolman, B., Konings, W.N. and Robillard, G.T. (1983) Eur.J.Biochem. 134, 41-46

PHYLOGENETIC RELATIONS OF SULFATE REDUCING BACTERIA

M. Bruschi
Laboratoire de Chimie Bactérienne, C.N.R.S.,
B.P. 71, 13277 Marseille Cedex 9, France

Sulfate reducing bacteria are strict anaerobes which are still able to perform today a very ancient process : the dissimilatory reduction (or respiration) of sulfate. It has been shown that this biological reduction of sulfates did occur as far back as 1 to 2.5 billion years. Reduction of sulfate into hydrogen sulfide necessitates the presence of a sophisticated set of electron carriers (different cytochromes c and non haem iron proteins).

Amino acid sequences and three dimensional structures comparisons obtained for some of these electron transport proteins are a powerful way of obtaining informations about the evolutionary history of the organisms.

Cytochrome c_3 can be defined as an original group of c type cytochromes characterized by a very low redox potential and the presence of four haems per molecule for a molecular weight comparable to mitochondrial cytochrome c. Alignment of the amino acid sequences of cytochrome c_3 from six different sulfate reducing bacteria shows only 25 % of homology (1). It should be postulated that, very early during the evolution of the molecule, duplication of the gene has taken place.

The three dimensional structure of two different cytochromes c_3 shows the same folding of the molecule with a core of non parallel haems (presenting a relatively high exposure to the solvent).

When compared to eucaryotic cytochrome c, this class of multihaemic cytochrome c probably represents evolutionary convergence with the same mode of attaching the haem group rather than divergence from a common ancestor.

A monohaemic cytochrome c, cytochrome c_{553} from Desulfovibrio vulgaris (Hildenborough) (2) seems to be evolutionary homolog to some monohaemic bacterial cytochrome c and to eucaryotic cytochrome c. An evolutionary tree has been proposed by Dickerson which indicates a common origin for electron transport chain in photosynthesis and respiration (3). The earliest bacteria were anaerobic fermenters resembling present day Clostridia then the topping of solar energy was carried out by the ancestors of green sulfur bacteria such as Chromatium and Chlorobium, with the use of H_2S and release of sulfate. With the availability of sulfate, a sulfur cycle could have developed between sulfide photosynthesizers and sulfate reducing bacteria.

Examination of the sequence of rubredoxins (4,5) and ferredoxins (6,7,8) has provided an independant set of evolutionary data between diverse groups of bacteria (9) including Clostridia, methanogenic and photosynthetic bacteria.

REFERENCES

(1) Bruschi M. (1981) Biochim. Biophys. Acta, 671, 219-226.
(2) Bruschi, M. and Le Gall J. (1972) Biochim. Biophys. Acta, 271, 48-60.
(3) Dickerson R.E., Timkovich R. and Almassy R.J. (1976) J. Mol. Biol. 100, 473-491.
(4) Bruschi, M. (1976) Biochim. Biophys. Acta, 434, 4-17.
(5) Bruschi, M. (1976) Biochem. Biophys. Res. Commun. 2, 615-621.
(6) Bruschi, M. (1979) Biochem. Biophys. Res. Commun., 91, 623-628.
(7) Bruschi M. and Hatchikian E.C. (1982) Biochimie, 64, 503-507.
(8) F. Guerlesquin, M. Bruschi, G. Bovier-Lapierre, Bonicel, J. and P. Couchoud (1983) Biochimie, 65, 43-47.
(9) Vogel, H., Bruschi M. and Le Gall J. (1977) J. Mol. Evol., 9, 111-119.

THE FUNCTION OF THE HYDROPHOBIC SUBUNITS (frdC AND frdD) OF
ESCHERICHIA COLI FUMARATE REDUCTASE
Joel H. Weiner, M. Lynn Elmes, David J. Latour and Bernard D. Lemire
Department of Biochemistry, University of Alberta
Edmonton, Alberta, Canada T6G 2H7

Fumarate reductase serves as the terminal electron transfer enzyme
when *Escherichia coli* is grown anaerobically on glycerol plus fumarate medium. The holoenzyme is a membrane-bound tetramer composed of
four non-identical subunits in equimolar ratios. The subunits are
encoded by an operon with four open reading frames corresponding to
the frdA, frdB, frdC and frdD genes which code for polypeptides of
69, 27, 15 and 13 Kd respectively (1). This operon has been cloned
into a multicopy vector and used to amplify high levels of reductase.
The 69 Kd and 27 Kd subunits form a membrane-extrinsic iron-sulfur
flavoenzyme. This catalytic dimer is capable of reducing fumarate
using artificial electron donors and has been purified by hydrophobic
exchange chromatography and characterized (2). We have previously
shown that this catalytic dimer does not associate with the membrane
or with phospholipids in the absence of the two small subunits.

Three lines of evidence have led us to propose that the two small
subunits are membrane intrinsic anchors for the catalytic dimer.
(i) The sequences of the small subunits are extremely hydrophobic and
analysis for hydrophobic stretches showed that each contained three
regions of about 20 amino acids capable of crossing the membrane as
α-helical segments (3). (ii) A plasmid coding for only the frdA and
frdB genes produces large amounts of soluble, cytoplasmic fumarate
reductase (4). (iii) Membranes which have been treated with 6M urea
to remove the catalytic dimer still contain the two small subunits
which are capable of binding fresh catalytic dimers and reconstituting
a functional membrane-bound tetramer.

In addition to serving an anchoring role we have shown that the
hydrophobic polypeptides serve to stabilize the catalytic activity of
the dimer. The dimer is rapidly denatured by incubation at alkaline
pH (pH 8.8) or temperatures above 40 C. The membrane-bound form and
tetramer, however are quite stable to these conditions. Further,
while the dimer requires anions for maximal activity, the tetramer
does not. Together these results indicate that the two small subunits
induce conformational changes in the catalytic dimer resulting in
optimal stability and activity.

Still another role for the small hydrophobic subunits is to participate in lipid-protein and protein-protein interactions which give
rise to a novel membranous organelle composed of a helical array of

tetrameric fumarate reductase and lipid. This structure appears as a result of dramatic overproduction of the reductase in plasmid carrying hosts.

To further probe the roles of the individual anchor subunits, we have isolated a deletion of the frdD gene generated by Bal31 digestion and a transposon Tn5 insertion into the frdD gene. These modifications result in accumulation of soluble fumarate reductase activity in the cytoplasm indicating that the frdD polypeptide plays a role in membrane attachment and that frdC is not sufficient to bind the catalytic dimer to the membrane.

A nitrosoguanidine generated mutation in the chromosomal frdC gene results in a nonfunctional fumarate reductase (cells do not grow anaerobically on fumarate). This mutation has been cloned into a multicopy vector and the cells contain high levels of active, membrane-bound reductase. Unlike the wild-type enzyme, this activity is thermolabile and very sensitive to protease degradation. Together these results indicate that the frdC subunit performs a role in stabilizing the catalytic dimer and is required for the enzyme to function in the glycerol-fumarate electron transport chain.

References:

(1) Cole, S.T., T. Grundstrom, B. Jaurin, J.J. Robinson and J.H. Weiner (1982) Eur. J. Biochem., 126, 211-216.
(2) Dickie, P. and J.H. Weiner (1979) Can. J. Biochem., 57, 813-821.
(3) Weiner, J.H., B.D. Lemire, R.W. Jones, W. Anderson and D.G. Scraba (1984) J. Cell. Biochem. (in press).
(4) Weiner, J.H., B.D. Lemire, M.L. Elmes, R.D. Bradley and D.G. Scraba (1984) J. Bacteriol. (in press).

Supported by the Medical Research Council of Canada and The Alberta Heritage Foundation for Medical Research.

FNR PROTEIN - THE GENE ACTIVATOR PROTEIN FOR ANAEROBIC ELECTRON TRANSPORT IN ESCHERICHIA COLI

G. Unden° and J.R. Guest*
° Mikrobiologie, Fachbereich Biologie, Philipps-Universi-
tät Marburg, 3550 Marburg, FRG
* Department of Microbiology, Sheffield University,
Western Bank, Sheffield S10 2TN, England

Anaerobic electron transport systems like nitrate and fumarate respiration in E. coli are expressed only under anaerobic conditions. Their expression depends on the presence of the Fnr protein, the product of the fnr gene (1). The fnr gene has been cloned and the primary structure has been deduced from the nucleotide sequence (2). The protein sequence revealed a striking homology between the Fnr protein and CRP (3), the cyclic AMP receptor protein, which mediates catabolite repression by binding to cata- bolite-sensitive genes. The homology suggests that the Fnr protein may function like CRP to activate genes con- cerned with anaerobic respiration.

The homology of Fnr to CRP implied the presence of a nucleotide binding site in Fnr. To investigate the possible involvement of different nucleotides we have tested rele- vant mutants for the effect on the expression of anaerobic electron transport systems. In mutants lacking adenylate cyclase the fumarate respiratory system was completely dependent on the presence of exogenous cyclic AMP, whereas CRP was not required (4). This may suggest that cyclic AMP acts as the effector of the Fnr protein, thereby assigning a second function to cyclic AMP apart from the control of the catabolite repression.

To obtain an in vitro system for studying the operation of the Fnr protein, we have purified the protein to homogeneity. The identity of the protein as the product of fnr was confirmed by determining the aminoterminal sequence. The isolated Fnr protein is a monomer of M_r 29 000, whereas it is most likely a dimer in the bacteria. In agreement with the postulated function as a gene acti- vator protein, Fnr showed DNA binding activity.

The proposed function of cyclic AMP as the effector

molecule of Fnr suggests the involvement of an additional
control system, which enables a different activation of
CRP and of the Fnr protein. Fnr contained 2-3 reactive
cysteine residues making the protein possibly itself redox
sensitive and its activity dependent on the redox state of
the bacterium. As a consequence the Fnr protein would
require anaerobic conditions plus cyclic AMP to activate
transcription of genes involved in anaerobic respiration.

1. Lambden, P.R. and Guest, J.R. (1976) J. Gen. Microbiol.
   97, 145-160
2. Shaw, D.J. and Guest, J.R. (1982) Nucleic Acids Res.
   6119-6130
3. Shaw, D.J., Rice, D.W. and Guest, J.R. (1983) J. Mol.
   Biol. 166, 241-247
4. Unden, G. and Guest, J.R. (1984) FEBS Lett., in press

ASPECTS OF ELECTRON FLOW DISTRIBUTION IN *PARACOCCUS DENITRIFICANS*

I. Kučera, R. Matyášek, and V. Dadák
Department of Biochemistry, J. E. Purkyně University,
61137 Brno, Czechoslovakia

The report deals with the distribution of electron flow among the terminal acceptors nitrite, nitrous oxide and oxygen in the cells of *P. denitrificans* grown anaerobically on succinate.

1) During the reduction of NO_2^- under anaerobic conditions N_2O is accumulated in the medium only when the pH of the reaction mixture is lower than 7.3-7.4. At pH 6.4 an almost stoicheiometric conversion of NO_2^- to N_2O was found followed by a rapid reduction of N_2O to N_2. If the pH value of the medium exceeds 7.4 N_2O does not accumulate. By the inhibition of nitrous oxide reductase with acetylene and by following the time course of the sum of NO_2^- and a double of N_2O concentrations it was proved that no other nitrogenous intermediate accumulates markedly. When using a natural respiratory substrate (succinate) the pH optimum of the nitrite reductase activity of cells was found between 7.8-8.2, this value differing clearly from the pH optimum at 6.2 observed in the case of using an artificial electron donor TMPD plus ascorbate. No similar shift of pH optimum depending on the used substrate was found with nitrous oxide reductase, which shows its optimum activity between 8.0-8.7 of pH. The degree of cyt *c* reduction at pH < 7.3 was higher with N_2O than with NO_2^- ; the opposite is true when working at pH > 7.3. The obtained experimental results can be explained :
i) by increased electron flow into the terminal part of the respiratory chain caused by increasing the pH,
ii) by different pH optima for nitrite and nitrous oxide reductases,
iii) by a mutual influencing of activities of both enzymes via the redox state of cyt *c* as a common electron donor.

2) In a previous work [1] we found that the preferential use of O_2 over NO_2^- was abolished in the presence of an uncoupler. On further examining this effect we found that it was also observable when working with an artificial electron donor - TMPD plus ascorbate. The effect of the uncoupler can be substituted by adding 20 mM tetraphenylphosphonium chloride into the reaction mixture thus indicating the necessity of the $\Delta\Psi$ decrease for switching the electron flow. The shape of the dependence of oxygen concentration on time is markedly influenced by the used concentration of nitrite. Under certain conditions two distinct inhibitory phases can be distinguished, separated by a phase of enhanced oxygen consumption. Anaerobically

grown cells of *P. denitrificans* exert an inhibitory effect on the oxidase activity (TMPD as substrate) of membrane vesicles derived from aerobically grown cells in the presence of nitrite and an uncoupler. The utilization of nitrite (and probably of nitrous oxide) under aerobic conditions accompanied by the inhibition of oxygen consumption can be achieved even at a small decrease in the activities of terminal oxidases caused by hydroxylamine similarly as it has been found in the case of nitrate reduction [2]. In interpreting the results we suppose:

i) a redistribution of nitrite between the cell and its surroundings when the $\Delta\Psi$ is lowered and the inhibition of terminal oxidases by nitrite (cf. [1]),

ii) the redistribution of the electron flow among oxygen, nitrite and nitrous oxide which is formed in the reaction mixture,

iii) an assumed inhibition of terminal oxidases due to the transiently formed nitrogenous intermediate of not yet identified nature.

References:

[1] Kučera,I. and Dadák,V. (1983) Biochem. Biophys. Res. Commun. 117, 252-258

[2] Kučera,I., Karlovský,P. and Dadák,V. (1981) FEMS Microbiol. Lett. 12, 391-394

MECHANISMS OF SELECTION OF ELECTRON TRANSFER PATHWAYS IN BACTERIA

S.J. Ferguson

Department of Biochemistry, University of Birmingham, P.O. Box 363, Birmingham, B15 2TT, England.

The electron transfer chains of many bacteria can use physiological electron acceptors other than oxygen. Such bacteria often are able to select one acceptor preferentially. This report summarises recent findings with Paracoccus denitrificans and Rhodopseudomonas capsulata.

The electron transfer chain of the denitrifying bacterium P. denitrificans can, after growth of cells under denitrifying conditions, use four electron acceptors, oxygen, nitrate, nitrite and nitrous oxide. Oxygen and nitrous oxide are used in strict preference to nitrate [1]. Oxygen has also been shown to be used before either nitrite or nitrous oxide when physiological substrates are the donors to the electron transfer chain. Electron transfer reactions in this organism therefore pose the question of how one electron acceptor can be selected preferentially. Before addressing this problem it is useful to recall that the pathway for electrons to oxygen is considered to proceed from ubiquinol either via an antimycin-sensitive cytochrome bc_1 segement, cytochrome c and cytochrome aa_3 or directly to an alternative oxidase, cytochrome o [2]. Nitrate reductase, a membrane protein, for which the active site is thought to face the cytoplasm, probably receives its electrons from ubiquinol via a distinct type of cytochrome b that is associated with the nitrate reductase enzyme [2]. Nitrate and nitrous oxide reductases are periplasmic enzymes which receive electrons from cytochrome c.

Recent work has shown that the inhibition of nitrate reduction when oxygen is available is not a consequence of a direct inhibitory effect of molecular oxygen [1]. This is concluded because nitrous oxide and ferricyanide mimic the effect of oxygen and antimycin partially relieves the inhibitory effect of oxygen. Hence nitrate reduction is controlled by the extent of oxidation of one or more components of the respiratory chain. Given its probable location at the branch point of the electron transfer chain, the oxidation state of the ubiquinone/ubiquinol couple is likely to be important. Evidence for this has been obtained [1]. It is thought that the control is exerted on the movement of nitrate across the plasma membrane to the active site of the reductase rather than on the flow of electrons to the reductase. Two observations support this view. First, inside-out membrane vesicles, in which nitrate would have direct access to its reductase, can reduce oxygen and nitrate simultaneously and at similar rates. Second, the nitrate reductase in cells of P.

denitrificans can only reduce chlorate, a substrate analogue, after a permeability barrier has been disrupted. Significantly, the titre of detergent that permits the appearance of chlorate reduction also removes the control by oxygen on nitrate reduction [1]. The control of nitrate reduction also raises interesting questions about the operation of a mobile ubiquinone pool [1].

Quite different mechanisms underlie the inhibition by oxygen of nitrate and nitrous oxide reduction. In the former case the reductase appears to be starved of electrons by competition from the oxidases, whereas the nitrous oxide reductase enzyme appears to be directly inactivated by oxygen [1].

The protonmotive force appears to play no direct role in controlling selection of electron transfer pathways in P. denitrificans, although in the presence of an uncoupler nitrite reduction takes precedence over oxygen reduction because of the highly inhibitory effect on oxidases of a product (possibly nitroxyl anion) of nitrite reduction. Only after collapse of the protonmotive force can this species reach its inhibitory sites on the cytoplasmic surface of the membrane. In contrast, the inhibitory effect of light upon electron flow to oxygen and nitrate in Rps. capsulata is mediated by the protonmotive force [3,4]. The newly discovered respiratory nitrate reductase in Rps. capsulata is periplasmic and therefore the mechanism by which oxygen inhibits nitrate reduction in this organism must differ from that in P. denitrificans, but the protonmotive force is not involved [3,4].

The following references are not meant to be comprehensive but to serve only as a guide to the relevant literature from Birmingham and other laboratories. Study of Rps. capsulata is a joint project with J.B. Jackson.

1. Alefounder, P.R., Greenfield, A.J., McCarthy, J.E.G., and Ferguson, S.J. (1983) Biochim. Biophys. Acta 724, 20–39.
2. Ferguson, S.J. (1982) Biochem. Soc. Trans. 10, 198–200
3. McEwan, A.G., Jackson, J.B. and Ferguson, S.J. (1984) Arch. Microbiol. 137, 344–349.
4. McEwan, A.G., Cotton, N.P.J., Ferguson, S.J. and Jackson, J.B. (1984) in Advances in Photosynthesis Research (Sybesma, C. ed.) pp. II.5. 449–452, Martinus Nijhoff, The Hague.

Halorhodopsin, the light driven chloride pump in Halobacterium Halobium

P.Hegemann and D.Oesterhelt
Max-Planck-Institut für Biochemie
D-8033 Martinsried/München, FRG

Recently a second retinal protein, called Halorhodopsin (HR), was demonstrated to exist in Halobacterium halobium. By light-scattering experiments (1) and electrical measurements on black lipid membranes (2) it was shown to function as an outwardly directed chloride pump. In light, this chromoprotein creates changes in the membrane potential and is able to drive phosphorylation, documenting light energy conversion (3,4). Because of the different ion specifities of the pump and the catalytic system for ATP-synthesis which only utilizes protons, HR is not capable of promoting anaerobic growth in mutants, containing HR as the only chromoprotein. Therefore we asssume that HR has an accessory bioenergetic function under conditions of illumination and low oxygen tension.

Isolation of the native chromoprotein HR is time consuming because this pigment is present in much smaller amounts than bacteriorhodopsin (BR) and is not concentrated in special areas of the cell membrane. Isolation is possible by a combination of membrane preparation, selective extraction with detergent, solubilisation and chromatography on phenylsepharose and hydroxyapatite (5,6). By the black lipid membrane technique it was shown that the isolated chromoprotein is the active chloride pump (7).

Upon light excitation HR is converted within 5 psec into a product with red shifted absorption. The maximum is between 590 and 670 nm and might well be similar to the absorption maximum of the intermediate K in the photocycle of BR (8). The subsequent dark reactions proceed within 20 msec. In contrast to BR no deprotonation and reprotonation reactions were observed during the HR-photocycle.

The formation of the dominant intermediate with an absorption maximum of 520 nm (HR_{520}) is accelerated by chloride, indicating that chloride binds during this step. The chloride is released upon conversion of HR_{520} to the following intermediate absorbing at 640 nm (HR_{640}). This finding is the basis for us to hypothesize that the irreversible transport step could take place by binding and release of chloride at the protonated Schiff base itself. A mechanistic model for the chloride translocation involving the positive charge of the Schiff base will be presented.

Under continuous illumination HR is partially converted into a species absorbing at 410 nm (HR_{410}) This species is not an intermediate of the ion transport cycle and returns to HR only within minutes. Its formation and decay is accompanied by a reversible deprotonation and reprotonation of the Schiff base. Furthermore blue light accelerates reformation of HR from HR_{410}. Therefore the photosteady state level of HR and HR_{410} are influenced by blue light. The function of the HR_{410}-form in the cell is still obscure but it could act as an blue light receptor which might be involved in regulation processes or phototactic reactions in addition to the slow cycling rhodopsin.

(1)   Schobert,B. and Lanyi,J.K. (1982) J.Biol.Chem. 257,10306-10313.
(2)   Bamberg,E., Hegemann,P. and Oesterhelt,D. (1984) Biochim.Biophys.Acta. (in press)
(3)   Mukohata,Y. and Kaji,Y. (1981) Arch.Biochem. Biophys. 206,72-76.
(4)   Wagner,G., Oesterhelt,D., Krippahl,G. and Lanyi,J.K. (1981) FEBS Lett. 131,341-345.
(5)   Steiner,M. and Oesterhelt,D. (1983) EMBO J. 2,1379-1385.
(6)   Taylor,M., Bogomolni,R.A. and Weber,H.J. (1983) Proc.Natl.Acad.Sci.USA 80,6172-6176.
(7)   Bamberg,E. Hegemann,P. and Oesterhelt,D. (1984) Biochem. (submitted)
(8)   Polland,H.-J., Franz,M.A., Zinth,W., Kaiser,W., Hegemann,P. and Oesterhelt,D. (1984) Biophys.J. (submitted)

CYTOCHROME C OXIDASE IN EUKARYOTES AND PROKARYOTES

B. Ludwig   Institute of Biochemistry, Medical University
D 2400 Lübeck   West Germany

Cytochrome c oxidase has long been appreciated as a key enzyme in mitochondrial electron transport, and recently gained particular attention because of its much-debated role in energy conservation. Over the past five years, the presence of this enzyme has been established in a number of bacteria as well.

In mitochondria, cytochrome c oxidase is composed of a number of non-identical subunits that ranges from 7 or 8 to up to 12 or 13 (for detais, see reviews 1-6) in mostly 1:1 stoichiometry. Sequences of all the polypeptides of the beef heart enzyme have been determined (7,8), and a mol.weight of slightly above 200 000 can be calculated for the monomer. Four metal centers are liganded by the protein: 2 copper ions and 2 hemes, a and a_3 , which were analyzed by a variety of spectroscopic techniques aiming at determining reaction mechanisms, kinetic intermediates, ligand surroundings etc (see reviews).

The best view of the 3-D arrangement of mitochondrial oxidase has been obtained from electron diffraction studies of crystalline arrays of reconstituted oxidase (see 1); it is seen as a Y - shaped complex spanning the membrane with an asymmetric mass distribution.

Wikström's suggestion that oxidase acts as a proton pump coupled to electron transport (7) has been confirmed by many laboratories, however, without unanimous acceptance: some critics claim it an artefact (for a discussion, see 10, 11), others assume even higher H^+/e^- ratios than 1 , but recent measurements of disappearance of protons from the vesicle inside nicely match the original suggestion (12).

What was the impact on our understanding of this mito-chondrial enzyme from studies of bacterial oxidases ? One of the major results has been that bacterial oxidases have a much simpler structure while being very similar or identical in most spectral properties, i.e. redox components and their chemical surroundings. Oxidases from a number of bacterial sources (for reviews, see 13,14) all appear to consist of 2 to 3 different polypeptides (with some carrying a further cytochrome c as the third component). Their

homology (where tested) extends even to clear sequence
similarities with the mitochondrially coded subunits I and
II of the eukaryotic oxidase and suggests that also in the
multi-subunit enzyme it is the two largest subunits that
bind the redox ligands.

Even though simple in structure, most of these bacterial
oxidases appear to be active as proton pumps just as the
mitochondrial enzyme, however again with no agreed stoichiometry for the proton/electron ratio, for which values
for different enzymes between 0.6 and 1.4 are reported.

These findings that only two polypeptides suffice to
catalyze the same functions as the far more complex eukaryotic enzyme has questioned the assumption that subunit III
of the beef heart enzyme is directly involved in proton
translocation, and certainly stimulates the search for
functional assignments to the remaining subunits.

For this and other questions the availability of a high
resolution 3-D map of the enzyme would be desirable, and
for the above reasons bacterial oxidases should be suitable
candidates in attempts to obtain 3-D crystals.

References:

(1) Capaldi, R. et al.(1983) Biochim.Biophys.Acta 726,135-48
(2) Capaldi, R.A. (1982) Biochim.Biophys.Acta 694, 291-306
(3) Papa, S. (1982) J.Bioenerg.Biomembr. 14, 69-86
(4) Wikström, M. et al.(1981) Ann.Rev.Biochem. 50, 623-655
(5) Kadenbach, B. and Merle, P.(1981) FEBS-L.135,-1-11
(6) Azzi, A. (1980) Biochim.Biophys.Acta 594, 231-252
(7) Buse, G. et al.(1982) 2.EBEC Abstr., p. 163-164
(8) Anderson, S. et al.(1982) J.Mol.Biol. 156, 683-717
(9) Wikström, M. (1977) Nature 266, 271.273
(10) Mitchell, P. and Moyle, J. (1983) FEBS-L.151, 167-178
(11) Casey, R.P.and Azzi, A. (1983) FEBS-L. 154, 237-242
(12) Wikström, M. (1984) Nature 308, 558-560
(13) Ludwig, B. (1980) Biochim.Biophys.Acta 594,177-189
(14) Poole, R.K. (1983) Biochim.Biophys.Acta 726, 205-243

FURTHER TESTS OF THE PROTONMOTIVE Q CYCLE AND ASSOCIATED PROTON
TRANSLOCATION IN THE CYTOCHROME bc$_1$ COMPLEX OF THE RESPIRATORY CHAIN

B. L. Trumpower[o]
Department of Biochemistry
Dartmouth Medical School
Hanover, NH 03756 U.S.A.

Available evidence indicates that electron transfer through the bc$_1$
complex of the respiratory chain proceeds by a protonmotive Q cycle
pathway (1,2). This pathway accounts for proton translocation by the
bc$_1$ complex and the generally agreed upon stoichiometries thereof by
transmembranous oxidation-reduction of ubiquinone. However, several
laboratories have reported that proton translocation, but not electron
transfer, is inhibited by DCCD (3-6). This inhibition and labeling of
cytochrome b by DCCD (7) have implicated this cytochrome as a "proton
pump" without prescribing how proton pumping is linked to electron
transfer and can be "uncoupled" therefrom by DCCD.

These considerations prompted us to further test the Q cycle pathway
in isolated mitochondrial succinate-cytochrome c reductase. We thus
examined the effects of antimycin and myxothiazol on electron transfer
from cytochrome b to fumarate and on the triphasic reduction of
cytochrome b which is observed with stoichiometric amounts of succinate.
Oxidation of cytochrome b by fumarate is inhibited by antimycin but not
by myxothiazol. This suggests that only one of the two pathways of b
reduction prescribed by the Q cycle is reversible.

Triphasic reduction of cytochrome b consists of an initial partial
reduction of b by an antimycin-insensitive, myxothiazol-sensitive
reaction. The rate of this reaction varies with pH and with succinate
concentration and appears to be equal to and linked to the rate of
cytochrome c$_1$ reduction, which is monophasic under all conditions.
The initial partial reduction of b is followed by an antimycin-sensitive
reoxidation, which is then followed by a slow myxothiazol-sensitive
reduction of b. Triphasic reduction of cytochrome b provides additional
evidence of two pathways of b reduction in support of the Q cycle, and
illustrates that, under certain conditions, reduction of cytochrome b
can switch from one pathway to the other.

As an additional approach to elucidating the pathway of electron
transfer and mechanism of proton translocation, we have purified a
ubiquinol oxidase complex from membranes of aerobically grown
Paracoccus denitrificans. This complex consists of only seven polypeptides, including a 2 peptide cytochrome oxidase, a novel bound
cytochrome c-552, and a 3 or 4 subunit bc$_1$ complex. Electron transfer

through this bc_1 complex is inhibited by antimycin and myxothiazol. The effects of these inhibitors, individually or together, indicate that the electron transfer occurs by a Q cycle pathway.

The ubiquinol oxidase complex of *P. denitrificans* provides a unique opportunity to evaluate H^+/e^- stoichiometries simultaneously in the bc_1 and c-aa_3 segments in a highly resolved system. Because of its structural simplicity and apparent functional identity to the mitochondrial complex, the bc_1 complex of *P. denitrificans* may be especially useful to elucidate the structural basis by which proton pumping is linked to the Q cycle and the role of cytochrome b therein. As a working hypothesis it is suggested that there are sequestered redox domains for ubiquinone on cytochrome b and DCCD sensitive proton conducting pathways between these domains and the aqueous interface.

(1) Mitchell, P. (1976) J. Theor. Biol. 62, 327-367
(2) Trumpower, B.L. (1981) Biochim. Biophys. Acta 639, 129-155
(3) Beattie, D.S. and Villalobo, A. (1982) J. Biol. Chem. 257, 14745-14752
(4) Clejan, L. and Beattie, D.S. (1983) J. Biol. Chem. 258, 14271-14275
(5) Lenaz, G., Esposti, M.D. and Castelli, G.P. (1982) Biochem. Biophys. Res. Commun. 105, 589-595
(6) Price, B.D. and Brand, M.D. (1982) Biochem. J. 205, 419-421
(7) Beattie, D.S. and Clejan, L. (1982) FEBS Lett. 149, 245-248

^{o}BLT gratefully acknowledges receipt of a "Humboldt Preis" from the Alexander von Humboldt-Stiftung. Research was supported by a National Institutes of Health Research Grant GM 20379.

STRUCTURE AND FUNCTION OF BACTERIORHODOPSIN

A.V.Kiselev, N.G.Abdulaev, R.G.Vasilov, I.R.Nabiev, Yu.A.Ovchinnikov
Shemyakin Institute of Bioorganic Chemistry USSR Academy of Sciences,
ul. Vavilova, 32, USSR

The report presents data on recent investigations into the bacteriorhodopsin structure and function.

The total amino acid sequence of the chromoprotein was determined.
It consists of 248 amino acid residues, 67% of them being of hydrophobic character (1). Delipidated bacterioopsin and its fragments
have a relatively low solubility in aqueous media, therefore special
methods were developed to overcome difficulties in their cleavage and
separation, in particular, amidation of all the accessible ε-amino
groups of the protein with dimethyl 3,3'-dithiobispropionimidate was
performed, the S-S bond of the modifying reagent was reduced by dithiothreitol and the resultant protein derivative with the selectively introduced SH-groups was immobilized on the controlled pore activated thiol-glass(CPG/Thiol):

$$|\text{-S-S-}\langle \rangle_{N=} + \text{HS-protein} \longrightarrow |\text{-S-S-protein} + \text{S} = \langle \rangle_{NH}$$

Cleavage of the immobilized delipidated bacterioopsin by chemicals
and proteases yields ($\sim$75%) various lysyl containing protein fragments necessary for the structural studies and for localization of
the introduced "reporter" groups.

By proteolysis of bacteriorhodopsin in purple membranes four regions
1-4, 65-73, 162-163 and 231-248, accessible to the enzyme action,
were found and localized on the membrane surface. Peptide bond 71-72
is located on the outer membrane surface and C-terminal region - on
the cytoplasmic one(2) Removal of regions 1-3, 66-72 and 232-248 of the
bacteriorhodopsin polypeptide chain does not change its ability of
proton translocation across the membrane (3).

Immunochemical methods were applied to identify the amino acid residues of various regions in the bacteriorhodopsin polypeptide chain
which interact with monoclonal antibodies of the known specificity.
They are of undoubt value refine the disposition of these residues on
the membrane surface. The residues are:<Glu-1, Ala-2, Gln-3; Asp-36
Asp-38, Phe-42,Phe-156 and Glu-194. The region between Pro-200 and
Leu-207 contacts apparently an aqueous phase (4). It is of special
interest to prove the exposure of residue Glu-194 since the aromatic
nucleous of the photosensitive retinal analog forming the aldimine
bond with the Lys-216 contacts polypeptide chain region 193-194 (5).
Amino acid residues Lys-40, Lys-41 and Lys-159, though composing the
identified antigenic determinants of bacteriorhodopsin do not influence antigenic properties of the protein molecule.

The results show that Lys-216 is the sole attachment site of retinal in bacteriorhodopsin and that its change does not occur during the
photocycle, proton translocation and light-dark adaptation (6). Six of
seven lysine residues of bacteriorhodopsin (Lys-30, Lys-40, Lys-41,
Lys-129, Lys-159 and Lys-172) do not participate in formation of proton conductivity across the membrane, as completely succinylated bacteriorhodopsin can generate a photopotential being incorporated into
the artificial phospholipid membrane. The positive charge on the
ε-amino group of lysines seems to be an essential element for self-
assembling of bacteriorhodopsin from the denatured state in the
presence of phospholipids and all-trans-retinal.

With the help of biosynthetically and chemically modified bacteriorhodopsin analogs the nature of groups located in the vicinity of
the aldimine bond was studied. The comparative analysis of kinetic resonance Raman spectra of biosynthetically fluorated and native chromoprotein shows that tyrosine residue catalizes the deprotonation of
the aldimine bond in $\sim 10\,\mu$s after light absorption. Bacterioopsin
modified in the SDS solution at pH 5,7 with tetranitromethane was reconstituted into the phospholipid membrane. It is the tyrosine residue
subjected to modification that catalizes deprotonation of the aldimine
bond and the hydroxyl group of this residue seems to form a hydrogen
bond with the negatively charged radical of aspartic or glutamic acid
in the protein molecule. We tentatively concluded that the amino acid
residues Tyr-26, Tyr-57 and probably Tyr-43, Tyr-131 and Tyr-133 are
not directly involved in the process of aldimine deprotonation.

R E F E R E N C E S

1. Ovchinnikov Yu.A., Abdulaev N.G., Feigina M.Yu., Kiselev A.V.
   Lobanov N.A., Nasimov I.V. (1978) Bioorg. Khim., 4, 1573-1574.
2. Ovchinnikov Yu.A., Abdulaev N.G., Feigina M.Yu., Kiselev A.V.,
   Lobanov N.A. (1979) FEBS Lett., 100, 219-224.
3. Abdulaev N.G., Feigina M.Yu., Kiselev A.V., Ovchinnikov Yu.A.,
   Drachev L.A., Kaulen A.D., Khitrina L.V., Skulachev V.P.(1978) FEBS
   Lett., 90, 190-194.
4. Ovchinnikov Yu.A., Vasilov R.G., Vturina I.Yu., Kuryatov A.B., Kiselev A.V. (1984) FEBS Lett. (in press).
5. Huang K.-S., Radhakrishnan R., Bayley H., Khorana H.G. (1982)
   J.Biol.Chem., 257, 13616-13623.
6. Abdulaev N.G., Dencher N.A., Dergachev A.E., Fahr A., Kiselev A.V.
   (1984) Biophys.Struct.Mech., 10, 211-227.

HIGH RESOLUTION X-RAY STRUCTURE OF PHOTOSYNTHETIC REACTION CENTRES FROM RHODOPSEUDOMONAS VIRIDIS

H. Michel, J. Deisenhofer, K. Miki and O. Epp
Max-Planck-Institut für Biochemie
D-8033 Martinsried, West Germany

Photosynthetic reaction centres from the purple photo-synthetic bacteria consist of three protein subunits which are called H (heavy), M (medium) and L (light), 4 bacterio-chlorophyll molecules, 2 bacteriopheophytin molecules, one non-heme-iron atom and two quinone molecules. In several cases they contain an additional cytochrome subunit. Two bacteriochlorophyll b molecules, which are non-covalently linked ("special pair") serve as primary electron donor upon excitation by light. The photooxidized primary donor is then rereduced by a cytochrome c molecule.

The photosynthetic reaction centre from the bacterio-chlorophyll b containing bacterium Rhodopseudomonas viri-dis has been crystallized in three dimensions in the presence of the small amphiphilic molecule heptane-1,2,3-triol by ammonium sulphate precipitation using N,N-di-methyldodecylamine-N-oxide as detergent. The crystals diffract X-rays to beyond 2.5 Å resolution (1). They are photochemically active (2).

A native data set was collected at 2.9 Å resolution comprising 55,000 independent reflections. Multiple heavy-atom-isomorphous replacement was used to phase the reflec-tions. Three independent heavy atom derivatives were ob-tained by soaking the crystals in solutions containing heavy atom compounds. The experimental phases were im-proved by solvent flattening. The resulting electron density map can be interpreted:

The reaction centre is an elongated complex of about 120 Å length probably perpendicular to the membrane. The central intramembraneous part is formed by the L- and M-subunit, each possessing five transmembrane helices. These two subunits are connected by the non-heme-iron atom. They share the pigments and possess very similar protein folding. Very interestingly the pigments and the L- and M-subunits are related by a non-crystallographic twofold rotation axis, despite the fact that the amino acid sequences of both subunits are different. This axis runs through the non-heme-iron atom and the special pair. The

special pair is formed by two nearly parallel bacteriochlorophyll b molecules, which have their planes perpendicular
to the membrane. Close to each bacteriochlorophyll b molecule of the dimer one more bacteriochlorophyll b molecule
and then a bacteriopheophytin molecule are found. This
arrangement suggests that the electron has two ways to get
to the electron-accepting quinone molecules on the opposite side of the membrane. The H-subunit possesses only one
transmembrane helix and a large (probable) cytoplasmic
domain. The cytochrome molecule is attached to the L and
M subunit from the (probable) extracellular side. 4 heme
groups are found in a linear chain. In this chain the last
heme group lays on the twofold non-crystallographic axis.
Its distance to the special pair is about 20 Å.

References:

1 H. Michel (1982). J. Mol. Biol. 158, 567-572

2 W. Zinth, W. Kaiser and H. Michel (1983). Biochim.
   Biophys. Acta 723, 128-131

ELECTRON MICROSCOPY AND IMMUNO-ELECTRON MICROSCOPY AS AN APPROACH TO THE QUATERNARY STRUCTURE OF F_1 AND CF_1

H. Tiedge°, H. Lünsdorf", G. Schäfer° and H. U. Schairer"

° Institut für Biochemie, Medizinische Hochschule Lübeck, D-2400 Lübeck 1
" Gesellschaft für Biotechnologische Forschung, D-3300 Braunschweig-Stöckheim

Subunit stoichiometry and iuxtaposition of F_1-type ATPase have long been a matter of dispute. On the basis of electron microscopic tilting experiments, we proposed an alternating sequence of three α- and three β-subunits, arranged in two layers, for the enzyme from beef heart mitochondria (1). By use of immuno-electron microscopy, it has been shown recently that the corresponding enzyme from E. coli contains three α- and three β-subunits in a symmetric arrangement (2).

In the study presented here, we investigated the quaternary structure of CF_1-ATPase from spinach chloroplasts, which for a long time has been a prominent candidate for a subunit stoichiometry of $\alpha_2\beta_2\gamma_{1-2}\delta_{1-2}\epsilon_2$ (for a review, see (3)). Monoclonal antibodies were raised against the large subunits (α and β) of the enzyme. The ATPase was incubated with a tenfold excess of monoclonal anti α-IgG; antigen-antibody complexes were isolated by gel filtration and analysed by electron microscopy. We observed complexes of the following kind: CF_1 labelled by one, two, or maximally three IgG molecules, complexes of two CF_1-molecules linked by two antibodies, and complexes of higher molecular mass. The angle spanned by two neighbouring IgG molecules bound to the same molecule of CF_1 was characteristically 120°.

These results confirm the hypothesis that the large subunits of F_1, CF_1, and BF_1 from E. coli form a rotary symmetrical arrangement. Three α- as well as three β-subunits are located at the vertices of equilateral triangles, thus forming the typical hexagonal image of F_1 when observed at a top view projection. This kind of structural symmetry seems to be a common feature of proton translocating ATPases of the F_1-type.

References

(1) Tiedge, H., Schäfer, G., and Mayer, F. (1983)
    Eur. J. Biochem. 132, 37–45
(2) Lünsdorf, H., Ehrig, K., Friedl, P., and Schairer, H. U. (1984)
    J. Mol. Biol. 173, 131–136
(3) Shavit, N. (1980)
    Annu. Rev. Biochem. 49, 111–138

Acknowledgement: This work is supported by Deutsche Forschungs-
               gemeinschaft, grant Scha 125/14-1

Fig. 1: Electron micrograph showing a complex of three monoclonal
        anti-αIgG molecules bound to one CF_1-ATPase molecule (left)
        and interpreting sketch (right).

50 nm

LABELING OF AMINO ACID RESIDUES EXPOSED AT THE LIPID PHASE BY [125I] TID IN THE ATP-SYNTHASE FROM NEUROSPORA CRASSA

J. Hoppe* and W. Sebald*
*Department of Cytogenetics, GBF - Gesellschaft für Biotechnologische Forschung mbH., D-3300 Braunschweig, F.R.G.

Membrane integrated proteins in mitochondria from Neurospora crassa were reacted with the carbene generated from the hydrophobic photoactivatable reagent 3-(trifluoromethyl)-3-(m-[125I]iodophenyl)diazirine ([125I]TID). After solubilization of the membranes in Triton X-100 the ATP synthase was isolated by precipitation with antisera raised against F_1 protein (1). SDS gelelectrophoresis followed by autoradiography revealed four labeled polypeptides: three F_0-subunits and the F_1 subunit ß. The proteolipid subunit of the F_0 part which was the most heavily labeled protein was isolated by chloroform/methanol extraction. Label was traced back to individual amino acid residues by Edman degradation of the whole protein (Fig. 1).

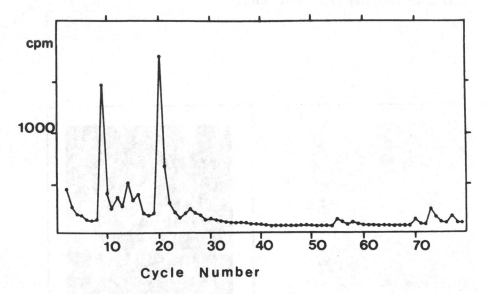

Figure 1. Edman degradation of the purified proteolipid subunit. The proteolipid subunit was immobilized on phenylendiisothiocyanato activated porous glass beads and sequenced in an automated solid phase sequenator using a short program of 50 min. per cycle. Released samples were analyzed for [125]-I-radioactivity. The performance of the Edman degradation was monitored by thin layer chromatography or high performance liquid chromatography.

Labeling was confined to five residues at the NH_2 -terminus and five residues at the C-terminus of the polypeptide chain (met-9, val-12, leu-16, ser-20, thr-26, and ser-55, ile-58, phe-79, met-73, and met-77). In a putative α-helix of the respective segments these labeled residues would lie on one side of the surface. The results are thus indicative for tightly packed α-helices of the proteolipid subunits in the F_0 -part of the ATP-synthase.

Furthermore, labeling occurred at similar positions compared to the homologous protein (subunit c) in the ATP synthase from E. coli (3). These findings are even more remarkable since labeling occurred in segments which are not homologous. As a consequence at corresponding positions different types of amino acids were labeled (Fig. 2). The results strongly suggest that $[^{125}I]$TID is a powerful reagent to detect residues exposed at the lipid phase. From the patterns of labeled residues possible conformations of the polypeptide can be derived.

YSSEJAQAMVEVSKNLGMGSAAJGLTGAGJGJGLVFAALLNGVARNP
MENLNMDLLYMAAAVMMGLAAJGAAJGJGJLGGKFLEGAARQP

ALRGQLFSYAJLGFAFVEAJGLFDLMVALMAKFT
DLJPLLRTQFFJVMGLVDAJPMJAVGLGLYVMFAVA

Figure 2. Comparison of $[^{125}I]$TID labeled residues in the sequence of the proteolipid subunit from Neurospora crassa (first line) and from E. coli (second line). Labeled residues are indicated by bold letters.

References

1 Sebald, W. and Wild, G. (1979) Methods Enzymol. 55, 344-351

2 Hoppe, J. and Sebald, W. (1980) Eur. J. Biochem. 107, 57-65

3 Hoppe, J., Brunner, J. and Jørgensen, B.B. (1984a) Biochemistry, in press

ARCHITECTURE OF Na,K-ATPASE IN THE MEMBRANE

N.N.Modyanov,A.N.Barnakov,V.V.Demin and K.N.Dzhandzhygazyan
Shemyakin Institute of Bioorganic Chemistry,USSR Academy of Sciences,
Moscow,117312,USSR

Immunochemically homogeneous Na,K-ATPase was isolated from outer medulla of pig kidneys. Subunits of two types,α and β, compose the enzyme molecule in an equimolar ratio. Their molecular masses are: 96 kDa - for the α-subunit, 40 kDa and 7 kDa - for protein and carbohydrate moieties of the β-subunit, respectively[1]. The sequence investigation of the subunits and their structural genes are now in progress.

Information on the spatial organization of Na,K-ATPase was derived from the following results. Exposed domains of the α-subunit were found on both sides of the membrane by different immunochemical methods using rabbit and pigeon antibodies with intact cells and "inside-out" proteoliposomes as model systems [2]. Chemical modification and immunochemical analysis revealed the hydrophilic part of the β-subunit, including the N-terminal glycosylated region [1], only on the outer membrane surface [2].

To determine the molecular mass and, consequently, the subunit composition of the functional unit of membrane-bound Na,K-ATPase the active site affinity labelling and freeze-fracture electron microscopy were applied. Covalent binding of γ-/4-(N-2-chloroethyl-N-methylamino)/ benzylamide ATP (ClR-ATP) to the α-subunit of the membrane-bound enzyme results in complete inhibition of the ATP-hydrolyzing activity[3]. ATPase does not hydrolyze ClR-ATP. This reagent is a functional analog of the substrate for Na,K-ATPase according to the following criteria: K_i=0.03mM; the effective prevention by ATP the reagent binding and enzyme inactivation as well as protection of the enzyme from ClR-ATP inhibition under conditions for the phosphoenzyme formation(10mM $MgCl_2$ 10mM P_i) or for increase of K_m for ATP (20mM K^+). Binding of one ClR-ATP molecule per 480-530 kDa of the protein is sufficient for complete inhibition as follows from measurements of the enzyme inactivation versus reagent incorporation. This value is a minimal molecular mass of the functional unit of membrane-bound Na,K-ATPase.

Na,K-ATPase was revealed by freeze-fracture electron microscopy as intramembrane particles (IP) of 9-12 nm in diameter [4]. Stereological analysis of micrographs allowed us to define the volume concentration of IP in the sample. The average molecular mass of IP (540 + 40 kDa) was calculated by comparison of IP and protein concentrations in the same sample [4]. Thus, a structure-functional unit of the membrane-bound enzyme consists of four protomers $\alpha\beta$.

This conclusion was supported by the investigation of 2D-crystals of Na,K-ATPase formed at prolonged incubation of the membranes (4°C) in the presence of $Mg^{2+}$and VO_3^- (5).After negative staining a lattice cell unit formed by protomer $\alpha\beta$ has the following parameters: a=5,4nm, b=6,9nm, γ=112°. Freeze-fracture microscopy of the crystals revealed a lattice with cell unit parameters: a=10,6nm, b=13,4nm, γ=111°; they are in accord with the IP size in noncrystaline membranes. Direct comparison of these crystal lattices (Fig.1) demonstrates that one freeze-fracture and respectively a single IP includes four units identified by negative staining.

All these findings evidence that a functional unit of membrane-bound Na,K-ATPase is a oligomer complex α_4 β_4.

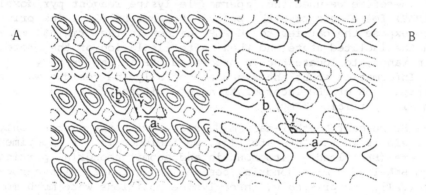

Fig.1.Filtered images of Na,K-ATPase crystals.
     A- negative staining, B- freeze-fracture.

References:

1.Dzhandzhugazyan,K.N.,Modyanov,N.N.,Ovchinnikov,Yu.A. (1981)
  Bioorgan.Khim., 7, 847-857
2.Dzhandzhugazyan,K.N.,Modyanov,N.N.,Ovchinnikov,Yu.A. (1981)
  Bioorgan.Khim., 7, 1790-1800
3.Modyanov,N.N.,Dzhandzhugazyan,K.N. (1983) J.Cell.Biochemistry,
  Suppl. 7B,375
4.Demin,V.V.,Barnakov,A.N.,Dzhandzhugazyan,K.N.,Vasilova,L.A.
  (1981) Bioorgan.Khim., 7, 1783-1789
5.Skriver,E.,Maunsbach,A.B.,Jørgensen,P.L. (1981) FEBS Letters
  131, 219-222

PRIMARY STRUCTURE OF THE MITOCHONDRIAL ADENINE NUCLEOTIDE CARRIER,
SPANNING IN THE MEMBRANE AND RELATION TO THE PRIMARY STRUCTURE OF
THE UNCOUPLING PROTEIN

H.Aquila, W.Bogner and M.Klingenberg

Institut für Physikalische Biochemie, Universität München,
Goethestrasse 33, 8000 München 2, FRG

According to the amino acid sequence of the ADP/ATP carrier its 22
lysine residues are distributed rather uniformly over the polypeptide
chain (1). Therefore we used the impermeable lysine reagent pyridoxal
phosphate (PLP) for elucidating the protein topography (2). The principle was to expose the carrier to ^{3}H-PLP either in the original membrane or in the isolated state and then, after isolation of the protein
from the membranes, to locate the incorporated PLP in the primary
structure. Information about the sidedness of a lysine residue was obtained by comparison of PLP incorporation into mitochondria and into
submitochondrial particles (SMP).

Since the SH-reagent N-ethyl-maleimide reacts with the carrier only
when the protein is in the "m-state" (3), we included in our experiments
also bonkrekate (BKA)-loaded mitochondria and submitochondrial particles
derived from BKA-loaded mitochondria, besides those loaded with carboxyatractylate (CAT). Surprisingly, three lysine positions were much more
strongly labeled when BKA was bound to the carrier. Position 22 is
accessible from the c-side, whereas positions 42 and 48 can be attacked
from the m-side. When mitochondria are loaded with atractylate, which
- in contrast to CAT - is removed by PLP treatment, position 22 is
additionally labeled by PLP. This suggests that lysine 22 is involved
in the binding of CAT. The different reactivity of lysine 42 and 48
may reflect a conformational change between the "c-" and "m-states".

Now we have also elucidated the primary structure of the uncoupling
protein from brown adipose tissue (BAT) of hamster. This protein is
responsible for uncoupling of respiration in BAT mitochondria (4).
Coupling can be restored by purine nucleotides such as GDP, GTP, ADP
and ATP, which bind to this protein (5). The sequence of the 306 residues of the protein was derived largely by automatic Edman degradation
of cyanogen bromide peptides and of peptides obtained by cleavage of
citraconylated protein with trypsin or the Glu-specific protease respectively.

Analysis of the sequence for internal homology by the DIAGON-plot
revealed that this protein, like the ADP/ATP carrier, contains three
related sequences of about 100 amino acids each, here the third repeat
is less homologous. Comparison of the sequence of the uncoupling protein with that of the carrier protein demonstrated a clear homology

of the two proteins, although the amino acid compositions differ strongly. The two proteins appear to have evolved from one ancestral gene by two gene duplications followed by diversification. Hydropathy plots according to Kyte and Doolittle (6) also show up the triplicated structure of both proteins. Only 3 to 4 hydrophobic segments that might form transmembrane helices of about 20 to 30 largely nonpolar residues are detectable. If a modified hydroplot is used, possible sided helices and sided ß-sheets are indicated, the two proteins diverge now.

(1) Aquila,H., Misra,D., Eulitz,M. and Klingenberg,M. (1982) Hoppe-Seyler's Z.Physiol.Chem. 363, 345-349.

(2) Bogner,W., Aquila,H. and Klingenberg,M. (1982) FEBS Lett.146, 259-261.

(3) Aquila,H. and Klingenberg,M. (1982) Eur.J.Biochem.122, 141-145.

(4) Heaton,G.M., Wagenvoord,R.J., Kemp,A. and Nicholls,D.G. (1978) Eur.J.Biochem.82, 515-521.

(5) Lin,C.S. and Klingenberg,M. (1980) FEBS Lett.113, 299-303.

(6) Kyte,J. and Doolittle,R.F. (1982) J.Mol.Biol.157, 105-132.

STRUCTURAL AND FUNCTIONAL ASYMMETRY OF THE RECONSTITUTED ADP/ATP CARRIER FROM MITOCHONDRIA

R. Krämer and M. Klingenberg

Institut für Physikalische Biochemie, Universität München, Goethestrasse 33, 8000 München 2, FRG

Structural and functional parameters of the adenine nucleotide carrier from mitochondria were investigated using the two possible types of orientation of the transport protein when reconstituted into phospholipid vesicles, i.e. right-side-out and inside-out orientation. These two populations can be discriminated in the reconstituted system by titration with side-specific inhibitors (1,2).

The two populations of carrier protein, oriented in opposite directions, are functionally equivalent with regard to several aspects:

a) The affinity for nucleotides is similar on both sides of the membrane (2).

b) The binding sites on both sides accept only free nucleotides and not the complexes with divalent cations.

c) Transport from both sides can be inhibited with acyl-CoA.derivatives.

d) Anions compete with the anionic substrates ADP and ATP on both sides, showing very similar K_i-values (3).

e) The electrophoretic regulation of nucleotide exchange by membrane potential is the same for the two oppositely oriented carrier populations.

In contrast to this symmetrical behaviour, a considerable number of parameters in the reconstituted system were found not to be equivalent at the cytosolic side and the matrix side of the ADP/ATP carrier.

a) There is firstly the wellknown side-specific binding of the inhibitors CAT and BKA (1).

b) Anions were found to be essential for transport activation. Binding of anions to the reconstituted ADP/ATP carrier leads to a sigmoidal dependence of transport activation on the concentration of added anions, with a Hill coefficient of n = 2. This activation, however, takes place only at the cytosolic side.

c) Modulation of transport parameters by the surface potential turned out to be very effective at the matrix side, but only weak at the cytosolic side of the reconstituted carrier protein.

d) Furthermore, for the first time, a side-specific stimulation of transport activity by negatively charged phospholipids was found, but only at the matrix side.

These results lead to a new conception of the nucleotide carrier protein embedded in the phospholipid membrane, with definite structural and functional asymmetry. The binding site at the cytosolic face of the translocator is situated at a considerable distance from the plane of the membrane and shows - in addition to the binding of nucleotides - a cooperative interaction with anions from the water phase. The binding site at the matrix side, however, is in close contact and effective interaction with the inner membrane surface.

(1) Klingenberg,M. (1976) in 'The Enzymes of Biological Membranes' (Martonosi,A.N.ed.) Vol.3, pp.383-438, Plenum Press, New York

(2) Krämer,R. (1983) Biochim.Biophys.Acta 735, 145-159

(3) Krämer,R. and Kürzinger, G. (1984) Biochim.Biophys.Acta, in

STRUCTURAL ORGANIZATION OF THE PHOTOSYSTEMS IN WILD-TYPE AND IN A
CHLOROPHYLL b-LESS MUTANT OF BARLEY

Birger Lindberg Møller, Landis E.A. Henry, Ursula Hinz,
David J. Simpson and Peter Bordier Høj
Department of Physiology, Carlsberg Laboratory,
DK-2500 Copenhagen Valby, Denmark.

The chloroplast lamellar system is composed of flattened membrane
sacs (thylakoids) which are either appressed (grana stacks) or non-
appressed (stroma lamellae). The appressed and non-appressed membrane
regions can be separated by differential centrifugation after disruption of the lamellar system. The disruption was initially carried out
using digitonin (1). More efficient disruption is obtained using
Triton X-100 (2,3,4,5) or a combination of French press treatment and
phase partitioning (6,7). Analyses of the polypeptide composition and
measurements of photochemical activities of such preparations show
that photosystem (PS) II is localized in the grana stacks whereas PS I
is restricted to the non-appressed membrane regions (5,6). Electron
microscopy reveals that the PS II preparations are composed of pairs
of appressed membrane sheets derived from the grana partition regions
and exposing the original luminal sides towards the medium (4,5,6).
When prepared by the phase-partitioning method, the sheets seal and
form inside-out vesicles (4,5,8). When washed with 200 mM NaCl, three
polypeptides of M_r 32,000, 23,000, and 13,500 are released and the
preparations are inactivated with respect to oxygen evolution. Both
preparations are reactivated by rebinding of the isolated M_r 23,000
polypeptide (4). The PS I preparation derived from the stroma
lamellae is composed of much smaller and non-appressed vesicles which
appear to contain the polypeptides of the lamellar system except the
nine or ten polypeptide components of the PS II preparation (9). These
stroma lamellae fragments provide a good starting material for isolation of PS I vesicles with a minimum number of polypeptides (10). The
appressed and non-appressed membrane regions appear to have very few
if any polypeptides in common.
   The major light-harvesting chlorophyll a/b-protein of the thylakoids, chlorophyll a/b-protein 2, has been shown to be involved in
membrane stacking (11,12). The barley mutant chlorina-f2-2800 and
other mutants in this gene cannot synthesize chlorophyll b and the
thylakoids are devoid of chlorophyll a/b-protein 2 (13). Nevertheless,
the mutant chloroplasts contain a normal number of grana which can be
isolated in the presence of 10 mM $MgCl_2$ and a freeze fracture particle
distribution analysis indicates a segregation of the two photosystems
(14). The absence of chlorophyll a/b-protein 2 may be why it is very
difficult or impossible to restack the mutant thylakoids after de-
stacking. Boardman and Highkin (15) had earlier reported on unsuccess-

ful attempts to fractionate the mutant membranes. Since this work was carried out at low salt conditions, a study was undertaken to fractionate the mutant thylakoids using the Triton X-100 as well as the French press/phase partitioning procedures and $MgCl_2$ concentrations ranging from 0.5 to 25 mM. It was not possible to separate PS I from PS II under any of the conditions used. When the grana fraction prepared with the French press in the presence of $MgCl_2$ concentrations at 10 mM or more was subjected to phase partitioning, inside-out vesicles showing reverse proton pumping were isolated. These vesicles have a similar size to those isolated from the wildtype but are less tightly appressed as evidenced by a number of unappressed membrane regions within each pair of membrane sheets. Thus the absence of the light-harvesting complex neither prevents grana formation nor formation of inside-out vesicles. The polypeptide composition of the inside-out vesicles as well as of the stroma thylakoid fragments of the mutant is similar to that of the total mutant thylakoids. It appears that the presence of the light-harvesting complex is essential to retain an energetically favorable exclusion of PS I from the appressed regions of the lamellar system during fractionation.

Literature cited:
(1) Wessels,J.S.C. & G.Voorn: Photosynth., Proc. 2nd Int. Congr. I, 833 (1972)
(2) Berthold,D.A., G.T. Babcock & C.F.Yocum: FEBS Lett. 134, 231 (1981)
(3) Kuwabara,T. & N.Murata: Plant Cell Physiol. 23, 533 (1982)
(4) Møller,B.L. & P.B. Høj: Carlsberg Res. Commun. 48, 161 (1983)
(5) Møller,B.L., P.B. Høj & L.F.A.Henry: Photosynth., Proc. 6th Int. Congr. III, 3, 219.
(6) Henry,L.E.A. & B.L.Møller: Carlsberg Res. Commun. 46, 227 (1981)
(7) Åkerlund,H.-E., B. Andersson & P.-Å.Albertsson: Biochim. Biophys. Acta 449, 525 (1976)
(8) Andersson,B., D.J.Simpson & G.Høyer-Hansen: Carlsberg Res. Commun. 43, 77 (1978)
(9) Henry,L.E.A. & B.L.Møller: Photosynth., Proc. 6th Int. Congr. III, 3, 203.
(10) Møller,B.L., G.Høyer-Hansen & R.G.Hiller: Photosynth., Proc. 5th Int. Congr. III, 245 (1981)
(11) Ryrie,I.J., J.M.Anderson & D.J.Goodchild: Eur.J.Biochem. 107, 345 (1980)
(12) Steinback,K.F., J.J.Burke & C.J.Arntzen: Arch. Biochem. Biophys. 195, 546 (1979)
(13) Machold,O., D.J.Simpson & B.L.Møller: Carlsberg Res. Commun. 44, 234 (1979)
(14) Simpson,D.J.: Carlsberg Res. Commun. 44, 305 (1979)
(15) Boardman,N.K. & H.R.Highkin: Biochim. Biophys. Acta 126, 189 (1966)

THE ROLE OF SURFACE ELECTRICAL CHARGES IN CONTROLLING THE ORGANISATION
AND FUNCTIONING OF THE CHLOROPLAST THYLAKOID MEMBRANE

J. Barber
Department of Pure and Applied Biology,
Imperial College, London SW7 2BB., U.K.

Like all biological membranes, the chloroplast thylakoid membrane
carries excess negative charge on its surfaces at neutral pH (1). The
negative charges on the outer surface are due to the carboxyl groups
of glutamic and aspartic acid residues of exposed polypeptides with
no obvious contribution from the low level of acidic lipids which
form a part of the general membrane matrix. Associated with these
fixed charges is a diffuse layer of counterions immediately adjacent
to the surface whose precise characteristics are governed by the nature of the electrolyte solution in which the membrane is suspended (2).
It is the distribution of positive charges within the diffuse layer
which determines the forces of interaction between protein complexes
in and on the membrane and between adjacent membrane surfaces (3).
Coupled with this basic concept is the fact that the thylakoid membrane is highly fluid as judged by fluorescence anisotropy measurements using the hydrophobic probe diphenylhexatriene. Therefore when
thylakoid membranes are suspended in different electrolyte solutions
the membrane arranges itself into preferred structures. At low levels
of electrolytes the membrane is totally unstacked with no appressed
regions and with the various intrinsic complexes randomised along its
plane. However, when the surface charges are well screened by adding
appropriate cations there is a reorganisation of the membrane into
appressed and non-appressed regions with corresponding segregation of
the functionally different intrinsic supermolecular complexes. Photosystem two (PS2), and its closely associated chlorophyll a/b light-
harvesting complex (LHC) are localized in the appressed regions while
photosystem one (PS1) and the coupling factor ATP synthetase (CF_O-CF_1
complex) are restricted to non-appressed regions (1). The position of
the cytochrome b_6-f complex, which acts as an intermediate component
in the electron/proton transfer between PS2 and PS1, is less certain
but could be at the interface of the appressed/non-appressed lamellae
(4). There is also some evidence that there is a partial separation of
lipid classes between the appressed and non-appressed region which is
indicative of specific lipid-protein interactions. Within the intact
organism the electrolyte level in the chloroplast stroma is such as to
maintain appressed and non-appressed thylakoids and thus a lateral
separation of PS1 and PS2. This lateral separation emphasises the need
for mobile redox carriers between complexes. Three mobile redox carriers can be identified; plastoquinone (PQ), plastocyanin (PC) and soluble ferredoxin (Fd). PQ acts as an electron/proton carrier between
PS2 and the cyt b_6-f complex, presumably by rapid lateral and trans-

membrane diffusion aided by the highly fluid nature of the lipid
matrix (see Millner and Barber, these proceedings). PC is a copper
containing protein held to the inner thylakoid surface and can diffuse
between the cyt b_6-f and PS1 complexes.On the other hand, Fd is on the
outer surface and can shuttle electrons from the reducing side of PS1
to cyt b_6-f complexes in order to facilitate cyclic electron flow.
All these mobile carriers are probably not stoichiometric with the
intrinsic complexes so that intercommunication between several complexes is possible.

The model described above is not static since the amount of appressed and non-appressed membrane, and therefore of the relative concentrations of the associated complexes, varies with growth conditions.
Moreover, on the short-time scale, phosphorylation of exposed portions
of the intrinsic complexes alters surface charge properties and leads
to reorganisation of components (1,5). It is this mechanism whereby
the phosphorylation of LHC induces lateral movements of LHC and LHC-
PS2 complexes from the appressed to the non-appressed regions with the
consequence of changing the quantal distribution between PS2 and PS1.
This phosphorylation is controlled via a membrane kinase which is activated when the PQ pool becomes over reduced. When the PQ pool is
oxidised the kinase is not operative and the LHC is dephosphorylated
by a phosphatase. In response to this, the LHC and LHC-PS2 diffuse
back to their original location in the appressed regions. The precise
extent of reorganisation associated with LHC phosphorylation/dephos-
phorylation is dependent on the electrolyte concentration of the suspension medium as expected for a phenomenon governed by surface charge
changes (see Telfer et al. these proceedings). The physiological expression of this reorganisation is known as the State 1-State 2
transition which exists to optimise photosynthetic efficiency when
changes in lighting conditions occur in the natural canopy or algal
bloom. However, for this mechanism to be effective, and also to aid
PQ diffusion, the lipid matrix of the thylakoid membrane must be fluid.
Optimization of thylakoid membrane fluidity has been detected in response to changes in growth temperature, which in the case of cold
resistant plants, seems to involve regulation of the protein to lipid
ratio rather than alterations in the degree of unsaturation of the
fatty acids of the lipid matrix.

1. Barber, J. (1982) Ann. Rev. Plant Physiol. 33, 261-295
2. Barber, J. (1980) Biochim. Biophys. Acta 594, 253-308
3. Barber, J. (1982) BioScience Reps. 2, 1-13
4. Barber, J. (1983) Plant, Cell and Environ. 6, 311-322
5. Barber, J. (1983) Photobiochem. Photobiophys. 5, 181-190

STRUCTURAL ASPECTS OF VECTORIAL REACTIONS IN PHOTOSYNTHETIC REACTION CENTERS

P. Mathis
Service de Biophysique, Département de Biologie
CEN Saclay, 91191 Gif-Sur-Yvette cédex, France

In photosynthetic organelles, primary photochemistry takes place in reaction centers. These membrane subunits are made of a few hydrophobic polypeptides, of some pigment molecules and of 4-6 redox centers. They are highly organized to permit efficient photochemistry, forward electron transfer and build-up of a membrane potential. Some aspects of the structure-function relation will be discussed : new developments, methods, comparison between different classes of reaction centers, major unanswered questions. Two levels of organization have to be considered :

Internal organization

The progressive mapping of reaction centers has involved many complementary methods (X-ray diffraction will not be considered ; its recent application to RC's should permit enormous progress, but other methods should remain important) :
- local environment (absorption spectroscopy, CD, resonance Raman spectroscopy ; NMR has not yet been used).
- relative orientation of the molecules (linear dichroism, fluorescence polarization).
- ESR provides a lot of information : local level, orientation, distances,...
- reaction kinetics, as studied by flash absorption, range from picoseconds to milliseconds. They seem to be in direct relation with distances between redox centers.
- specific information is derived from more specialized methods : triplet transfer, effect of temperature.
- the relation between proteins and redox centers is approached through specific inhibitors, photo affinity labels, primary sequencing,...

The emerging figure is that of a protein matrix holding pigment molecules and redox centers. This ensemble has a well-defined geometry : fixed orientations, increasing distances between redox centers when going from 1ary to 2ary partners, great similarities between classes of RC's. The synthesis of artificial models helps understanding the factors which are important for the primary photochemistry. Our knowledge remains weak in two respects :
- intramolecular distances,
- sites of binding of redox centers on the protein moities.

The reaction center in the membrane

Several methods showed that the reaction centers have a specific positioning in the photosynthetic membrane :

- differential access to artifical redox mediators, inhibitors, proteolytic enzymes, antibodies,... permits to localize various components at the interior or on either side of the membrane.
- most reaction centers components are oriented with respect to the membrane plane, as shown by spectroscopic experiments (linear dichroism, EPR) with oriented membranes.
- electrical measurements by means of electrochromism or directly by electrodes give an estimate of the depth of various charged species within the membrane.
- the general shape of bacterial reaction centers, their position in artificial bilayers, have been determined by diffraction studies.

Reaction centers are coupled to secondary electron carriers. These may be tightly (like cytochrome c_{553} in Chromatium vinosum) or loosely coupled (like ferredoxins). The coupling with quinones in PS-II and in purple bacteria, presents unique characters.

The transverse orientation of RC's in the photosynthetic membrane is a general property. How is it determined ? Is it associated with a transverse asymmetry of lipid composition ? This transverse orientation is probably favorable for the efficiency of electron transfer ; it is a pre-requisite for the build-up of a membrane electrochemical potential.

References

1. OKAMURA M.Y., FEHER G. and NELSON, N (1982) in Photosynthesis Energy conversion by plants and bacteria, Vol. 1 (Govindjee, ed) pp. 195-272, Academic Press
2. MATHIS P. and PAILLOTIN G. (1981) in Biochemistry of plants, Vol. 8 (M.D. Hatch and N.K. Boardman, eds) pp. 97-161, Academic Press
3. BRETON J. and VERMEGLIO A. (1982) in Photosynthesis : Energy conversion by plants and bacteria, Vol. 1 (Govindjee, ed.) pp. 153-194
4. TREBST A. (1974) Ann. Rev. Plant Physiol. 25, 423-458

THE POLYPEPTIDE SUBUNITS OF THE CYTOCHROME SYSTEM OF MITOCHONDRIA AND
THEIR FUNCTION

S. Papa, M. Lorusso, R. Capitanio, M. Buckle and D. Gatti
Institute of Biological Chemistry, Faculty of Medicine, University of
Bari, Italy

The cytochrome chain of mitochondria is composed of numerous polypeptides and attempts are being made in various laboratories to clarify their exact quantity, structure, topography and function.

SDS-PAGE analyses generally reveal seven major polypeptide bands in
mammalian oxidase preparations (1). Using high-resolution conditions
up to twelve polypeptides can, however, be identified (2,3). Comparison
of oxidases preparations from eukaryotes and prokaryotes indicate that
subunits I and II are essential for electron flow and energy conservation. In fact the functional binding site for cytochrome c is located
in subunit II which apparently binds also haeme a and Cu_A.

Subunit III of the beef-heart oxidase seems also to be involved in
the binding of cytochrome c. Its presence favours the formation of dimers and appears to be essential for a redox linked acidification,exhi-
bited by the oxidase reconstituted into phospholipid vesicles. This is
considered by various authors as being involved in energy-linked proton pumping by the enzyme (4; see however 5).

There is evidence for cooperative linkage between the redox state
of haemes aa_3 and protolytic equilibria elsewhere in the enzyme (6).

Proton conduction in polypeptides of cytochrome oxidase could be
involved in proton transfer from the matrix (N) aqueous phase to the
catalytic center of the oxidase (which is apparently displaced towards
cytochrome c at the outer side) for protonation of reduced oxygen to
H_2O. It may just be the specific aminoacid sequence in the vicinity of
the reaction center that gives rise to preferential proton access from
the N side of the membrane. The possibility that analogous proteace-
ous pathways may further result in active transfer of the protons,deri-
ving from the N aqueous space, to the P space is under investigation.

Electrophoretic analysis of cytochrome c reductase reveals eight polypeptides bands (7), which can be resolved under suitable conditions,
into eleven polypeptides. Studies with chemical modification followed
by cross-linking, or selective proteolytic digestion indicate that low
M.W. polypeptides can be involved in redox-linked proton pumping by
the reductase. In a model favoured by Papa et al. (8) polypeptide

subunits would constitute a pathway for transmembrane electrogenic proton conduction from the matrix aqueous phase to the a semiquinone/quinol system in the membrane and from this to the outer phase. In this model polarizability of hydrogen-bonded aminoacid residues co-operatively linked to protonmotive catalysis by the quinone system can result in the proper adjustment in space and time of the proton accepting and proton donating capabilities of residues so to result in asymmetric protonation/deprotonation of the quinone center.

References

1. Capaldi, R.A., 1982, Biochim. Biophys. Acta 694, 291.
2. Kadenbach, B. and Merle, P., 1981, FEBS Lett. 135, 1.
3. Buckle, M. and Papa, S., 1984, submitted.
4. Wikström, M., Krab, K. and Saraste, M., 1981, Ann.Rev.Biochem. 50, 623.
5. Papa, S., Lorusso, M., Capitanio, N. and De Nitto, E., 1983, FEBS Lett. 157, 7.
6. Papa, S., 1982, J.Bioenerg.Biomembr. 14, 69.
7. Lorusso, M., Gatti, D., Boffoli, D., Bellomo, E. and Papa, S., 1983, Eur.J.Biochem. 137, 413.
8. Papa, S., Lorusso, M. and Guerrieri, F., 1982, in: Cell Function and Differentiation, Part B (G. Akoyounoglou et al., eds), p. 423, Alan R. Liss Inc., New York.

THE Fo·F1 ATP SYNTHASE-ATPase COMPLEX: CHARACTERIZATION OF ITS
CATALYTIC β-SUBUNIT.

Z. Gromet-Elhanan, D. Khananshvili and S. Weiss
Department of Biochemistry, Weizmann Institute of Science, Rehovot,
Israel.

The β-subunit has been removed from the membrane bound Fo·F1-ATP
synthase of *Rhodospirillum rubrum* (1) and purified to homogeneity (2).
The resulting β-less chromatophores have lost all their ATP synthesis
and hydrolysis activities, but these could be fully restored upon
rebinding the isolated, purified β-subunit. This isolated β-subunit
has no catalytic activity by itself (1), although, when assembled
into the Fo·F1 complex, the β-subunit has been suggested to function
as the catalytic subunit and to contain binding sites for nucleotides
and for inorganic phosphate (3,4). The *R. rubrum* system provides,
therefore, an ideal system for the possible direct identification,
on the isolated β-subunit, of substrate binding sites as well as the
catalytic site of the Fo·F1 complex and the amino-acid residues
involved in them. Such studies have recently been carried out by
using a number of approaches:
a. by following the effect of various chemical modifiers, known to
    interact with specific amino-acid residues (5), on the capacity
    of the purified, reconstitutively active β-subunit to bind to
    β-less chromatophores and restore their lost activities (6-8).
b. by measuring the capacity of the isolated β to bind labeled
    ATP, ADP and inorganic phosphate (9,10).
c. by comparing the effect of the various chemical modifiers on the
    reconstitutive activity of the isolated β-subunit and on the
    properties of the substrate binding sites identified on this
    subunit.

Two nucleotide binding sites, a high affinity site not affected
by $MgCl_2$ and a low affinity $MgCl_2$-dependent site, have been
demonstrated on the isolated β-subunit. Furthermore, the low affinity
site has been found to bind also inorganic phosphate and a number of
tests seem to identify it with the catalytic site of the Fo·F1
complex.

Since the β-less chromatophores can rebind the β-subunit with
restoration of their lost ATP-linked activities they provide a
unique system for testing the possible formation of active hybrid
Fo·F1 complexes. In search of such hybrids we have recently tested
the capacity of the β-less *R. rubrum* chromatophores to bind, in a
functionally active form, the β-subunit that has been isolated and
purified from the *E. coli* Fo·F1 complex (11). A hybrid system capable

of carrying out photophosphorylation as well as ATP hydrolysis has
been obtained. The properties of this heterologous *R. rubrum - E. coli*
hybrid system will be discussed in comparison with those of the
homologous *R. rubrum* system.

References:

1. Philosoph, S., Binder, A. and Gromet-Elhanan, Z, (1977) J. Biol.
    Chem. 252, 8747-8752.
2. Khananshvili, D. and Gromet-Elhanan, Z. (1982) J. Biol. Chem.
    257, 11377-11383.
3. Futai, M. and Kanazawa, H. (1983) Microbiol. Rev. 47, 285-312.
4. Senior, A.E. and Wise, J.G. (1983) Membr. Biol. 73, 105-125.
5. Cross, R.L. (1981) Annu. Rev. Biochem. 50, 681-714.
6. Khananshvili, D. and Gromet-Elhanan, Z. (1982) Biochim. Biophys.
    Res. Commun. 108, 881-887.
7. Khananshvili, D. and Gromet-Elhanan, Z. (1983) J. Biol. Chem.
    258, 3714-3719.
8. Khananshvili, D. and Gromet-Elhanan, Z. (1983) J. Biol. Chem.
    258, 3720-3725.
9. Gromet-Elhanan, Z. and Khananshvili, D. (1984) Biochemistry
    23, 1022-1028.
10. Khananshvili, D. and Gromet-Elhanan, Z. (1984) Poster presented
    at this meeting.
11. In colaboration with M. Futai and H. Kanazawa.

STRUCTURAL BASIS OF H[+]-TRANSFER IN BACTERIORHODOPSIN

D. Kuschmitz[o], M. Engelhard[o], K.-D. Kohl[o], K. Gerwert",
F. Siebert" and B. Hess[o]
[o]Max-Planck-Institut für Ernährungsphysiologie,
Rheinlanddamm 201, 4600 Dortmund 1, FRG
"Institut für Biophysik und Strahlenbiologie der Univer-
sität Freiburg, Albertstraße 23, 7800 Freiburg, FRG

The overall process of proton pumping by bacteriorhodopsin (bR) can be subdivided into the following steps (1):
a. proton activation at the active site, following light
excitation of the retinal chromophore, b. proton transfer
within the protein from the active site to the outside of
the membrane, c. proton release into the medium across the
outside membrane bulk interface, d. proton uptake from the
cytoplasma across the inside membrane bulk interface and e.
proton rebinding to the active site. Two mechanisms were
proposed for the proton transfer through the protein: a.
two hydrogen bonded pathways serving for proton conduction,
one leading from the active site to the outside medium,
the other leading from the cytoplasma to the active site
(2), b. a proton transfer mechanism composed of transfer
in discrete steps through proton donor-acceptor sites constituted by the amino acid side chains of the protein such
as tyrosyl, lysyl, aspartyl and glutamyl residues (1).

In order to define the proton pathway through the bR molecule and to discriminate between the two mechanisms of
proton transfer we have investigated the chemical behaviour
of tyrosine residues as well as carboxyl side chains using
UV (3,4) and IR (5) techniques. In addition, the properties of proton binding sites at the two membrane surfaces
as a function of pH and ionic strength (1) as well as a
function of specific cations (6) during the photocycle
were investigated.

Proton pumping is initiated by a reversible trans-cis
isomerisation of the retinal chromophore as indicated
spectrophotometrically (4). Following this event, during
the K-L transition two aspartyl residues become protonated
and deprotonated respectively (5). Three proton donating
groups can be identified in the L-M transition time: the
chromophore, donating the proton from a Schiff's base configuration as well as one tyrosine and one aspartic acid

(3,4,5). Also two proton accepting groups, namely two aspartate residues, have been identified in the L-M transition (5). These proton transfer reactions occur in the hydrophobic part of the opsin moiety of bR, as indicated by the inaccessibility of these groups to changes in pH, ionic strength as well as to $H_2^{18}O$ exchange and in addition by the IR absorption properties. Proton binding groups of pK 7-7.5 (high ionic strength) titratable at the membrane surface appear and disappear during the M-bR ground state transition (1). These groups have not been identified in terms of amino acid residues, they might be related to the proton transfer process across the membrane bulk interfaces. In addition it has been found that an unknown binding site is specifically occupied by mono- respectively divalent cations (6). They are obligatory for the purple bacteriorhodopsin state. One cation is reversibly released and rebound during the photocycle and suggests a control function for the proton transfer through the bR molecule (6).

Our results indicate that proton transfer through bacteriorhodopsin takes place stepwise, correlated in first approximation with the discrete steps of the photocycle of the retinal chromophore. So far we have identified by different techniques one tyrosyl and four aspartyl residues and we have excluded the participation of any glutamyl residues. The complete spatial sequence of the pathway can only be given if more actual structural data are available.

1. Hess, B., Kuschmitz, D. and Engelhard, M. (1982) in Membranes and Transport 2 (Martonosi, A.N., ed.) pp. 309-318, Plenum Press, New York.
2. Nagle, J.F. and Morowitz, H.J. (1978) Proc. Natl. Acad. Sci. U.S.A. 75, 298-302.
3. Hess, B. and Kuschmitz, D. (1979) FEBS Lett. 100, 334-340.
4. Kuschmitz, D. and Hess, B. (1982) FEBS Lett. 138, 137-140.
5. Engelhard, M., Hess, B. Gerwert, K., Kohl, K.-D., Mäntele, W. and Siebert, F. this meeting.
6. Kohl, K.-D., Engelhard, M. and Hess, B. this meeting.

The Structures of the Light-harvesting Pigment-Protein Complexes and their Polypeptides from Cyanobacteria and Purple photosynthetic Bacteria

H. Zuber, W. Sidler, P. Füglistaller, R. Brunisholz, R. Theiler and G. Frank.

Institut für Molekularbiologie und Biophysik, Eidgen. Technische Hochschule, ETH-Hönggerberg, CH-8093 Zürich, Switzerland

In the primary processes of photosynthesis, light-energy is absorbed by a number of pigment molecules, mainly tetrapyrroles. Early evidence strongly points to a specific arrangement of these pigment molecules in the light-harvesting (or antenna) pigment-protein complexes of various photosynthetic organisms. Within these complexes the pigments are conjugated with a few types of polypeptides. The principal objective of our investigations in Zürich is to determine the structures of the light-harvesting polypeptides and, based on this of the antenna complexes (1). The idea is further to provide by these studies a structural interpretation of their light-harvesting and energy transfer function.

In the photosynthetic apparatus of cyanobacteria and eucaryotic red algae the main light-harvesting complexes of photosystem II are the phycobiliproteins (1). They are located as extramembrane entities at the surface of the thylacoid membrane in form of the phycobilisome antennae. The primary structures of the α- and β-polypeptide chains of the three phycobiliproteins of the phycobilisome from the cyanobacterium Mastigocladus laminosus were determined (2,3,4). A typical structural principle of two highly organized (α- and β-) light-harvesting polypeptides carrying pairs of excitonically coupled phycobilin (pigment) molecules and forming cyclic (disc-shaped) hexamers and stacks of hexamers for heterogeneous energy transfer, became obvious.

In the group of the intramembrane light-harvesting BChl-protein complexes of purple photosynthetic bacteria a similar structure principle for energy transfer has been found in our laboratory. From a series of Rhodospirillacene (Rs. rubrum (6,7), Rp. sphaeroides (8), Rp. viridis, Rp. gelatinosa, Rp. capsulata) the primary structures of the light-harvesting polypeptides were analysed. In the B 800-850 or the B 870 (B 890, B 1015) antenna complexes of these bacteria two types of polypeptides (α-, β-) form pairs of transmembrane α-helices within the hydrophobic domain, bordered by polar/charged N- and C-terminal domains at or near the membrane surface. In the polypeptide pair two BChl molecules are exciton coupled. The helix pairs which are the basic structural units of the light-harvesting BChl-protein complexes aggregate further to build up a two-dimensional array of cyclic systems for heterogeneous energy transfer to the reaction centers within the membrane.

References

1) Zuber H (1983) Structure and function of the light-harvesting phycobiliproteins from the cyanobacterium mastigocladus laminosus. In: Papageorgiou GC, Packer L (eds) Photosynthetic Procaryotes: Cell differentiation and function. Elsevier Science publishing Co, New York, Amsterdam, Oxford, pp 23-42.

2) Frank G, Sidler W, Widmer H, Zuber H (1978) The complete amino acid sequence of both subunits of C-phycocyanin from the cyanobacterium mastigocladus laminosus. Hoppe Seyler's Z. Physiol Chem 359: 1491-1507.

3) Sidler W, Gysi J, Jsker E, Zuber H (1981) The complete amino acid sequence of both subunits of allophycocyanin a light harvesting protein pigment complex from the cyanobacterium mastigocladus laminosus. Hoppe Seyler's Z Physiol Chem 362: 611-628.

4) Füglistaller P, Suter F, Zuber H (1983) The complete amino acid sequence of both subunits of phycoerythrocyanin from the thermophilic cyanobacterium mastigocladus laminosus. Hoppe Seyler's Z Physiol Chem 364: 691-712.

5) Zuber H (1982) Structural studies on the light-harvesting pigment-protein complexes from cyanobacteria and photosynthetic (purple) bacteria. IVth Int Symp Photosynthetic procaryotes Bombannes, France, abstracts C 40.

6) Brunisholz RA, Cuendet PA, Theiler R, Zuber H (1981) The complete amino acid sequence of the single light harvesting protein from chromatophores of rhodospirillum rubrum G-9$^+$. Febs letters 129: 150-154.

7) Brunisholz RA, Suter F, Zuber H (1984) The light-harvesting polypeptides of rhodospirillum rubrum. The amino acid sequence of the second light harvesting polypeptide B 880-β (B 870-β) of rhodospirillum rubrum S 1 and the carotenoidless mutant G-9$^+$. Aspects of the molecular structure of the two light-harvesting polypeptides B 880-α (B 870-α) and B 880-β (B 870-β) and of the antenna complex B 880 (B 870) from rhodospirillum rubrum. Hoppe Seyler's Z Physiol. Chem, in press (June 1984).

8) Theiler R, Suter F, Wiemken V, Zuber H (1984) The light-harvesting polypeptides of rhodopseudomonas sphaeroides R-26.1. I Isolation, purification and sequence analysis. Hoppe Seyler's Z Physiol Chem, in press (June 1984).

REACTION CENTER-$\underline{bc}_1$ COMPLEX INTERACTION: THE Q POOL AND ROUTES TO CYTOCHROME $\underline{b}$ REDUCTION

P.L. Dutton, S. DeVries, K.M. Giangiacomo, C.C. Moser and D.E. Robertson
Department of Biochemistry & Biophysics, University of Pennsylvania,
Philadelphia, PA 19104, U.S.A.

The photosynthetic reaction center from $\underline{Rps.}$ $\underline{sphaeroides}$ serves, following flash activation, to generate an oxidized $\underline{c}$-type cytochrome and ubiquinol-10 (QH_2). The quinol and the ferricytochrome $\underline{c}$ are substrates for the $\underline{bc}_1$ complex. This system has been studied in two forms: a) in the native chromatophore membranes of $\underline{Rps.}$ $\underline{sphaeroides}$, in which there are approximately 25 quinones per reaction center, (~50 per $\underline{bc}_1$ complex); and b) in a minimal detergent suspension of a hybrid system comprised of reaction centers, $\underline{bc}_1$ complex from beef heart and cytochrome $\underline{c}$ from horse heart in which (excluding Q_A of the reaction center) there is a total of 2-3 quinones per reaction center.

The hybrid system promotes the single turnover flash activated reduction of cytochromes $\underline{b}_{562}$ or $\underline{b}_{566}$ in milliseconds $\underline{via}$ a route that demands both QH_2 and oxidized Rieske FeS. This route involves: a) occupancy by the QH_2 of the Q_z site; and b) the Q_zH_2 induced reduction of both Rieske FeS and cytochrome $\underline{b}_{562}$ or $\underline{b}_{566}$ depending on the conditions. Myxothiazol, considered to interfere with the Q_z site, inhibits this route, while antimycin has no significant effect.

The hybrid system clearly displays an additional, distinctly different route to reduction of the b-cytochromes. This route requires QH_2, but does not require an oxidized Rieske FeS; hence this route proceeds in the absence of added cytochrome $\underline{c}$ and under conditions that maintain Rieske FeS and cytochrome $\underline{c}_1$ reduced throughout the experiment. This route most likely involves: a) occupancy by QH_2 of the Q_c binding site; b) the Q_cH_2 induced reduction of cytochrome $\underline{b}$ yielding perhaps the Q_c semiquinone. Cytochrome $\underline{b}$ reduction $\underline{via}$ this sequence is inhibited by antimycin, considered to interfere with the Q_c site, while myxothiazol has little effect.

While the two routes to cytochrome $\underline{b}$ reduction display high discrimination for antimycin and myxothiazol, both routes are inhibited by $\underline{o}$-hydroxy quinones such as UHDBT and UHNQ. The distinct and independent nature of the two routes (possibly owing their existence to discrete Q_c and Q_z catalytic sites associated with the function of the $\underline{bc}_1$ complex) is supported by UHDBT titrations of the inhibition of cytochrome $\underline{b}$ reduction either $\underline{via}$ the Q_c route in the presence of myxothiazol or the Q_z route in the presence of antimycin. We find in each case that close to one UHDBT molecule is required per $\underline{bc}_1$ complex for completed inhibition. Further, in the absence of antimycin or myxothiazol close to two UHDBT molecules are required to block cytochrome b reduction.

The hybrid system lends itself to easy manipulation of the environmental conditions which modulate the communication between the oxidized cytochrome c and QH_2 substrates and the various reaction sites on the bc_1 complex. The halftimes of ferricytochrome c mediated oxidation of cytochrome c_1 and Rieske FeS, and of QH_2 mediated cytochrome b reduction via the Q_c or Q_z routes are all linearly related to the solution viscosity (changed by glycerol). This is expected of diffusion events. Although QH_2 is a smaller molecule than cytochrome c, its diffusion appears slower, suggesting that the QH_2 in the hybrid system is never found free in the aqueous phase. The observed difference in b reduction via the Q_c and Q_z routes at high viscosity may be explained tentatively by a larger "target size" offered by the Q_c site relative to the Q_z site for a diffusing QH_2.

The chromatophore system clearly displays the myxothiazol sensitive, Q_z mediated route. The other, antimycin sensitive, Q_c mediated route is not directly observable in native chromatophores and has not been reported. However, we have now revealed it in chromatophores after extraction of the majority of the Q pool to approximate the pool size encountered in the hybrid system. This indicates that the Q pool size is important for the operation of this route. Under experimental conditions which poise the b-cytochrome and the Q pool oxidized prior to the flash activation, it is clear that the large pool encountered in vivo totally suppresses this antimycin sensitive route. Whether the source of the suppression is simply a matter of a higher E_h that will be apparent on entry of a QH_2 from the reaction center into a large oxidized Q pool will be determined. Further, it remains to be established what part this route plays, if any, under these and lower potential conditions, in the forward direction of the catalytic acts of energy conversion processes in the bc_1 complex. However, since it can be now resolved in the coupled chromatophore membrane system, new opportunities are available for the measurement of proton binding/release and electrogenic reactions uniquely associated with this route.

The extracted chromatophores provide a continuous link, then, between the Q rich native membrane and the Q poor hybrid system while still maintaining a membrane environment for Q mobility.

HYDROPHOBIC MARKERS FOR THE DETERMINATION OF THE ENVIRONMENT OF MEMBRANE PROTEINS

R. Bachofen, J. Brunner[*] and H.P. Meister
Institute for Plant Biology, University of Zürich, CH-8008 Zürich, Switzerland and
[*] Laboratorium für Biochemie der Eidgenössischen Technischen Hochschule, ETH Zentrum, CH-8092 Zürich, Switzerland

In order to obtain information concerning the structural and topological organization of the proteins of chromatophores from the photosynthetic bacterium R. rubrum G-9$^+$, chemical labeling using a variety of hydrophobic reagents has been evaluated.
Dicyclohexylcarbodiimide (1) and fluorescamine were found to react exclusively with subunit M of the bacterial reaction center implying the presence of reactive and accessible carboxyl groups and amino groups within a hydrophobic environment. The light-activable carbene-generator 2-[8-14 C] naphthyl 2-diazo-3,3,3,-trifluoropropionate (NADIT) (2) is a promising reagent for the general labeling of the membranous polypeptide segments of integral proteins. NADIT and another carbene generating reagent, 3-(trifluoromethyl)-3-(m-[125 I] iodophenyl)-diazirine (TID) (3) were both determined to label all of the intrinsic pigment proteins, especially the light harvesting polypeptides B-870, the primary structures of which have been elucidated recently (4). When chromatophores labeled with TID or NADIT were treated with proteinase K, up to 30 % of the label covalently bound to the membrane proteins was released upon washing of the membranes after the digestion. These labeling reagents were therefore concluded to not only bind within the hydrophobic core of the membrane but also to protein segments located at the cytoplasmic surface, or in the region of the polar phospholipid head groups of the chromatophores.
Sequencing of the TID-labeled B 870-α chain showed substantial amounts of label in residues of the N-terminal segment, thought to be located in the cytoplasmic space or in the bilayer interphase. In the membrane spanning part the periodicity of the radioactivity pattern from the Edman degradation analysis corresponds to two amino acid residues between labeled sites (see fig. 1). These data suggest (i) a helical structure with tight packing to other polypeptides, (ii) that only a fraction of the helix is exposed to the lipid phase, and (iii) the arrangement of light harvesting polypeptides into a polymeric unit structure.
Furthermore, 1-palmitoyl-2-[10-[4-(trifluoromethyl)diazirinyl]-

phenyl]-[9-3 H]-8-oxodecanoyl]-sn-glycero-3-phosphocholine (5)
(PTPC) has been successfully incorporated into the lipid
bilayer of chromatophores. Upon light activation, integral proteins were labeled. The detailed determination of the labeling
pattern in the primary structure of the light harvesting
polypeptides is currently in progress.
Besides chemical labels which partition into the membrane,
intrinsic markers may provide more information on the environment of the membrane peptides. Growth of the cells on medium
supplemented with F-phenylalanine results in labeling of the
different phenylalanine of the peptides under study. This
allows the investigation of the environment of these aminoacids
by ^{19}F NMR. Such experiments (in collaboration with Drs. Ghosh
and Hauser from the ETH Zurich) are under way.

Fig. 1: Radioactivity profile of the Edman degraded TID-labeled
B 870-α . Cpm values are not corrected for decreasing yields.

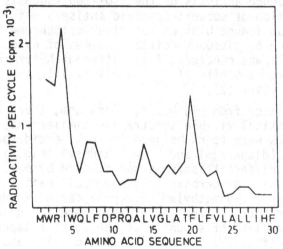

1) Zürrer, H. et al. (1983) FEBS Lett. 153, 151-155

2) Meister, H.P. and Bachofen, R. (1984) Biochim. Biophys. Acta
771, 103-106

3) Brunner, J. and Semenza, G. (1981) Biochemistry 20, 7174-
7184

4) Brunisholz, R.A. and Zuber, H. (1981) FEBS Lett. 129, 150-
154
Brunisholz, et al. (1984) Hoppe-Seyler's Z. Physiol. Chem.
365, 689-701

5) Brunner et al. (1983) Biochemistry 22, 3812-3820.

ISOZYMES AND POLYMORPHISM OF MAMMALIAN CYTOCHROME C OXIDASE

B. Kadenbach, A. Stroh and L. Kuhn-Nentwig
Fachbereich Chemie, Biochemie, Universität Marburg,
Hans-Meerwein-Straße, D-3550 Marburg, FRG

Cytochrome c oxidase is composed of various subunits, the exact number depending on the evolutionary stage of the organism. Two to three subunits have been found in bacteria, seven in yeast and thirteen in mammalian tissues (review in 1). In eucaryotic cells three subunits are of mitochondrial origin, the remainder are determined by the nuclear genome. The mitochondrially coded subunits I-III contain the two heme a and copper ions and the proton pump. The function of nuclear coded subunits is yet unknown. They are suggested to have a regulatory role (1). The number of nuclear coded subunits in mammalian cytochrome c oxidase has been determined by SDS-gel electrophoresis (2). The identity of ten different nuclear coded subunits was demonstrated by N-terminal amino acid sequence analysis of the isolated subunits (3), and by the exclusive reaction of subunit-specific antisera with the corresponding subunit on an immuno blot of rat liver cytochrome c oxidase (4). The occurrence of tissue-specific isozymes of cytochrome c oxidase in higher animals was concluded from different N-terminal amino acid sequences between subunits VIa, VIIa, VIIb, VIIc and VIII from pig (beef) heart and liver (3).

Cytochrome c oxidase, isolated from pig heart, diaphragm, liver and kidney, showed almost identical visible spectra and similar enzymatic activity. Small differences were found between the polypeptide pattern of the enzymes from heart (diaphragm) and liver (kidney) (fig. 1). The functional meaning of different isozymes was studied by measuring the protection by cytochrome c of carboxylic group modification in various subunits by 1-ethyl-3(3-dimethylaminopropyl) carbodiimide (EDC) and ^{14}C-glycine ethylester (GEE). Large differences were found in the cytochrome c-protected labelling of subunits of the heart (diaphragm) and liver (kidney) enzyme (fig. 2). Subunits VIa, VIbc, VIIa and VIII showed a much stronger and subunit II a weaker cytochrome c-protected labelling in liver and kidney than in heart and diaphragm. The data indicate a different involvement of nuclear coded subunits in the binding of cytochrome c to the enzyme from different tissues.

1) Kadenbach, B. (1983) Angew. Chem. 95, 273-281; Angew. Chem. Int. Ed. Engl. 22, 275-282.
2) Kadenbach, B., Jarausch, J., Hartmann, R. and Merle, P. (1983) Anal. Biochem. 129, 517-521
3) Kadenbach, B., Ungibauer, M., Jarausch, J., Büge, U. and Kuhn-Nentwig, L. (1983) Trends Biochem. Sci. 8, 398-400
4) Kuhn-Nentwig, L. and Kadenbach, B. (1984) FEBS-Lett. in press

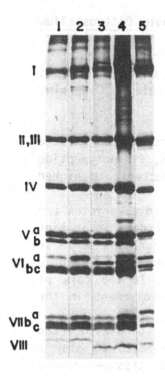

Fig. 1: SDS-polyacrylamide gel electrophoresis (2) of isolated cytochrome c oxidases from different tissues. 1: Pig heart; 2: pig diaphragm; 3: pig liver; 4: pig kidney; 5: rat liver. Subunits VIb and c are not separated in the pig enzymes. Subunit VIa from pig liver and kidney shows a double bond (as demonstrated by a subunit-specific antiserum). The lower band of subunit IV represents a proteolytic product of the upper band. Subunit III is not visible due to aggregation.

Table 1: Labelling of cytochrome c oxidase subunits from different tissues by EDC and GEE in the presence and absence of cytochrome c. 9 nmoles of the indicated isolated enzyme were incubated for 20 h at 0°C with or without 9 nmoles of cytochrome c with 2 mM EDC and 0.5 mM GEE in 10 mM NaP_i, pH 7.0 and 5 mM lauryl maltoside. The acetone precipitated enzymes were separated by SDS-PAGE and stained. Radioactive bands were visualized by fluorography, cut out and counted in a scintillation counter.

Enzyme source		heart	diaphragm	liver	kidney
		\- cpm -			
Total labelling of all subunits *		7587	6401	9339	12823
Cytochrome c- protected labelling of subunit	II	2290	1924	1884	1574
	VIa	129	100	405	459
	VIbc	445	396	670	922
	VIIa	41	57	461	413
	VIIbc	0	0	91	154
	VIII	54	47	496	412

*without cytochrome c

Cytochrome c oxidase from the cellular slime mould Dictyostelium discoideum

R. Bisson, G. Schiavo and E. Papini

Centro C.N.R. per la Fisiologia dei Mitocondri e Laboratorio di Biologia e Patologia Molecolare, Istituto di patologia Generale, Padova, Italy

Although the known catalytic functions of cytochrome c oxidase are essentially identical in all aerobic organisms, its polypeptide composition shows an increasing complexity from bacteria to higher eukaryotes (1,2).

For two of the three largest subunits, which in eukaryotes are coded for by mt-DNA, there are clear experimental data indicating their function. Subunit II binds cytochrome c and contains at least one of the four redox centers present in the enzyme (3-5) while subunit III is implicated in proton pumping (6). In the case of subunit I only indirect evidence suggest its involvement in the oxidase catalytic process (7).

The additional nuclear DNA coded polypeptides, found associated to eukariotic cytochrome c oxidase, have been proposed to play a regulative role on the base of the occurrence of tissue-specific isoenzymes (8).

D. discoideum is a slime mould which grows as single ameboid cell when supplied with appropriate nutrients. Under starvation conditions, the amoebae aggregate into multicellular bodies that eventually differentiate in a fruiting body consisting of a stalk supporting a balloon like structure filled with spores (9). Hence D. discoideum may be an useful model to study possible changes in cytochrome c oxidase as function of the different cellular processes.

In this investigation D. discoideum cytochrome c oxidase has been purified with a procedure based on hydrophobic and affinity chromatography. Under different SDS-PAGE conditions the Coomassie blue staining pattern of the enzyme shows six polypeptides apparently lacking subunit III.

The composition of the enzyme purified from cells in exponential growth and in stationary phase are clearly different. In this transition the smallest subunit is replaced by a larger polypeptide in a process that is strictly correlated with the two different

conditions.
The possible role of this change is under investigation.

References

1. Azzi, A. (1980) Biochim. Biophys. Acta 594, 231-252
2. Ludwig, B. (1980) Biochim. Biophys. Acta 594, 177-189
3. Bisson, R., Azzi, A., Gutweniger, H., Colonna, R., Montecucco, C. and Zanotti, A. (1978) J. Biol. Chem. 253, 874-880
4. Millet, F., De Jong, G., Paulson, L. and Capaldi, R.A. (1983) Biochemistry 22, 546-552
5. Darley-Usmar, V.M., Capaldi, R.A. and Wilson, M.T. (1981) Biochem. Biophys. Res. Commun. 103, 1223-1330
6. Prochaska, L., Bisson, R. and Capaldi, R.A. (1980) Biochemistry 19, 3174-3179
7. Winter, D.B., Bruyninckx, W.J., Foulke, F.G., Grinich, N.P. and Mason, H.S. (1980) J. Biol. Chem. 255, 11408-11414
8. Kadenbach, B. (1983) Angew. Chem. Int. Ed. Engl. 22, 275-283
9. Loomis, W.F. (1975) in Dyctyostelium Discoideum: A Developmental System, Academic Press, New York.

GENES FOR IMPORTED MITOCHONDRIAL PROTEINS: INSIGHTS INTO PROTEIN STRUCTURE, IMPORT AND EVOLUTION

L.A.Grivell, A.C.Maarse, C.A.M.Marres, M.De Haan, P.Oudshoorn, H.Van Steeg and A.P.G.M.Van Loon. Section for Molecular Biology, Laboratory of Biochemistry, University of Amsterdam, Kruislaan 318, 1098 SM Amsterdam, The Netherlands

Biogenesis of mitochondria is dependent on information provided by both nuclear and mitochondrial DNA. The application of rapid DNA sequencing techniques to the relatively simple organellar DNAs of animals, yeasts and plants has yielded a wealth of information on the structure of many proteins of the respiratory chain and on the ways in which they have changed during evolution (1,2). As a result, it has sometimes been possible to identify functional domains and to assess minimal requirements for function. Extension of this analysis to the mtDNAs of evolutionarily distant organisms, such as those of Trypanosomes, in which overall homologies of mtDNA-encoded proteins to their counterparts in other organisms are significantly lower (cf. ref.3) should allow further advances to be made.

In order to obtain similar information for those proteins that are imported by the mitochondrion, we have cloned yeast nuclear genes coding for a number of proteins belonging to different compartments of the organelle. These genes include those for the manganese-superoxide dismutase, subunit IV of cytochrome c oxidase and four subunits of the ubiquinol-cytochrome c reductase. Nucleotide sequences of these genes have been determined and wherever possible they have been compared with amino acid sequences of the mature proteins from both yeast and other organisms. Salient features are as follows:

Mn-superoxide dismutase (ref.4): the gene encodes a protein of 233 aa, whose sequence corresponds with that of the protein (5), except for extensions of 27 and 4 aa at the N- and C-termini respectively. The N-terminal extension is consistent with the observation that superoxide dismutase is synthesized as a precursor about 2 kDa larger than the mature protein (5). Processing at the C-terminus cannot be formally excluded, but seems unlikely because two of the extra residues are conserved in the mature proteins of other species and aspecific proteolysis is a common event in yeast. The protein has 38% homology with the Mn-enzyme of E. coli and the predicted secondary structures of both proteins are highly similar. Like other imported mitochondrial proteins, the presequence of superoxide dismutase contains several basic amino acids characteristically interspersed with neutral polar residues (mainly Ser and Thr). Acidic residues and long hydrophobic stretches are absent. Two basic residues just upstream of the mature N-terminus may form part of the recognition site for the chelator-sensitive protease that is involved in the maturation of this and other imported proteins (4,6).

Cytochrome oxidase subunit IV (ref.7): the gene encodes a protein of 155 aa. Comparison with the amino acid sequence of the mature subunit reveals an N-terminal extension of 25 aa, which is rich in basic and neutral polar residues. Its N-terminal moiety shows short stretches of homology with the pre-sequences of other imported inner membrane proteins. The mature protein is 37% homologous to bovine subunit VIa, which has a length of only 98 aa. Most of this difference in length is accounted for by an insertion of 25 aa at the N-terminus of the yeast protein.

UQ-cytochrome c reductase: genes coding for the 40, 17, 14 and 11kDa subunits of this complex have been isolated and complete nucleotide sequences for the latter three are available. As yet, amino acid sequence data on the mature proteins are lacking. The 17kDa subunit is predicted to be an unusual protein, rich in charged residues, most of which are acidic (8). Its most striking feature is a stretch of 25 acidic residues between positions 48 and 72. We have identified this protein as a homologue of the so-called "hinge-protein" of the bovine complex (9) by virtue of acidic stretches in both proteins and a 35-40% homology between the two proteins on the C-terminal side of the acidic blocks. Preferential conservation of acidic residues in this area suggests a role in function.

The 14kDa (ref.10) and 11kDa subunits of the complex are predicted to be relatively polar. They are nevertheless firmly associated with the complex, co-purifying with cytochrome b on sub-fractionation. In mutants lacking a functional cytochrome b, levels of both subunits are reduced as a result of increased turnover, suggesting that this cytochrome, either directly or indirectly mediates their interaction with the complex (10).

1. Senior, A.E. (1983) Biochim.Biophys.Acta 726, 81-95.
2. Capaldi, R.A., Malatesta, F. and Darley-Usmar, V.M. (1983) Biochim. Biophys.Acta 726, 135-148.
3. Hensgens, L.A.M., Brakenhoff, J., De Vries, B.F., Sloof, P., Tromp, M.C., Van Boom, J.H. and Benne, R. (1984) Nucl.Acids Res., submitted.
4. Marres, C.A.M., Van Loon, A.P.G.M., Oudshoorn, P., Van Steeg, H., Grivell, L.A. and Slater, E.C. (1984) Europ.J. Biochem.
5. Autor, A.P. (1982) J.Biol.Chem. 257, 2713-2718.
6. Bohni, P.C., Daum, G. and Schatz, G. (1983) J.Biol.Chem. 258, 4937-4943.
7. Maarse, A.C., Van Loon, A.P.G.M., Riezman, H., Gregor, I., Schatz, G. and Grivell, L.A. (1984) EMBO J. in press.
8. Van Loon, A.P.G.M., De Groot, R.J., De Haan, M., Dekker, A. and Grivell, L.A. (1984) EMBO J. 3, 1039-1043.
9. Wakabayashi, S., Tadeka, H., Matsubara, H., Kim, C.H. and King, T.E. (1982) J.Biochem. (Tokyo) 91, 2077-2085.
10. De Haan, M., Van Loon, A.P.G.M., Kreike, J., Vaessen, R.T.J.M. and Grivell, L.A. (1984) Europ.J.Biochem. 138, 169-177.

THE ATP SYNTHASE OPERON OF <u>ESCHERICHIA</u> <u>COLI</u>: STRUCTURE AND EXPRESSION.

Kaspar von Meyenburg[+], Jørgen Nielsen[+], Birgitte B. Jørgensen[+], Ole Michelsen[+], Flemming G. Hansen[+], and Bo van Deurs[°].

[+]Department of Microbiology, The Technical University of Denmark, 2800-Lyngby, Copenhagen, Denmark.
[°]Department of Anatomy, The Panum Institute, University of Copenhagen, 2200-Copenhagen N, Denmark.

The eight subunits of the membrane-bound ATP synthase of E.coli are coded by 8 genes in the <u>atp</u> (<u>unc</u>) operon (Fig. 1; for review see ref. 1, 2); a ninth gene in the operon, <u>atpI</u>, codes for a polypeptide which is dispensable, yet may have a role in ATP synthase function at extreme pH (3).

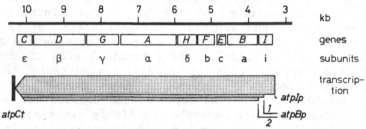

Fig. 1. Genetic organization of the E.coli atp operon. The atp genes are located between 3 and 11 kilobases counterclockwise of oriC. Arrow indicates position of promoters and termination site.

Transcription of the <u>atp</u> operon can start at three promoters one of which, <u>atpIp</u> (Fig. 1), is the most prominent one (4,5). Expression of the <u>atp</u> operon was found not to be affected by a variety of growth conditions (e.g. anaerobiosis, carbon source); thus, it appeares to be constitutive resulting in rather constant levels of ATP synthase. In order to obtain cells with increased levels of ATP synthase a series of high copy number plasmids was constructed by molecular cloning of the eight structural genes resulting in 2 fold, 4-5 fold and 10-12 fold overproduction, respectively. The ATP synthase was calculated to represent 3%, 6-7% and 18-23%, respectively, of total protein in cells with these plasmids. In wild type cells ATP synthase represents 1.5-2% of total protein equivalent to appr. 3000 enzyme complexes per average cell. While 2 or 4-5 fold wild type level of the ATP synthase had only minor effects, the 10-12 fold overproduction (strain CM2788) resulted in a pronounced inhibition of cell division and growth and in the formation of membrane cisterns and vesicles within the cells (Fig. 2).

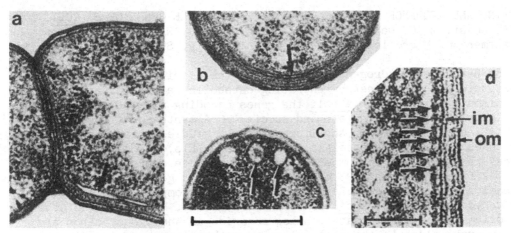

Fig. 2. Thin section electron micrographs of cells of strain CM2788 showing (a,b,c) cisterns and vesicles (bar:0.5μm) and (d) regularly spaced globular structures (arrows) on the cytoplasmic aspect of the inner membrane (bar:0.1μm).

The membrane vesicles and cisterns in strain CM2788 are like the cytoplasmic membrane covered with an appr. 100 Å thick layer of protein (Fig. 2). The globular substructure of the proteinlayer on the vesicles and the cytoplasmic membrane suggested that this layer represents the F_1 sectors attached to F_0 sectors in the membrane. Insertion of the hydrophobic F_0 sector subunits into the cytoplasmic membrane presumably results in an expansion of the total membrane area leading to the spontaneous invagination of the cytoplasmic membrane at random places. These invaginations may seal themselves off yielding closed cytoplasmic vesicles and cisterns, the lumen of which would represent periplasmic space, while their outer side would be equivalent to the inner side of the cytoplasmic membrane.

Rendering the expression of the atp genes inducible by linking them to a regulated promoter allows us now to study not only the fine kinetics of expression but also the time course of biogenesis of membrane vesicles.

References:
(1) Futai, M. and Kanazawa, H. (1983) Microbiol Rev. 47:285-312.
(2) von Meyenburg, K. et al. (1982) Tokai J. Exp. Clin. Med. 7:23-31.
(3) Michelsen, O. (1984) this volume.
(4) von Meyenburg, K. et al. (1982) Mol. Gen. Genet. 188:240-248.
(5) Nielsen, J. et al. (1984) Mol. Gen. Genet. 193:64-71.

CLONING AND SEQUENCE ANALYSIS OF CITRIC ACID CYCLE GENES OF E. COLI

J. R. Guest, M. G. Darlison, M. E. Spencer, R. J. Wilde, and D. Wood
Department of Microbiology, Sheffield University, Sheffield, U.K.

The pyruvate dehydrogenase complex and the citric acid cycle constitute the major energy-generating catabolic pathway of aerobic organisms. In Escherichia coli the genes encoding all the steps in this pathway except aconitase and succinyl CoA synthetase, have been identified and located in the linkage map by studies with mutants [1]. The genes encoding the specific dehydrogenase (E1p, aceE) and transferase (E2p, aceF) components of the pyruvate dehydrogenase complex form an operon at 3 min, close to the lpd gene encoding the lipoamide dehydrogenase (E3) components of two complexes (Fig.1C). The dehydrogenase (Eo, sucA) and transferase (E2o, sucB) genes of the 2-oxoglutarate dehydrogenase complex comprise an analogous operon at 17 min, very close to the gltA (citrate synthase) and sdh (succinate dehydrogenase) genes (Fig.1A). Because the chemical reactions of the cycle form two successive groups of five analogous reactions, it has been suggested that the corresponding pairs of enzymes may have evolved by gene duplication and diversification [2]. It has also been suggested that the cycle may have evolved from pre-existing oxidative and reductive pathways leading to 2-oxoglutarate and succinate, by coupling them into an oxidative cycle with 2-oxoglutarate dehydrogenase [3]. Under anaerobic conditions, the primitive non-cyclic situation is restored because the 2-oxoglutarate dehydrogenase complex is repressed; succinate dehydrogenase is also repressed but the interconversion of succinate and fumarate is catalysed by a derepressed fumarate reductase. The reductase is encoded by the frd genes at 94 min (Fig.1B).

The advent of gene cloning and nucleotide sequence analysis has promoted the elucidation of enzyme primary structures and facilitated

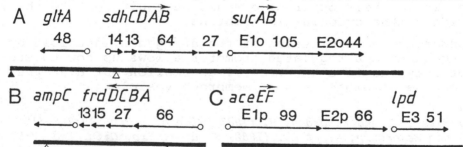

Fig. 1. Organisation of citric acid cycle and related genes. Segments of E. coli chromosome (A, 11 kb; B, 5 kb; C, 7 kb), sizes of gene products (kd), and HindIII (Δ) and EcoRI (▲) targets are shown.

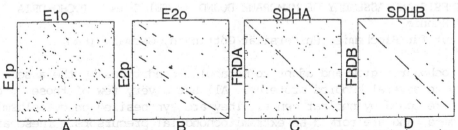

Fig. 2. Homology between enzyme subunits. Comparison matrices by DIAGON [9]: probability < 0.001, span = 15-21 and scale = 100 residues.

the search for insights into evolutionary relationships between specific enzymes. So far the primary structures of the E. coli pyruvate and 2-oxoglutarate dehydrogenase complexes [4,5], citrate synthase [6], fumarase [7], succinate dehydrogenase [5] and fumarate reductase [8] have been defined by sequencing the corresponding genes. Segments of DNA encoding specific functions were isolated in phage and plasmid vectors by screening for their ability to complement mutant lesions and the genes were located by genetic analysis, transcription-translation studies, and sequence analysis (Fig. 1).

The E1 components of the pyruvate and 2-oxoglutarate dehydrogenase complexes exhibit very little sequence homology (Fig.2A), suggesting that they are not evolutionarily related, but the E2 components are strikingly homologous (Fig.2B).

The sdh and frd operons each contain four genes but the gene orders differ (Fig.1). The flavoprotein subunits (SDHA and FRDA) are remarkably homologous (Fig.2C), as are the iron-sulphur protein subunits (SDHB and FRDB; Fig. 2D). The hydrophobic subunits (C and D) are similar in size and composition but not sequence. This suggests that the major subunits have evolved from common ancestors and that their functional diversification is primarily due to the regulatory mechanisms controlling the two operons.

1. Bachmann, B. J. (1983) Microbiol. Rev. 47, 180-230
2. Weitzmann, P. D. J. (1981) Adv. Microb. Physiol. 22, 185-244
3. Gest, H. (1981) FEMS Lett. 12, 209-215
4 Stephens, P. E., Lewis, H. M., Darlison, M. G. and Guest, J. R.
        (1983) Eur. J. Biochem. 135, 519-527
5. Guest, J. R. et al.(1984) Eur. J. Biochem. & Biochem.J. in press
6. Ner S. S., Bhayana, V., Bell, A. W., Giles, I. G., Duckworth,
        H. W. and Bloxham, D. P. (1983) Biochemistry 22, 5243-5249
7. Miles, J. S. and Guest, J. R. (1984) N. A. Res. 12, 3631-3642
8. Cole, S. T. (1984) Biochem. Soc. Trans. 12, 237-238
9. Staden, R. (1982) Nucl. Acid. Res. 10, 2951-2961

BIOGENESIS AND ASSEMBLY OF MEMBRANE BOUND PROTEINS IN MITOCHONDRIA.
Bernd Hennig
Institut für Biochemie, Universität Göttingen, Germany (FRG)

Mitochondria grow and adapt to metabolic conditions by the assembly of up to several hundred proteins. All but a very few of those proteins are coded by nuclear genes. They are synthesized on cytoplasmic ribosomes and are formed as extramitochondrial precursors. These are taken up from a pool of free precursors into mitochondria by posttranslational mechanisms. This is in strong contrast to the assembly of proteins adressed to other cellular membranes, because the biogenesis of those proteins is in general governed by a cotranslational mechanism at the endoplasmic reticulum.

Some sixty mitochondrial protein precursors have been identified so far. With rare exceptions they were found to exhibit aminoterminal extensions varying in size between 0.5 to 10 kD. The few precursors without any extension (e.g. porin, cytochrome c, ATP/ADP carrier) differ in other covalent modifications and in molecular conformation from the mature proteins.

The posttranslational processes by which mitochondrial proteins become assembled at their functional sites were experimentally resolved and have conceptually been discriminated into the following steps: 1) Specific interaction of precursors with the mitochondrial surface. 2) Translocation of precursors into or across the outer or/and inner mitochondrial membrane. 3) Processing of the precursors to the mature proteins.

The first posttranslational step, i.e. the recognition between precursor and mitochondria, is mediated by receptor sites at the mitochondrial outer membrane (1,2). Additional cytoplasmic components are not required for selective and specific binding of purified precursors to mitochondria. The affinity constant and the number of high-affinity binding sites specific for apocytochrome c was determined at mitochondria by Scatchard analysis (1). Since only apocytochrome c has so far been found to interact with this particular receptor, several types of receptors must be present for the various precursors though the number of different types is expected to be rather limited.·

Binding of precursors to mitochondria decreases substantially when the mitochondria have been pretreated with proteases under conditions which do not completely disrupt the outer membrane. Simultaneously with this decrease of receptor activity mitochondria loose also the ability to import precursors. This not only indicates that proteins

are involved at receptor sites, but also demonstrates strong functional interrelationship between receptor sites and subsequent translocation of precursors into mitochondria.

Translocation of precursors into mitochondria proceeds by at least
two basically different mechanisms: Import of those precursors which
either become inserted into outer membrane or are translocated solely
across the outer membrane, is not prevented by deenergization of the
inner membrane. However, import of most precursors actually involves
also permanent or transient uptake into the inner membrane. Translocation of these precursors strictly depends on the membrane potential of
the inner membrane (3). Furthermore, it has been suggested that contact sites between outer and inner membrane are involved in translocation of these latter precursors. Yet the validity of this concept is
not experimentally proven so far.

Processing of precursors to the mature proteins proceeds not necessarily concomitant with translocation, but it occurs eventually subsequent to translocation: Processing can be inhibited by certain metal-
ion chelating reagents (e.g. o-phenanthroline) without affecting the
proper translocation step (4). The processing protease resides in the
mitochondrial matrix. During gel filtration of a mitochondrial extract the processing activity is recovered in fractions corresponding
to proteins of a molecular size larger than 100 kD. The possibility is
not excluded that the processing activity is composed of a mixture of
different proteases, since no unique consensus sequence is apparent at
the sites where precursors are proteolytically processed.

Several proteins are processed by two successive events in order to
be finally assembled at their functional sites (5). However, this feature and equally other unique details of the diverse import pathways
do not appear to be directly related to any particular submitochondrial site where percursors become finally assembled.

(1) Hennig, B., Koehler, H. and Neupert, W. (1983) Proc.Natl.Acad.
    Sci.USA 80, 4963-4967
(2) Riezmann, H., Hay, R., Witte,C., Nelson, N. and Schatz, G.
    (1983), The EMBO J. 2, 1113-1118
(3) Schleyer, M., Schmidt, B. and Neupert, W., (1982) Eur. J.
    Biochem. 125, 109-116
(4) Zwizinski, C. and Neupert, W. (1983) J. Biol. Chem. 258, 13340-
    13346
(5) Ohashi, A., Gibson, J., Gregor, I. and Schatz, G. (1982) J. Biol.
    Chem. 257, 13042-13047

R. SPHAEROIDES, GA b/c$_1$ COMPLEX GENES: IDENTIFICATION AND SEQUENCING.
N. Gabellini*, U. Harnisch[+], J.E.G.McCarthy*, G. Hauska°, and W. Sebald*

*Abt. Cytogenetik, GBF-Gesellschaft für Biotechnologische Forschung
mbH., D-3300 Braunschweig, F.R.G.
Institut für Biochemie, Universität Düsseldorf, D-4000 Düsseldorf,
F.R.G.
°Institut für Botanik, Universität Regensburg, D-8400 Regensburg,
F.R.G.

The photosynthetic and respiratory electron transport chain of
R. sphaeroides includes a membrane bound ubiquinol-cytochrome c
oxidoreductase (b/c$_1$ complex). The enzyme isolated from photosynthetic membranes is composed of cyt b (40 kd), cytochrome c$_1$ (33kd),
Fe-S-protein (25 kd) and a 10 kd polypeptide (1). This minimal structure catalyzes electron transport from ubiquinol to cyt c and gene-
rates membrane potential and pH gradient when reconstituted into liposomes (2).

The isolation and analysis of the genes of the b/c$_1$ complex subunits
represents a possibility to elucidate the primary structure of the enzyme and to understand the regulation and the coordination of the subunit synthesis.

The cloned gene of the Fe-S-protein of R. sphaeroides b/c$_1$ complex,
was identified by cross hybridization, using a 100 bp probe of cDNA derived from the homologous Fe-S-protein gene of N. crassa (3). Fourteen
positive clones with cross-hybridizing sequences were identified,
among 10.000 E. coli clones containing partial SAU3A fragments (5-9 kb)
of R. sphaeroides genomic DNA. Two plasmids chosen for further analysis covered a segment of 13 kb of genomic DNA surrounding the Fe-S-
gene. The restriction map (Fig. 1) of the cloned DNA is colinear with
the map of the genuine genomic DNA established by Southern blott analysis.

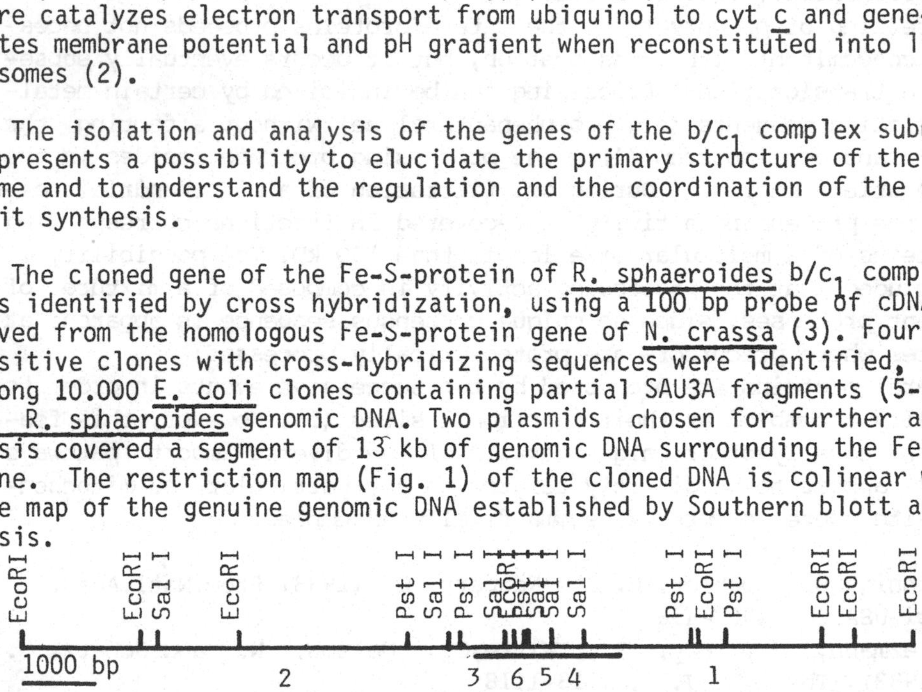

Figure 1: Restriction map of R. sphaeroides genomic DNA including
the Fe-S-protein and cyt b genes (underlined).

Partial DNA sequences were determined for Sal I restriction fragments 1, 3, 4, 6. The FE-S-protein gene of R. sphaerodies b/c$_1$ complex
is contained in Sal I fragments 3 and 6. It has an approximate size
of 600 bases. The terminal region of the gene containing the sequence

encoding for the 4 cysteins, which most likely bind the 2 Fe 2S cluster, is highly homologous with the correspondent nuclear gene of the mitochondrial Fe-S-protein (Fig. 2). The remaining sequence shows much lower homology.

Immediately after the Fe-S-protein gene, the gene for cytochrome b was detected, as deduced from the high homology of the coded aminoacid sequence with the mitochondrial cytochrome b(4,5)(Fig. 2). An intergenic distance of only 12 nucleotides including a Shine-Dalgarno consensus sequence, is deduced assuming that the cyt b starts with the first methionine detected after the stop codon of the Fe-S-protein gene. This short spacing suggests that the genes for the two subunits of the b/c$_1$ complex of R. sphaeroides are organized in an operon.

Fe-S-protein

R.sphaer. EWLVILGVCSHLGCVPMGDKSGDFGGWFCPCHGSHYDSAGHIRKGPAPRNLDIP
                140   150   160   170   180
N.crassa   EWLVMLGVCTHLGCVPIGE-AGDYGGWFCPCHGSHYDISGRIRKGPAPLNLEIP

cyt b

R.sphaer. PTPKNLNWWWLWGIVLAFTLVLQIVTGIVLAMHYTPHVACR
               20   30   40   50   60
yeast   PQPSSINYWWNMGSLLGLCLVIQIVTGIFMAMHYSSNIELA

Figure 2: Comparison of homologous aminoacid sequences of the Fe-S-protein and cyt b of R. sphaeroides with the correspondent mitochondrial proteins, which are numerized above.

References

1) Gabellini, N., Bowyer, J.R., Hurt, E.C., Melandri, B.A., and Hauska, g. (1982) Eur. J. Biochem. 126, 105-111.

2) Hurt, E.C., Gabellini, N., Shahak, Y., Lokau, W., and Hauska, G. (1983) Arch. Biochem. Biophys. 225, 879-885.

3) Harnisch, U. et al. (1984) 3rd. EBEC abstract .

4) Widger, W.R., Cramer, W.A., Herrmann, R.G., and Trebst, A. (1984) Proc.Natl.Acad.Sci. 81, 674-678.

5) Nobrega, F.G., and Tzagoloff, A. (1980) J. Biol. Chem. 255, 9828-9837.

ORGANIZATION AND STRUCTURE OF GENES ENCODING α AND β SUBUNITS OF
MITOCHONDRIAL ATPASE IN YEAST AND HIGHER PLANTS

Marc Boutry

Laboratory of Enzymology, University of Louvain, 1348 Louvain-la-
Neuve, Belgium
Department of Biochemistry, UTHSCSA, San Antonio, TX 78284, and
Laboratory of Plant Molecular Biology, The Rockefeller University,
New York, NY 10021-6399.

We have isolated and characterized several mutants of the fission
yeast Schizosaccharomyces pombe defective in mitochondrial ATPase
activity. Among these are two mutants which lack specifically the α
or the β subunits of the F_1ATPase (1). These two mutants were used
to select by in vivo complementation two S. pombe genomic clones
which contain the corresponding structural genes (2). Classical
genetic studies showed that the genes encoding the α and β subunits
are unlinked (3) in contrast to the bacterial ATPase genes which are
organized in an operon.

A plasmid carrying the β subunit gene of the budding yeast Saccharomyces cerevisiae was used to transform the S. pombe mutant lacking
the β subunit. Transformation restored partial ATPase activity as
well as growth on a mitochondria-dependent carbon source (4). These
results indicate that the genetic defect of the S. pombe mutant can
be complemented by the gene from S. cerevisiae. The transformant
contains a normal amount of a hybrid ATPase consisting of the mature
β subunit of S. cerevisiae and the remaining subunits of the S. pombe
ATPase (4). These data indicate that the mechanism of mitochondrial
import and the assembly of F_1ATPase subunits are similar in these
evolutionary divergent yeasts.

Higher plant mitochondrial F_1ATPase has a structure similar to
other coupling factors (5). However, in contrast to the situation in
other organisms where all the F_1ATPase subunits are encoded by nuclear genes, the plant α subunit is synthesized by isolated mitochondria, implying that the gene is localized within the organelle (5).
In contrast, the β subunit is not made by isolated mitochondria,
suggesting that the gene is nuclear encoded. By heterologous hybridization with the yeast β subunit gene, we have selected from a plant
(Nicotiana plumbaginifolia) genomic library, clones carrying nuclear
genes for the β subunit. The nucleotide sequence of one of them
has been determined. Unlike the yeast gene, the plant gene contains
several introns; nevertheless, the coding sequence is well conserved

in comparison to the yeast gene (75% homology in the amino acid sequence). Southern blot analysis indicated that 2 or 3 β subunit genes are present in the N. plumbaginifolia genome. Partial nucleotide sequence of a second β subunit gene indicated a sequence conservation in the exons but a divergence in the introns. S_1 mapping and Northern analysis are underway to determine the level of expression of the different genes.

1. M. Boutry and A. Goffeau (1982) Eur. J. Biochem. 125, 471-477.
2. M. Boutry, A. Vassarotti, M. Ghislain, M. Douglas and A. Goffeau (1984) J. Biol. Chem. 259, 2840-2844.
3. A. Vassarotti, M. Boutry, A.-M. Colson and A. Goffeau (1984) J. Biol. Chem. 259, 2845-2849.
4. M. Boutry and M. Douglas (1983) J. Biol. Chem. 258, 15214-15219.
5. M. Boutry, M. Briquet and A. Goffeau (1983) J. Biol. Chem. 258, 8524-8526.
6. M. Boutry and N.-H. Chua, manuscript in preparation.

This work has been performed in the laboratories and under the kind supervision of Drs. A. Goffeau (Louvain-la-Neuve), M. Douglas (San Antonio) and N.-H. Chua (New York).

PRIMARY STRUCTURE AND FUNCTION OF H^+-ATPase SUBUNITS OF E.COLI:ESSENTIAL RESIDUES AND DOMAINS

H. Kanazawa and M. Futai

Department of Microbiology, Faculty of Pharmaceutical Sciences, Okayama University, Okayama, Japan

To obtain information relating structure to function of the H^+-ATPase, we have determined complete amino acid sequence of F_1F_0 of E. coli (1). To date, the altered amino acid residues have only been determined in mutant strains defective in the c subunint of F_0. Knowledge of essential residues determined by chemical modification of F_1F_0 has been limited for the β subunit (2). In the present study, we estimated mutations of the α, β, γ and a subunits in a defined portion of the subunits and analyzed altered properties of the membranes and F_1F_0 for relating the primary structure to function of the subunits.

A mutant AN120 defective in the α subunit (2) has no catalytic activity of F_1ATPase at the steady state but has normal hydrolytic activity at a single turnover process. The mutant F_1 also lacks normal conformational transmission from the α to β subunit, essential for the steady state catalysis, from the α to β subunit. The mutation was mapped between residues 370 and 387 by genetic recombination test using plasmids carrying various portions of the α subunit gene (4). Finally the mutated gene was cloned and substitution of cytosine to thymine at residue 1118 resulting in substituion of Phe for Ser at residue 373 was identified. Thus we concluded that Ser-373 is essential for steady state catalysis by F_1ATPase (5).

Of 35 mutations sixteen were identified to be defective in the β subunit. Twelve of these mutants were defective in between residues 287 and 489 (domian II), while three mutants were defective in between residues 17 and 286 (domain I). All the mutants defective in domain II showed imcomplete assembly of F_1F_0 and lacks proton channel activity, suggesting that this domain is essential for the assembly of F_1F_0 (6). A mutant KF11 defective in domain I has a normal complex of F_1F_0. The mutant F_1 retains only 10 % of Mg^{2+}-ATPase activity of the wild type, but 60 % of Ca^{2+}-ATPase activity (7). The mutated gene of KF11 was cloned and altered sequence was determined: cytosine at residue 524 was changed to thymine, resulting in substitution of Phe for Ser at residue 174. These results indicated that Ser-174 is essential for Mg^{2+}-ATPase activity but not for Ca^{2+}-ATPase activity (8).

Three mutations (strains KF1, KF10 and NR70) and two mutations (strains KF12 and KF13) defective in the γ subunit were mapped between residues 1 and 167, and between 168 and 287, respectively. The former group has no assembly of F_1ATPase on the membranes, whereas the latter group has unstable but normal complex of F_1, suggesting that the amino terminal half of the γ subunit is essential for assembly of F_1 (9).

A mutation allele of strain KF3 defective in the a subunit was cloned and substitution of thymine for cytosine at residue 58 was determined by sequencing DNA. The mutation caused replacement of Gln at residue 20 by termination codon, resulting in no synthesis of the normal a subunit. Loss of the a subunit in the mutant resulted in no assembly of F_1F_0 subunits in the membranes, giving the first evidence that the a subunit is required for the first step of F_1F_0 assembly as suggested by Cox et al (10).

REFERENCES

1. Kanazawa, H., and Futai, M. (1982) Ann.N.Y.Acad.Sci. 402, 45-64
2. Futai, M., and Kanazawa, H. (1983) Microbiol. Rev. 47, 285-317
3. Kanazawa, H., Saito, S., and Futai, M. (1978) J. Biochem. 84, 1513-1517
4. Kanazawa, H., Noumi, T., Matsuoka, I., Hirata, T., and Futai, M. (1984) Arch. Biochem. Biophys. 228, 258-269
5. Noumi, T., Mosher, M., Natori, S., Futai, M., and Kanazawa, H. (1984) J. Biol. Chem. in press
6. Kanazawa, H., Noumi, T., Oka, N., and Futai, M. (1983) Arch. Biochem. Biophys. 227, 596-608
7. Kanzawa, H., Horiuchi, Y., Takagi, M., Ishino, Y., and Futai, M. (1980) J. Biochem. 88, 695-703
8. Noumi, T., Futai, M., and Kanazawa, H. (1984) J. Biol. Chem. in press
9. Kanazawa, H., Noumi, T., Futai, M. and Nitta, T. (1983) Arch. Biochem. Biophys. 223, 521-532
10. Kanazawa, H., Oka, N., Noumi, T., and Futai, M. Submitted
11. Cox, G.B., Downie, J.A., Langman, L., Senior, A.E., Ash, G., Fayle, D.R.H., and Gibson, F. (1981) J. Bacteriol. 148, 30-42

PURIFICATION, AMINO ACID SEQUENCE AND SECONDARY STRUCTURE OF AN HYDROPHOBIC SUBUNIT OF YEAST ATP-SYNTHASE

J. VELOURS[°], J. HOPPE[''], W. SEBALD[''], P. DAUMAS['''], F. HEITZ[''']
and B. GUERIN[°]

[°] Institut de Biochimie Cellulaire, 1 rue Camille St Saëns
33077 BORDEAUX, FRANCE

[''] Department of cytogenetics, GBF - Gesellschaft für Biotechnologische Forchung mbH, D 3300 BRAUNSCHWEIG, FRG

['''] Laboratoire de Physicochimie des Systèmes Polyphasés
34033 MONTPELLIER, FRANCE

The purification (I) and the aminoacid sequence of a proteolipid translated on mitoribosomes in yeast mitochondria is reported. This protein was found associated to the immunoprecipitated ATP synthase (2). It was extracted from whole mitochondria with chloroform/methanol (2/1) and purified by chromatography on phosphocellulose and reverse phase h.p.l.c. A molecular weight of 5500 was estimated by chromatography on Bio-Gel P-30 in 80 % formic acid. The complete aminoacid sequence of this protein was determined by automated solid phase Edman degradation of the whole protein and of fragments obtained after cleavage with cyanogen bromide. The sequence analysis indicates a lenght of 48 aminoacid residues. The calculated mol. wt. of 5870 corresponds to the value found by gel chromatography. This polypeptide contains three basic residues and no negatively charge side chain. The three basic residues are clustered at the C terminus. The primary structure of this proteolipid is identical to the predicted sequence deducted from an URF located on yeast mitochondrial DNA between Oli 2 gene and the Cytochrome oxidase subunit I gene (3). Evidences occur that this protein is also encoded in mitochondrial DNA of Aspergillus nidulans (4), since the predicted sequence revealed 50 % homology with the primary structure of the protein isolated from mitochondria. An analogous gene (URF A6L) could be found in mammalian mitochondria near the subunit 6 gene (5). As a matter of fact, a stricking similarity with the proteolipid of fungus was observed in the N terminal region of the protein which exhibits the common sequence : Met - Pro - Gln - Leu -

The secondary structure prediction methods indicate that the protein traverses the membrane once by an α helical segment very likely from residue 15 to 35. The C terminal part of the protein is predicted to be outside the membrane. The α helix structrure was confirmed by CD spectroscopy in various conditions. When introduced into lipid bilayers, the protein was shown to increase the membrane conductivity through discrete fluctuations of the transmembrane current. Such a behavior is reminiscent of that of Alamethicin (6) and suggests that the pores are built by assembly of α helices.

References

1. VELOURS, J., ESPARZA, M., HOPPE, J., SEBALD, W. and GUERIN, B. (1984) EMBO J., 3, 207-212.

2. ESPARZA, M., VELOURS, J. and GUERIN, B. (1981) FEBS Lett., 134, 63-66.

3. MACREADIE, I.G., NOVITSKI, C.E., MAXELL, R.J., JOHN, U., OOI, B.G., Mc MULLEN, G.L., LUTKINS, H.B., LINNANE, A.W. and NAGLEY, P. (1983) Nucleic Acids Res., 11, 4435-4451.

4. GRISI, E., BROWN, T.A., WARING, R.B., SCAZZOCCHIO, C. and DAVIS, R.W. (1982) Nucleic Acids Res., 10, 3531-3539.

5. ANDERSON, S., BANKIER, A.T., BARRELL, B.G., de BRUIJN, M.H.L., COULSON, A.R., DROUIN, J., EPERON, I.C., NIERLICH, D.P., ROE, B.A. SANGER, F., SCHREIER, P.H., SMITH, A.J.H., STADEN, R. and YOUNG, I.G. (1981) Nature, 290, 457-465.

6. FOX, R.O. and RICHARDS, F.M. (1982) Nature, 300, 325-330.

REQUIREMENT FOR ENERGY DURING EXPORT OF ß-LACTAMASE IN ESCHERICHIA
COLI IS FULFILLED BY THE TOTAL PROTONMOTIVE FORCE

Evert P. Bakker[1] and Linda L. Randall[2]
[1] Fachgebiet Mikrobiologie, Universität Osnabrück, D-4500 Osnabrück, FRG.
[2] Biochemistry/Biophysics Program, Washington State University,
Pullman, Washington 99164-4660, USA.

Proteins that are exported into the periplasmic space or to the outer membrane of E. coli are synthesized on membrane-bound polysomes in
precursor forms which contain extra amino-terminal sequences of about
25 amino acids. During export, these amino-terminal sequences, termed
leader or signal sequences, are proteolytically removed by an enzyme,
leader peptidase (review in (1)). Export of proteins requires energy
(2-4). It is vital to know whether this requirement is specific for
ATP, for total protonmotive force or for the membrane potential. We,
therefore, investigated the mechanism of export of plasmid pBR322-en-
coded ß-lactamase in valinomycin-treated cells of E. coli. The magnitude of the membrane potential of these cells was manipulated by varying the KCl concentration in the medium, and the pH-gradient was varied by incubating the cells at different pH values. Both components
were simultaneously affected by the addition of the protonophore FCCP.

Table I shows that in the control strain KCl-induced depolarization
inhibited processing and decreased the ATP level, whereas in the ATP-
synthase negative strain KCl inhibited processing and slightly increased the ATP level. This result indicates that the inhibition of processing was not a result of decreased cytoplasmic ATP concentration.

TABLE I: Inhibition of processing of ß-lactamase is independent of ATP.
Valinomycin-treated cells were chased with the KCl or NaCl
concentration indicated.

condition	strain G6/pBR322 (control)		strain BH273/pBR322 (unc)	
	processing	ATP content (nmol/mg)	processing	ATP content (nmol/mg)
0.6 mM KCl	proceeds	8.5	proceeds	6.0
100 mM KCl	inhibited	1.9	inhibited	7.9
100 mM NaCl	proceeds	n.d	n.d.	n.d.

Half-maximal accumulation of precursor of ß-lactamase was observed
in all cases when the level of protonmotive force was decreased to a-
bout 150 mV. Under these conditions the membrane potential varied from
65 to 140 mV and the pH-gradient from 95 to 25 mV (Fig. 1). Thus, the
energy requirement is satisfied by the total protonmotive force, rather
then by either the membrane potential, the pH-gradient, or the value

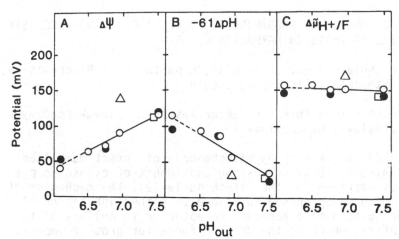

Fig. 1: Export of ß-lactamase requires total protonmotive force.
Values of $\Delta\Psi$ (A), $-61\Delta pH$ (B) or $\Delta\tilde{\mu}_H+/F$ at which half-maximal processing was observed are given as a function of pH_{out}. Symbols: cir-
cles and triangles, strain G6/pBR322; squares, strain BH273/pBR322;
triangles, FCCP-treated cells;open symbols, control cells; closed symbols, the chase solution contained chloramphenicol (100 µg/ml).

of cytoplasmic pH (5). We,therefore, reject a mechanism, according to
which export of protein is triggered by the membrane potential that
pulls negative domains of the exported protein across the cytoplasmic
membrane (4). The role of protonmotive force may either be direct, in
that a neutral domain of the protein is exported across the cytoplasmic membrane in symport with hydroxyl ions. Alternatively, protonmotive force may have a indirect effect, in that it activates the proteins involved in export of other proteins. It should be investigated
whether the import of protein into mitochondria also requires total
protonmotive force, rather than membrane potential alone (6).

REFERENCES:
1) Michaelis,S. and Beckwith, J.(1982) Annu. Rev. Microbiol. 36,435-
   465.
2) Date,T., Goodman, J. and Wickner, W.(1980) Proc. Natl. Acad. Sci.
   U.S.A. 77, 4669-4673.
3) Enequist, H.G., Hirst, T.R., Hardy, S.J.S., Harayama, S. and
   Randall, L.L. (1981) Eur. J. Biochem. 116, 227-233.
4) Daniels, C.J., Bole, D.G., Quay, S.C. and Oxender, D.L. (1981)
   Proc. Natl. Acad. Sci. U.S.A. 78, 5396-5400.
5) Bakker, E.P. and Randall, L.L. (1984) EMBO J. 3,895-900.
6) Schleyer, M., Schmidt, B. and Neupert, W. (1982) Eur. J. Biochem.
   125,109-116.

THE MECHANISM OF ACTION OF RHODAMINE 6G ON THE IMPORT AND PROCESSING
OF MITOCHONDRIAL PROTEINS IN MAMMALIAN CELLS

B.D. Nelson, J. Kolarov[1] and S. Kuzela[1], Department of Biochemistry,
University of Stockholm, Stockholm, Sweden

[1]permanent address: Institute for Cancer Research, Slovak Academy of
Sciences, Bratislava, Czechoslovakia

Rhodamine 123 is taken up by mitochondria of intact mammalian
cells (1). It also inhibits processing and import of cytosolic precursor peptides destined for the mitochondria (2). The mechanism of
Rhd 123 inhibition was proposed to be due to either inhibition of
precursor binding to a mitochondrial receptor or to collaps of the
membrane potential, which is the driving force for protein import.
During the course of our studies on the import of inner membrane peptides in mammalian cells, we found that Rhd 6G is also a potent inhibitor of procursor processing. We show, however, that Rhd 6G inhibition is due to a direct effect on the matrix processing enzyme.
  In vivo studies. Rat Zajdela hepatoma ascites cells were labeled in vitro with 35-S methionine, fractionated with digitonin into
a soluble and a pellet fraction and then immunoabsorbed with monospecific antibodies raised against subunits of rat liver cytochrome
b-c_1 complex. After a 10 min pulse, small amounts of pFeS protein,
(Fig. 2) and pcyto c, were detected in both fractions, but the bulk of
the protein was already processed. Rhd 6G inhibited processing completely and led to accumulation of the precursors in both the soluble
and pellet fractions. Association of precursor with the pellet fraction indicated that Rhd 6G does not inhibit binding to a mitochondrial receptor. Rhodamine 6G was compared to an uncoupling agent, CCCP,
and found to have the following advantages: a) Rhd 6G does not appear
to uncouple oxidative phosphorylation at concentrations which inhibit
processing, b) it has little effect on cytoplasmic protein synthesis,
whereas CCCP reduces this process drastically, and c) Rhd 6G appears
to be a more effective inhibitor of processing in whole cells than
CCCP. Based upon these findings, our in vivo experiments suggest that
the most likely mode of action of Rhd G6 is on matrix processing.
  In vitro studies. Free polysomes from rat liver were translated
in a reticulocyte lysate system and immunabsorbed with antibodies
against rat liver inner membrane proteins. Precursors were detected
for core protein 1, cytochrome c_1 and the iron sulfur protein of
Complex III, the α and β subunits of F_1-ATPase, and subunit IV of
cytochrome oxidase. The precursor nature of these radio-labeled
peptides was confirmed by competition with unlabeled antigens and

83

by processing with a crude matrix fraction from rat liver mitochondria. Rhodamine 6G and o-phenanthroline inhibited matrix processing, (Fig. 2) thus confirming our conclusion from whole cell experiments that Rhd 6G acts by inhibiting the matrix protease rather than by collapsing the mitochondrial membrane potential.

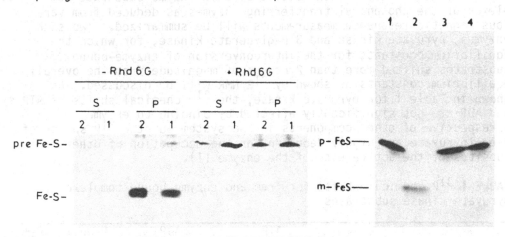

Fig. 1. Synthesis of the FeS protein in hepatoma cells. Cells labeled in the presence or absence of Rhd 6G and were fractionated into soluble (S) or pellet (P) fractions. Lanes 1, 10 min pulse, lanes 2, 10 min pulse + 10 min chase.

Fig. 2. In vitro translation of the FeS protein. Reticulocyte lysates were programmed with free polysomes from rat liver. FeS protein was immunoabsorbed from: 1) untreated lysate, 2) lysate processed with matrix fraction, 3) processed in the presence of 2 mM o-phenanthroline or 4) 2 mM Rhd 6G.

References.

1) Johnson, L.V., Walsh, M.L., Bockus, B.J. and Chen, L.B. (1981) J. Cell Biol. 88, 526-535.

2) Morits, T., Mori, M., Ikeda, F., and Tatibana, M. (1982) J. Biol. Chem. 257, 10527-10550.

NMR STUDIES OF ENZYME SUBSTRATE INTERACTIONS IN ATP REACTIONS
M. Cohn
Institute for Cancer Research
Philadelphia, Pennsylvania 19111, U.S.A.

The changes induced at the active site in enzyme-substrate complexes of the phosphoryl transferring enzymes as deduced from various magnetic resonance measurements will be summarized. Two such enzymes, pyruvate kinase and 3-P-glycerate kinase, for which the equilibrium constants for the interconversion of enzyme-bound substrates shifted more than 2 orders of magnitude from the overall equilibrium constants as shown by ^{31}P NMR will be discussed. As shown in Table 1 for pyruvate kinase, the ^{31}P chemical shifts of ATP and ADP are not significantly affected by binding to enzyme irrespective of other components in the system but the ^{31}P shifts of P-enolpyruvate are highly dependent on the occupation of other subsites of the active site of the enzyme (1).

TABLE 1 ^{31}P chemical shifts for free and enzyme-bound complexes of pyruvate kinase substrates

Complex	ATP			ADP		P-enol-pyruvate	Pi+ AMP
	α	β	γ	α	β		
Free substrates	10.8	21.4	5.8	10.4	6.2	-0.9	
Free substrates + Mg^{2+}	10.6	19.2	5.5	10.1	5.9		
Bound substrates	11.1	21.7	6.4	10.7	6.0	-0.2	
Bound substrates + Mg^{2+a}	10.5	19.0	5.8	10.0	5.7	-2.3	
E.P-enolpyruvate (-K$^+$)						-0.1	
Equilibrium mixture	10.7	19.2	5.8	10.7	5.8	-3.6	-3.6
Equilibrium mixture + EDTA	10.9	21.6	6.2	10.9	6.2	-3.1	-3.7

^{a}[Mg^{2+}] is ∿ 35% in excess of the sum of enzyme and nucleotide concentrations.

In the case of 3-P-glycerate kinase, in addition to observing the effect of binding on the ^{31}P chemical shifts of the substrates (2), the effect of binding of the substrates on the ^1H-NMR of the yeast and NMR of the yeast and horse enzymes has been investigated (3). A photo-CIDNP technique (4) has been used to observe the highly enhanced resonance signals of the histidine, tyrosine and tryptophan residues which are accessible to a photo-excited flavin dye in the yeast and horse 3-P-glycerate kinases and their substrates complexes. Interpretation of a chemical shift change in a protein upon ligand binding in the usual NMR experiment is ambiguous in that it is not possible to distinguish between 1) an indirect conformational change at a distance from the active site and 2) a direct contact between the ligand and the affected residue. In CIDNP spectra, the appearance of a new resonance, i.e. an inaccessible residue becoming accessible, may be unequivocally attributed to a conformational change. Assignment of some of the resonances in the CIDNP spectra to specific amino acids in the protein sequence is possible in many cases since both the sequence (5) and the crystallographic structure (6) of the yeast and horse enzymes have been determined. In yeast 3-P-glycerate kinase, the CIDNP enhancement of a tyrosine residue is decreased upon binding of Mg ATP indicating lowered accessibility of that tyrosine to the flavin dye. In the horse enzyme, histidine 390 which lies in the hinge region between the two domains of the enzyme becomes accessible when 3-P-glycerate and MgATP are bound to the enzyme, indicating a significant change in conformation in the hinge region which presumably is responsible for bringing the two substrates into proximity for reaction.

1 Nageswara Rao, B.D., Kayne F.J., Cohn M. (1979) J. Biol. Chem. 254, 2689-96
2 Nageswara Rao, B.D., Cohn M. and Scopes, R.K. (1978) J. Biol. Chem. 253, 8056-8060
3 Scheffler J. and Cohn M. (in preparation)
4 Kaptein, R. (1982) in Biological Magnetic Resonance, (Berliner, L.J. and Reuben, J. Eds.), Vol. IV, pp. 145-191, Plenum/New York
5 Perkins R.E., Conroy, S.C., Dunbar B., Fothergill L.A., Tuite M.F., Dobson M.J., Kingsman, A.J. (1983) Biochem J. 211, 199-218
6 Watson H.C., Walker N.P.C., Shaw P.J., Bryant T.N., Wendells P.L., Fothergill L.A., Perkins R.E., Conroy S.C., Dobson M.J., Tuite M.F., Kingsman A.J. and Kingsman S.M. (1982) EMBO J. 1, 1635-1640

THE COUPLING OF ACTOMYOSIN ATPASE ACTIVITY WITH STRUCTURAL CHANGES
DURING MUSCLE CONTRACTION

R.S. Goody, Max-Planck Institut für Medizinische Forschung,
           Heidelberg, F.R.G.

Kinetic investigations on actomyosin ATPase have led to a detailed
understanding of rates and equilibria of steps involved in the interaction of myosin with nucleotides and actin. Thus steps in the mechanism associated with large negative free energy changes can be identified, and are often regarded as candidates for the "power-stroke" in
the contraction mechanism. However, it is not clear in which manner
the coupling between enzymatic and structural events occurs. Mechanisms can be envisaged in which a step involving a large free-energy
is directly coupled, indeed is identical with the crucial structural
change leading to tension development and filament sliding, but it is
also conceivable that the coupling is less direct, and that the power-
-stroke occurs as a consequence of a particular step in the cycle (1).
This difference is not trivial, since the former type of mechanism
should be identifiable in simplified systems, such as actin mixed
with proteolytic subfragments of myosin which retain the enzymatic
properties of the whole molecule, whereas in the second (indirect)
type of mechanism this may be impossible, for instance because the required region of myosin is no longer present. There is at present no
compelling evidence distinguishing between these two classes.

The mode of interaction of ATP, ADP and phosphate with the myosin
active site, as well as certain details of the hydrolysis mechanism,
are reasonably well understood. However, this information does not yet
offer any clues concerning the mechanism of energy transduction.

1 Geeves, M.A., Goody, R.S. and Gutfreund, H. (1984) J. Muscle Res.
Cell Motility, in press.

ROLE OF WATER IN PROCESSES OF ENERGY TRANSDUCTION

L. de Meis
Instituto de Ciências Biomédicas, Dept. de Bioquímica
Universidade Federal do Rio de Janeiro
Rio de Janeiro, 21910 - Brasil

In conditions similar to those found in the cytosol, the hydrolysis of acetyl phosphate, ATP or inorganic pyrophosphate(PPi)is accompanied by a large change in free energy. During the catalytic cycle of energy converting enzymes there are steps in which the hydrolysis of these compounds is accompanied by only a small energy change. This includes the Ca^{2+}-dependent ATPase of sarcoplasmic reticulum (acylphosphate residue of low and high energy), F_1-ATPase (tightly bound ATP) and inorganic pyrophosphatase (tightly bound PPi). At present it is not known why the energy of hydrolysis of these compounds decreases to low values when they are trapped in the catalytic site of the enzyme. For the sarcoplasmic reticulum ATPase, conversion of the acylphosphate residue from low into high energy seems to be associated with a change of water activity at the catalytic site (de Meis,L. et al. 1980, Biochemistry 19, 4252; de Meis,L., et al, 1982, J.Biol.Chem. 257,4993; de Meis,L. and Inesi,G., 1982, J.Biol.Chem. 257, 1289). This finding has led us to explore further the possible role of water structure in determining the ΔG^o of hydrolysis of a compound. For this purpose the equilibria between pyrophosphate and orthophosphate catalyzed by soluble yeast inorganic pyrophosphatase were measured at pH values varying between 6.0 and 8.0; in the presence of different Mg^{2+} and Ca^{2+} concentration; and in the presence of either polyethyleneglycol or different organic solvents. When $MgCl_2$ is present, organic solvents and polyethyleneglycol promote a significant decrease of the observed equilibrium constant (Kobs) for PPi hydrolysis. Thus, depending on the conditions Kobs may vary from a value higher than $4x10^3$ M (ΔG^oobs more negative than -5.1 Kcal/mol) to a value smaller than 0.04 (ΔG^oobs more positive than + 2.0 Kcal/mol). The relevance of these observation to the formation of high energy compounds by a membrane-bound enzyme was explored. Chromatophores of the photosynthetic bacteria Rhodospirillum rubrum retain a membrane-bound pyrophosphatase which is able to drive the synthesis of PPi from Pi when a H^+ electrochemical gradient is formed across the chromatophore membrane (Baltscheffsky, M. et al., 1966 Science 153, 1120). Spontaneous synthesis of PPi, in amount similar to or higher than those attained with chromatophores of R. rubrum during illumination (i.e., when a H^+ gradient is formed), can be attained with soluble yeast inorganic pyrophosphatase (without H^+ gradient) if the water activity of the medium is decreased by the addition of either ethyleneglycol (60% v/v) dimethylsulfoxide (30% v/v) or polyethyleneglycol (40-50g %).

The data presented are interpreted according to the concept that the energy of hydrolysis of a compound depends on the difference in solvation energy of reactants and products (George,P. et al., 1970 Biochem. Biophys. Acta 223, 1).

It may be that in the two systems discussed, energy is used to create at the catalytic site of the enzyme microenvironments having different solvent structures. In one environment the product would be of low energy and could be formed spontaneously. Energy might be used by the enzyme to change its conformation, forming a new microenvironment at the catalytic site where the product would became of high energy.

MECHANISM OF ATP HYDROLYSIS BY BEEF HEART MITOCHONDRIAL ATPase.
H.S. Penefsky, Department of Biochemistry, The Public Research
Institute, New York, N.Y., 10016, U.S.A.

Previous studies on the mitochondrial ATPase (F_1) support a reaction mechanism containing at least two interacting catalytic sites on
each molecule of enzyme (1,2,3,4). Uncer appropriate conditions,
such as a 3-fold molar excess of enzyme over substrate (conditions
which result primarily in the occupation of only one catalytic site),
it is possible to study single site catalysis (3,4). A minimal mechanism for single site catalysis is described as follows:

$$F_1 \underset{(K_1)}{\overset{\text{ATP}}{\rightleftharpoons}} F_1 \cdot \text{ATP} \underset{H_2O \ (K_2)}{\overset{H_2O}{\rightleftharpoons}} F_1 \cdot \text{ADP} \cdot \text{Pi} \underset{(K_3)}{\overset{\text{Pi}}{\rightleftharpoons}} F_1 \cdot \text{ADP} \underset{(K_4)}{\overset{\text{ADP}}{\rightleftharpoons}} F_1$$

Two key experimental observations are of considerable relevance to the
role of the ATPase in the mechanism of ATP formation in oxidative
phosphorylation. First, the equilibrium constant K_1, which is an
association constant, is 10^{12} M^{-1} and second, the equilibrium constant in the catalytic site, K_2, is near unity (3). Additional evidence that ATP hydrolysis is reversible in the high affinity catalytic site is provided by experiments with ATP labeled in the γ-phos-
phoryl group with ^{18}O. Under the conditions of single site catalysis,
the rate of washout of ^{18}O-labeled ATP bound in the high affinity
site was properly described by the forward and reverse rate constants
for step 2 (5). The proposed model and supporting experimental observations are consistent with reports that [γ-^{32}P]ATP may be formed on
the chloroplast ATPase (6) and on the beef heart enzyme (7) from ADP
and ^{32}Pi. The model also is consistent with the major proposals of
the binding change mechanism (8).

Further developments in an understanding of the reaction mechanism
of the ATPase depend upon a more detailed description of its elementary steps, particularly additional or as yet undetected steps and on
a determination of the extent to which the details of mechanism for
the soluble enzyme apply to the membrane-bound ATPase. Recent experiments deal with the reaction mechanism of the ATPase in coupled
submitochondrial particles.

Single site catalysis as defined for soluble F_1 (3,4) is demonstra-
ble in submitochondrial particles. Addition of [γ-^{32}P]ATP (about 1/3

of a catalytic site equivalent) to submitochondrial particles is accompanied by a very slow rate of formation of free ^{32}Pi. Hydrolysis
of bound [γ-^{32}P] is accelerated by a cold chase of nonradioactive ATP.
About 80 to 90% of added [γ-^{32}P]ATP is bound in catalytic sites of
the membrane-bound enzyme. Similar to soluble F_1, the equilibrium
constant in the catalytic site (K_2) is near unity since the ratio of
ADP + ^{32}Pi to [γ-^{32}P]ATP in the particles remained constant with time
of incubation even though the total amount of radioactivity bound
decreased.

1. Grubmeyer, C. and Penefsky, H.S. (1981) J. Biol. Chem.
   256, 3718-3727.
2. Grubmeyer, C. and Penefsky, H.S. (1981) J. Biol. Chem.
   256, 3728-3734.
3. Grubmeyer, C., Cross. R.L. and Penefsky, H.S. (1982) J. Biol.
   Chem. 257, 12092-12100.
4. Cross, R.L., Grubmeyer, C. and Penefsky, H.S. (1982) J. Biol.
   Chem. 257, 12101-12105.
5. Cross. R.l., Penefsky, H.S. and Boyer, P.D. (1982) Unpublished experiments.
6. Feldman, R. and Sigman, D.S. (1982) J. Biol. Chem. 257,
   2676-1683.
7. Sakamoto, J. and Tonomura, Y. (1983) J. Biochem. 93,
   1601-1614.
8. Gresser M.J., Myers, J.A. and Boyer, P.D. (1982) J. Biol.
   Chem. 257, 12030-12038.

MODULATION OF THE CHLOROPLAST ATP SYNTHETASE: CONFORMATIONAL STATES AND PHOTOAFFINITY LABELING OF NUCLEOTIDE BINDING SITES

N. Shavit
Ben Gurion University of the Negev,
P. O. Box 653, Beer Sheva 84105, Israel.

Molecular mechanisms of energy transduction being considered propose that upon energization of mitochondrial, bacterial and chloroplast membranes, the H^+-translocating membrane-bound ATP synthetase undergoes conformational changes linked to the interaction between nucleotide binding sites. The binding and release of nucleotides has been suggested to involve three tight binding sites on the ATP synthetase complex that alternately act as catalytic sites (1). However, the behavior of tightly bound ADP and ATP indicates that these binding sites are rather different from the catalytic site(s) since they are kinetically incompetent to be directly involved in catalysis (2,3). Regulation of the ATP synthetase by the binding of nucleotides at noncatalytic sites has also been suggested (3,4). We have proposed that ATP binds preferentially to noncatalytic sites on the energized form of the ATP synthetase while the noncatalytic sites on the ATPase form that operates in the dark have a higher affinity for ADP. Upon deenergization, the active enzyme·ADP complex undergoes a conformational change to produce the inactive enzyme form (4,5).

The mechanism of modulation of the ATP synthetase activities is rather complex. Nevertheless, the participation of different enzyme conformations in the catalytic process and the existence of several classes of nucleotide binding sites on the synthetase seems warranted. Studies on the tightly bound ATP accumulated during photophosphorylation revealed that the ATP formed and released by the catalytic site binds tightly to proximal noncatalytic sites (2). The kinetic behavior of this class of tightly bound ATP suggests that the newly formed $AT^{32}P$ exists as a free form in a space near the active site before it diffuses to the bulk medium or rebinds to noncatalytic sites.

Photolabeling experiments of both the soluble ATPase from lettuce chloroplasts (CF_1) and that of the thermophilic bacterium PS3 (TF_1) or the membrane-bound $CF_0 \cdot CF_1$ with the nucleotide analog, 3'-0-(4-Benzoyl)benzoyl ADP (BzADP), show differences that might be due to the different structures of these complexes. BzADP binds noncovalently to both CF_1- and TF_1-ATPase and is a reversible inhibitor of ATP hydrolysis, changing the kinetics of the reaction from noncooperative to cooperative. Photoactivation of the bound BzADP results in the incorporation of the analog to the β subunit with the parallel inhibition of the ATPase activity. Complete inactivation of both CF_1- and TF_1-

ATPases occurs after covalent binding of about 2 mol BzADP/mol enzyme. Photoinactivation by BzADP is prevented by ADP with CF_1 while with TF_1 both ADP and ATP are effective. BzADP also interacts with the membrane-bound ATP synthetase but is not a substrate for photophosphorylation. It is a strong competitive inhibitor with respect to ADP and ATP in phosphorylation or hydrolysis and the exchange reactions, respectively. BzADP competes with ADP for the tight-binding site on the enzyme but unlike ADP does not induce the inactivation of the enzyme. However, photoactivation of the $Bz[^3H]ADP$ noncovalently bound to the membrane-bound CF_1 results in its covalent attachment to both the α and β subunits and not only to the β subunit as obtained with the soluble CF_1 and TF_1 enzymes (6).

These findings indicate that binding of the analog to the α subunit requires that the α and β subunits be rather spatially proximal, a condition fulfilled apparently only by the structure of the membrane-bound $CF_0 \cdot CF_1$ complex. Therefore, the conformation of the soluble enzymes, at least regarding the interactions between the nucleotide binding sites on the two large subunits, appears to be different from that of the membrane-bound enzyme. More structural information is needed to define whether a direct transfer of nucleotides between sites on different subunits occurs or whether such a transfer involves a free nucleotide species distinct from that existing in the bulk medium.

References

1. Boyer, P. D. (1979) in Membrane Bioenergetics (Lee, C. P., Schatz, G., and Ernster, L., eds) pp. 461-478. Addison-Wesley Publishing Company, Reading, PA.
2. Aflalo, C. and Shavit, N. (1982) Eur. J. Biochem. 126, 61-68.
3. Bruist, M. and Hammes, G. (1981) Biochemistry 20, 6298-6305.
4. Bar-Zvi, D. and Shavit, N. (1982) Biochim. Biophys. Acta 681, 451-458.
5. Shavit, N., Aflalo, C., Bar-Zvi, D. and Tiefert, M. A. (1983) Proc. Sixth Int. Congr. Photosynth. (C. Sybesma, ed.) in press.
6. Bar-Zvi, D., Tiefert, M. A. and Shavit, N. (1983) FEBS lett. 160, 233-238.

MECHANISM AND REGULATION OF MITOCHONDRIAL F_1-ATPase: ASYMMETRY IN
  STRUCTURE AND FUNCTION.

R.L. Cross, F.A.S. Kironde, and D. Cunningham
Biochem. Dept., SUNY Upstate Med. Ctr., Syracuse, N.Y. 13210 U.S.A.

Previous studies have shown the mitochondrial F_1-ATPase to contain
a total of six nucleotide binding sites (1). Three of these sites
behave as catalytic sites in that they exchange rapidly with medium
nucleotides during ATP hydrolysis. The other three sites do not exchange with medium nucleotide in buffers of neutral pH containing Mg^{2+},
and unlike the catalytic sites, they are highly specific for adenine
nucleotides. These noncatalytic sites may have a structural or regulatory role.

Until recently, another criteria for distinguishing between nucleotide sites was based on affinity, with noncatalytic sites defined as
"tight sites" and catalytic sites considered to be "loose sites".
Using these definitions, Harris et al. (2) showed that nucleotides
that remain bound through repeated ammonium sulfate precipitations
("tight sites"), will exchange rapidly with medium ligand at low pH in
the presence of EDTA.

Assuming that nucleotides at catalytic but not at noncatalytic
sites will readily chase in the presence of MgATP, we have used the
following assay to examine the EDTA-dependent exchange at noncatalytic
sites. F_1 is incubated in the presence of 0.2 mM EDTA and 0.7 mM
(^{14}C)ADP at pH 8.0. At various times the exchange is stopped by
addition of excess Mg^{2+} and unbound ligand is removed on Sephadex centrifuge columns. Since catalytic sites also retain tightly-bound
(^{14}C)ADP through the column, labeled ligand is chased from these sites
by addition of sufficient MgATP to allow several thousand turnovers.
Unbound nucleotide is removed on a second centrifuge column and the
mol of ^{14}C label per mol of F_1 in the column effluent is taken as a
measure of EDTA-dependent nucleotide exchange at noncatalytic sites.
Under these conditions, we find that one site exchanges nucleotide rapidly with a half-time of less than one minute (Fig. 1, squares),
one site exchanges slowly over a period of one to two hours, and one
site does not exchange. Consistent with these sites being noncatalytic sites are the findings that GDP and IDP do not affect the binding of (^{14}C)ADP and bound label is not detected when (^{3}H)GDP is used
in place of ADP.

It was of interest to determine whether the capacity for fast exchange is randomized among the three noncatalytic sites during catalytic turnover. If this were to occur, only 1/3 of the bound label
might be expected to dissociate rapidly during a second incubation
with EDTA and unlabeled ADP. However, such an experiment shows this

is not the case. The same unique noncatalytic site undergoes rapid exchange with medium ligand during a second incubation in the presence of EDTA (Fig. 1, circles) despite an intervening period of ATP hydrolysis at near Vmax rates.

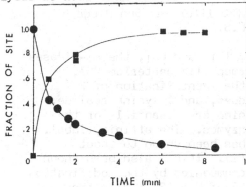

TIME (min)

Fig. 1. PERMANENT ASYMMETRY OF NONCATALYTIC SITES ON SOLUBLE F_1. Solid squares: EDTA-dependent labeling of a single fast-exchanging noncatalytic site. Solid circles: EDTA-dependent loss of label from the fast exchanging noncatalytic site following catalytic turnover. See text for further details.

This permanent asymmetry does not appear to be caused by heterogeneity of adenine nucleotides bound at the noncatalytic sites. We have prepared F_1 in which these sites are nearly uniformly occupied by ADP (< 0.2 mol ATP/mol F_1), yet their asymmetric properties remain unchanged. Factors which could contribute include asymmetric interactions of nucleotide-binding subunits either with the single copies of the smaller subunits (3) or with EDTA-resistant, tightly-bound Mg^{2+} (4). It is also possible that this asymmetric behavior is altered when F_1 interacts with membrane-bound F_o.

The catalytic sites on F_1 also behave asymmetrically, but their properties appear to change rapidly during turnover. It was previously reported that the affinities for MgATP binding at the first and second catalytic sites on F_1 may differ by a factor of 10^6 (5). We have recently shown that the single high affinity site can be readily detected during steady-state turnover over a wide range of ATP concentrations. During this work, properties of ADP binding at catalytic sites were re-examined. We have measured a site with higher affinity ($K_d = 10^{-8}$ M) and with more rapid rates of binding (10^5 M^{-1}s^{-1}) and dissociation (10^{-3}s^{-1}) than previously reported (6).

1. Cross, R.L., et al. (1982) J. Biol. Chem. 247, 2874-2881
2. Harris, D.A., et al. (1978) Biochim. Biophys. Acta. 504, 364-383.
3. Amzel, L.M., et al. (1982) Proc. Natl. Acad. Sci. USA 79, 5852-56.
4. Senior, A.E., et al. (1980) J. Biol. Chem. 255, 7211-7217.
5. Cross, R.L., et al. (1982) J. Biol. Chem. 257, 12101-12105.
6. Grubmeyer, C., et al. (1982) J. Biol. Chem. 257, 12092-12100.

IDENTIFICATION AND FUNCTION OF ESSENTIAL AMINO ACIDS IN THE F_1-ATPASES

W.S. Allison, W.W. Andrews, D.A. Bullough, P.K. Laikind, and M. Yoshida
Department of Chemistry, University of California, San Diego
La Jolla, CA 92037 U.S.A.

Chemical modification studies on MF_1, TF_1, and EF_1, the F_1-ATPases from bovine heart mitochondria, the thermophilic bacterium PS3, and E. coli, respectively, have led to the identification of 2 tyrosine residues, 2 glutamic acid residues, and 1 lysine residue, all of which reside in the β subunit, which are essential for the hydrolytic reactions catalyzed by the enzymes. The affinity label, 5'-p-F-sulfonylbenzoyladenosine (FSBA) has been shown to react with Tyr-β-368 when it inactivates MF_1 (1). It has also been shown that inactivation of TF_1 with DCCD is accompanied by the modification of the Glu which corresponds to Glu-β-188 of MF_1 (2), while the inactivation of MF_1 and EF_1 with DCCD proceeds with the modification of Glu-β-199 (3).

We have recently found that the inactivation of MF_1 with 7-chloro-4-nitro[^{14}C]benzofurazan ([^{14}C]Nbf-Cl) at pH 7.5, as first described by Ferguson et al. (4), followed by the reduction of the nitro-group of the incorporated reagent at pH 6.0 with $S_2O_4^=$ leads to the modification of Tyr-β-311, one of the residues shown by Hollemans et al. (5) to be modified when the mitochondrial enzyme is photo-inactivated with 8-N_3-[^{14}C]ATP. When MF_1 is inactivated with [^{14}C]Nbf-Cl at pH 7.4 and then brought to pH 9.0 to promote the O $\longrightarrow$ N migration, first described by Ferguson et al. (6), Lys-β-162 is labeled. Walker et al. (7) have pointed out that the amino acid sequence around Lys-β-162 is homologous with sequences found in various proteins which are known to bind purine nucleotides. Crystallographic analysis has shown that Lys-21 of adenylate kinase, which is one of these homologous sequences, probably interacts with the phosphate which is transferred in the reaction catalyzed by the enzyme (8). Therefore, it is possible that Lys-β-162 might participate in electrostatic interactions with adenine nucleotides which bind to the catalytic site of MF_1.

The functions of the DCCD reactive carboxyl groups of TF_1 and MF_1 are not clear. Whereas modification of Glu-β-188 of TF_1 abolishes both ATP hydrolysis and the synthesis of enzyme bound ATP, modification of Glu-β-199 of MF_1 only abolishes ATP hydrolysis. This suggests that Glu-β-188 might have a direct catalytic role while Glu-β-199 might not. Chlorpromazine inhibits the ATPase activity of MF_1, protects Glu-β-199 against modification by DCCD and markedly protects the enzyme against cold inactivation. These effects suggest that Glu-β-199 might reside at the binding site for the MF_1 inhibitor protein.

The pH-inactivation profile for the modification of Tyr-β-368 with FSBA suggests that this residue has a direct catalytic role (1). The inactivation of MF_1 with FSBA is cooperative in the sense that the modification of Tyr-β-368 in at least two copies of the catalytic site occurs at different rates. Chlorpromazine accelerates the rate of inactivation of MF_1 by FSBA and also affects the cooperativity of the inactivation.

References

1. Esch, F.S., and Allison, W.S. (1978) J. Biol. Chem. 253, 6100-6106.
2. Yoshida, M., Poser, J., Allison, W.S., and Esch, F.S. (1981) J. Biol. Chem. 256, 148-153.
3. Yoshida, M., Allison, W.S., Esch, F.S., and Futai, M. (1982) J. Biol. Chem. 257, 10033-10037.
4. Ferguson, S.J., Lloyd, W.J., Lyons, M.H., and Radda, G.K. (1975) Eur. J. Biochem. 54, 117-126.
5. Hollemans, M., Runswick, M.J., Fearnley, I.M., and Walker, J.E. (1983) J. Biol. Chem. 258, 9307-9313.
6. Ferguson, S.J., Lloyd, W.J., and Radda, G.K. (1975) Eur. J. Biochem. 54, 127-133.
7. Walker, J.E., Saraste, M., Runswick, M.J., and Gay, N.J. (1982) EMBO Journal 1, 945-951.
8. Pai, E.G., Sachsenheimer, W., Schirmer, R.H., and Schultz, G.E. (1977) J. Mol. Biol. 114, 37-45.

Supported by Grant NIH GM-16974 of the United States Public Health Service.

FUNCTIONAL AND REGULATORY ASPECTS OF F1-FO-ATP SYNTHASE IN
RECONSTITUTED SYSTEMS.

D.C. Gautheron, A. Di Pietro, G. Deléage, C. Godinot, G. Fellous,
M. Moradi-Ameli and F. Penin
Université Claude Bernard de Lyon, LBTM-CNRS
43, Bd du 11 novembre 1918. 69622 Villeurbanne, France

With the view of understanding the mechanism of ATP synthesis in mitochondrial inner membrane, our experimental approach has been to compare the properties of isolated F1 to those of F1 integrated in a vesicular form of F1-FO-ATP synthase complex [1], in inverted submitochondrial particles or in reconstituted systems from Pig heart mitochondria. Kinetic studies of ATPase activity in the presence of ADP revealed cooperativity between ATP and ADP sites suggesting a regulatory
role of nucleotides [2] studied in this report.

Hysteretic inhibition and nucleotides sites. Preincubation of F1 with
ADP and Mg induces an hysteretic inhibition which slowly develops during Mg-ATP hydrolysis [3]. The inhibition is due to the binding of at
least 1 mole of ADP per mole of F1 at a site distinct from catalytic
sites since labeled ADP is not rapidly removed during a rapid Mg-ATP
hydrolysis, by gel filtration in the presence of Mg, or by pyruvate kinase used as an auxilliary enzyme. The regulatory site is very specific
of adenine nucleotides or analogues while F1 can hydrolyze many XTP. In
the presence of Mg, ADP thus appears entrapped at regulatory site. The
ADP-induced inhibition and the concomitant binding of ADP could be reversed only by Am_2SO_4 precipitation or gel filtration in the absence of Mg indicating that the regulatory site is also distinct from
tightly-bound nucleotide sites.

Conformational changes. During the setting up of hysteretic inhibition,
slow conformational changes can be detected. After the binding of ADP
to regulatory site, F1 is no longer inactivated by trypsin. A monoclonal antibody anti β-subunits 5G11, preincubated with F1, partially prevents the hysteretic inhibition while it does not affect the ATPase
activity, probably by hindering the conformational change. Once the
hysteretic inhibition established, anions no longer activate F1.
Finally, measurements of circular dichroism demonstrate that the final
step of hysteretic inhibition, after substrate addition starting Mg-ATP
hydrolysis, leads to another slow conformational change toward a more
compact structure characteristic of inhibited F1 that was termed F1*.

Localization of regulatory site on a β-subunit of F1. The arylazido-β-
alanyl ADP (NAP3-ADP) behaves similar to ADP to produce the hysteretic inhibition of F1 with the same concentration dependence and the

same binding stoichiometry and is therefore a very good photoaffinity label to localize the regulatory site. Catalytic sites being saturated by GMP-P(NH)P, a non hydrolysable analogue of the substrate that cannot bind to regulatory site, the photoirradiation induces the covalent binding of NAP_3-ADP specifically to β-subunits of F1 that is directly correlated to the hysteretic inhibition while there is no correlation with the limited binding to α. About 1 mol NAP_3-ADP bound is specifically related to the maximal hysteretic inhibition. Therefore, there is a regulatory site on one β-subunit only of F1 while the two other β-subunits would contain one catalytic site each. The three β-subunits do not appear equivalent and this suggests that the third subunit with the regulatory site could control the whole conformation of F1.

Effects of Pi on the time dependence of hysteretic inhibition. One can wonder what event triggers the final slow conformational change leading to F1*. When F1 is preincubated with Pi in addition to ADP andMg, no lag time is observed after addition of Mg-ATP, indicating that F1* conformation is immediately reached. As shown by pH-dependence the monoanionic Pi is compulsory to obtain the inhibition and to reach the F1* conformation. Also, the monoclonal antibody 19D3 only inhibits the ATPase activity of F1 in the presence of monoanionic Pi at pH 6.7.

Comparison of the various forms of F1. The inhibited form F1* has been compared to isolated F1 and F0-F1 integrated in the isolated F1-F0 complex or in the membrane. Its properties and conformation appear very close to those of the enzyme competent for ATP synthesis. Therefore F1* with a low ATPase activity seems a much better model than F1 with a high ATPase activity to understand the mechanism of ATP synthesis. Other arguments come from the use of monoclonal antibodies that preferentially recognize F1 conformations different for ATP synthesis or hydrolysis.

Reconstituted systems of ATP synthesis. F1* in association with OSCP is fare more efficient than F1 to reconstitute ATP synthesis. Preparations of F0-F1, membrane and reconstituted systems were all assessed by their good capacities for ATP synthesis and well coupled proton translocation.

1 - Penin, F., Godinot, C., Comte, J. and Gautheron D.C. (1982) Biochim. Biophys. Acta 679, 198-209.
2 - Godinot, C., Di Pietro, A. and Gautheron D.C. (1975) FEBS Lett. 60, 250-255.
3 - Di Pietro, A., Penin, F., Godinot, C. and Gautheron, D.C. (1980) Biochemistry 19, 5671-5678.

AUROVERTIN AND NUCLEOTIDE BINDING SITE INTERACTIONS ON MITOCHONDRIAL
F_1-ATPase

J. LUNARDI, G. KLEIN and P.V. VIGNAIS
DRF/Biochimie (CNRS/ERA 903 et INSERM U.191), Centre d'Etudes
Nucléaires, 85X, 38041 Grenoble cedex, France.

Aurovertin D, a fluorescent probe of F_1-ATPase was introduced by
Lardy et al. (1) as a tool to investigate the mechanism of oxidative
phosphorylation in mammalian mitochondria. Binding to F_1 enhances up
to 100 fold the weak fluorescence intensity exhibited by aurovertin D
in aqueous solution. Beef heart soluble F_1 contains 3 aurovertin binding sites of different affinities located on the 3 β subunits as shown
by direct titration with (^{14}C)aurovertin D (2). The fluorescent intensity of the F_1-aurovertin complex is modified upon addition of nucleotides, P_i and Mg^{++}, probably due to modification in the rigidity of
the bound fluorophore (3). Consequently, aurovertin D was extensively
used to probe conformational changes of F_1 induced by the F_1
substrates.

ADP and ATP have opposite effects on the fluorescence level of the
F_1-aurovertin complex in a medium containing EDTA, ADP inducing a
slight stimulation whereas ATP promotes a rapid quenching. Under these
conditions, addition of Mg^{++} results in strong quenching of the fluorescence and subsequent addition of ADP produces an important stimulation. This situation is comparable to that observed with *E. coli* F_1 in
the presence of EDTA. The differences between the fluorescence responses observed in the presence of EDTA or $MgCl_2$ could be related to the
nature of the divalent cation binding sites present on the enzyme.
Senior et al. (4) reported that beef heart mitochondrial F_1 contains
2 Mg^{++} binding sites, one of which being a tight binding site, whereas
soluble F_1-ATPase from *E. coli* possesses 2 mol of tightly bound Mg^{++}.

In an experiment where the aurovertin fluorescence changes were
measured together with the amount of bound nucleotides, it was clear
that maximal stimulation of the fluorescence of the F_1-aurovertin
complex was correlated with the binding of 1 mol (^{14}C)ADP per mol F_1
under conditions where the ratio of bound aurovertin to F_1 (mol/mol)
was close to 1. Conversely maximal quenching was attained upon binding
of 1 mol (^{3}H)ATP per mol F_1. Further binding of (^{14}C)ADP or (^{3}H)ATP
did not change the level of fluorescence. In a double labeling experiment performed with (^{14}C)ADP and (^{3}H)ATP, the fluorescence levels of
the F_1-aurovertin complex were correlated with the amount of bound
(^{14}C)ADP and (^{3}H)ATP. The filling of the high affinity ADP binding
site was achieved upon incubation with small concentrations of (^{14}C)-
ADP and resulted in the maximal increase of the fluorescence level.

Then (^{3}H)ATP was added at increasing concentrations ; binding of
(^{3}H)ATP occured concomitantly with the release of the previously
bound (^{14}C)ADP and the decrease in the fluorescence level, suggesting
a 1 to 1 competition for binding between ADP and ATP at low concentration of added ATP. In the presence of EDTA, ADP and ATP induced
opposite effects with respect to the fluorescence level of the F_1-
aurovertin complex ; furthermore ATP binding was accompanied by ADP
release in a 1/1 stoichiometry. Two different explanations are
consistent with these results : i) ADP and ATP bind to the same site
which may assume two different conformations (ADP and ATP conformations), interacting differently with the aurovertin site.
ii) ADP and ATP bind to two different interacting and specific sites.
These considerations are illustrated in the Schemes I and II respectively.

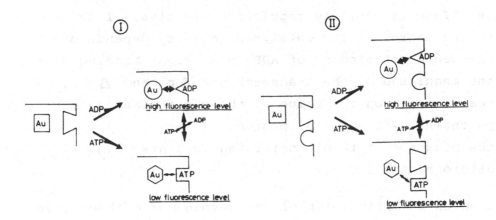

References

(1) Lardy, H.A., Connelly, J.L. and Johnson, D. (1964) Biochemistry
    3, 1961-1968.
(2) Issartel, J.P., Klein, G. and Vignais, P.V. (1983) Biochemistry
    22, 3492-3497.
(3) Chang, T.C. and Penefsky, H.S. (1973) J. Biol. Chem. 248, 2746-
    2754.
(4) Senior, A.E. (1979) J. Biol. Chem. 254, 11319-11322.

ACTIVATION AND MODIFICATION OF THE CHLOROPLAST ATPase

Peter Gräber

Max-Volmer-Institut für Physikalische und Biophysikalische Chemie, Technische Universität Berlin, D-1000 Berlin 12, Germany

The proton transport-coupled ATP synthesis/hydrolysis in chloroplasts is catalyzed by a membrane-bound ATPase, CF_oF_1. Recent work has shown that the catalytic activity of the APTase is strictly regulated. In class II chloroplasts the degree of the catalytic activity depends on:

(a) the degree of binding of ADP to a "tight binding site"

(b) the magnitude of the transmembrane ΔpH and $\Delta\Psi$, resp., i.e., on the value of the electrochemical potential difference of protons,

(c) the presence of thiol-containing reagents, e.g. dithiothreitol.

Based on the principles of the chemiosmotic theory, an enzyme kinetic model has been formulated which describes the steady state kinetics of ATP synthesis/hydrolysis with explicit formulation of the activation process of the ATPase (1). Under non-energized conditions the ATPase is in an inactive state and is converted into an active state upon energization. Only in the active state is the ATPase able to catalyze ATP synthesis/hydrolysis. On the other hand, a high energization favors the ATP synthesis direction. Since the energization required for activation is higher than that required for ATP synthesis, usually no

ATP hydrolysis is observed in chloroplasts.

In the presence of dithiothreitol, the ATPase can be modified, and it is well known that under these conditions ATP hydrolysis can be observed. In the model described above, this modification decreases the ΔpH required for activation: depending on the reaction conditions, the activation step can be removed either completely or shifted by about 1 ΔpH unit to lower values.

(1) Gräber, P. and Schlodder, E. in: Photosynthesis II
    (Akoyunoglou, G., ed.), Balaban Intern. Sci. Serv.,
    Philadelphia, Pa., pp. 867-879

IS CYTOCHROME b REDUCTION TOO FAST FOR A Q-CYCLE ?

P.R. Rich
Department of Biochemistry, University of Cambridge,
Cambridge CB2 1QW, United Kingdom

In a Q-cycle, cytochrome b is reduced by a semiquinone produced after a quinol has donated an electron to the Rieske centre/cytochrome c_1 region of the bc_1 complex. It therefore follows that the rate of cytochrome b reduction should not be faster than the rate of Rieske centre/c_1 reduction. However, several recent studies, both with mammalian cytochrome bc_1 complex and with chloroplast cytochrome bf complex, have highlighted situations in which cytochrome b reduction appears to be too fast to be accommodated within this model (reviewed in ref. 1). An experimental examination of this possible anomaly is made.

The experimental system consisted of reaction centres and bc_1 complex, either free in solution (2) or reinserted into purified soyabean phosphatidylcholine vesicles (3), together with cytochrome c in a buffer containing 50mM potassium phosphate, 2mM EDTA and 1mM KCN at pH 7.2 and 25°C. An actinic flash of less than 5μs half-width was provided by a xenon flashlamp. Cytochrome c redox changes were monitored at 550nm - 1.15 x 542nm and cytochrome b at 562nm - 0.91 x 542nm. Either ascorbate, succinate/fumarate or dithionite were used as reductants to redox poise the system before flash activation and myxothiazol and antimycin A were used to test for the involvement of centres o and i respectively. Enough ubiquinone-50 was present in the purified preparations to provide a quinone at centre o.

In the presence of 1mM ascorbate, the cytochromes b and any endogenous ubiquinone were essentially fully oxidised. Flash activation caused a rapid cytochrome c oxidation and rapid cytochrome b reduction. Very little cytochrome c rereduction was seen on the timescale of cytochrome b reduction since centre o (often termed the Q_z site) was not loaded with a quinol. Myxothiazol, but not antimycin A, prevented this rapid cytochrome b reduction.

In the presence of 5mM succinate/5mM fumarate, the cytochromes b were approximately 50% reduced and the Q_z site was loaded with a quinol. Flash activation caused a rapid cytochrome c oxidation and rereduction. Cytochrome b was transiently reduced, and then almost fully oxidised on the same timescale as c rereduction. Either antimycin A or myxothiazol prevented this cytochrome c rereduction and cytochrome b reoxidation.

An attempt was made to kinetically poise the system with both cytochromes b reduced but with the reaction centres still operative. This was achieved by adding a small aliquot of sodium dithionite and waiting for a sufficient time so that the cytochromes had just become reduced but with Q_A and Q_B of the reaction centre still oxidised. When such a system was flash activated very little cytochrome b redox change was observed even although cytochrome c oxidation confirmed that the reaction centres were still operative.

Similar results were obtained with the vesicle-reincorporated system. The only difference was that the lipid competed with Q/QH_2 for the centre o (Q_z) site. In order to observe Q_z-like behaviour, it was necessary to use vesicles into which ubiquinone-50 had been previously inserted.

The conclusions of these studies are :-
1) a rapid electron transfer is possible from Q_B^- of the reaction centre to cytochrome b via centre o. It is equivalent to the second half-reaction of oxidant-induced reduction;
2) when the complex turns over rapidly, the oxidant at centre i is produced at centre o and travels between the centres without equilibrating with the quinone pool (see ref 1);
3) although the above experiment with dithionite is by no means definitive, the evidence that fully-reduced cytochromes b can be oxidised in an antimycin-sensitive manner by a pulse of oxidant (e.g. see discussion in ref. 4) is questioned. After examination of the original data quoted in support of the claim, it is not clear whether the cytochromes b are actually fully reduced before the addition of the oxidant;
4) the Q/QH_2 at centre o can rapidly exchange with other quinones or lipid.

References

1   Rich, P.R. (1984) Biochim. Biophys. Acta 768, 53-79
2. Packham, N.K., Tiede, D., Mueller, P. and Dutton, P.L. (1980) Proc. Natl. Acad. Sci. USA 77, 6339-6343
3. Rich, P.R. and Heathcote, P. (1983) Biochim. Biophys. Acta 725, 332-340
4. Bowyer, J.R. and Trumpower, B.L. (1981) in Chemiosmotic Proton Circuits in Biological Membranes (Skulachev,V.P. & Hinkle,P.C., eds.), pp. 105-122, Addison-Wesley Publishing Co..

This work is funded by the BP Venture Research Unit.

EPR ANALYSIS OF IRON SULFUR CLUSTERS IN THE NADH-MQ OXIDOREDUCTASE
SEGMENT OF T. THERMOPHILUS HB-8

Tomoko Ohnishi*, Koyu Hon-nami", and Tairo Oshima"
*Dept. of Biochem.& Biophys., Univ. of Penn., Phila., PA 19104, USA
"Mitsubishi-Kasei Institute of Life Sciences, Machida, Tokyo, JAPAN

NADH-UQ oxidoreductase of beef heart mitochondria is extremely
complex system which contains 26 polypeptides (1) with three
tetranuclear and 5 to 6 binuclear iron-sulfur clusters (2,3). In
our efforts to search for simpler NADH-UQ oxidoreductase systems, we
have examined a rotenone sensitive thermophilic bacterial strain T.
thermophilus HB-8. The NADH oxidation of the cytoplasmic membrane
preparation is inhibited maximally about 80 % by rotenone with I_{50}
value of 15 nmoles/mg protein. Upon addition of NADH to the rotenone
pretreated uncoupled membrane under aerobic condition, an EPR
spectrum with an axial symmetry with $g_{//}=2.023$ and $g_{\perp}=1.931$ was
obtained, at the sample temperature of 25 K. In the FCCP
non-treated membrane, an additional spectral feature was seen at the
same temperature which suggested the presence of another N-1 type
cluster with a very low midpoint redox potential, as in case of
clusters N-1b and N-1a, respectively, in the beef or pigeon heart
mitochondrial membrane. In the latter system, lowering of the
sample temperature to 10 K revealed the presence of an additional
iron-sulfur cluster with $g_y=1.945$ and $g_x=1.89$.

These results indicate the presence of at least 2 binuclear and 1
tetranuclear type iron-sulfur clusters in the NADH-MQ oxidoreductase
segment of the respiratory chain in the T. thermophilus membrane,
and the two iron-sulfur species have relatively low midpoint redox
potentials.

Succinate addition to the uncoupled or rotenone treated membrane
gave rise to an EPR spectrum of rhombic symmetry with $g_z=2.027$,
$g_y=1.931$, and unresolved g_x which considerably differs from the
spectral characteristics of the above described N-1 type clusters.

Potentiometric titrations of the "g=1.94" at 25 K showed E_m
values of approximately +70, -260, < -360 mV for the Fd-type cluster
in the succinate dehydrogenase (S-1), N-1b, and N-1a, respectively.
The titration at 10 K indicated E_m values of approximately < -360
and -260 mV for the tetranuclear cluster in the NADH-MQ
oxidoreductase segment and S-2 in the SDH, respectively.

Thus this thermophilic bacterial system has simpler iron sulfur composition in the NADH-MQ oxidoreductase segment than the counter part in all rotenone sensitive mitochondria (4) or in the P. denitrificans cytoplasmic membrane (5) and may serve as a useful model system for the study of the Site I energy coupling.

References:

1 Heron, C., Smith, S. and Ragan, C.I. (1979) Biochem. J. 181, 435-443
2 Paech, C., Reynolds, J.G., Singer, T.P., and Holm, R.H. (1981) J. Biol. Chem. 256, 3167-3170
3 Ohnishi, T., Ragan, C.I., and Hatefi, Y. (1984) in press for J. Biol. Chem.
4 Ohnishi, T. & Salerno, J.C. (1982) in Iron Sulfur Proteins (T.G. Spiro, ed.) Vol. 4, pp. 285-327, J. Wiley & Sons, Inc., New York
5 Albracht, S.P.J., von Verseveld, H.W., Hagen, W.R., and Kalkman, M.L. (1980) Biochim. Biophys. Acta 593, 173-186

STUDIES ON THE MECHANISM OF PHOTOSYNTHETIC WATER OXIDATION

G. Renger, W. Weiss and M. Völker

Max-Volmer-Institut für Biophysikalische und Physikalische Chemie,
Technische Universität Berlin, D-1000 Berlin 12, Germany

The crucial step of photosynthetic water cleavage into O_2 and meta-bolically bound hydrogen is the realization of water oxidation. It is energetically driven by photooxidation of a special chlorophyll-a com-plex (P680) within the reaction center of system II (RC II) and takes place via a four-step univalent reaction sequence catalyzed at a man-ganese containing enzyme system Y. RC II and system Y are functionally connected via redox component D_1 (for review see ref. 1).

In order to attack unresolved problems referring to the nature of the different redox states S_i (i = 0-4) in system Y as well as to the structure and the mechanistic details of the function of proteins acting as apoenzymes flash-induced absorption changes and oxygen yield measurements were performed in selectively modified thylakoids. The following results were obtained:

(1) Based on kinetic data (2) and using the difference spectrum as-cribed to the turnover of the primary plastoquinone acceptor of system II, Q_A^-/Q_A (3), measurements of absorption changes induced in dark adapted inside-out thylakoids by the 3rd flash of a train of single turnover flashes permit the determination of the D_1^{ox}/D_1-difference spectrum in systems with high oxygen-evolving capacity. Within the experimental error, the spectrum obtained is identical to that report-ed for Tris-washed PS II-particles (4).

(2) Applying this D_1^{ox}/D_1-difference spectrum to previously measured data (5) allowed the separation of the difference spectrum between the redox states S_3 and S_0 of the water oxidizing enzyme system Y.

(3) An analysis of flash-induced absorption changes in dark adapted thylakoids did not reveal striking differences in the spectra character-

izing the redox transitions $S_i \rightarrow S_{i+1}$ for $i = 0,1,2,3$.

(4) The destruction of the oxygen-evolving capacity in inside-out thylakoids by trypsin exhibits a remarkable pH-dependence between pH=6.5 and pH=7.4 that cannot only be due to differences in the proteolytic activity of the enzyme itself. The phenomenon rather reflects pH dependent structural changes of polypeptides that are closely related to the water oxidizing enzyme system Y or are part of it.

(5) Trypsination of inside-out thylakoids causes manganese release. The pH-dependence of this effect correlates with that of the impairment of the oxygen-evolving capacity.

(6) Trypsination of inside-out thylakoids at pH=7.4 retards the $P680^+$-reduction by D_1 in a similar manner as Tris-washing, but the internal D_1^{ox}-reduction in the dark is significantly faster after the former treatment. Trypsination of Tris-washed inside-out thylakoids does not affect the D_1-oxidation but greatly enhances its dark recovery. This trypsin effect does not reveal the above-mentioned strong pH-dependence.

Based on the above-mentioned data and previous results, a model for the functional and structural organization of photosynthetic water oxidation will be proposed and discussed.

References

1. Renger, G., Eckert, H.J. and Weiss, W. (1983) in: The Oxygen Evolving System in Photosynthesis (Inoue, Y., Crofts, A.R., Govindjee, Murata, N., Renger, G. and Satoh, K., eds.), pp. 73-82, Academic Press, Japan
2. Boska, M. and Sauer, K. (1984) Biochim. Biophys. Acta 765, 84-87
3. van Gorkom, H.J., Thielen, A.P.C.M. and Gorren, A.C.F. (1982) in: Function of Quinones in Energy Conserving Systems (Trumpower, B., ed.), pp. 213-225, Academic Press, New York
4. Decker, J.P., van Gorkom, H.J., Brok, M. and Ouwehand, L. (1984) Biochim. Biophys. Acta 764, 301-309
5. Renger, G. and Weiss, W. (1983) Biochim. Biophys. Acta 722, 1-11

OPTICAL STUDIES ON THE OXYGEN EVOLVING COMPLEX OF PHOTOSYNTHESIS

H.J. van Gorkom, J.P. Dekker, J.J. Plijter and L. Ouwehand
Department of Biophysics, Huygens Laboratory of the State University,
P.O. Box 9504, 2300 RA Leiden, The Netherlands

The oxygen evolving complex of photosynthesis was studied by absorbance changes upon flash illumination of photosystem II. During photosynthesis this complex releases an oxygen molecule once every 4 turnovers of photosystem II. Between photoreactions, the complex can be in any of the 4 successive redox states, called S_0, S_1, S_2, and S_3. In S_3, a subsequent photoreaction is followed by reduction to S_0 in about a millisecond, with concomitant oxygen release. Since the low S-states, S_0 and S_1, are stable and the high S-states, S_2 and S_3, are reduced to S_1 in minutes, the turnover of the complexes can be synchronized by a period of darkness. Subsequent illumination by a series of single turnover flashes leads to an oscillating amount of oxygen evolution, with maxima on flash numbers 3, 7, 11, etc., and allows the study of the successive S-state transitions [1].

One of the phenomena which have been found to oscillate with a periodicity of 4 in these conditions is an absorbance change in the ultraviolet part of the spectrum [2]. With system II particles isolated from spinach according to the method of Berthold et al. [3], we have recently been able to determine the spectrum of this change on the first flash after dark adaptation [4]. It consists of a broad and asymmetric band appearing around 300 nm, which could be due to an oxidation of Mn(III) to Mn(IV) on the $S_1 \rightarrow S_2$ transition.

Next, we set out to accurately determine the oscillation pattern of this absorbance change throughout the spectrum, in the hope that absorbance changes due to the other S-state transitions might turn up. To this end, it was first established that the period 2 oscillations at the acceptor side of system II, as measured in the presence of the electron donor tetraphenylboron, could be prevented completely by the addition of the lipophilic electron acceptor dichlorobenzoquinone at a concentration of 100 µM, using 200 µM chlorophyll. Then the oscillating absorbance changes without artificial electron donor were measured, with sufficient time resolution to detect the 1 ms transient which accompanies oxygen evolution. The initial S_0/S_1 ratio and the percentages of misses and double hits were determined from the phase, period and damping of the oscillation pattern of this transient. For the contributions of each S-state on each flash calculated with these parameters, the difference spectra of the four S-state transitions were calculated from the observed flash-induced difference spectra.

It turned out that the difference spectra thus obtained were composed of only two components, one of which was obviously due to the acceptor side. An irregular period 2 oscillation of a plastosemiquinone anion was observed, indicating that in the absence of tetraphenylboron the secondary quinone acceptor Q_B^- was not fully reoxidized between flashes. The oscillation pattern suggested that only in S_O some Q_B^- oxidation by dichlorobenzoquinone occurred.

The other absorbance change had the same spectrum and amplitude as determined previously for the first flash, and was the same for each S-state advance. If this change is due to an Mn(III) to Mn(IV) transition, three different Mn(III) ions are oxidized successively and all three are rereduced during the $S_3 \rightarrow S_O$ transition. Independently, the spectrum and amplitude of the 1 ms transient confirmed this conclusion: it consists of the three equivalent Mn difference spectra plus the difference spectrum of the secondary donor Z. If the fourth Mn ion, which is known to be present, is transiently oxidized as well, the rapid reduction of the complex by water prevents its accumulation. The kinetics of the Mn oxidation on successive flashes were in agreement with the reported Z^+ reduction times [5].

1. Joliot, P. and Kok, B. (1975) in Bioenergetics of Photosynthesis (Govindjee, ed.), pp. 387-412, Academic Press, New York
2. Pulles, M.P.J., van Gorkom, H.J. and Willemsen, J.G. (1976) Biochim. Biophys. Acta 449, 536-540
3. Berthold, D.A., Babcock, G.T. and Yocum, C.F. (1981) FEBS Lett. 131, 231-234
4. Dekker, J.P., van Gorkom, H.J., Brok, M. and Ouwehand, L. (1984) Biochim. Biophys. Acta 764, 301-309
5. Babcock, G.T., Blankenship, R.E. and Sauer, K. (1976) FEBS Lett. 61, 286-289

EFFECT OF CHLOROFORM ON PHOTOPHOSPHORYLATION AND ON THE
GENERATION OF ΔP IN BACTERIAL CHROMATOPHORES.

G. Mandolino, R. Tombolini, B. A. Melandri
Institute of Botany,Chair of Molecular Biology - Univ. of Bologna,
40126 Bologna, Italy

The inhibition of ATP formation in biological membranes by
uncouplers is believed to be obligatorily paralleled by the
dissipation of the protonic gradient. Nevertheless in rat liver
mitochondria chloroform and halotane were shown to inhibit
oxidative ATP synthesis, and to accelerate the rates of respiration
and of ATP hydrolysis, as all uncouplers do, without affecting the
extent of $\Delta\bar{\mu}_{H^+}$(1). The addition of these substances was also shown
to increase the fluidity of the membrane lipid phase (1). This
behavior is consistent with the existence of alternative modes of
coupling between electron flow and phosphorylation, requiring close
functional interactions between energy trasducing complexes (2,3).

The effect of $CHCl_3$ on the generation of $\Delta\bar{\mu}_{H^+}$ and on some energy
linked reactions by chromatophores of Rhodopseudomonas capsulata
Kb1 was studied. Chloroform, at a concentration, of 25 mM severely
inhibited photo- phosphorylation; this phenomenon was accompanied
by a marked inhibition of the proton gradient, which was decreased
from 370 mV of the control to about 200 mV (Fig. 1). This
inhibitory effect on the proton gradient was more significant for
the pH component than for $\Delta\psi$. It was also somewhat dependent on the
ionic composition of the assay, being more severe at low ionic
strengths. The variation of $\Delta\bar{\mu}_{H^+}$ in the steady state corresponded to
a small increase in the ion conductivity of the membrane: the decay
half time of was lowered from 3.5 s to 1.3 s at 22 mM $CHCl_3$, and
the acceleration of the reversal of the light-induced quenching of
9 amino-acridine fluorescence was linearly related to the $CHCl_3$
concentration. Neither the decrease of the steady state $\Delta\bar{\mu}_{H^+}$,
always well above the minimal threshold for ATP synthesis, nor the
variations in ion conductivity, however, seemed to be sufficient to
account fully for the inhibition of ATP synthesis.

The functionality of ATPase was also examined. In the presence
of $CHCl_3$ up to 35 mM, the rate of ATP hydrolysis was stimulated; the

activity became considerably insensitive to oligomycin (Fig. 2), indicating possibly a decoupling of F_1 from the membrane. At all $CHCl_3$ concentrations, however, ATPase was still capable of sustaining a pH gradient ranging from 1.7 to 1.2 units. Again therefore the effect of $CHCl_3$ on the H^+-ATPase did not seem fully consistent with a classical uncoupling effect.

REFERENCES
1 Rottenberg, H. (1983) Proc. Natl. Acad. Sci.U.S.80,3313-3317
2 Venturoli, G. and Melandri B.A. (1982)Biochim. Biophys. Acta 680, 8-16
3 Mandolino, G. and Melandri,B.A. (1983) Biochim. Biophys. Acta 723, 428-439

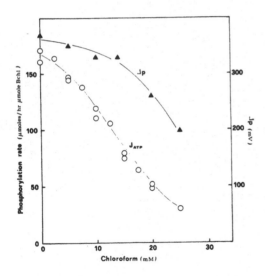

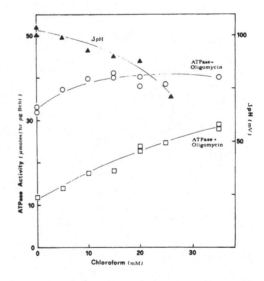

Figure 1. The effect of chloroform on light-induced phosphorylation and on the formation of the H^+ gradient.

Figure 2. The effect of chloroform on ATPase activity and on the ATP dependent H^+ translocation in the presence of K^+ and valinomycin.

Poster Presentations

THE NEGATIVE COOPERATIVE INTERACTION BETWEEN CYTOCHROME c AND CYTOCHROME OXIDASE

R. Bolli, K. A. Nałęcz, and A. Azzi
Medizinisch-Chemisches Institut der Universität Bern,
Bühlstrasse 28, 3012-Bern, Switzerland

The biphasic kinetics of the reaction between bovine cytochrome c oxidase and its substrate, cytochrome c, has been explained on the basis of two catalytic sites with different affinity for the substrate (1). When the enzyme was dispersed in /B-D-dodecylmaltoside as the detergent, the kinetic pattern and the molecular state depended on ionic strength. At low salt concentrations monomeric oxidase and a monophasic Eadie-Hofstee plot were found, whilst at high salt concentration biphasic kinetics and dimers were present (2). Since the concentrations of the enzyme used in the kinetic assay were much smaller than those used for the molecular studies, the question of the possibility of monomerization of the dimeric enzyme on dilution was investigated.

Gel filtration of 1-5 nM oxidase (detectable only by activity) on Ultrogel AcA 34 at 50 mM KCl gave only dimers, excluding the possibility of monomerization with dilution. When, on the contrary, at low ionic strength the enzyme concentration was increased to 50 nM, by adding oxidase inhibited with KCN to permit the activity measurement, biphasic kinetics were observed, suggesting dimerization with increasing protein concentration. With high enzyme concentration (3-30 /uM) the sedimentation velocity studies in /B-D-dodecylmaltoside revealed sedimentation coefficients of 13-15 S, for low and high ionic strength, what could correspond to the dimeric species (3).

Monomerization of bovine heart cytochrome c oxidase, has been obtained in cholate (4), and at high pH, high salt concentration in the presence of Triton X-100 (5). Under these conditions the kinetics of the reaction were monophasic but a biphasic kinetic pattern was obtained by lowering the pH or removing cholate.

Fig. 1 shows an Eadie-Hofstee plot of the kinetics of cytochrome c oxidase reconstituted into phospholipid vesicles according to Casey et al. (6). The plot is

monophasic at high and low ionic strength supporting the
conclusion (6) that the reconstituted enzyme was mainly
monomeric.

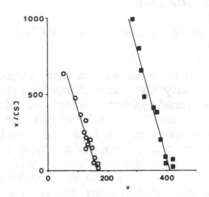

Figure 1. Eadie-Hofstee plot of the activity of
cytochrome c oxidase reconstituted into phospholipid
vesicles. Assay conditions: 10 mM buffer, pH 7.4 (Tris,
Hepes), 0.25 /uM valinomycin, 2.5 /uM CCCP in the absence
(O) or presence (■) of 50 mM KCl; v, s⁻¹; (S), /uM.

The present results suggest a model of cytochrome c
oxidase with one catalytic site per monomer. The electric
charge and spatial interaction between two molecules of
cytochrome c in the dimeric enzyme may produce the
apparent negative cooperativity.

REFERENCES:
1. Ferguson-Miller, S., Brautigan, D.L., and Margoliash,
   E. (1976) J. Biol. Chem. **251**, 1104-1115
2. Nałęcz, K.A., Bolli, R., and Azzi, A. (1983)
   Biochem. Biophys. Res. Commun. **114**, 822-828
3. Darley-Usmar, V.M., Alizai, N., Al-Ayash, A.I., Jones,
   G.D., Sharpi, A., and Wilson, M.T. (1980) Comp.
   Biochem. Physiol. **68B**, 445-456
4. Robinson, N.C., and Capaldi, R.A. (1977) Biochemistry
   16, 375-381
5. Georgevich, G., Darley-Usmar, V.M., Malatesta, F., and
   Capaldi, R.A. (1983) Biochemistry **22**, 1317-1322
6. Casey, R.P., O'Shea, P.S., Chappell, J.B., and Azzi,
   A. (1984) Biochim. Biophys Acta **765**, 30-37

118

STRUCTURAL FACTS FOR THE SITE OF e^- AND H^+ TRANSLOCATION IN CYTOCHROME C OXIDASE

G. Buse°, R. Biewald°, G. Raabe* and J. Fleischhauer*
°Abteilung Physiologische Chemie, RWTH Aachen
 D-5100 Aachen, West Germany
*Lehr- und Forschungsgebiet Theoretische Chemie, RWTH Aachen
 D-5100 Aachen, West Germany

From the 12 polypeptide pattern of a mammalian type cytochrome c oxidase, respiratory complex IV, only the three largest mitochondrially synthesized proteins are functional subunits which are directly involved in substrate binding, e^- and H^+ translocation and O_2 reduction. The main evidence for this statement comes from the investigation of bacterial oxidases of the two copper-, heme a, a_3-type (1). Thus the two subunits found in the *Paracoccus denitrificans* oxidase show close sequence homology to subunits I and II of the mitochondrial enzymes, which therefore must bind the copper and heme and perform the oxidative function (2). In the bovine enzyme the large hydrophobic subunit I with possibly 12 membrane penetrating sections represents the core of a membrane domain (M_1) which also contains the two N-terminal membrane sections of subunit II besides cytoplasmic chains. The hydrophilic part of subunit II is a main constituent of the cytoplasmic domain (C) of the enzyme and forms the negatively charged high affinity cytochrome c binding site (3,4). Previously we detected its remote homology to electron conducting copper proteins of the azurin family (5). The structure of subunit II from bacteria to men contains an invariant cluster of tryptophan and tyrosine residues which possibly is involved in electron transfer between the metal centers of the enzyme and may contribute to the coupling of e^- and H^+ translocation (6). Quantum mechanical MNDO (Modified Neglect of Differential Overlap) calculations show that e^- transfer within the structures given below may facilitate a coupled H^+ transfer and vice versa.

-Trp-Tyr-Trp- x -Tyr- x -Tyr-

Evidence for the function of subunit I is presented from experiments with Li-Dodecylsulfate, which allow for a protective dissociation of the oxidase complex. The isolation of subunit I with spectroscopically active heme is demonstrated and some spectroscopic data are given.

Alkali treatment and reduction of the entire oxidase with borohydride on the other hand leads the covalent fixation of more than 75% of the chromatographically detectable heme of the oxidase to subunit I via Schiff bases. Subunit I therefore is assumed to bind both heme a and a_3 in the intact oxidase. These results together suggest an arrangement of subunit II with copper A as the first electron acceptor of the enzyme. Electrons may then be donated probably via quantum mechanical tunneling in a cluster of tryptophan and tyrosine residues to subunit I with heme a, and the a_3/copper B O_2-activating center.

References

1 Ludwig, B. and Schatz, G. (1980) Proc. Natl. Acad. Sci (USA) 77, 196-200.
2 Steffens, G.C.M., Buse, G., Oppliger, W. and Ludwig, B. (1983) Biochem. Biophys. Res. Comm. 116, 335-340.
3 Bisson, R., Steffens, G.C.M. and Buse, G. (1982) The Journ. of Biol. Chem. 257, 6716-6720.
4 Bisson, R., Steffens, G.C.M., Capaldi, R.A. and Buse, G. (1982) FEBS Letters 144, 359-363.
5 Steffens, G.J. and Buse, G. (1979) Hoppe-Seyler's Z. Physiol. Chem. 360, 613-619.
6 Buse, G., Steffens, G.C.M. and Meinecke, L. (1983) in Structure and Function of Membrane Proteins (Quagliariello, E. and Palmieri, F. eds.) pp. 131-138, Elsevier/North-Holland Biomedical Press, Amsterdam.

NON-STEADY-STATE KINETIC STUDIES ON THE MECHANISM OF ELECTRON FLOW IN
THE MITOCHONDRIAL bc_1 COMPLEX

M.Degli Esposti and G.Lenaz
Institute of Botany,University of Bologna,40126 Bologna,Italy.

We have investigated the non-steady-state kinetics of the mitochondrial bc_1 complex by monitoring the rapid reduction of the fully oxidized enzyme by ubiquinol-1 (Q_1H_2) in a crude succinate-cytochrome c reductase fraction. We have selected this preparation because it retains,
contrary to the isolated bc_1 complex, the same kinetic behaviour of the
ubiquinol-cytochrome c reductase activity as in the mitochondrial mem-
brane |1|. The reduction of cytochrome c_1 is usually biphasic; the fi-
rst rapid phase (accounting for more than 70% of the overall reaction)
follows a non-linear saturation, apparently with negative-cooperativity,
as a function of Q_1H_2 concentration. At high quinol concentrations, an
apparent half-saturation constant (Km) ranging between 7 and 14 μM can
be extrapolated. Such behaviour has been previously reported also for
the steady-state activity of the enzyme when assayed at low cytochro-
me c concentrations |1,2|, and is parallel to the analogous non-linear
saturation of the rates of cytochrome b reduction under non-steady-sta-
te conditions |2|. Therefore, the half-reaction of the catalytic cycle
of the bc_1 complex, constituted by the electron donation by ubiquinol
into the oxidized enzyme, contains some rate-determining steps conferring the non-linear saturation to the Q_1H_2 titration.

Two explanations can be advanced: 1) The electron input in the complex is governed by a one-electron transfer step, so that the rate of
bc_1 reduction by the two-electrons donor quinol is mainly proportional
only to the square root of Q_1H_2 concentration; 2) Conformational rate-
determining isomerisations and/or monomeric-oligomeric equilibria of
the enzyme are involved in the reduction phase of the bc_1 catalysis.

We have tested the effects of a number of specific inhibitors of
the bc_1 complex on the Q_1H_2 titration of cytochrome c_1 reduction. The
rapid phase of the reaction and its contribution extent are generally
decreased by antimycin |3|. Such effect is however dependent on the u-
biquinol concentration, since the inhibitor also markedly modifies the
saturation pattern of the ubiquinol titration, becoming linear and with

a much lower Km (Table I). At certain ratios of the electron donor to the enzyme, even a slight stimulation of the rate can be observed. The effects of many other inhibitors resemble that of antimycin. They include not only antimycin-like inhibitors such as HQNO, but also other compounds such as UHDBT, which are usually believed to act at a different site than that of antimycin |4|. The set of the experimental evidences obtained on the effects of inhibitors is summarized in Table I. Myxothiazol appears the only compound capable to completely block the electron transfer to cytochrome c_1 by ubiquinol, but without any signignificant modification of the quinol titration (Table I).

TABLE I- Effects of inhibitors of the bc_1 complex on the non-steady-state reduction of cytochrome c_1 by ubiquinol-1.

Inhibitor	Km (μM)	Q_1H_2 saturation pattern
none	7-14	non-linear
antimycin A	1.2	linear
HQNO	3.2	linear
UHDBT	1.8	linear
DBMIB	4.5	linear
BAL + O_2	4	linear
myxothiazol	4.8	non-linear

The finding that the effects of specific inhibitors such as antimycin on the rate of cytochrome c_1 reduction depend on the ratio between the electron donor and the enzyme may explain the large discrepancy existing in the literature about the kinetic effect of the antibiotic on the first turnover of the bc_1 complex |5|.Moreover, the above evidences outline that it could be hazardous to deduce reaction mechanisms for the bc_1 complex merely based on the effects of its inhibitors, since all of them can induce per se changes in the electron transfer pathways catalyzed by the enzyme.

|1| Degli Esposti,M. and Lenaz,G. (1982) BBA 682,189-200.
|2| Degli Esposti,M. and Lenaz,G. (1982) Arch.Biochem.Biophys.216,727-735.
|3| Degli Esposti,M. and Lenaz,G. (1982) FEBS Lett.142, 49-53.
|4| De Vries,S. (1983) ph.D. Thesis, University of Amsterdam.
|5| Rich,P.R. (1983) BBA 722, 271-280.

ELECTRON TRANSFER AFTER LOW TEMPERATURE FLASH PHOTOLYSIS OF MIXED VALENCE CO-BOUND CYTOCHROME c OXIDASE IN PLANT MITOCHONDRIA.

M. Denis and P. Richaud*
Faculté des Sciences de Luminy, Case 901, Laboratoire de Physiologie Cellulaire, ERA 619, 13288 Marseille Cédex 9, France.
* Present address : C. E. N. Cadarache, DRF, ARBS, B. P. 1, 13115 Saint Paul lez Durance, France.

The study of ligand binding to the active site of cytochrome c oxidase has proved very efficient in the building of our knowledge on its configuration in the absence of the three dimensional structure of this complex enzyme. Indeed, such an approach avoids, in principle, the interference of the intricated electron transfer mechanism still to be solved. In this prospect, carbon monoxyde has been widely used due to its property to be photodissociable at any temperature(1). We have recently reported the dynamic properties of CO rebinding to the fully reduced cytochrome c oxidase in plant mitochondria(2) using a low temperature-flash photolysis-spectrometric technique(3). In contrast with the mammalian system which has been found to give rise to a monophasic CO recombination(4), the plant complex is characterized by a multiphasic CO rebinding process observed in analogous conditions in the 160-200 K temperature range(2). The kinetics of CO rebinding to the mammalian cytochrome c oxidase in the mixed valence state have also been found monophasic in this temperature domain(5). However, room temperature observations yielded evidence of a reversible electron transfer between presumably a_3 and Cu_A after flash photolysis of the carboxy-mixed valence form(6). It was therefore of great interest to investigate, with the plant system, the events following flash photolysis of the CO-bound mixed valence cytochrome c oxidase in comparison with observations related to the fully reduced form and known results from the mammalian enzyme.

The low temperature flash photolysis technique and the purified potato mitochondria involved in this study were as described previously(2). Absorbance changes were recorded at 586/630 nm and 445/460 nm and referred to the absorbance level of the initial CO-bound state to follow the rates of CO recombination in the temperature range 160-210 K.

Flash photolysis, as observed at 586/630 nm, was followed by a multiphasic CO recombination which presented the additional peculiarity to stop far from completion. The succession of about 5 flash-rebinding cycles yielded up to 80% of unbound form, the remaining 20% keeping the property to be photolysed and bound repetitively. Wavelength scanning in the α range after flash photolysis confirmed the partial rebinding of CO and failed in resolving different optical forms which could be correlated to the phases of the rebinding process.

Observations in the Soret region after a first flash looked contradictory as the absorbance level came back to the initial one. However, a second flash revealed that the amount of photodissociable species was only 50% of the starting material and this amount decreased flash after flash accordingly with observations in the α region.

By running experiments in the conditions which provided the resolution in time and wavelength of compound A_2 and compound $C(7)$, we could rule out the responsability of O_2 for this lack of CO rebinding and we identified the optical properties of the unbound species, very similar to the ones of compound C.

The obtention of this unbound form after flash photolysis of the CO-bound mixed valence state has been interpreted as the product of an irreversible(in our time scale) electron transfer from Cu_B to Cu_A, assigning to the metal atoms of cytochrome $\underline{c}$ oxidase the same redox level as in compound $C(7)$. This interpretation could stand for the electron transfer observed by Boelens et al. at room temperature(6) if one takes into account the optical interference in the Soret region from the unbound form. On the basis of this interpretation, the observed electron transfer would be the first one reported as being restricted to the copper atoms of cytochrome $\underline{c}$ oxidase. These results support the idea of the coexistence of several conformers for the active site of cytochrome $\underline{c}$ oxidase in plant mitochondria, a system which deserves more attention.

REFERENCES

1. YONETANI, T., IIZUKA, T., YAMAMOTO, H. and CHANCE, B.(1973) in Oxidases and Related Redox Systems(KING, T. E., MASON, H. S. and MORRISON, M. eds.) vol. 1, pp 401-405, University Park Press, Baltimore.
2. DENIS, M. and RICHAUD, P.(1982) Biochem. J. 206, 379-385.
3. CHANCE, B., GRAHAM, N. and LEGALLAIS, V.(1975) Anal. Biochem. 67, 552-579.
4. SHARROCK, M. and YONETANI, T.(1976) Biochim. Biophys. Acta 434, 333-344.
5. De FONSEKA, K. and CHANCE, B.(1979) Biochem. J. 183, 375-379.
6. BOELENS, R., WEVER, R. and VAN GELDER, B. F.(1982) Biochim. Biophys. Acta 682, 264-272.
7. DENIS, M. and BONNER, W. D. Jr(1978) in Plant Mitochondria(DUCET, G. and LANCE, C., eds.) pp 35-42, Elsevier/North-Holland Biomedical Press, Amsterdam.

STRUCTURAL AND FUNCTIONAL ASPECTS OF CYTOCHROME C OXIDASE FROM
BACILLUS SUBTILIS W23

W. de Vrij°, B. Poolman°, A. Azzi" and W.N. Konings°
°Department of Microbiology, University of Groningen, Kerklaan 30,
9751 NN HAREN, The Netherlands
"Medizinisch-Chemisches Institut, University of Bern, Bühlstraße 28,
CH-3000 BERN 9, Switzerland

Eurkaryotes and many prokaryotes contain cytochrome c oxidase as
the terminal component of the electron-transport chain. Although
bacterial cytochrome c oxidases are structural less complex than
eukaryotic oxidases, the functional properties of both type of oxidases show striking similarities. In this report we will describe
some characteristics of Bacillus subtilis cytochrome c oxidase,
purified to homogeneity by cytochrome c affinity chromatography
(1). The purified enzyme showed absorption maxima at 414 nm and 598
nm in the oxidized form and at 443 nm and 601 nm in the reduced
form. Upon reaction with carbon monoxide of the reduced purified
enzyme the absorption maxima shifted to 431 nm and 598 nm. SDS-PAA
gelelectrophoresis indicated that the purified enzyme is composed
out of three subunits with apparent molecular weights of 57, 37 and
21 kD. The reaction catalyzed by this oxidase was strongly inhibited
by cyanide, azide and carbon monoxide, characteristic for an aa_3
type oxidase. Furthermore immunoabsorption experiments indicated a
transmembranal localization of the protein in the cytoplasmic membrane. Eukaryotic cytochrome c oxidase has been described to exhibit
mutli-phasic steady state kinetics. This kinetic behaviour is determined by the aggregation state of the enzyme. The monomeric form has
monophasic kinetics and the dimeric form shows multi or biphasic
kinetics with strong negative cooperativity between the binding
sites. The ratio of the mono- and dimeric cytochrome c oxidase in a
detergent solubilized state could be varied with the ionic strength
of the medium. We investigated the kinetics of cytochrome c oxidase
from B. subtilis at different ionic strengths. The results show that
the maximum turnover number of the purified enzyme in solubilized
state is independent of the ionic strength. However, increase of the
ionic strength results in an increase of the K_m for cytochrome c,
indicating electrostatic interaction between enzyme and substrate.
The bacterial enzyme exhibits positive cooperativity at low ionic
strength, which is lost at high ionic strength. Furthermore there is
no relation between the steady state kinetics and the molecular
weight of the enzyme, which independently of ionic strength is
present in a dimeric form. Under more physiological conditions i.e.
in the cytoplasmic membrane the enzyme exhibited monophasic kinetics.

Cytochrome c oxidase serves as a terminal member of the electron transport chain, transferring electrons from reduced cytochrome c to oxygen and functions in the generation of a proton motive force. Reconstitution experiments of B. subtilis cytochrome c oxidase showed that it is possible to generate an electrical potential across the artificial membrane (inside negative). The $\Delta\psi$'s measured with an ion-selective electrode sensitive to tetraphenylphosphonium (TPP^+) varied from -30 to -40 mV. A slight increase of the $\Delta\psi$ was observed with increasing pH, possibly due to changes in ion permeability of the artificial membrane. Important for optimal reconstitution is the incorporation of the enzyme in the artificial membrane. Several factors influence the net-incorporation of the oxidase (i.e. cytochrome c binding site facing outwards). Decrease of the net charge of the liposome results in higher percentage right-side out orientated oxidase molecules. A similar effect is observed when the protein to lipid ratio is increased.

REFERENCES

1. Vrij, W, de, Azzi, A. and Konings W.N. (1983) Eur.J.Biochem. 131, 97-103

LATERAL DIFFUSION OF UBIQUINONE IN MODEL AND MITOCHONDRIAL MEMBRANES
R.Fato,M.A.Battino and G.Lenaz,Istituto di Chimica Biologica and Istituto Orto Botanico, Università di Bologna,40I26 Bologna.

The classical pool function of Q as a mobile component of the respiratory chain(I,2)in contrast with its more recently postulated function as a prosthetic group in a protein-bound form has been tested in our laboratory.We have applied the method of fluorescence quenching(3)of fluorophore I2-(9-antroylstearic acid)(AS)by different ubiquinones in phospholipids bilayers:diffusional collisions depend on D_L and on local quencher concentrations,therefore quenching reveals both diffusion rates and lipid-water partition coefficients of the quencher.We have applied the equation:

$$I/K_{app} = \alpha_m(I/K_m - I/K_mP) + I/K_mP$$

where K_m is the bimolecular quenching costant in the membrane,α_m is the volume factor of the membrane in the suspension,K_{app} is the observed quenching costant and P is the partition coefficient of the quencher.Aplot of I/K_{app} vs α_m for different ubiquinones allows calculation of partition coefficients and of lateral diffusion coefficients of probe plus quencher.

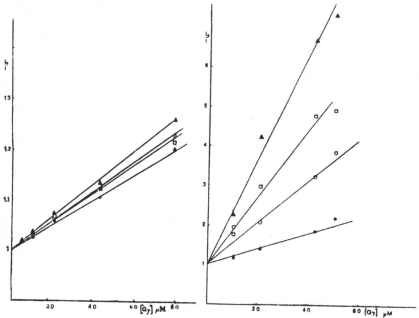

Fig.1 - Quenching of AS fluorescence by Q_7.Left,Q_7 added from ethanolic solution; rigth,Q_7 cosonicated with lipids.Phospholipids concentrations :△—△ 0.25 mg/ml; ▫—▫   0.5 mg/ml; o—o   1mg/ml; ●—● 2 mg/ml.

Whereas for short chain ubiquinones the way of addition does not appear to modify the quenching characteristics,long chain ubiquinones appear to substantially quench AS fluorescence only when sonicated together with the phospholipids(Fig.I). Such behavior is showed also by other fluorescent probes,such as perylene and diphenylexatriene. Since long chain ubiquinones appear to be incorporated into lipid vesicles when added from ethanolic solutions(4),the lack of quenching might be due to a non-monomeric physical disposition of the quinones in the bilayer, also leading to apparent low partition coefficients measured with the quenching technique. The values found for different ubiquinones are reported in table I

Table I

Quencher	$K_m(M^{-1}s^{-1})$	P	$D_L(cm^2/s)$
Q_1 (added)	3.259×10^9	9.143×10^4	1.58×10^{-7}
Q_2 (")	5.936×10^9	17.576×10^4	2.88×10^{-7}
Q_3 (")	7.061×10^9	6.568×10^4	3.43×10^{-7}
Q_5 (")	7.526×10^9	7.490×10^4	3.65×10^{-8}
Q_7 (")	1.086×10^9	1.114×10^4	5.30×10^{-8}
Q_7 (cosonicated)	3.870×10^9	2.560×10^5	2.66×10^{-7}
Q_{10}(cosonicated)	3.260×10^9	5.960×10^4	1.58×10^{-7}

(Pvalues are reported in $(Mol_Q/Mol_{PL})/(Mol_Q/Mol_W)$;the diffusion coefficients,D_L,are calculated from K_m by the equation of Smoluchowsky(3)). Lateral diffusion has also been calculated theoretically(5):if the quinone has an orientation parallel to the lipid molecules and assuming viscosity around I P at 25°C,it can be calculated that Q_{10} in the mitochondrial membrane diffuses with D_L of $10^{-8}cm^2/s$.By the relation $r^2=4D_Lt$,the resulting displacement rate is of 64 nm/ms for $D_L=10^{-8}cm^2/s$. It can be calculated that at such rate the average distance between complex I and III of 30nm can be covered in 0.46ms,well below the turnover number of the respiratory chain.Under normal conditions the diffusion of Q in mitochondria cannot therefore be rate-limiting,in accordance with a collection of kinetic data.

I. Green D.E.,Comp.Biochem.Physiol. 4,8I, I962
2. Kröger A. and Klingenberg M.,Eur.J.Biochem. 34,358,1973
3. Lakowicz J.R.,Hogen D.and Omann G.,Biochim.Biophys.Acta 47I,40I,1977
4. M.Degli Esposti,E.Bertoli,G.Parenti Castelli,R.Fato,S.Mascarello and G.Lenaz, Arch.Biochem.Biophys.,2IO,2I,I98I
5. Saffmann P.G. and Delbrück M.,Proc.Nat.Acad.Sci. USA 72,3III,I975

RECENT PROGRESS IN THE CHARACTERIZATION OF THERMUS RESPIRATORY PROTEINS

J.A. Fee[‡], T. Yoshida[‡], D. Kuila[‡], B.H. Zimmermann[‡], J. Cline[°], and B. Hoffman[°]
Biophysics Research Division, The University of Michigan[‡], Ann Arbor, Michigan 48109 U.S.A.
Department of Chemistry, Northwestern University[°], Evanston, Illinois 60201 U.S.A.

The extremely thermophilic aerobe Thermus thermophilus has been shown to possess respiratory proteins having redox centers remarkably similar to those observed previously in the proteins of mammalian mitochondria. We are focussing our efforts on two of these systems:

Rieske-type iron sulfur protein. This protein was purified from Thermus and characterized in great detail [Fee et al, J. Biol. Chem. 259, 124 (1984)]. The essential finding was that the two identical [2Fe 2S] centers must be ligated to amino acids other than cysteine. We have recorded the ENDOR spectrum of the Thermus Rieske-type protein, and this has provided solid evidence for the binding of nitrogen atoms to the iron cluster. The spectrum has been interpreted in terms of coordinated histidine.

We are also measuring the redox potentials of the [2Fe 2S] clusters, and data on this subject will be presented as it becomes available.

Cytochrome $c_1 aa_3$. We purified this protein from Thermus and found it to consist of two polypeptides. The ~33 Kdal protein binds the heme and has spectral properties ($g_z \cong 3.3$) similar to mammalian cytochrome c_1. The ~55 Kdal protein is thought to bind all the redox components of aa_3, i.e. 2 Cu and 2 heme A [Yoshida et al, J. Biol. Chem. 259, 112 (1984)]. The cytochrome $c_1 aa_3$ complex was also shown to pump protons, suggesting that this activity may be part of a very simple peptide composition [Yoshida and Fee, J. Biol. Chem. 259, 1031 (1984)].

At the present meeting, we will report results of on-going efforts to determine the redox properties of the enzyme and its ability to bind peroxides.

THE AMINO ACID SEQUENCE OF THE IRON-SULFUR SUBUNIT OF THE CYTOCHROME REDUCTASE FROM NEUROSPORA MITOCHONDRIA DERIVED FROM THE cDNA-SEQUENCE

Uwe Harnisch*, Hanns Weiss*, Walter Sebald°

*Institut für Biochemie, Universität Düsseldorf, FRG
°Gesellschaft für Biotechnologische Forschung,
3300 Braunschweig-Stöckheim, FRG

The Rieske-iron-sulfur subunit of ubiquinol:cytochrome c reductase is an amphiphilic protein, which mainly extends into the intermembrane space of mitochondria and is anchored to the inner mitochondrial membrane only by a small hydrophobic protein strech (1,2).

The amino acid sequence of the subunit was derived from the cDNA-sequence. A first cDNA clone was identified from a cDNA bank cloned in E.coli by hybridization-selection of mRNA and subsequent cell-free protein-synthesis and immunoabsorption. Further clones were identified by colony-filter-hybridization. The complete cDNA-sequence was determined from five overlapping cDNA-fragments.
The sequence of residues 1 to 24 was confirmed by automated Edman-degradation of the isolated protein.

The subunit consists of 199 amino acid residues; 90 are hydrophobic, 23 acidic and 27 basic. The molecular weight is 21800. The sequence contains a small and a large hydrophilic domain and a hydrophobic region from amino acid 30 to 54 (Fig.).

This region probably is the only part of the subunit
which lies in the bilayer. The protein contains only
4 cysteine residues located at position 142,147,161
and 163 in a region of moderate hydrophobicity. These
cysteines most likely bind the 2Fe-2S cluster.

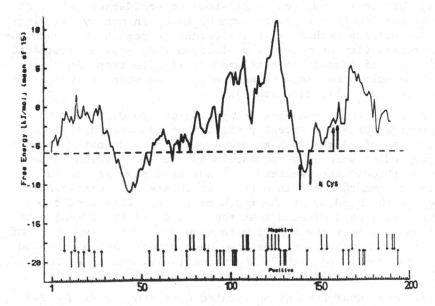

Fig.: Profile of polarity of the iron-sulfur subunit. The
area below the dotted line indicates an increased probability for the amino acid to be located in the membrane. The
arrows indicate the position of the 4 cysteine residues.

References

(1) Li,Y., de Vries,S., Leonhard,K. and Weiss,H., (1981),
    FEBS Lett. 135, 277-280
(2) Karlsson,B., Hovmöller,S., Weiss,H. and Leonhard,K.,
    (1983), J.Mol.Biol. 165, 287-302

INHIBITION OF CYCLIC AND NON-CYCLIC ELECTRON FLOW AT THE Q_c SITE OF THE CYTOCHROME b/f-COMPLEX

A. Hartung, H. Wietoska, and A. Trebst

Abt. Biologie, Ruhr-Universität, D-4630 Bochum 1, FRG

In non-cyclic photosynthetic electron flow from water to NADP the the two photosystems are connected with each other via two mobile redox carriers (plastoquinone and plastocyanin) and the integral cytochrome b_6/f-complex. Recent investigations on the role of cytochrome b_6 indicated that it is involved in plastoquinol-oxidation. Although this assigns an obligatory role for cytochrome b_6 in non-cyclic electron flow, the precise mechanism of its oxidation remains unclear. The antimycin insensitivity of non-cyclic electron flow - as against antimycin sensitivity of ferredoxin catalyzed cyclic electron flow - argues against an obligatory role of an electrogenic step in cytochrome b_6 reactions in non-cyclic electron flow.

Cytochrome b_6 contains two heme groups, which are difficult to distinguish spectroscopically. Recently the DNA of the chloroplast gene for cytochrome b_6 of 23 kD has been sequenced. The hydropathy analysis of the corresponding aminoacid sequence predicts the folding pattern of the peptide through the membrane. It indicated several helical spans across the membrane and in particular located the histidines responsible for the binding of the two heme groups. This prediction of the peptide folding indicated that the histidines are arranged on two helical spans across the membrane in such a way that two pairs of histidines face each other such that two heme groups can be alligned between his 96 and 181 and his 82-196 respectively, one on either side of the membrane - as required for the Q-cycle hypothesis.

Accordingly one inner heme group located close to the Rieske FeS center would be involved in plastoquinol oxidation - a site, Q_z sensitive to DBMIB and DNP-INT. The other, outer heme group of cytochrome b_6 could be involved in plastoquinone reduction - Q_c and antimycin sensitive site. Because of the known inhibitor sensitivity the Q_z site would be involved in both non-cyclic and cyclic electron flow, the Q_c site only in cyclic electron flow. However, we wish to show that the Q_c site may also participate in a non-cyclic system and if it does, increases energy conservation.

We reinvestigated the effect of new inhibitors at the Q_c site - in comparison with antimycin - on certain cyclic electron flow and on the duroquinol donor system that are DBMIB sensitive and therefore include the cytochrome b/f-complex. The results strengthen reports already in the literature that indeed only ferredoxin dependent cyclic systems are antimycin sensitive as well as to the other inhibitors. To the known ferredoxin dependent cyclic system we have added a ferredoxin duroquinol catalyzed cyclic system with high rates. Antimycin and the new types of inhibitors inhibit these systems down to the "basal" rate i.e. that rate obtained either in the absence of DCMU (DCMU poises and this way stimulates cyclic electron flow) or of ferredoxin.

Interesting is the effect of the Q_c site inhibitors on the duroquinol non-cyclic donor system to PS I and MV. The effect on electron flow is small - in order of 20 % - i.e. electron flow is almost insensitive to antimycin. But significantly the stoichiometry of ATP formation to electron flow is changed from 0.6 to 0.4. In accordance with the literature the slow rise of the field is not affected by antimycin or the other inhibitors indicating that the field is built up, but not necessarily net electron flow through it. We interpret these results as the existence of Q_c-antimycin sensitive site that binds quinones close to the outer heme group of cytochrome b_6. This Q_c site is obligatory in ferredoxin dependent cyclic systems, as ferredoxin provides just one electron for PQ reduction on the outside of the membrane. But the Q_c site also participates in the duroquinol donor system. In the latter case the Q_c site is not obligatory. But electron flow through the Q_c site increases coupling efficiency, as shown by the higher stoichiometry in the absence of the inhibitors.

The conclusion is that a quinol is oxidized by the cytochrome b/f-complex at the Q_z site (Rieske center and the inner heme group of cytochrome b_6) and both electrons are transfered to photosystem I without the participation of the Q_c site and without the outer heme group of cytochrome b_6. It is proposed that just one heme group of cytochrome b_6 is able to catalyze dismutation of plastosemiquinone: cytochrome b_6 is reduced by the semiquinone/quinone couple and is oxidized by the quinol/semiquinone couple. The two reactions may occur on opposite sides of the membrane as in the classical Q-cycle hypothesis, but in the dismutase reaction in non-cyclic electron flow they are on the same site and side of the membrane.

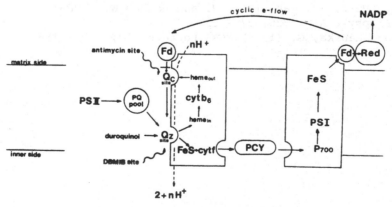

FURTHER STUDIES ON THE H^{+}-PUMPING
CYTOCHROME bc-COMPLEXES

G. Hauska

Department of Biology and Preclinical Medicine,
Institute of Botany, University of Regensburg,
84 Regensburg, W-Germany

The dominant types of electron transport chains in photosynthesis and
respiration contain cytochrome bc-complexes which function as H^{+}-trans-
locating quinol-cytochrome c (or plastocyanin) oxidoreductase. Although
various functionally active isolates show different polypeptides
compositions (Fig. 1), they have a universal redox center composition -
1 haem c, 2 haems b and one Fe2S2-center -, and catalyze the energy
conserving reaction by the same, complex mechanism, manifested by
"oxidant-induced reduction" of cytochrome b. The features of these
complexes have recently been reviewed from the comparatives aspects
(1). In Fig. 1 the polypeptide patterns of these isolates are shown
in addition to a cytochrome bc-preparation from the green sulfur bacterium Chlorobium (2). Cytochrome c (c1 or f) is presented by subunits
3, 1 ab, and 2 in the cases of mitochondria, chloroplasts and the
cyanobacterium Anabaena, and the photosynthetic bacteria, respectively.
The double-haem protein cytochrome b is presented by 4, 2 and 1 in
the same order. The Fe2S2-center (Rieske FeS-protein) has been identified with 5, 3 and 3ab (proteolytic artefact!) in the same order
(the Chlorobium preparation does not contain this center). Improvements on the isolation and reconstitution and progress with regard to
the elucidation of the structure (genes and primary structures, crys-
tallization, proteolysis and labelling, ect.) emphasizing the photosynthetic complexes will be summarized.

1) Hauska, G., Hurt, E., Gabellini, N. and Lockau W. (1983)
   Biochim. Biophys. Acta 726, 97 - 133
2) Hurt, E. and Hauska, G. (1984) FEBS Letters 168, 149 - 154

Fig. 1 – SDS-PAGE Patterns of Quinol-Cytochrome c Oxidoreductases.

The gel (15 % after Laemmli) was stained with Coomassie blue. Subunits
of the oxidoreductases are numbered on the right of each track.
A) standard. B) bc1-complex from beef heart mitochondria (courtesy of
Dr.s Engel and von Jagow, Munich). C) b6f-complex from spinach chloroplasts. D) b6f-complex from Anabaena variabilis. E) bc1-complex from
Rhodopseudomonas sphaeroides GA. F) bc-preparation from Chlorobium
limicola f. thiosulfatophilum.

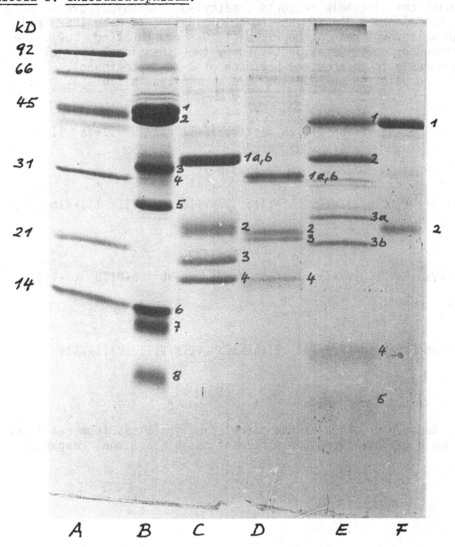

STRUCTURE AND ORIENTATION OF CYTOCHROME C OXIDASE INVESTIGATED BY CROSS-LINKING AND PROTEASE TREATMENT

J. Jarausch and B. Kadenbach
Fachbereich Chemie, Biochemie, Universität Marburg,
Hans-Meerwein-Straße, D-3550 Marburg, FRG

Mammalian cytochrome c oxidase is composed of 13 different polypeptides, as demonstrated by a SDS-gel electrophoretic system of high resolution (1). The isolated complex of rat liver was cross-linked with the cleavable reagents dimethyl-3,3'-dithiobispropionimidate (DTBP) and 3,3'-dithiobis(succinimidyl)propionate (DSP) and with the sulfhydryl reagent cupric di(1,10-phenanthroline (CuP). The cross-linked products were identified by two-dimensional SDS-gel electrophoresis in absence and presence of 2-mercaptoethanol. Cross-linked pairs of polypeptides are summarized in table 1.

agent

	I+II	I+Va	I+Vb	I+VIIb	I+VIII	II+II	II+IV	II+Va	II+Vb
DTBP		+	+	+	+		+	+	
DSP							+	+	
CuP	+	+	+	+	+	+!		+	+

	II+VI	II+VIb	II+VIc	II+VIIb	II+VIIc	II+VIII	III+VIb
DTBP	+	+	+		+		
DSP		+					+
CuP				+		+	

	IV+Va	IV+Vb	IV+VIa	IV+VIb	Va+Va	Va+Vb	Va+VIIb	Va+VIII
DTBP	+	+	+	+		+		
DSP	+			+		+		
CuP					+!	+	+	+

	Vb+Vb	Vb+VIIb	Vb+VIII	VIIb+VIIb	VIIb+VIII	VIII+VIII
DTBP						
DSP						
CuP	+!	+	+	+!	+	+!

Table 1: Summary of cross-linked pairs of polypeptides from rat liver cytochrome c oxidase formed by different homobifunctional reagents.

Cross-linked products of the large, mitochondrially synthesized poly-peptides were formed with 9 of the 10 nuclear coded polypeptides. Pairs between identical polypeptides and complete loss of enzymatic activity was only observed with CuP, indicating larger conformational changes and aggregation of enzyme complexes. Therefore only cross-linked products with DTBP and DSP were used to draw up a model for the structural arrangement of polypeptides.

Seven polypeptides of rat liver cytochrome c oxidase were found cross-linked by treatment with CuP. The same polypeptides were labeled with the sulfhydryl reagent ^{3}H-N-ethylmaleimide (I,II,III,Va,Vb,VIIb,VIII).

The orientation of the native enzyme towards the cytoplasm was inves-tigated by protease treatment. Mitoplasts were incubated with trypsin, subtilisin and pronase, washed and dissolved in sample buffer. After separation by gel electrophoresis the mitoplast proteins were trans-ferred on nitrocellulose (2) and the amount of proteolytic degradation was studied with an antiserum against rat liver cytochrome c oxidase. Results are presented in table 2.

polypeptides with orientation towards cytoplasm	protease trypsin	subtilisin	pronase
II		+	+
IV	+		+
Vb			+
VIa	+	+	+
VIIa			+
VIII			+

Table 2: Orientation of rat liver cytochrome c oxidase polypeptides towards the cytoplasm, as studied by protease treatment of mitoplasts.

1) Kadenbach, B., Jarausch, J., Hartmann, R. and Merle, P. (1983) Anal. Biochem. 129, 517-521
2) Jarausch, J. and Kadenbach, B. (1982) Hoppe-Seyler's Z. Physiol. Chem. 363, 1133-1140

LOCALIZATION OF CYTOCHROME b-562 HEME AT THE M-SIDE OF THE MITOCHONDRIAL MEMBRANE

A.A. Konstantinov, W.S. Kunz, P. Berndt, E. Popova
A.N. Belozersky Laboratory of Molecular Biology and Bioorganic Chemistry, Moscow State University, Moscow W-234, 119899, USSR

We have examined interaction of the respiratory chain of mitochondria and inverted submitochondrial particles (SMP) with 2 membrane impermeable redox compounds: $Fe(CN)_6^{3-/4-}$ (FeCy) and $Ru(NH_3)_6^{3+/2+}$ (RuAm). Among the results obtained, the following two may be interesting.

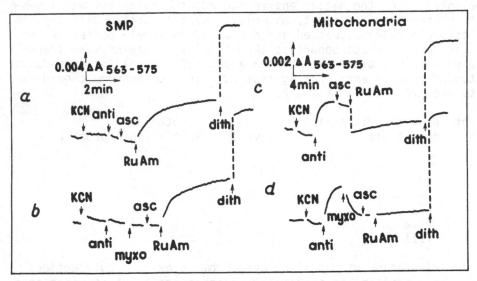

Fig. 1. Reduction of cytochrome b by hexaamminoruthenium. (a,b) beef heart SMP, 1.2 mg/ml; (c,d) pigeon heart mitochondria, 0.7 mg/ml. 0.3 M sucrose, 20 mM HEPES pH=7.5, 1 mM EDTA, 1 μM CCCP, 3 μM rotenone. Additions. KCN, 4 mM; antimycin, 1 μM; myxothiazol, 1 μM; ascorbate, 1 mM; RuAm; 20 μM.

1. In SMP RuAm mediates rapid reduction of cytochrome b-562 by ascorbate (Fig. 1a). The reaction is not inhibited by antimycin + myxothiazol (Fig. 1b) and therefore is likely to involve direct interaction of the mediator with heme b-562. In the antimycin + KCN-inhibited mitochondria RuAm brings about relaxation of the cytochrome b extrareduction followed by a very sluggish re-reduction of b-562 (Fig. 1c), which is further inhibited by myxothiatol (Fig. 1d). These results (see also [1]) indicate that ferric heme b-562 is localized on the M-side of the mitochondrial membrane, whereas the myxothiazol-sensitive centre o may be moderately accessible from its C-side.

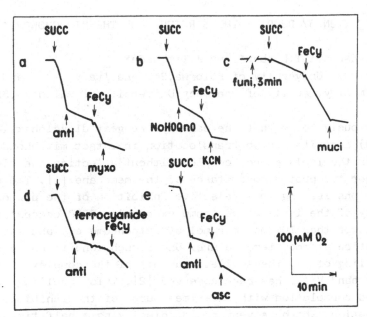

Fig. 2. Ferricyanide-induced stimulation of succinate oxidase activity in beef heart SMP. Conditions as in Fig. 1, but 5 mM MgSO4 added and EDTA omitted. FeCy, 1 mM; succinate 20 mM; NoHOQnO, 10 μM; funiculosin, 4 μM; mucidin, 6 μM; ascorbate 2 mM.

When succinate oxidation by SMP is inhibited by antimycin, funiculosin or 2-nonyl-4-hydroxyquinoline N-oxide, FeCy partially restores oxygen consumption (Fig. 2). The FeCy-stimulated succinate oxidase activity is coupled to $\Delta \psi$ -generation [2,3] and is abolished by myxothiazol, mucidin or KCN, as well as by ascorbate which reduces ferricyanide to ferrocyanide (Fig. 2). As discussed in [3] the data strongly support the Q-cycle hypothesis and show that at least one of the ferrous cytochrome b hemes can serve as electron donor to FeCy at the M-side of the membrane. Most likely, FeCy is able to accept electrons directly from b-562 heme and thus uncouples QH_2 oxidation by FeS-Rieske and b-566 in centre o of the Q-cycle from the antimycin-sensitive step of b-562 oxidation by CoQ in centre i.

[1] Konstantinov, A.A. and Kunz, W.S. (1984) Biochemistry (USSR) 49, 1046-1048.
[2] Liberman, E.A., Vladimirova, M.A. and Tsofina, L. (1977) Biophysics (USSR) 22, 255-259.
[3] Kunz, W.S., Konstantinov, A.A., Tsofina, L. and Liberman, E.A. (1984) FEBS Lett. in press.

CHEMICAL MODIFICATION BY DCCD OF THE STRUCTURE OF THE MITOCHONDRIAL bc$_1$
COMPLEX.

G.Lenaz,M.Degli Esposti,T.Barber* and J.Timoneda*
Institute of Botany, University of Bologna,Bologna,Italy,   and
*Dept.of Biochemistry,Faculty of Pharmacy,University of Valencia,Spain.

We have previously found that the carboxyl reagent dicyclohexylcarbodiimide (DCCD) inhibits proton translocation in intact mitochondrial
particles and in the isolated bc$_1$ complex,without affecting the electron transfer nor the proton conductance of the membrane [1]. The expe
rimental conditions leading to a selective inhibition of the proton
pumping activity of the isolated bc$_1$ complex require DCCD concentrations generally lower than 100 mol per mol of cytochrome c$_1$, and short
times of incubation at room temperature. Under such conditions, a preferential labelling of   the smallest subunit of the complex of about 9kDa, namely band VIII, has been observed [2]. This labelling appears also in good correlation with the time-course of the inhibition of
proton translocation, which is very rapid, displaying a half-time shor
ter than 1 min at room temperature[1,2].

The time of incubation with the reagent has been found to be quite
critical in the labelling experiments. At times longer than 10 min, in
fact, a broad labelling of many subunits of the complex has been detec
ted by both gel-slices counting and fluorographic studies. Even some
contaminant bands seem to bind DCCD in a relevant manner. Similar results can be obtained using DCCD concentrations higher than 100 mol per
mol of the enzyme. Moreover, clear cross-linking phenomena between
bands V,VII and VIII occur either by prolonging the time of incubation
or increasing the DCCD concentration. Such unspecific labelling pattern
and structural modifications can be correlated to the onset of inhibition of the redox activity of the bc$_1$ complex and to a loss of its antimycin binding. Therefore, we believe they reflect a less specific,or
secondary,action of the ddimide, occurring after the primary action has
been already accomplished on the proton extrusion activity [2].

In Fig.1 is shown the DCCD concentration dependence of the labelling
of subunit VIII and subunit III (the apoprotein of cytochrome b) at two
different times of incubation. The behaviour of the binding at 10 min
(Fig.1A) correlates well with the titration of the inhibition of the

proton-translocating function of the enzyme at short times of incubati
on. Since the labelling of band III is much lower under such conditi
ons (cf.Fig.1A), and displays a slow time-dependence (half-time around
30 min), it is difficult to consider cytochrome b as the major site of
the DCCD modification of the bc_1 complex, as reported by other Authors
|3,4|.

On the basis of our results, it is tempting to speculate that the
proton extrusion device in the bc_1 complex involves non-redox proteins
such as band VIII.

Fig.1- DCCD titration of the labelling of Band VIII and III at 10 min
(A) and 4 hours (B) of incubation at room temperature.

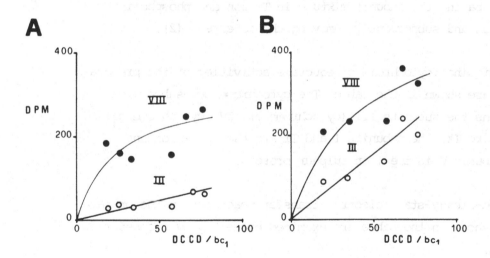

|1|Degli Esposti,M.,Meier,E.M.M.,Timoneda,J., and Lenaz,G. (1983) BBA
   725,349-360.
|2|Degli Esposti,M.,Saus,J.,Timoneda,J.,Bertoli,E. and Lenaz,G. (1982)
   FEBS Letters 147,101-105.
|3|Beattie,D.S. and Clejan,L. (1982) FEBS Letters 149,245-249.
|4|Naɫecz,M.J.,Casey,R.P. and Azzi,A. (1983) BBA 724,75-82.

RECONSTITUTION OF MITOCHONDRIAL UBIQUINOL:CYTOCHROME C REDUCTASE FROM ISOLATED SUBUNITS

Petra Linke, Hanns Weiss

Institut für Biochemie der Universität

Düsseldorf, FRG

In order to enlighten the function of the subunits I and II of cytochrome reductase the enzyme is dissociated into isolated subunits and subunit complexes (1) and reconstituted in steps. Partially and totally reconstituted preparations are obtained by incubating the subunit mixture in Triton and phospholipid solution and subsequently removing the detergent (2).

The ubiquinol:cytochrome c reductase activities of the preparations are shown in the table. The cytochrome bc_1 subcomplex contains the subunits III (cytochrome b), IV (cytochrome c_1) and VI to IX, the subunits I and II are the core-proteins, the subunit V is the iron sulphur protein.

The pre-steady-state electron transfer reactions in the preparations shown in the table are examined presently by stopped-flow method.

Preparation	turnover number in Triton solution* (3) $\left[\text{min}^{-1}\right]$
Cytochrome reductase	400
Cytochrome bc_1-subcomplex	10
Cytochrome bc_1-subcomplex + subunit I, II	10
Cytochrome bc_1-subcomplex + subunit V	10
Cytochrome bc_1-subcomplex + subunit V + subunit II	10
Cytochrome bc_1-subcomplex + subunit V + subunit I	100
Cytochrome bc_1-subcomplex + subunit V + subunit I, II	250

* The turnover numbers of the preparations in phospholipid bilayers are five times higher.

References

(1) Hovmöller, S., Leonard, K., Weiss, H., (1981) FEBS Lett. 123, 118 - 122
(2) Wingfield, P., Arad, T., Leonard, K., Weiss, H., (1979) Nature 280, 696 - 697
(3) Weiss, H., Wingfield, P., (1979) Eur.J.Biochem. 99, 151 -160

PROTON TRANSLOCATION IN THE $b-c_1$ COMPLEX FROM BEEF-HEART MITOCHONDRIA
RECONSTITUTED INTO PHOSPHOLIPID VESICLES

M. Lorusso, D. Gatti and S. Papa
Institute of Biological Chemistry, Faculty of Medicine, University of
Bari, 70124 Bari, Italy

Electron flow from reduced quinone to ferricytochrome c along the
$b-c_1$ complex of the respiratory chain, either in the native mitochondrial membrane or in the reconstituted system, results in the effective translocation of two protons from the inner space (N phase) to
the outer space (P phase); two more protons, deriving from the oxidation of the hydrogen donor, are released in the P phase, thus giving
an overall H^+/e^- stoicheiometry of 2 (1-2).

General models so far proposed for vectorial proton translocation
in the $b-c_1$ complex involve either a diffusion of hydrogen-carrying
quinone molecules along specifically oriented conduction pathways (3)
or cooperative phenomena in the polypeptides, linked to electron flow
at the redox centers (2,4).

We have recently proposed (5), on the basis of various observations,
a mechanism based on a combination in series of proton transfer reactions at a protein-bound semiquinone-quinone center in the membrane
together with protolytic effects in the apoproteins, which provide
asymmetric proton exchange between the catalytic center and the two
aqueous-phases.

To verify and possibly identify a role of polypeptides of the $b-c_1$
complex in its redox and proton-motive activity, we are persuing two
general approaches, i) studies with chemical modifiers of specific aminoacid residues and ii) studies with proteolytic enzymes.

Treatment of $b-c_1$ complex reconstituted into liposomes vesicles
with DCCD results in suppression of the redox-linked vectorial proton
translocation, with no effect on the rate of uncoupled electron flow
and enhancement of the rate of electron flow in the coupled state. The
inhibition of proton translocation is not due to enhanced rate of passive proton back-flow.

Treatment of $b-c_1$ vesicles with papain for short time incubation
and relatively low concentrations of the proteolytic enzyme, results
also in a marked inhibition of proton translocation and decrease of
H^+/e^- stoicheiometry. Separate experiments show that papain-treated
$b-c_1$ vesicles exhibit even a lower passive proton conductance.

From these observations it is concluded that either modification by DCCD of critical glutamic residue in the hydrophobic region of the membrane or cleavage of externally exposed polypeptide segments of the b-c$_1$ complex, leads to loss of proton translocation, leaving the electron flow at the redox centers unaffected.

These observations are consistent with the idea that specific proton conduction pathways in the apoproteins are involved in the protonmotive activity of the b-c$_1$ complex.

^{14}C-DCCD binding experiment supports an involvement of low molecular weight polypeptides (the 8kDa subunits) and possibly of the Fe-S protein in the redox-linked proton translocation in this segment of the cytochrome system. Experiments are in progress to specify the subunits involved in the proteolytic digestion by papain.

---

1) Papa, S. (1982), J.Bioenergetics and Biomembranes 14, 69-86.
2) Wikström, M.K.F., Krab, K. and Saraste, M. (1981) Ann.Rev.Biochem. 50, 623-625.
3) Mitchell, P. (1976) J. Theor. Biol.62, 327-367.
4) Von Jagow, G., Engel, W.D. and Shagger, H. (1981) in "Vectorial Reactions in Electron and Ion Transport in Mitochondria and Bacteria (Palmieri, F. et al.eds.), pp.149-161, Elsevier/North Holland, Amsterdam.
5) Papa, S., Lorusso, M., Boffoli, D. and Bellomo, E. (1983), Eur. J. Biochem. 137, 405-412.

STEADY-STATE KINETICS OF ELECTRON TRANSFER THROUGH THE MITOCHONDRIAL
UBIQUINOL: CYTOCHROME c OXIDOREDUCTASE

C.A.M. Marres and J.A. Berden
Laboratory of Biochemistry, B.C.P. Jansen Institute, University of
Amsterdam, Plantage Muidergracht 12, 1018 TV Amsterdam, The Netherlands

It is generally assumed that the degree of non-linearity of the relation between inhibition of succinate or NADH oxidation and saturation
of the ubiquinol: cytochrome c oxidoreductase with inhibitors like antimycin, HQNO, myxothiazol and BAL is a measure for the overcapacity of
the quinol oxidase system (V_2) above the quinone reductase system (V_1)
[1,2] in agreement with the model of Kröger and Klingenberg [3]. According to this model a pool of diffusible ubiquinone molecules can accept
electrons from all ubiquinone-reducing enzymes and can donate them to
all ubiquinol-oxidizing enzymes (in the same membrane continuum) and
the inhibition of the ubiquinol oxidase activity is assumed to be proportional to the saturation of the inhibitor-binding site.

However, it has been shown previously that the quinol oxidation itself may be hyperbolically inhibited [2] and we have demonstrated that
the relation between inhibition of quinol oxidation or the quinol:
cytochrome c oxidoreductase reaction and saturation of inhibitor-binding site depends on both the type of inhibitor and the concentration
of the substrate [4] : a hyperbolic inhibition curve is obtained only
at low substrate concentrations while high substrate concentrations
result in linear inhibition curves (see Figure).

To account for these observations Marres [4] has proposed a model in
which a rapid substrate transfer can take place between different enzyme molecules so that substrate bound to inhibited enzyme molecules is
available to non-inhibited molecules without equilibrating with the free
substrate pool (see also ref.5.). It is likely that the substrate is
transferred in its active form, i.e. as QH^- (see ref. 6).

With the introduction of this rapid substrate transfer the following
rate equation is obtained for the quinol oxidation in the presence of
inhibitors that make the enzyme-substrate complex catalatically inactive
without affecting the binding of substrate

$$v = \frac{k_1 \, k_2 \, e(1-n) \, S}{k_{-1} + k_2 \, (1-n) + k_1 \, S}$$

This equation describes a hyperbolic relation between overall rate and
inhibitor saturation (n) at low S and a linear relation at high S. Within this model many observations concerning the kinetics of the ubiquinol
oxidase reaction can be explained:
1. The inhibition by the SH-reagent 5,5'-dithiobis(2-nitrobenzoate) is
linearly related to the saturation of the inhibitory site, also at low

substrate concentrations [4,7]. This is understandable if DTNB prevents the binding of ubiquinol, as is suggested by EPR data [7].

2. Lineweaver-Burk plots for the oxidation of QH_2 and duroquinol by cytochrome c in which QH_2 or duroquinol is the variable substrate reveal an apparent negative cooperativity which disappears at high concentrations of cytochrome c (see also ref.8). The biphasicity at low concentrations of cytochrome c may be explained by substrate transfer between cytochrome c-free(i.e. non-active) and cytochrome c-containing (i.e. active) enzyme molecules, resulting in relatively higher activities at non-saturating concentrations of substrate.

3. The HQNO-induced increase in the steady-state reduction level of cytochrome b does not parallel the inhibition of overall electron transfer from succinate to oxygen [1] and the ratio between V_2 and V_1 for the oxidation of NADH or succinate, as calculated from inhibitor titrations is much larger (≥ 7, refs. 1 and 3) than the ratio between the measured values of succinate of NADH: Q oxidoreductase activity and quinol: cytochrome c oxidoreductase activity, which is only 1-2. The present model suggests that V_1 and V_2 derived from inhibitor titrations do not simply constitute the acitvities of the dehydrogenase and cytochrome chain, respectively, unless corrections have first been made for the hyperbolic inhibition of the QH_2 oxidation itself.

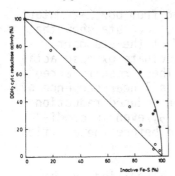

Figure: Inhibition of DQH_2:cytochrome c reductase activity in mitoplasts (obtained by hypotonic-isotonic treatment of mitochondria) by BAL(+ O_2) treatment, which destroys the Fe-S cluster [9]. The degree of destruction of the Fe-S cluster was measured by EPR. The enzyme activity with 400 μM (0) and 40 μM (●), DQH_2, respectively, was measured in 50 mM Mops-Tris buffer (pH 7.6) containing 0.25 M sucrose, 2 mM EDTA, 1 mM KCN and 9 M cytochrome c (the same results were obtained with other inhibitors [4].

REFERENCES
1. Van Ark, G. and Berden,J.A. (1977) Biochim.Biophys.Acta 459, 119-137
2. Zhu,Q.S., Berden,J.A., De Vries,S. and Slater,E.C.(1982) Biochim. Biophys. Acta 680, 69-79
3. Kröger,A. and Klingenberg,M.(1970) Vitamins and Hormones 28, 533-574
4. Marres,C.A.M.(1983),Ph.D.Thesis University of Amsterdam,Rodopi,Amsterdam
5. Ragan,C.J. and Heron, C (1978) Biochem. J., 174, 783-790
6. Rich,P.R.(1981) Biochim. Biophys. Acta 637, 28-33
7. Marres,C.A.M., De Vries,S. and Slater,E.C. (1982) Biochim. Biophys. Acta 681, 323-326
8. Esposti,M.D. and Lenaz,G.(1982) Biochim. Biophys.Acta 682, 189-200
9. Slater,E.C. and De Vries,S. (1980) Nature 288, 717-718

THE CYTOCHROMES b AND THE Q CYCLE IN THE MITOCHONDRIAL RESPIRATORY CHAIN OF SACCHAROMYCES CEREVISIAE : USE OF A mit⁻ MUTANT

D. Meunier-Lemesle and P. Chevillotte-Brivet
Laboratoire de Chimie Bactérienne, C.N.R.S., B.P. 71,
13277 Marseille Cedex 9 (France)

1) In the recent litterature on cytochromes b, the behaviour of these components has always been described as "anomalous" : indeed, all these experiments show an oxidant-induced reduction of these cytochromes, concerning either the cytochrome b_{565} alone (with succinate as substrate) or the two cytochromes b (1). At the present time, these phenomena are explained in the frame work of the Q cycle established by Mitchell (2).

However, in yeast mitochondria, with succinate as substrate, we have obtained experimental conditions, depending on pH, where there is no oxygen induced reduction of the cytochrome b_{565}, but oxidation by oxygen, and reduction in anaerobiosis (3). These observations cannot accounted for by a single Q cycle mechanism as described by Mitchell (1).

2) A cryosensitive revertant PS408/WR4 obtained from box 2-2 mutant is described. It is devoid of cytochrome b_{565} (4). The data show that the disappearance of this cytochrome does not affect the standart redox potential of the cytochrome b_{562} and the succinate oxidase activity, if we compare to the wild type strain (3). These results agree better with a disposition in "parallel" than with a linear sequence of the two cytochromes. In this revertant, the oxygen induced reduction of cytochrome b does not affect the cytochrome b_{562}, even in conditions where this cytochrome is not totally reduced before the addition of oxygen.

3) Our experimental results agree with a Q cycle modified in the following fashion : a) the cytochrome b_{565} and the cytochrome b_{562} are arranged in parallel and not in a linear sequence. b) a direct reaction of succinate with quinone, as proposed by Van Ark (5), leading to the formation of QH_2 diffusible from the inside to the outside surface of the inner membrane, is necessary to explain the reduction of cytochrome b in anaerobiosis. c) the inhibitor Antimycin A must acts on the two sites of complexe III involving the oxidation $(Q°H \rightarrow Q)$ and the reduction $(Q'H \rightarrow QH_2)$ of the semiquinone. Its inhibitory effect is effective alternatively on one site or on the other, at a given time, depending on the sate of mitochondria (aerobiosis or anaerobiosis).

In these conditions, the oxido-reduction state of cytochrome b_{565} depends on the relative rate of the reactions involving Q'H, Q°H and the cytochromes b (3).

REFERENCES

(1) Trumpower, B.L. and Katki, A.G. (1979) in "Membrane proteins in energy transduction" (R.A. Capaldi, ed.) pp. 89-200. M. Dekker Inc., New York and Basel.

(2) Mitchell, P. (1976) J. Theor. Biol., 62, 327-367.

(3) Meunier-Lemesle, D. and Chevillotte-Brivet, P. (submitted for publication).

(4) Meunier-Lemesle, D., Chevillotte-Brivet, P. and Pajot, P. (1980) Eur. J. Biochem., 111, 151-159.

(5) Van Ark, G. (1980) Thesis, Amsterdam.

THE ROLE OF CYTOCHROME b-563 IN CHLOROPLAST ELECTRON TRANSPORT

D. A. Moss and D. S. Bendall
Department of Biochemistry, University of Cambridge, Tennis Court Rd.,
Cambridge, CB2 1QW, U. K.

Most current models of electron transport in the thylakoid membrane
incorporate a Q-cycle or similar mechanism (1). The central feature
of such models is the role of cytochrome b-563 in carrying an electron
across the membrane from a quinone species near the inner surface to
one near the outer surface. It has also been suggested (2) that under
appropriate conditions plastoquinol may be oxidized to the quinone by
two sequential turnovers of cytochrome f, rather than by one turnover
of each of cytochromes b-563 and f. This would allow the chloroplast
Q-cycle to be facultative, accounting for the observed variability of
the H^+/e^- ratio in chloroplast electron transport (3).

The model set out above predicts that where b-563 turnover is
prevented by direct inhibition or by low ambient redox potential the
Q-cycle will not occur. The loss of the Q-cycle would be observable
experimentally as a partial inhibition of photophosphorylation under
steady state illumination, or as an inhibition of the slow phase of
the membrane potential-indicating carotenoid bandshift (slow ΔA_{518})
under flash illumination.

The effect of direct inhibition of b-563 turnover can be observed
using 2-heptyl-4-hydroxyquinoline N-oxide (HQNO) at 10 μM (4). We
have measured the effect of HQNO on cyclic photophosphorylation in the
reconstituted systems described earlier (5), using broken chloroplasts
in the presence of DCMU and photoreduced anthraquinone-2-sulphonate
(AQS) or ferredoxin. In saturating steady state illumination HQNO was
found to inhibit the AQS-catalyzed system by 28.3±4.4% and the
ferredoxin-catalyzed system by 16.5±5.7%: while at limiting intensity
the values were 50.8±0.8% and 44.8±7.3% respectively. These data are
consistent with the facultative Q-cycle model (2) in which a low light
intensity (and thence a low rate of reoxidation of f) would favour the
reduction of b-563 by plastosemiquinone, causing a greater proportion
of the observed phosphorylation to be due to HQNO-sensitive electron
transport through the Q-cycle.

There have been several recent reports (4,6,7) of redox titrations
of the slow ΔA_{518}, but the results have been conflicting. We have
investigated the dependence of the results of such experiments on the
mediators used. The ΔA_{518} was measured as described previously (5),
using the reconstituted AQS-catalyzed cyclic system described above
but varying the ambient potential by varying the length of
preillumination. The slow ΔA_{518} was found to be unaffected by ambient

potential between 0 and -200 mV; but the addition of 10 μM 2-hydroxy-naphthoquinone (HNQ) or 1 μM anthraquinone (AQ) completely removed the slow phase at -200 mV, while having no effect at 0 mV. This is in good agreement with references 6 and 7, though reference 4 reported observation of the slow ΔA_{518} down to -300 mV in the presence of HNQ. One possible explanation of the present observations would be that b-563 is not in equilibrium with the ambient potential until HNQ or AQ is added. However, direct observation of the poise of b-563 under identical conditions showed approximately 90% reduction at around -200 mV, roughly consistent with a midpoint for b_L of -140 mV: and no change was detectable on adding AQ. Both HNQ and AQ are effective cofactors of DBMIB-sensitive cyclic electron transport (8,9) so DBMIB-like inhibition of b-563 turnover would appear to be unlikely, though a HQNO-like effect cannot be ruled out.

Thus while the circumstantial evidence for a model along the general lines of a Q-cycle remains compelling, a substantial modification of the classical Q-cycle is necessary to account for the presence of the electrogenic reaction with b-563 highly reduced and for its inhibition by AQ and HNQ.

References

1 Bendall, D. S. (1982) Biochim. Biophys. Acta 683,119-151
2 Rich, P. R. (1984) Biochim. Biophys. Acta 768,53-79
3 Cox, R. P. and Olsen, L. F. (1982) in Electron Transport and Photophosphorylation (Barber, J., ed.), pp. 49-79, Elsevier Biomedical Press, Amsterdam.
4 Hind, G., Clark, R. D. and Houchins, J. P. (1983) Proceedings of the 6th International Congress on Photosynthesis, in the press.
5 Moss, D. A. and Bendall, D. S. (1983) Proceedings of the 6th International Congress on Photosynthesis, in the press.
6 Giorgi, L., Packham, N. and Barber, J. (1983) Proceedings of the 6th International Congress on Photosynthesis, in the press.
7 Girvin, M. and Cramer, W. A. (1983) Proceedings of the 6th International Congress on Photosynthesis, in the press.
8 Hauska, G., Reimer, S. and Trebst, A. (1974) Biochim. Biophys. Acta 357,1-13
9 Binder, R. G. and Selman, B. R. (1980) Biochim. Biophys. Acta 592,314-322

The authors would like to thank the Science and Engineering Research Council for financial support.

FUNCTION OF THE 33 kDa PROTEIN IN THE PHOTOSYNTHETIC OXYGEN EVOLUTION

N. Murata[o], M. Miyao[o], and T. Kuwabara"
[o]Department of Biology, College of Arts and Sciences, University
of Tokyo, Komaba, Meguro-ku, Tokyo 153, Japan
"Department of Chemistry, Faculty of Science, Toho University,
Miyama, Funabashi 274, Japan

A protein having a molecular mass of 33 kDa was purified from
photosystem (PS) II particles of spinach chloroplasts, and chemically
and physicochemically characterized [1,2]. It has been proved to
be one of the three water-soluble proteins (33 kDa, 24 kDa, and
18 kDa) involved in the photosynthetic oxygen evolution system [3-5].
This study was intended to disclose the function of the 33 kDa protein in oxygen evolution.

As previously reported [4,6], 2.6 M urea removed this protein
from PS II particles with concomitant inactivation of oxygen evolution. Preservation of Mn in the particles during the urea treatment
depended on Cl^- in the medium; in 10 mM NaCl, two of the four Mn
atoms in the oxygen evolution system were lost, whereas in 200 mM
NaCl almost all the four Mn atoms were left bound. These (urea +
NaCl)-treated particles lacking all the three water-soluble proteins
were very similar to those prepared with 1.0 M $CaCl_2$ by Ono and Inoue
[7].

When the (urea + NaCl)-treated PS II particles depleted of the
33 kDa protein were suspended in 10 mM Cl^-, two of the four Mn atoms
were gradually lost. In 200 mM Cl^-, the loss of Mn was suppressed.
In PS II particles containing the 33 kDa protein, on the other hand,
the Mn was not lost in either 10 mM or 200 mM Cl^-. These findings
suggest that the 33 kDa protein is essential for preserving Mn in
the oxygen evolution system, and that 200 mM Cl^- can partially substitute for it.

Oxygen evolution activity was high in PS II particles containing
the 33 kDa protein and low in PS II particles depleted of this protein. 200 mM Cl^- could partially restore oxygen evolution activity
in the particles depleted of the 33 kDa protein.

In conclusion, the 33 kDa protein is necessary for both oxygen evolution and binding of Mn. 200 mM Cl$^-$ can partially replace the function of the 33 kDa protein.

A part of this work will be published elsewhere [8].

References
1. Kuwabara, T. and Murata, N. (1979) Biochim. Biophys. Acta 581, 228-236
2. Kuwabara, T. and Murata, N. (1982) Biochim. Biophys. Acta 680, 210-215
3. Kuwabara, T. and Murata, N. (1982) Plant Cell Physiol. 23, 533-539
4. Kuwabara, T. and Murata, N. (1983) Plant Cell Physiol. 24, 741-747
5. Miyao, M. and Murata, N. (1983) FEBS Lett. 164, 375-378
6. Miyao, M. and Murata, N. (1984) Biochim. Biophys. Acta, in press
7. Ono, T. and Inoue, Y. (1983) FEBS Lett. 164, 255-260
8. Miyao, M. and Murata, N. (1983) FEBS Lett., in press

SUBUNIT III OF CYTOCHROME c OXIDASE: STRUCTURAL AND REGULATORY ROLE

K. A. Nałęcz[O], R. Bolli[O], B. Ludwig[X], and A. Azzi[O]
[O] Medizinisch-Chemisches Institut der Universität Bern,
   Bühlstrasse 28, 3012-Bern, Switzerland
[X] Institut für Biochemie, Medizinische Hochschule, D-2400
   Lübeck, West Germany

In previous studies the steady-state kinetics of bovine heart cytochrome c oxidase was correlated with its molecular form. The monophasic kinetics was ascribed to a monomer, whilst the biphasic Eadie-Hofstee plot was explained as the occurence of homotropic negative cooperativity, caused by the interactions of two cytochrome c molecules in the dimeric enzyme (1). Since it was known from the cross-linking studies (2) that the back face of cytochrome c is in close proximity to the 3rd largest subunit of the enzyme, the studies on the role of this subunit were undertaken, comparing the kinetics of the native bovine enzyme with an oxidase depleted of subunit III (3) and the two-subunit Paracoccus denitrificans oxidase (4-5).

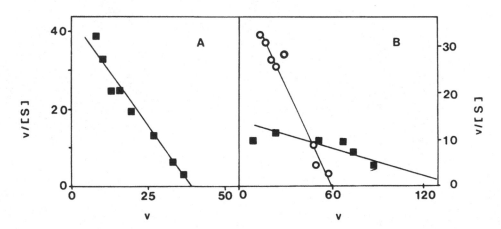

Figure 1. Eadie-Hofstee plot of spectrophotometrically measured reaction of the subunit III depleted bovine oxidase (A) or Paracoccus denitrificans enzyme (B) with cytochrome c. Assay conditions according to Nałęcz et al. (1) without (O) or with 50 mM KCl (■); v, s^{-1}; (S), /uM.

The two enzymes which did not contain subunit III gave monophasic kinetics in both, low and high ionic strength (Fig. 1) in the presence of /B-D-dodecylmaltoside, a detergent known to give very high enzymatic activities and to allow separation of oxidase monomers and dimers (1).

The molecular form of both preparations was checked by filtration in Ultrogel AcA 34 under conditions facilitating dimerization of the control enzyme (high ionic strength). Bovine oxidase depleted of subunit III revealed the presence of both molecular forms, dimeric and monomeric, with a tendency of the latter form to increase by decreasing the protein concentration. In the case of Paracoccus denitrificans oxidase one species occurs at high ionic strength (dimer) and two (monomer and dimer) at low salt concentration.

As it was proposed, subunit III in mammalian oxidase may have a role in the transition of monomers to dimers (6) and thus in the regulation of the interaction with cytochrome c in a cooperative way. The removal of this subunit would facilitate monomerization and loss of cooperative kinetics. In the bacterial oxidase the negative cooperative interaction could not be observed even in the dimers, suggesting that the absence of subunit III would prevent the cytochrome c - cytochrome c interaction responsible for the negative cooperativity.

REFERENCES:
1. Nałęcz, K.A., Bolli, R., and Azzi, A. (1983) Biochem. Biophys. Res. Commun. **114**, 822-828
2. Birchmeier, W., Kohler, C.E., and Schatz, G. (1976) Proc. Natl. Acad. Sci. USA **73**, 4334-4338
3. Bill, K., and Azzi, A. (1982) Biochem. Biophys. Res. Commun. **106**, 1203-1209
4. Ludwig, B., and Schatz, G. (1980) Proc. Natl. Acad. Sci. USA **77**, 196-200
5. Bill, K., Casey, R.P., Broger, C., and Azzi, A. (1980) FEBS Lett. **120**, 248-250
6. Georgevich, G., Darley-Usmar, V.M., Malatesta, F., and Capaldi, R.A. (1983) Biochemistry **22**, 1317-1322

STRUCTURAL-FUNCTIONAL ANALYSIS OF THE MITOCHONDRIAL CYTOCHROME bc1 COMPLEX IN MONOMERIC AND DIMERIC FORM

Maciej J. Nałęcz, Reinhard Bolli and Angelo Azzi,
Medizinisch-chemisches Institut, Universität Bern,
Bühlstrasse 28, 3012 Bern, Switzerland

The M_r of the bc1 complex, estimated from its subunit composition, varies between 190 and 260,000, depending on the material, resolution into eight or more subunits and their assumed stoichiometry (1,2). Hydrodynamic measurements of the enzyme solubilized in Triton X-100 gave a M_r of 440,000 to 550,000 (3,4), suggesting its dimeric form. Similar conclusion came from studies on two dimensional crystals of the bc1 complex in Triton X-100 (5). In addition, the recent proposal of the so called "double Q-cycle" (6) assumes a dimeric structure of the bc1 complex as the only functional form of the enzyme.

It has been recently shown in our laboratory that monomers of cytochrome c oxidase may be prepared from enzyme's dimers using filtration on Ultrogel AcA 34 (7). With a similar technique we were also able to isolate active monomers and dimers of the bc1 complex.

RESULTS

Bovine heart cytochrome bc1 complex dispersed in 0.1% dodecylmaltoside, 10 mM Tris/HCl (pH 7.4) was subjected to filtration on Ultrogel AcA 34 columns in the presence of different concentrations of KCl. The typical elution profile contained three peaks of apparent M_r about 400,000, 230,000 and 170,000. In the medium containing no added KCl the species of the lowest M_r dominated always over the two others. In the medium containing 50 mM KCl the result was reversed and the species of the highest M_r dominated. When the intermediate concentrations of KCl were used, the elution profile showed an increasing amount of species having M_r of about 400,000, in correlation with increasing salt concentration. The apparent M_r and the amount of species forming the "middle" (230,000) peak was not influenced by the ionic strenght of the medium. When the 400,000 M_r species eluted with 50 mM KCl were collected, dialysed, concentrated and re-applied to the column in the absence of KCl, the elution profile showed only one peak of M_r about 170,000. The reverse was also true, i.e. collected 170,000 species, when treated with KCl, formed only one peak of M_r about 400,000. All this suggests that both species, of 170,000 and 400,000 respectively, are composed of the same material undergoing a salt-dependent molecular conversion between two different aggregation states. Indeed, SDS-gel electrophoresis showed identical polypeptide pattern of both species, with all subunits of the bcl complex in the same proportions. Material collected as 230,000 species contains additionally a high M_r polypeptide (around 70,000), most likely representing the largest subunit of the succinate dehydrogenase.

Virtually the same results were obtained also with an alternative method for estimating apparent M_r values, sucrose density gradient centrifugation.

No substantial difference was found in the rates of cytochrome c reduction measured at low or high ionic strenght of the medium, thus under conditions known to induce different molecular states of the enzyme. The same was true for the isolated low and high M_r species of the bcl complex. The apparent K_m value for cytochrome c was found slightly higher in the medium containing 50 mM KCl (0.34/uM) than in the medium containing no added KCl (0.28/uM), but the V_{max} value was virtually the same (about 65 nmols cytochrome c reduced/s per nmol haem b, with 34.5/uM DBH as electron donor).

CONCLUSIONS

We conclude to have isolated monomeric and dimeric species of the bcl complex. The monomeric enzyme seems to be stabilized by the strong interaction with the largest subunit of the succinate dehydrogenase. Both aggregation states of the enzyme appear to be equally active, a small difference in K_m towards cytochrome c being due to the effect of salt. Our data do not support functional models based exclusively on a dimeric form of the bcl complex.

1. Rieske,J.S. (1976) Biochim.Biophys.Acta 456,195-247
2. Wikström,M.,Krab,K. & Saraste,M. (1981) Ann.Rev. Biochem. 50,623-655
3. Von Jagov,G.,Schägger,H.,Riccio,P.,Klingenberg,M. & Kolb,H.J. (1977) Biochim.Biophys.Acta 462,549-558
4. Weiss,H. & Kolb,H.J. (1979) Eur.J.Biochem. 99,139-149
5. Leonard,K.R.,Wingfield,P.,Arad,T. & Weiss,H.(1981) J.Mol.Biol. 149,259-274
6. De Vries,S.,Albracht,S.P.J.,Berden,J.A. & Slater,E.C. (1982) Biochim.Biophys.Acta 681,41-53
7. Nałęcz,K.A.,Bolli,R. & Azzi,A. (1983) Biochem. Biophys.Res.Commun. 114,822-828

158

PLASTOQUINONE BINDING PROTEINS

W. Oettmeier, K. Masson, and H.J. Soll
Lehrstuhl Biochemie der Pflanzen, Ruhr-Universität, D-4630 Bochum 1, FRG

In photosynthetic electron transport plastoquinone is reduced at the acceptor side of photosystem II and plastohydroquinone reoxidized at the cytochrome b_6/f-complex. Therefore, at least two different plastoquinone binding proteins must be present within the thylakoid membrane. We have identified them by use of photoaffinity labels.

1. Plastoquinone binding proteins in the cytochrome b_6/f-complex

Certain halogen-substituted dinitrodiphenylethers like DNP-INT are efficient inhibitors of plastohydroquinone oxidation. An azido-derivative of DNP-INT, DNP-ANT, out of the five polypeptides of isolated spinach cytochrome b_6/f-complex (cytochrome f, 33, 34; cytochrome b_6, 23.5; Rieske Fe S protein, 20; small subunit, 17.5 kDa) exclusively labels the cytochrome b_6 and the Rieske Fe S protein (1). An identical result was obtained by using an azido-plastoquinone (2). Thus, cytochrome b_6 and the Rieske Fe S protein as well simultaneously take part in plastohydroquinone oxidation.

2. Plastoquinone binding proteins in the photosystem II complex

Like plastohydroquinone oxidation, also plastoquinone reduction can be specifically blocked by inhibitors, some of which are utilized as efficient herbicides. So far, two different types of inhibitors are recognized: "DCMU-type" and phenolic (3). For both types of inhibitors photoaffinity labels are available: azido-atrazine ("DCMU-type") and azido-dinoseb (phenolic) (4). Azido-atrazine exclusively labels a 32-34 kDa protein (4); whereas azido-dinoseb preferentially tags the photosystem II reaction center proteins (spinach: 43 and 47 kDa; Chlamydomonas reinhardtii: 47 and 51 kDa) (4). To establish, whether indeed all "DCMU-type" inhibitors bind to the 32-34 kDa "herbicide binding protein", an azido-triazinone has been synthesized (5). The labeling experiments demonstrate that in fact azido-atrazine and azido-triazinone bind to an identical 32-34 kDa protein (5).

Presently, for the mode of action of photosystem II inhibitors a competition for binding between plastoquinone and inhibitors is assumed. Indeed, a competitive displacement between radioactively labeled DCMU and plastoquinone could be verified (6). To ascertain, whether inhibitors and plastoquinone bind to an identical protein, labeling patterns for azido-plastoquinone, azido-atrazine, and azido-triazinone have been compared in thylakoids and photosystem II preparations. Fig. 1 shows an azido-plastoquinone labeling pattern in a spinach photosystem II preparation (2 nmol; LDS-PAGE gel, 10-15 %). Indeed, azido-plastoquinone labels a protein in the 32 kDa molecular weight region. However,

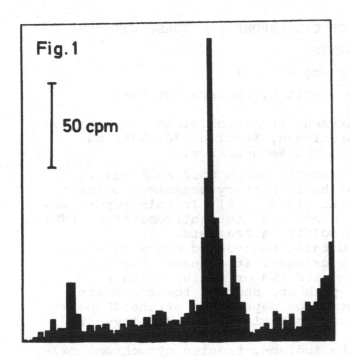

Fig. 1

50 cpm

the proteins labeled by azido-plastoquinone and by either azido-atrazine or azido-triazinone differ by 0.05 in their R_f-values (the R_f-values are identical, if 4 M urea is included in the gel) (7). This can be interpreted in terms of two distinct proteins or only one protein, whose mobility is differentially affected by the attachment of chemically distinct labels.

References

1. Oettmeier, W., Masson, K., and Olschewski, E. (1983) FEBS Lett. 155, 241-244.
2. Oettmeier, W., Masson, K., Soll, H.J., Hurt, E., and Hauska, G. (1982) FEBS Lett. 144, 313-317.
3. Oettmeier, W., and Trebst, A. (1983) in "The Oxygen Evolving System of Photosynthesis" (Inoue, Y. et al., eds.) pp. 411-420, Academic Press, Tokyo.
4. Johanningmeier, U., Neumann, E., and Oettmeier, W. (1983) J. Bioenerg. Biomembr. 15, 43-66.
5. Oettmeier, W., Masson, K., Soll, H.J., and Draber, W., in preparation.
6. Oettmeier, W., and Soll, H.J. (1983) Biochim. Biophys. Acta 724, 287-290.
7. Oettmeier, W., Soll, H.J., and Neumann, E. (1984) Z. Naturforsch. 39c, in press.

CHEMICAL MODIFICATION OF CYTOCHROME c OXIDASE WITH
TYROSINE SPECIFIC REAGENTS

B. Poolman[*], G. Petrone and A. Azzi

Medizinisch-Chemisches Institut, Universität Bern,
3012 Bern, Switzerland.
[*]present address: Department of Microbiology, University
                of Groningen, Kerklaan 30, 9751 NN
                Haren, The Netherlands.

The involvement of aromatic amino acid residues in
the catalytic action of beef heart cytochrome c oxidase
has been proposed several times (1,2). In this report we
present the results obtained with tetranitromethane (TNM)
and iodine as tyrosine modifying reagents.

Figure 1 shows the dithionite-reduced minus oxidized
spectra of cytochrome c oxidase. It appears that upon
treatment of the enzyme with TNM or iodine (data not
shown) the soret and α-peak are shifted towards lower
wavelength. The concentration dependence of the α-peak
blue shift and the enzymatic activity are displayed in
the inset of figure 1. Further, it is observed that in
the TNM-, but not in the iodine-, treated cytochrome oxidase bleaching of the heme a_3 occurred on reduction of
the enzyme with dithionite. Since the effect is largely
vanished when ascorbate/TMPD is the reducing agent, an
opening of the heme a_3 binding-site to the action of oxygen radicals or dithionite produced hydrogen peroxide may
be postulated.

Studies in which in the presence of ascorbate/TMPD and
oxygen the levels of heme a and heme a_3 reduction were
measured, indicated that TNM and iodine modify cytochrome
oxidase in the domain where electrons are transferred
from heme a_3 to oxygen. With cytochrome oxidase vesicles
the electron transfer reaction and also the proton pump
activity of the enzyme were not affected by the TNM treatment. However, the spectral changes induced by TNM were
the same as observed in the detergent solubilized oxidase,
suggesting that labeling of more than one amino acid had
occurred. To obtain more information about the localization of the modified amino acid residues, the amino acid
composition of the isolated subunits were analyzed. Preliminary data on TNM-treated oxidase show that subunit I,
II and IV contain nitro-tyrosine (less than one residue
per subunit). Control experiments excluded the possibili-

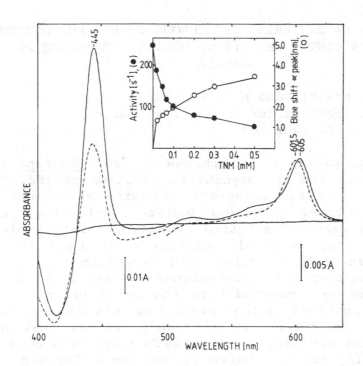

Figure 1. Beef heart cytochrome c oxidase (5 μM heme aa₃) was incubated with TNM in 50 mM Tris/HCl, pH 7.8, 50mM KCl and 0.1% (v/v) tween 80 for 60 min. at 28°C. After stopping the reaction with excess cysteine, the enzyme was dialyzed for 2 hours.

ty of nitration of the heme moiety of cytochrome c oxidase

References

1. Poulos, T.L. and Kraut, J. (1980), J. Biol. Chem. 255, 10322-10330.
2. Callahan, P.M. and Babcock, G.T. (1983), Biochem. 22, 452-461

BINDING SITES ON BACTERIAL CYTOCHROME c_2 FOR PHOTOSYNTHETIC REACTION CENTER AND MITOCHONDRIAL CYTOCHROME bc_1 COMPLEX

R. Rieder, E. Staubli and H.R. Bosshard
Biochemisches Institut der Universität Zürich,
CH-8057 Zürich, Switzerland

In vitro complexes of cytochrome c_2 from Rhodospirillum rubrum with the photosynthetic reaction center (the physiological electron acceptor) and with mitochondrial ubiquinol:cytochrome c oxidoreductase (cytochrome bc_1 complex) have been investigated by differential chemical modification. The chemical reactivity of each lysine residue (measured by reaction with radio-labeled acetic anhydride) was compared in cytochrome c_2 free in solution and cytochrome c_2 complexed with the redox partner. The surface area involved in the interaction was deduced from the decrease in reactivity of those lysines that are shielded in the complex. The same method had been used to map the binding site of horse cytochrome c for many of its electron transfer partners and it was found that a common recognition site is formed by a cluster of positively charged lysine residues around the heme edge of mitochondrial cytochrome c (Margoliash, E. and Bosshard, H.R. (1983) Trends Biochem. Sci. 8, 316-320).

From the structural similarity of cytochrome c_2 and mitochondrial cytochrome c, we expected high shielding of lysine residues 9, 12, 13, 72, 75, 94 and 97 of cytochrome c_2 since these residues are homologous to those at the known binding site of cytochrome c. Surprisingly, we now find that the homologous lysine residues of cytochrome c_2 are not involved in binding to the reaction center. Instead, the free N-terminal amino group and lysines 109 and 112 are highly shielded, suggesting a binding site on the right hand side of the molecule, away from the exposed heme edge. The reduced reactivity of lysines 56, 58 and 81 at the bottom part of the molecule

might be due to either a second area of binding or to a
conformational change. The increased reactivity of lysine
43 points to a conformational change. Essentially the
same shielding pattern is observed also with mitochondrial bc_1-complex. It is now difficult to imagine the
same mechanism of electron transfer via the exposed heme
edge for the cytochrome c_2 - reaction center system as it
has been proposed for the mitochondrial cytochrome c
system.

The Figure gives shielding factors R representing the
ratio "reactivity in free state/reactivity in complexed
state" of individual lysine residues of cytochrome c_2.

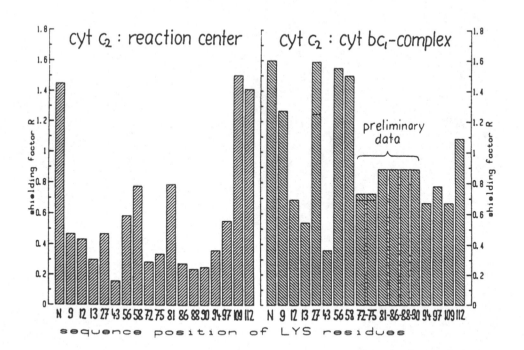

THE FUNCTION OF THE QUINONE POOL IN BACTERIAL PHOTOSYNTHETIC ELECTRON TRANSPORT

M. Snozzi
Institute for Plant Biology, University of Zurich, CH-8008 Zurich

A new model for the cyclic photosynthetic electron transport has been postulated. In contrast to earlier models the whole quinone pool of the membrane has ben integrated into this model.

The flash induced electron transport in the presence of antimycin has been reinvestigated, paying special interest to the function of the quinone pool. Variations of the quinone concentration in the membrane were achieved either by fusion of chromatophores with liposomes or by synchronous growth of the cells. The measured changes of the kinetics of the electron transport agree very well with the model. It could be demonstrated that the ubiquinol:cytochrome c_2 oxidoreductase reacts with quinol from the pool in a second order reaction. The oxidation of the quinone itself is a concerted reaction of a one electron reduction of the b-cytochromes with a one electron reaction of the Rieske type iron sulphure center.

Membranes from synchronously grown cells have been useful for the determination of thermodynamic parameters, due to their good homogeneity. In this way a k_2 of 1.4×10^5 M^{-1} s^{-1} for the reaction of the quinol with the ubiquinol:cytochrome c_2 oxidoreductase has been determined. The second order reaction becomes saturated , when the quinol concentration in the lipid part of the membrane reaches a concentration greater than 10 mM.

THE NATURALLY OCCURRING ELECTRON CARRIERS IN H_2-RECYCLING FOR PHOTOSYNTHETIC NITROGEN FIXATION IN RHODOPSEUDOMONAS CAPSULATA

Song Hongyu, Chen Hancai, Zhu Changxi and Wu Monggan

Shanghai Institute of Plant Physiology, Academia Sinica, Shanghai, China

In vivo the hydrogenase of Rhodopseudomonas capsulata is capable of recycling molecular hydrogen, which is coupled to the nitrogenase for acetylene reduction (1). A ferredoxin from R. capsulata can be reduced by native hydrogenase with molecular hydrogen as electron donor (2). The reducing power generated by H_2-hydrogenase could couple to nitrogenase-dependent acetylene reduction via ferredoxin (3). A natural fraction from crude extracts of R. capsulata has been separated, which functions as an active electron carrier between H_2-hydrogenase system and acetylene reduction reaction by native nitrogenase. A low midpoint redox potential component was identified in this natural fraction. The evidences indicate that the component effective for the coupling of electron might be a cytochrome C_3. A methyl viologen linked diaphorase activity specific to NADPH has been identified. The possible role of ferredoxin-cytochrome C_3 complex as an electron carrier system in the hydrogen evolution and hydrogen recycling process by R. capsulata was discussed.

1 Song Hongyu, Chen Hancai, Wu Monggan, Chen Bingjian and Yu Baolin (1980) Scientia Sinica, 23, 252-260
2 Zhu Changxi, Wu Monggan, Chen Hancai, Yu Baolin and

Song Hongyu (1980) Acta Phytophysiol. Sinica, 6, 299-305

3 Song Hongyu, Chen Hancai, Wu Monggan and Zhu Changxi (1981) in Current Perspectives in Nitrogen Fixation (Gibson, A.H. and Newton, W.E., eds.) pp.370, Australian Academy of Science, Canberra, Australia

Intra-vesicular pH changes in cytochrome c oxidase vesicles as a result of ferrocytochrome c oxidation.

M. Thelen, P. S. O'Shea, G. Petrone and A. Azzi
Med.-Chem. Institut, Bühlstr.28, CH-3012 Bern, Switzerland

The existence of a proton pump associated with bovine cytochrome c oxidase (E.C. 1931) (CO) has over the last few years been a matter of considerable dispute. In an attempt to resolve some of the problems with the measuring systems we have synthesised fluorescein-phosphatidylethanolamine (FPE) (1) which when reconstituted with CO into phospholipid vesicles provided, under suitable experimental conditions, a reliable indicator of the intravesicular pH. Attempts have been made to measure respiration induced intra-vesicular pH changes by using entrapped water soluble pH-indicating dyes (2). However, the results of these experiments were often more qualitative then quantitative. Using FPE phospholipid-vesicles it proved possible to directly relate intra-vesicular absorbance changes to externally added aliquots of acid or base (1). Applying this technique; measurements were made of the H^+ uptake from the intra-vesicular space upon addition of externally added ferrocytochrome c (i.e. the H^+/e^--ratio). This method, as with all the respiratory pulse experiments, may only be applied if an optimal charge compensation system ($K^+/$valinomycin) is present. In the absence of such a system (cf.3,4) there exists a small but influential residual $\Delta\Psi$ (5). This $\Delta\Psi$ was expressed as a change in the order of the pH relaxation rate after the turnover of CO and was held (6,4) to argue against the true H^+-translocating activity of the enzyme. It is clear from the figure however that a first order process is responsibe for the pH-relaxation rate after the turnover of CO. We therefore ascribe the high-decay kinetics previously obtained (3,4) to an insufficient charge compensation system (a homeopathic valinomycin concentration).
With ferrocytochrome c pulse experiments a constant H^+/e^--ratio close to 2.0 was found even for high numbers of turnovers (up to 15). Similar results were obtained in parallel experiments by measuring potentiometrically the extra-vesicular pH changes upon ferrocytochrome c pulses. The constant H^+/e^--ratio measured on both sides of the membrane unequivocally demonstrated a H^+-translocating activty of CO and excludes a H^+ release from the outer COV surface upon binding of cytochrome c as the origin of the

169

turnover induced extra-vesicular acidification (6).
When a subunit-III depleted preparation of CO was
reconstituted into FPE-vesicles a constant H^+/e^--ratio
close to 1.0 was observed. These observations are
consistent with previous results in which the involvement
of subunit-III in the H^+-translocating activity of CO has
been demonstrated (7,8).

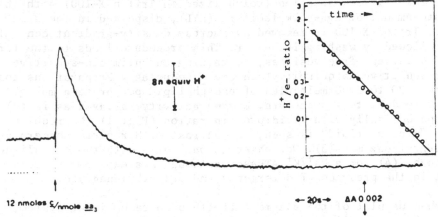

Figure: Experimental trace of the intra-vesicular pH change
with FPE-COVs and (inset) semi logarithmic evaluation of
the pH excursion. Experimental conditions: 84µl COVs
prepared as in (1), 0.287nmoles aa_3 in an active
orientation (90% of total), 1.5ml 50mM K^+Hepes, 100mM KCl,
1.9mM $CaCl_2$, pH 7.3 and 2.5nmoles valinomycin; T=15°C,
A=504.5-556.2nm. Dotted line: Trace corrected for dilution
and colour changes upon ferrocytochrome c addition.

1. Thelen, M., Petrone, G., O'Shea, P.S. and Azzi, A.
(1984) Biochim. Biophys. Acta 'in press'
2. Wrigglesworth, J.M. and Nicholls, P. (1979) Biochim.
Biophys. Acta 547,36-46
3. Casey, R.P., Chappell, J.B. and Azzi, A. (1979) Biochem.
J. 182,149-156
4. Casey, R.P. and Azzi, A. (1983) FEBS Lett. 154,237-242
5. Casey, R.P., O'Shea, P.S., Chappell, J.B. and Azzi, A.
(1984) Biochim. Biophys. Acta 765,30-37
6. Mitchell, P.D. and Moyle, J. (1983) FEBS Lett.
151,167-178
7. Casey, R.P., Thelen, M. and Azzi, A. (1980) J. Biol.
Chem. 255, 3994-4000
8. Penttila, T. (1983) Eur. J. Biochem. 133, 355-361.

STUDIES ON RECONSTITUTED COMPLEX III FROM BEEF-HEART MITOCHONDRIA

José M. Valpuesta and Félix M. Goñi
Department of Biochemistry, Faculty of Science, University of the
Basque Country, P.O. Box 644, Bilbao, Spain.

We have purified beef-heart mitochondrial complex III containing
about 80 phospholipids per molecule (1). Reconstitution is carried out
by mixing the enzyme suspension (solubilized in Triton X-100) with the
appropriate amount of egg-yolk lecithin (EYL), dispersed in the same
detergent. Triton X-100 is removed by sucrose density-gradient centrifu
gation, followed by washing in buffer. This procedure leads to the formation of detergent-free vesicles, containing an antimycin-sensitive,
uncoupler-sensitive ubiquinol:cytochrome c reductase. Preparations containing from 30 to 3000 molecules of phospholipid per protein molecule
may be obtained by this procedure. Enzyme activity, assayed as in (2),
varies hyperbolically with lipid:protein ratios (Fig. 1). No requirement of a "minimal lipid" is seen, in contrast with reports concerning
other integral enzymes (3). Our observations agree with those published
by Ottolenghi (4) for a Na^+/K^+ ATPase, although his experiments were ca
rried out in the presence of detergents and polyethyleneglycol.

An Arrhenius plot of the enzyme activity of a recombinant system,
containing 80 phospholipids per protein molecule is shown in Fig. 2. A
discontinuity is observed near 25°C; above 37°C the enzyme activity de
creases abruptly. An Arrhenius plot of the intrinsic Trp fluorescence
of the same recombinant system may also be seen in Fig. 2. A discontinuity is also seen at 37°C, suggesting a conformational change implying
thermal denaturation. However, no change in fluorescence is seen in the
region around 25°C. Recombinant systems containing between 60 and 1000
phospholipids/protein molecule behave in a similar way with respect to
fluorescence and enzyme activity measurements. In contrast, no breaks
are seen in the corresponding Arrhenius plots of the detergent-solubili
zed complex III preparation, suggesting that protein conformation changes upon incorporation into the lipid bilayer.

1. Engel, W.D., Schägger, H. and Von Jagow, G. (1980) Biochim. Biophys.
   Acta 592, 211-222.
2. Barbero, M.C., Valpuesta, J.M., Rial, E., Gurtuaby, J.I.G., Goñi, F.
   M. and Macarulla, J.M. (1984) Archiv. of Biochem. Biophys. 228, 560-
   -568.
3. Hesketh, T.R., Smith, G.A., Houslay, M.D., McGill, K.A., Birdsall,
   N.J.M., Metcalfe, J.C. and Warren, G.B. (1976) Biochemistry, 10,
   4145-4151.
4. Ottolenghi, P. (1979) Eur. J. Biochem. 99, 113-131.

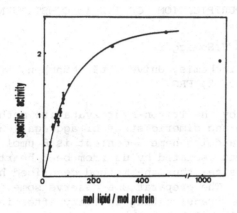

Fig. 1. Ubiquinol:cytochrome c reductase activity of EYL/complex
III recombinant systems as a function of phospholipid/pro
tein molar ratios.

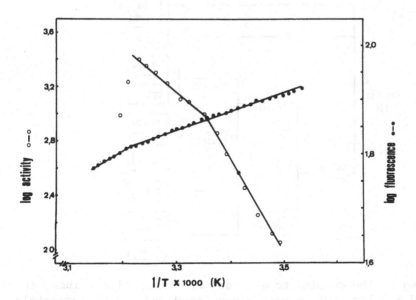

Fig. 2. Arrhenius plots of (O) enzyme activity, and (●) intrinsic
Trp fluorescence of a 80:1 EYL:complex III recombinant system.

THE bc_1 COMPLEX ISOLATED BY THE TRITON-HYDROXYAPATITE METHOD, A PREPARATION SUITABLE FOR PURIFICATION OF ITS 10 CONSTITUENT PROTEINS AND A c_1 SUBCOMPLEX

G. von Jagow, W.D. Engel and H.Schägger

Institut für Physikalische Biochemie, Universität München, Goethestrasse 33, 8000 München 2, FRG

The bc_1 complex purified by the Triton-hydroxyapatite method (1,2) is present in a monodisperse and dimeric state of aggregation. Its molecular weight is 500 000 and its heme b content is 8 µmol/g protein. The multiprotein unit has been isolated by us from beef heart, calf liver, N.crassa and S.cerevisiae; thus the method described has a wide range of applicability. The preparations deserve some interest, since they reveal good proton translocating activity after incorporation into phospholipid vesicles although they lack endogenous ubiquinone. Due to its homogeneous state of aggregation, the beef-heart complex is an ideal starting material for preparation of all the subunits, i.e. the constituent 10 proteins.

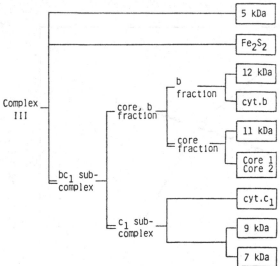

After binding of the complex to a second hydroxyapatite column, the FeS protein and a 5 kDa protein were dissociated and eluted separately from the column by buffers containing urea. Subsequently the bc_1 subcomplex, still bound to the hydroxyapatite column, was cleaved by a buffer containing guanidine. While 5 proteins - a 12 kDa protein, cytochrome b, an 11 kDa protein and the core proteins - were directly eluted by this buffer, the cytochrome c_1 subcomplex remained bound. It is a real subcomplex, resisting even 2 M guanidine. However, 3 M guanidine did cleave it into a heme-carrying 27.8 kDa protein, a 9.2 kDa protein and a 7.2 kDa protein. Cytochrome c_1 eluted first, and the 9.2 kDa and 7.2 kDa proteins were subsequently eluted separately by high phosphate concentrations. The core,b fraction containing 5 proteins was applied to a third hydroxyapatite column to separate the cytochrome b fraction from the core fraction. Cytochrome b, eluting unbound, was accompanied by a 12 kDa protein; the 2 core proteins and the 11 kDa protein were eluted together subsequently. These proteins could be separated from each other by gel chromatography, only the 2 core proteins stayed together.

The isolated FeS protein could be rebound to the bc_1 subcomplex and showed reactivation of electron flow and proton translocation (3). Reconstitution attempts with the other isolated proteins have been unsuccessful so far.

The primary sequences of cytochrome b (4), cytochrome c_1 (5), the 9.2 kDa (6) and the 7.2 kDa proteins (7) have been elucidated. The isolated cytochrome b was present as a monomer, revealing a Mr of 40 000 in ultracentrifugal studies. It had a heme b content of 40-45 µmol/g protein, indicating a protein endowed with two heme b centers. The redox potentials of these two centers in isolated cytochrome b were 0 mV and -100 mV respectively. The protein showed an abnormal migration behaviour in SDS gel electrophoresis, comparable to that of the lactose carrier protein: although possessing a Mr of 43.5 kDa, an apparent Mr of 30 000 was simulated in SDS gel electrophoresis (8).

The heme c center of the cytochrome c_1 subcomplex was non-reactive with carbon monoxide and possessed a redox potential of +265 mV. The c_1 subcomplex is regarded as a valuable candidate for kinetic studies of electron transfer from cytochrome c to cytochrome c_1.

(1) Engel,W.D., Schägger,H. and Von Jagow,G. (1980) BBA 592, 211-222.
(2) Engel,W.D., Schägger,H. and Von Jagow,G. (1983) Z.Physiol.Chem.364, 1753-1763.
(3) Engel,W.D., Michalski,C. and Von Jagow,G. (1983) Eur.J.Biochem.132, 395-402.
(4) Anderson,S., De Bruijn,H.L., Coulson,A.R., Eperon,I.C., Sanger,F. and Young,I.G. (1982) J.Mol.Biol.156, 683-717.
(5) Kim,C.H. and King,T.E. (1981) BBRC 101, 607-614.
(6) Wakabayashi,S., Takeda,H., Matsubara,H., Kim,C.H. and King,T.E. (1982) J.Biochem.91, 2077-2085.
(7) Schägger,H. and Von Jagow,G. (1983) Z.Physiol.Chem.364, 307-311.
(8) Loderbauer,S. (1983) Doctor's thesis, University of Munich.

DIFFERENCE IN REDOX EQUILIBRATION OF THE TWO REACTION CENTERS OF CYTOCHROME b WITH THE UBIQUINONE POOL

G. von Jagow and Th.A. Link

Institut für Physikalische Biochemie, Universität München, Goethestrasse 33, 8000 München 2, FRG

A new class of inhibitors - we have named them MOA inhibitors, as they carry ß-methoxyacrylate as crucial structural segment - has helped to characterize the two reaction sides of cytochrome b for ubiquinone (1,2). These are the b-566 center, which is believed to face the outer side of the mitochondrial inner membrane, and the b-562 center, believed to face the inner side of the mitochondrial inner membrane. While antimycin binds to the b-562 side of cytochrome b, myxothiazol - one of the MOA inhibitors - binds to the b-566 center. The combined application of these two inhibitors has proved clearly that cytochrome b possesses two reaction centers for ubiquinone which act independently of each other and in different ways: cytochrome b is only cut off from electron transfer after it has bound one molecule of antimycin per b-562 as well as one molecule of myxothiazol per b-566.

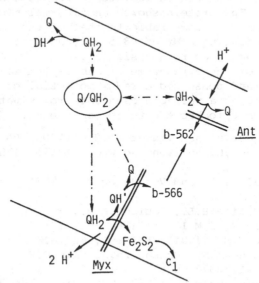

Even though myxothiazol was bound to the b-566 domain of cytochrome b, it exerted a drastic influence on the iron-sulfur protein, apart from inhibiting electron entry into heme b-566 (3) :
  i) it blocked the electron transfer from ubiquinol onto the Fe_2S_2 center;

ii) it shifted the g_z resonance in the EPR spectrum of the iron-sulfur protein;

iii) it abolished the midpotential shift induced by undecylhydroxyl-dioxobenzothiazol.

Keeping in mind that myxothiazol is bound only to cytochrome b, its multiple action on cytochrome b and on the iron-sulfur protein indicates the possible existence of a reaction center formed partly by the b-566 domain of cytochrome b and partly by a certain domain of the iron-sulfur protein. In this common reaction center ubiquinol is oxidized to ubiquinone during a two-step reaction sequence.

The reaction mechanism at the b-562 side seems to be less intricate than that at the b-566 side. Both on reduction and on oxidation of the ubiquinone pool, there was immediate equilibration of the b-562 center with the ubiquinone pool, provided the b-566 center was blocked by myxothiazol. Good redox equilibration was also indicated by comparable pseudo-first-order reduction kinetics of the heme b-562 center and the ubiquinone pool on succinate reductant pulses. The b-566 reaction side, on the other hand, did not enter into a redox equilibrium with the ubiquinone pool : when a reductant pulse was performed in the presence of antimycin, a 90% reduction of cytochrome b ensued while 85% of the ubiquinone pool was still in oxidized state (4). While the b-562 center was blocked by antimycin, reduced cytochrome b-566 did not alter its redox level on reoxidation of the ubiquinone pool. These findings are due to the fact that during ubiquinol oxidation the first electron was transferred onto the Fe_2S_2 center, while the second one was transferred onto the b-566 center, with a semiquinone molecule - stabilized in the reaction center - as intermediate. The virtual irreversibility of the b-566 side has to be ascribed - at least in the uncoupled state - to the fact that electrons situated in the b-566 center at an energy level of -50 mV cannot be retransferred onto the $Q/Q^{\cdot}$ couple that has a reduction potential of -240 mV.

(1) Becker,W.F., Von Jagow,G., Anke,T. and Steglich,W. (1981) FEBS Lett. 132, 329-333.
(2) Von Jagow,G. and Engel,W.D. (1981) FEBS Lett. 136, 19-24.
(3) Von Jagow,G., Ljungdahl,P.O., Graf,P., Ohnishi,T. and Trumpower,B.L. (1984) J.Biol.Chem., in press
(4) Von Jagow,G. and Link, Th.A (1984) in: Biomedical and Clinical Aspects of Coenzyme Q. Vol. 4 (K.Folkers and Y.Yamamura, eds.) Elsevier Science Publishers, Amsterdam, in press

SUBUNIT STOICHIOMETRY OF BEEF HEART CYTOCHROME C OXIDASE

L. Meinecke and G. Buse
Abteilung Physiologische Chemie, RWTH Aachen
D-5100 Aachen, West Germany

Cytochrome c oxidase, complex IV of the respiratory chain, is an oligomeric protein. The enzyme contains from 2 protein components in the case of bacteria (1) to 12 in the case of mammals (2). On the basis of the primary structures analysed in our laboratory (3) the subunit stoichiometry of the oxidase from beef heart was determined. In the table these data are shown together with the Mr, number of residues and N-terminal sequences of the constituting polypeptides. The stoichiometry was obtained in first instance from a direct Edman degradation of the entire enzyme and HPLC-quantitation of the resulting PTH-amino acids through several cycles (Figure; first cycle).

On the other hand the UV-absorption at 280 nm of gelchromatographically separated polypeptides after dissociation of the enzyme in 3% SDS solutions allowed for an exact determination of this stoichiometry on the basis of known tyrosine and tryptophan contents in the case of components I → VIa.

Additional evidence was obtained from quantitative determination of the C-terminal residues by hydrazinolysis as well as carboxypeptidase A and B digestion experiments. The results allow for a calculation of the exact Mr of the functional monomer of cytochrome c oxidase (Mr = 202.900) and show that all of the 12 protein components are stoichiometric constituents (stoichiometry 13).

Polypeptide	Mr	Synthesis	Stoichiometry	N-terminal Sequences
I	56993[a]	mit.	1	Formyl-Met-Phe-Ile-Asn-
II	26049	mit.	1	Formyl-Met-Ala-Tyr-Pro-
III	29918[a]	mit.	1	(Met)-Thr-His-Gln-
IV	17153	cyt.	1	Ala-His-Gly-Ser-
V	12436	cyt.	1	Ser-His-Gly-Ser-
VIa	10670	cyt.	1	Ala-Ser-Gly-Gly-
VIb	9419	cyt.	1	Ala-Ser-Ala-Ala-
VIc	8480	cyt.	1	Ser-Thr-Ala-Leu-
VII	10068	cyt.	1	Acetyl-Ala-Glu-Asp-Ile-
VIIIa	5441	cyt.	1	Ser-His-Tyr-Glu-
VIIIb	4962	cyt.	2	Ile-Thr-Ala-Lys-
VIIIc	6244	cyt.	1	Phe-Glu-Asn-Arg-

[a] from mtDNA sequence

Table: The protein components of cytochrome c oxidase from beef heart.

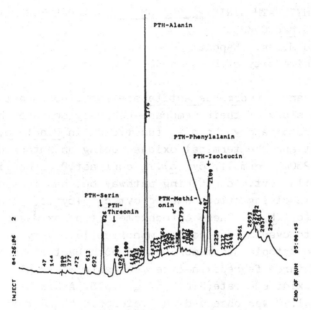

Figure: HPLC-quantification of the PTH-amino acids from the first Edman degradation cycle of cytochrome c oxidase from beef heart.

Although only the mitochondrial components I → III appear to be the
catalytic subunits of the enzyme the integral stoichiometry of the
cytoplasmic polypeptides indicates them to be constitutive components
of the respiratory enzyme complex, the function of which in the context of tissue specificity, respiratory regulation, biosynthesis and
structure of mitochondria remains to be established.

References

1 Ludwig, B. (1980) Biochem. Biophys. Acta 594, 177-189.
2 Buse, G., Steffens, G.C.M. and Meinecke, L. (1983) in: Structure
  and Function of Membrane Proteins (E. Quagliariello and F. Palmieri,
  eds.) pp. 131-138, Elsevier/North-Holland, Biomedical Press,
  Amsterdam.
3 See series of papers "Studies on Cytochrome c oxidase" I-X in Hoppe
  Seyler's Z. Phys. Chem. Last issue:
  Meinecke, L., Steffens, G.J. and Buse, G. (1984) Vol. 365, 313-320.

THE REDOX CHAINS OF PHYTOPATHOGENIC PSEUDOMONADS: A MODEL FOR ALTERNA-
TIVE ROUTES TO THE CYT.b/c COMPLEX.

D.Zannoni,S.Cocchi and M.Degli Esposti
Institute of Botany,University of Bologna,40126 Bologna(I)

Pseudomonas cichorii and Pseudomonas aptata are two fluorescent phytopathogenic bacteria.A study of their membrane-bound cytochromes has revealed that a linear respiratory chain is functional in P.aptata,cyto-
chromes c being absent and the terminal oxidase being an autooxidable
cyt.b with $Em_{7.0}$ of +250mV(formally cyt.o).In contrast,P.cichorii appears to retain a similar cyt.o-containing pathway but has in addition a second pathway involving cytochromes c(cyt.c_{240}+cyt.c_{335})to a
distinct cytochrome b(cyt.b_{380}).These cytochrome b type oxidases differ
in their sensitivity to cyanide and carbon monoxide(1).A very recent
e.p.r. analysis of the cytoplasmic membranes of both bacteria,demonst-
rated the presence of three ferredoxin-like centres(g_y=1.93)plus a centre detected in the oxidized-state(HiPIP)at g=2.015. A Rieske-type
iron-sulphur centre(g=1.90)was observed in P.cichorii but not in P.ap-
tata.The absence of the Rieske-centre in P.aptata parallels the absence of cytochromes c in this bacterium(2). The energy transduction by
membranes from P.cichorii and P.aptata has recently been examined(3).
It is concluded that the membrane-bound NADH-dehydrogenases of both
bacteria,which are rotenone-insensitive,are also completely uncoupled.
Proton gradient generating steps linked to respiratory substrates have been identified at the UQ/cyt.b levels of both bacterial strains.
As expected,the species deficient in cyt.c,i.e.P.aptata,did not perform ATP synthesis coupled to endogenous cyt.c-dependent respiration,
whereas P.cichorii presented both ATP and ΔpH generation linked to
cyt.c-oxidase activity(3).
  In this present communication we report on the effects of myxothiazol
and 5-undecyl-6-hydroxy-4,7-dioxobenzothiazole(UHDBT)in membranes fro-
m P.cichorii and P.aptata.In other bacterial respiratory chains,UHDBT
has been suggested to interact with both the Rieske-centre and cyt.b566
one of the two haems of b type potentiometrically resolved at the UQ/
cyt.b-c level.The effect of UHDBT is therefore different from that of
myxothiazol which seems to specifically bind cyt.b566(4).Myxothiazol
has moreover been shown to have a synergistic effect with antimycin A,
this latter antibiotic acting on the oxidation side of cyt.b562(4,5).

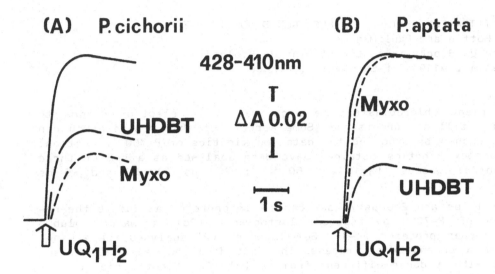

(A) P.cichorii   **428-410nm**   **(B) P.aptata**

UHDBT $\Delta A\ 0.02$

Myxo

Myxo

UHDBT

1 s

⇧ UQ₁H₂   ⇧ UQ₁H₂

The traces shown here,represent the patterns of cyt.b-reduction in membranes from P.cichorii(traces in A)and P.aptata(traces in B)using UQ_1H_2 as electron donor. It is apparent that both types of membranes are significantly affected by UHDBT whereas the myxothiazol sensitivity is only retained by P.cichorii membranes.This finding is therefore clearly inconsistent with previous data on other respiratory systems in which the presence of the Rieske-centre is considered a prerequisite for the UHDBT-inhibitory effect.In contrast,the absence of the Rieske-centre in P.aptata parallels the absence of myxothiazol-effect on this latter bacterial species.These data seem to agree with steady-state respiratory experiments in which NADH dependent respiration is biphasically affected by UHDBT.Indeed,the cyt.o-containing pathway in P.cichorii is 50% inhibited by 1μM UHDBT whereas the cyt.c reduction is affected only by higher UHDBT-concentration(Ki=10μM).

REFERENCES
(1)Zannoni D.(1982)Arch.Microbiol.133:267-273
(2)Zannoni D. and Ingledew J.W.(1984)J.Gen.Microbiol.(in press).
(3)Zannoni D.(1984)Arch.Microbiol.(in press).
(4)Crofts A.R.(1983)In:The enzymes of biological membranes(Martonosi
    A.N. eds)Plenum Publ.Comp.New York(in press).
(5)von Jagow G. and Engel W.D.(1981)FEBS Lett.136:19-24

REDOX TITRATION OF THE FAST ELECTRON DONOR TO P-700

Hervé Bottin and Paul Mathis
Service de Biophysique, Département de Biologie
CEN Saclay, 91191 Gif-Sur-Yvette, France

Introduction

In plant chloroplasts, the organization of electron donors to PS I is still not understood. Some aspects of this organization can be approached by studying the detailed kinetics of $P-700^+$ reduction. The complex kinetics obtained have been analysed as a sum of three exponential decays ($t\frac{1}{2}$=10 μs (50 %) : 200 μs (30 %) and >5 ms (20 %)).

The 10 μs and 200 μs phases can be interpreted as due to the reduction of $P-700^+$ by reduced plastocyanin (Pc), a water soluble "blue copper protein" (1-4). Some authors (2) analysed these kinetics as a second order process. The fast decay has also been associated with a donor different from Pc (3). Our first series of results (4) strongly support the proposal that the fast component is due to plastocyanine (more or less strongly) bound to the reaction center, the medium phase being related to a pool of mobile molecules. We have pursued these investigations by studying the effect of redox potential on the kinetics of $P700^+$ rereduction, following a single laser flash.

Material and methods

Spinach chloroplasts were osmotically broken and resuspended in 0.4 M sucrose, 10 mM NaCl, 20 mM tricine pH 7.8, 10 mM $MgCl_2$ (chlorophyll conc. 50 μg/ml). The redox titration was performed in presence of 2 μM DCMU, 20 μM ferrocene, 20 μM N,N,N',N'-tetramethyl-p-phenylenediamine, 2 mM $K_4Fe(CN)_6$ and increasing amounts of $K_3Fe(CN)_6$. Redox potential was continuously monitored. Kinetics of flash induced absorption changes at 820 nm were measured at 17°C, with excitation by a ruby laser pulse (694 nm, 6 ns) attenuated to be slightly sub-saturating.

Results

After its photooxidation by one flash, $P-700^+$ returns to the reduced state following a multiphasic exponential decay ($t\frac{1}{2}$=10 μs, 200 μs, 5 ms). When increasing the redox potential from 200 to 430 mV, the kinetics are continuously slowed down, mostly by the decrease of the amplitude of the fast phase, showing that the electron donor to P-700 becomes oxidized ; the $t\frac{1}{2}$ of this component remains constant. Through this potential range, the initial ΔA (i.e. the amount of photooxidized P700) is constant. This confirms that P-700 has a midpoint potential higher than +430 mV (5). The decay analysis can

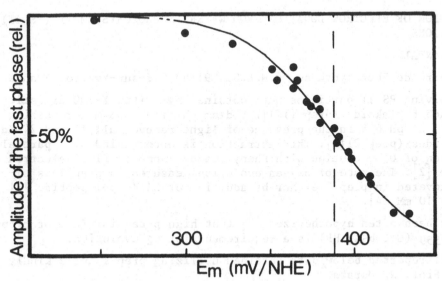

be done either by exponential decomposition or by direct measurement
of the amplitude after 50 μs. Both ways give the same titration
curve, showing the strict relation between the amplitude of the fast
phase and the redox poising. This result accounts for a molecule
titrating with one electron and a midpoint potential $E_{m_{7.8}}$=390 mV.
This value must be compared to the E_m of purified Pc (370-380 mV) or
Cyt f.

Conclusion

The present work provides a good argument for identifying Pc as
the immediate electron donor to P-700. One must also consider that
all through the titration the halftime of the fast phase remains
constant. This is in favour of a first order process, i.e. the existence of a complex between the reaction center of PS I and its electron donor, most likely plastocyanin.

References
1. HAEHNEL, W., PROPPER, A. and KRAUSE, H. (1980) Biochim. Biophys.
   Acta 593, 384-399
2. OLSEN, L., COX, R. and BARBER, J. (1980) FEBS Lett. 122, 13-16
3. BOUGES-BOCQUET, B. and DELOSME, R. (1978) FEBS Lett. 94, 100-103
4. BOTTIN, H. and MATHIS, P. (1984) in Proc. 6th Int. Photosynthesis, Brussels (C. Sybesma, ed.) Nijhoff/Junk Publishers, in press
5. SETIF, P. and MATHIS, P. (1980) Arch. Biochem. Biophys. 204,
   477-485

RESTORATION OF ELECTRON DONATION FROM WATER IN PHOTOSYSTEM II (PS II) PREPARATIONS

J-M. BRIANTAIS

Laboratoire de Photosynthèse, C.N.R.S., 91190 Gif-sur-Yvette, France

O_2 evolving PS II particles were obtained by Triton X-100 digestion of spinach thylakoids (prep 1)[1]. Sodium chloride one molar washing of prep 1 at pH 6.5 in the presence of light removes all 18 and 24 Kd polypeptides (prep 2)[2]. This extraction is accompanied by a partial inhibition of O_2 evolution with Phenylbenzoquinone as final electron acceptor [2]. The rate of oxygen evolution measured in prep 1 is almost recovered in prep 2 either by addition of 24 Kd polypeptide [3] or $CaCl_2$ 10 mM [4].

It has been often hypothetized [5] that high potential form of cytochrome b_{559} (Cyt b_{559} HP) is a requirement for O_2 evolution.

We will reported below, data obtained utilizing prep 1 and 2 kindly sent by Prof. N. Murata.

1. We observed that in prep 2 compared to prep 1

a) there is a decrease of the amount of ferrocyanide reducible form of cyt b_{559}; respectively 11 to 13 % and 35 % of total cyt b_{559}. It is accompanied by an increase of the lower potential forms,

b) there is a decrease of the reduction rate of the acceptor pool measured by chlorophyll fluorescence induction half rise time. This is partly reversed by incubation of prep 2 with NH_2OH 10 mM confirming an inhibition of water donation in this preparation,

c) $CaCl_2$ addition to prep 2 is able to stimulate O_2 evolution as in [4], and to decrease the half time of the fluorescence rise; it decreases slightly the amount of cyt b_{559} ferrocyanide reducible (6 %).

2. In order to check correlations between those variations induced by NaCl treatment, prep 1 have been incubated as in [2] but in the presence of intermediate concentrations of NaCl, from 10 to 1000 mM. The treated sample are not centrifuged in contrast to [2]. In the Table are reported the effects of these treatments and of subsequent addition of $CaCl_2$ or $MgCl_2$ 10 mM.

NaCl (mM)	10	150	250	500	1000	1000 then MgCl$_2$	1000 then CaCl$_2$
O$_2$ evolution	100	100	75	63	44	–	–
	100	73	–	58	40	30	77
Percentage of cyt b$_{559}$ Ferrocyanide red.	35	35	35	25	13	–	–
1/t$^{1/2}$ of fluorescence induction : minus NH$_2$OH	1.05	0.75	–	0.45	0.32	0.37	0.67

plus NH$_2$OH

O$_2$ evolution initial rate : 100 = 175 µmole O$_2$/mg Chl/hr, first line, values in same samples than cyt b$_{559}$ determination, second line same samples than for fluorescence induction. We noticed that larger is the inhibition, faster is a decrease of O$_2$ evolution developed during the time course of Hill reaction measurements.

It emerges a satisfactory correlation between progressive inhibition versus NaCl concentration of O$_2$ evolution and rate of reduction of the acceptor pool, reduction which is limited as in prep 2 by electron donation from water. CaCl$_2$ but not MgCl$_2$, even in the presence of 1000 mM NaCl$_2$ restores the donation by water.

There does not seem to be a correlation between the amount of cyt b$_{559}$ HP present and oxygen evolution, since this activity is inhibited by NaCl 250 mM whereas the proportion of cyt b$_{559}$ HP begins to decrease at higher concentration.

One may suggest that cyt b$_{559}$ HP is involved in the mechanism of oxygen evolution only for a part of PS II centers. This is doubtfull, indeed there is an anticorrelation as, in prep 2, CaCl$_2$ addition restores oxygen evolution, decreases the half-time of the fluorescence rise but slightly decreases the amount of cyt b$_{559}$ ferrocyanide reducible.

References

1 Kuwabara, T. and Murata, N. (1982) Plant Cell Physiol. 23, 533-539
2 Miyao, M. and Murata, N. (1983) Biochim. Biophys. Acta 725, 87-93
3 Miyao, M. and Murata, N. (1983) FEBS Lett. 164, 375-378
4 Miyao, M. and Murata, N. (1984) FEBS Lett. 168, 118-120
5 Butler, W.L. (1978) FEBS Lett. 95, 19-25.

THE PHOTOSYSTEM II REACTION CENTER OF THE THERMOPHILIC CYANOBAC-
TERIUM MASTIGOCLADUS LAMINOSUS

R. Frei, R. Rutishauser and A. Binder
Institute for Plant Biology, University of Zurich, CH-8008 Zurich,
Switzerland

Mastigocladus laminosus is a thermophilic cyanobacterium with a
broad spectrum of growth conditions, i.e. it can be cultivated from 35
to 65°C and pH values from 5 to 9.5. The growth performance under those
conditions have been investigated [1] In recent years several membrane
protein complexes from M. laminosus related to photosynthesis and
respiration have been isolated: phycobiliproteins [2], coupling factor
[3], cytochrome c [4] and PS I reaction centers [5,6].

Here we report now the isolation and characterization of the PS II
reaction center from M. laminosus. The complex was isolated from washed
membranes with 0.35% Sulfobetain SB-12 in the presence of 30 mM Tricine
pH 7.8, 10 mM $MgCl_2$ and 25% glycerol. It was purified on a sucrose gradient 15-60% and the presence of 10 mM $CaCl_2$ and on a Sepharose 6B
column. The activities are shown in Table 1.

Step	Activity (ueq. e/mg chl. x h)
Washed membranes	600
Crude PSII-reaction center (SB-12 sup.)	2600
After sucrose gradient	2400
no semicarbazide (H_2O as donor)	1800
+ DCMU 10 uM	900
20 uM	0
After Sepharose 6B	2200

Table 1 Purification and activities of PS II reaction centers. The
activities were measured spectrophotometrically at 420 nm or with o-
phenantrolin at 510 nm with ferricyanide as acceptor and, if not stated
otherwise, with semicarbazide as electron donor.

It has been shown earlier, that PS II dependent electron transport
in isolated membranes from cells grown under limiting CO_2 concentrations
is much lower than from those grown with sufficient CO_2 [7]. PS I
activities were demonstrated to remain constant under all conditions.

As a continuation of this work we present here the influence of the
conditions during cell-growth on the activities of isolated PS II reaction centers. In a controlled fermenter as described earlier [1], cells
were grown with different CO_2 concentrations in the air and PS II complexes were isolated as described above, but with 0.35% LDAO.

Table 2 demonstrates that cells grown with 400 ppm CO_2 show maximal PS II reaction center activities. These activities derease drastically when the cells continue growing at 0 ppm CO_2. These low activities are also observed, when cells grow from the beginning at 100 ppm CO_2. In contrast to PS II, PS I reaction centers are not influenced by the CO_2 concentration. The table also demonstrates, that the phycobiliprotein content decreases at low CO_2 concentrations, i.e. the cells degrade their phycobiliproteins which are not necessary anymore as antenna pigments and use them as carbon source. Thus, a decrease of PS II reaction center activity is coupled with a decrease of the phycobiliprotein content. At the same time, the chlorophyll content remains unchanged.

Measurements	Growth time and CO_2 conc. during growth			
	3 days 400 ppm	3 days 400 ppm	2 days 0 ppm	6 days 100 ppm
Activity of PS II reaction centers (LDAO extract)	(uequ.e/mg chl.x h)			
	471		20	60
PS I reaction centers (in washed membranes)	(chl./P700)			
	190		200	220
Yield (in whole cells)	(mg/mg protein)			
Chlorophyll	0.035		0.038	0.039
Phycobiliprotein	0.320		0.150	0.210

Table 2 Influence of CO_2 concentrations during cell-growth on the activity of extracted PS II and PS I reaction centers as well as on the yield of chlorophyll and phycobiliproteins in whole cells.

According to the results presented here, CO_2 concentration during growth regulates the activity of the PS II reaction center. It is feasible that under low CO_2 conditions, where the formation of redox equivalents for CO_2 assimilation is saturated, light is solely used to drive phosphorylation in a bacterial-like cyclic electron transport with PS I, while the PS II reaction center is shut off or degraded.

[1] Muster, Binder, Schneider and Bachofen (1983) Plant Cell Physiol. 24, 273-280
[2] in the group of H.Zuber, Zurich
[3] M.Wolf, A.Binder and R.Bachofen (1981) Eur.J.Biochem. 118, 423-427
[4] in the group of D.Krogmann, West Lafayette
[5] R.Nehushtai, P.Muster, A.Binder, R.Liveanu and N.Nelson (1983) Proc.Natl.Acad.Sci 80, 1179-1183
[6] P.Muster, A.Binder and R. Bachofen (1984) FEBS Lett. 166, 160-164
[7] A.Binder, R.Hauser and D.Krogmann (1984) Biochim.Biophys.Acta 765, 241-246

INHIBITOR AND PLASTOQUINONE INTERACTION AT THE ACCEPTOR SIDE OF
PHOTOSYSTEM II

H.J. Soll, and W. Oettmeier
Lehrstuhl Biochemie der Pflanzen, Ruhr-Universität, D-4630 Bochum 1,
FRG

1. Inhibitor/inhibitor interaction

Inhibitors of photosynthetic electron transport which block electron flow at the acceptor side of photosystem II can be divided into
two different classes: "DCMU-type" and phenolic. Their mechanism of
action is not yet fully understood. In their biochemical behaviour,
there exist similarities and differences as well (1). Both types of
inhibitors displace themselves mutually and in a competitive way from
the thylakoid membrane (2). This might indicate an identical binding
site for "DCMU-type" and phenolic inhibitors. Most recently, photoaffinity labels for both types of inhibitors have been developed: azido-atrazine (1), azido-triazinone (3) ("DCMU-type"), and azido-dinoseb
(phenolic) (4). Wherease the "DCMU-type" labels bind to a 32-34 kDa
protein, the phenolic labels tag the two photosystem II reaction center proteins (see Oettmeier, W., Masson, K., and Soll, H.J., this meeting). Contrary to the displacement experiments (2), the labeling patterns by the azido-compounds implicate two different binding sites
for "DCMU-type" and phenolic inhibitors. For further clarification,
binding studies with radioactively labeled inhibitors on thylakoids
have been performed, which have been pre-labeled by azido-atrazine.
After covalent binding of increasing concentrations of azido-atrazine,
analysis of subsequent atrazine binding reveals an identical binding
constant in all experiments. Conversly, the number of binding sites
drops for increasing concentrations of azido-atrazine. The same result
is obtained using DCMU. For the phenolic inhibitor 2-iodo-4-nitro-6-
isobutylphenol (4) the results are different: at increasing concentrations of azido-atrazine the number of binding sites remain constant, whereas the binding constants increase. This is interpreted in
terms of an identical binding site for azido-atrazine, atrazine and
DCMU, but a different one for the phenolic inhibitor. If a binding
site is occupied by covalently bound azido-atrazine, atrazine or DCMU
cannot bind to this site any longer. Contrary, the phenolic inhibitor
can still reach its particular binding site, although its accessibility is impaired by the bound azido-atrazine. This implicates the existence of close binding sites, but not necessarily on the same protein.
This notion is further corrobated by binding experiments of inhibitors
of both types to thylakoids after UV-treatment. UV-treatment, which is
known to selectively destroy the 32-34 kDa protein, diminishes binding of all "DCMU-type" inhibitors, whereas binding of phenolic inhibitors is either unchanged or even slightly enhanced.

2. Inhibitor/quinone interaction

For the mechnanism of action of photosystem II inhibitors most recently a model is discussed which implements direct competition between inhibitor and 'pool plastoquinone' at the acceptor side of photosystem II. Indeed, we could demonstrate that DCMU is competitively displaced from the thylakoid membrane in plastoquinone-depleted thylakoids by plastoquinone-1 and the photosystem II acceptor dichlorophenol-indo-phenol (5). Furthermore, DCMU can also be displaced by a series of 1,4-benzoquinones. In a quantitative structure activity relationship, the displacement behaviour of the quinones and their inhibitory activity in photosystem II reactions as well could be correlated to their redox and steric properties (6).

As reported elsewhere in this meeting (Oettmeier, W., Masson, K., and Soll, H.J.), azido-plastoquinone labels a protein in the 32 kDa molecular weight range, which seems to be different from the protein where azido-atrazine or azido-triazinone bind. In order to establish whether this 32 kDa protein is involved in plastoquinone/inhibitor competition, labeling experiments have been performed in the presence of a tenfold excess (with respect to azido-plastoquinone) of either DCMU, 2-iodo-4-nitro-6-isobutylphenol or tetraiodobenzoquinone (the best quinone type photosystem II inhibitor; pI_{50}-value >7). In all cases, no specific labeling in the 32 kDa molecular weight range can be observed, i.e. addition of the inhibitor prevents binding of azido-plastoquinone. As already stressed, binding of "DCMU-type" inhibitors and plastoquinone might not occur at the same protein. This would be similar to the mechnanism of "DCMU-type" and phenolic inhibitor interaction.

References

1. Oettmeier, W., and Trebst, A. (1983) in "The Oxygen Evolving System of Photosynthesis" (Inoue, Y., et al., eds.), pp. 411-420, Academic Press, Tokyo.
2. Laasch, H., Pfister, K., and Urbach, W. (1982) Z. Naturforsch. 37c, 620-631.
3. Oettmeier, W., Masson, K., Soll, H.J., and Draber, W., in preparation.
4. Oettmeier, W., Masson, K., and Johanningmeier, U. (1982) Biochim. Biophys. Acta 679, 376-383.
5. Oettmeier, W., and Soll, H.J. (1983) Biochim. Biophys. Acta 724, 287-290.
6. Soll, H.J., and Oettmeier, W. (1984) in "Advances in Photosynthesis Research" (Sybesma, C., ed.), Vol. 4, pp. 5-8, Martinus Nijhoff/Dr. W. Junk Publishers, The Hague, Boston, Lancaster.

HERBICIDE RESISTANCE AND KINETICS OF ELECTRON TRANSFER

G.F. Wildner and W. Haehnel
Lehrstuhl Biochemie der Pflanzen, Ruhr-Universität
D-4630 Bochum, West Germany

Mutants of <u>Chlamydomonas</u> <u>reinhardii</u> were isolated with altered herbicide binding properties against metribuzin, atrazine and DCMU (1). The impact of the mutants on the electron transport from Q_A to Q_B was measured by the flashphotometry technique at 335 nm. Intact cells were excited by 2 μs xenon flashes of saturating intensity with a repetition rate of 2.2 Hz and by additional background light at 716 nm with an intensity of 45 W.m^{-2}; 2000-3000 consecutive signals were averaged at a time resolution of 50 μs. Details of the equipment used in these studies were described elsewhere (2).

The light induced absorbance change of X-320 is attributed to the primary acceptor of photosystem II, to the dismutation of the semi-quinones $Q_A^{\cdot -} + Q_B^{\cdot -}$ to $Q_A + Q_B^{2-}$, catalyzed by the herbicide binding B-protein. The kinetic measurements with the wild-type cells and the four mutants are shown in Fig. 1, the results and the properties of the mutants were described in Table I.

Table I

strain	R/S values			$t_{1/2}$ (decay) (ms)
	metribuzin	atrazine	DCMU	
WT	1	1	1	0.4
MZ-1	9120	126	200	9.4
MZ-2	910	25	1	0.4
MZ-3	4570	100	126	8.4
MZ-4	36	1	1	0.4

These experiments suggest that two groups of mutants can be distinguished. The conclusions are summarized in the following model - an adaptation of the overlapping domain model by Trebst (3). The B-protein has beside the binding site for plastoquinone (A,B) additional binding sites for herbicides (AC, AD, AE). The mutagenic alteration of the C, D, E sites causes resistance but has no impact on the electron transfer kinetic. An alteration of the A-site alters the binding properties of the herbicide molecules as well as those of plastoquinone Q_B, resulting in

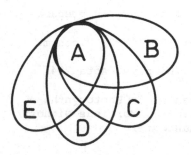

an significantly reduced electron transfer rate. The hypothesis of
competitive binding of plastoquinone Q_B and inhibitors of photosystem II(4)-sharing the site A- was substantiated by binding and replacement studies (5). The analysis of the B-protein sequence will
tell which amino acid residue causes differences in the herbicide and
plastoquinone binding, and it will promote the identification of each
site (in progress).

Fig. 1 Kinetic of X-320 in intact cells

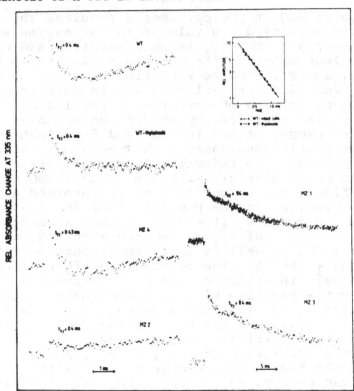

References

1 Pucheu, N., Oettmeier, W., Heisterkamp, U., Masson, K. and Wildner,
  G.F. (1984) Z. Naturforsch. in press
2 Andersson, B. and Haehnel, W. (1982) FEBS Lett. 146, 13-17
3 Trebst, A. (1980) Methods in Enzymology 69, 675-715
4 Velthuys, B.R. (1981) FEBS Lett. 126, 277-281
5 Vermaas, W.F.J. and Arntzen, C.J. (1983) Biochim. Biophys. Acta
  725, 483-491

CORRELATION BETWEEN THE NANOSECOND REDUCTION KINETICS OF
Chl-a$_{II}^{+}$ (P680^{+}) IN SINGLE FLASHES AND THE e-PATHWAY,
CHARGING OF THE S-STATES AND H^{+}-RELEASE IN THE O$_2$-EVOLVING
SYSTEM OF PHOTOSYNTHESIS

K. Brettel, E. Schlodder, and H.T. Witt
Max-Volmer-Institut für Biophysikalische und Physikalische
Chemie, Technische Universität Berlin, Strasse des 17. Juni
135, 1000 Berlin 12, FRG

The cleavage of H$_2$O in photosynthesis requires the accumulation of four oxidizing equivalents in the enzyme system
S. In the primary act, Chl-a$_{II}$ is photooxidized and extracts
in 4 turnovers four electrons from S via at least two electron donors, D$_1$ and D$_2$. The re-reduction kinetics of Chl-a$_{II}^{+}$
after the 1st, 2nd, 3rd, etc. flash given to dark-adapted
samples of oxygen-evolving photosystem II particles or spinach
thylakoids have been monitored by absorption changes at
824 nm in the ns range. After the 1st and 5th flash, the
re-reduction is nearly monophasic with $\tau \sim$ 23 ns. After
the 2nd and 3rd flash, the reduction is significantly slower
and can be adapted bi-phasicly with τ = 50 ns + 260 ns.
After the 4th flash, the reduction time is intermediate
(see Fig. 1). Considering the population of the S-states
of the O$_2$-evolving complex, it was shown that in state S$_0$
as well as in state S$_1$ Chl-a$_{II}^{+}$ is reduced within 23 ns;
whereas, in state S$_2$ as well as in S$_3$ the reduction takes
place with 50 ns + 260 ns. The retardation of the electron
transfer to Chl-a$_{II}^{+}$ in states S$_2$ and S$_3$ is assumed to be
caused by Coulomb attraction by a positive charge located
in the O$_2$-evolving complex. Such a charge in states S$_2$ and
S$_3$ can be explained, if the electron release pattern (1,1,1,1)
is accompanied by an intrinsic proton release pattern
(1,0,1,2) for the transitions S$_0 \to$ S$_1$, S$_1 \to$ S$_2$, S$_2 \to$ S$_3$, S$_3 \to$ S$_0$.
The values for the free energy change coupled with the reduc-
tion sequence can be derived, on one hand, from the kinetics
and equilibrium constant K, resp. On the other hand, this
free energy change can be explained by the Coulomb potential
created by the positive charges of the S$_2$ and S$_3$ states.
Thereby the possible distances for Chl-S and D-S can be also
evaluated (see Fig. 2). A kinetic model based on a linear
electron transfer from the S-states to Chl-a$_{II}^{+}$ via at least
two carriers, D$_1$ and D$_2$, can serve as quantitative explanation of the experimental results (see Fig. 2, bottom).
Under repetitive excitation, Chl-a$_{II}^{+}$ is multiphasicly reduced in the ns-time range. The ns decay can

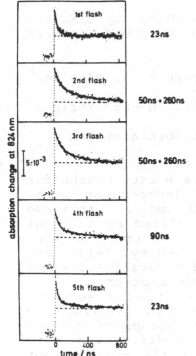

Fig. 1. Time course of the 824 nm absorption change in a series of 5 saturating flashes given to dark-adapted samples of spinach thylakoids.

be described by three-exponential phases. This result can be explained quantitatively by a superposition of the different kinetics obtained in single flash excitation, thereby taking the S-state population under repetitive excitation into consideration (25% each). For details see Brettel, Schlodder, Witt (1984) BBA (in press).

Fig. 2. Reaction sequence in the 1st and 2nd flash and the free energy difference, $\Delta\Delta G$, between both

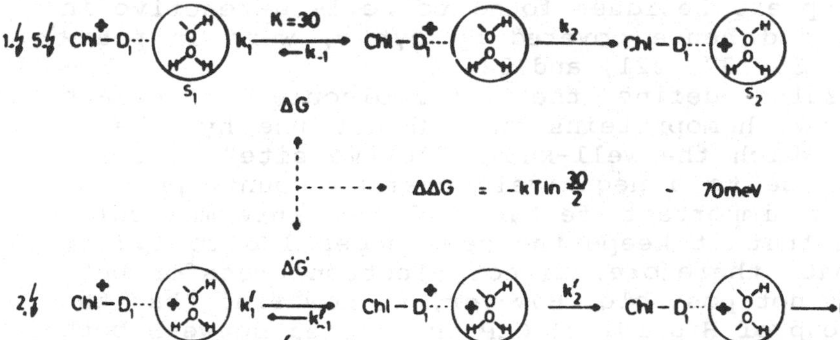

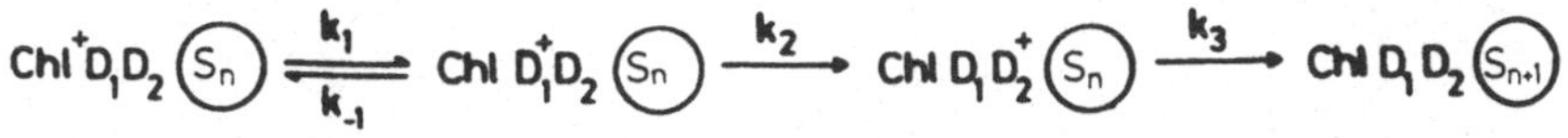

THE CYTOCHROME c PEROXIDASE - CYTOCHROME c ELECTRON TRANSFER COMPLEX AND THE MECHANISM OF ELECTRON TRANSFER BETWEEN HEMES

H.R. Bosshard[0], R. Bechtold[0], B. Waldmeyer[0], J. Bänziger[0]
T. Hasler[0], and T.L. Poulos[1]
[0]Biochemisches Institut der Universität, CH-8057 Zürich, Switzerland
[1]Dept. of Chemistry, University of California at San Diego, La Jolla, Cal 92093

The intermolecular interface of this electron transfer complex has been deduced in three independent ways. First, the site recognized by cytochrome c peroxidase (CCP) on cytochrome c (cyt c) was localized by chemical modification of lysines of cyt c (1,2). Second, the non-covalent complex was treated with EDC (1-ethyl-3-(3-dimethylaminopropyl)carbodiimide) to yield a covalent complex which had 16% residual activity towards exogenous cyt c. In this complex, crosslinking occurred to the sequence regions 30-48, 76-90 and 213-226 of CCP. Third, the chemical reactivity of all 48 carboxyl groups of CCP was measured, by differential chemical modification with EDC/taurine, in the free enzyme and in the non-covalent CCP-cyt c complex. Residues found to be less reactive in the complex, and hence covered by cyt c, were in positions 33, 34, 35, 37, 221, and 224.

These results define the intermolecular interface between the two hemoproteins and support the hypothesis according to which the well-known "active site" of cyt c has to juxtapose to a negatively charged counterpart on CCP (3,4). An important feature of the intermolecular interface is that it keeps the heme edges 1.6 to 1.8 nm apart and that, therefore, direct electron transfer between hemes is not possible. However, we now find that the imidazole group of His 181 of CCP is located between both hemes and near to the center of the intermolecular interface where it may serve as a bridging group in electron transfer. Indeed, destruction of this residue by photo-oxidation blocks electron transfer between hemes but does not interfere with the formation of the CCP-cyt c complex (5).

The figure shows a model of CCP (3,5). Sequence regions marked by a dotted line are either crosslinked to cyt c in the covalent CCP-cyt c complex or are less accessible towards EDC plus taurine (modification of carboxyl groups) in the non-covalent CCP-cyt c complex. The crucial His 181, photooxidation of which blocks the electron transfer between hemes, is indiacted by an arrow.

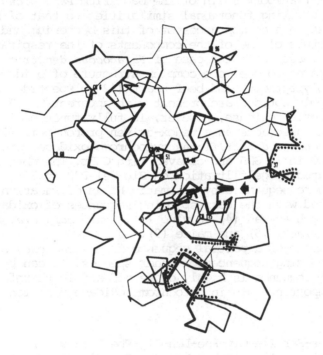

(1) Pettigrew, G.W. (1978) FEBS Lett. 86, 14-16
(2) Kang, C.H. et al. (1978) J. Biol. Chem. 253, 6502-6510
(3) Poulos, T.L. and Kraut, J. (1980) J. Biol. Chem. 255, 10322-10330
(4) Margoliash, E. and Bosshard, H.R. (1983) Trends Biochem. Sci. 8, 316-320
(5) Bosshard, H.R. et al. (1984) J. Biol. Chem. 259, in press.

194

SUBUNITS I AND II OF BACTERIAL AND MITOCHONDRIAL CYTOCHROME OXIDASES
ARE HOMOLOGOUS
G.C.M. Steffens*, G. Buse*, W. Oppliger° and B. Ludwig°
*Abteilung Physiologische Chemie, RWTH Aachen
 D-5100 Aachen, West Germany
°Biozentrum der Universität Basel
 CH-4056 Basel, Switzerland

The aerobic electron transport chain of the bacterium *Paracoccus denitrificans* shows striking functional similarities to that of mitochondria (1), suggesting a common ancestor of this bacterium and the organelle. The structure of one of the components of the respiratory chain, cytochrome oxidase is in the case of *Paracoccus denitrificans* much simpler as compared to the more complex structure of oxidases from yeast (ca. 9 polypeptides) and beef heart (12 polypeptides) mitochondria: only two polypeptides are present with apparent M_r 45000 and 28000 (2). This bacterial oxidase, however, is fully functional in electron transport and also acts as a redox-driven proton pump (3). The two subunits of *Paracoccus denitrificans* cytochrome oxidase were separated on Bio-Gel P200 in 3% SDS and cleaved with CNBr. The obtained fragments were isolated by gelfiltration on Bio-Gel P30 in 70% acetic acid. Some of them were sequenced by automatic Edman degradation and the sequences compared with the corresponding sequences of oxidases from other species such as man (4), bovine (5), *Neurospora crassa* (6), *Saccharomyces cerevisiae* (7,8) and maize (9).
Table 1 gives an example of a partial sequence from *Paracoccus* subunit I aligned to corresponding sequences of other species. It can be seen that this particular sequence is highly conserved and is therefore likely to play an important role in cytochrome oxidase function.

Table 1

277

Man	I	-Met-Met-Ser-Ile-Gly-Phe-Leu-Gly-Phe-Ile-Val-Trp-Ala-His-
Bovine	I	-Met-Met-Ser-Ile-Gly-Phe-Leu-Gly-Phe-Ile-Val-Trp-Ala-His-
Sacch. c.	I	-Met-Ala-Ser-Ile-Gly-Leu-Leu-Gly-Phe-Leu-Val-Trp-Ser-His-
Neuro. cr.	I	-Met-Met-Ser-Ile-Gly-Ile-Leu-Gly-Phe-Ile-Val-Trp-Ser-His-
Parac. d.	I	-Met-Ala-Ala-Ile-Gly-Ile-Leu-Gly-Phe-Val-Val-Trp-Ala-His-

Table 2 exemplifies the alignment of a partial sequence from *Paracoccus* subunit II to corresponding sequences of the other species. The first part of this sequence is not so much conserved, which might reflect the adaptation to different cytochromes c. Whereas the second part is characterized by the presence of a remarkable cluster of aromatic residues which is completely invariant in all cytochrome oxidases. This special structure is assumed to be essential for the function of the enzyme (10).

Table 2

89
Man	II	-Glu-Val-Asn-	-Asp-Pro-Ser-Leu-Thr-Ile-Lys-Ser-Ile-
Bovine	II	-Glu-Ile-Asn-	-Asn-Pro-Ser-Leu-Thr-Val-Lys-Thr-Met-
Saccharomyces c.	II	-Glu-Val-Ile-	-Ser-Pro-Ser-Ile-Thr-Ile-Lys-Ala-Ile-
Neurospora cr.	II	-Glu-Val-Ser-	-Asp-Pro-Ser-Met-Ser-Val-Leu-Ala-Glu-
Maize	II	-Gly-Val-Leu-Val-Asp-Pro-Ala-Ile-Thr-Ile-Lys-Ala-Ile-
Paracoccus d.	II	-Met-Pro-Asn-	-Asp-Pro-Asp-Leu-Val-Ile-Lys-Ala-Ile-

-Gly-His-Gln-Trp-Tyr-Trp-Thr-Tyr-Glu-Tyr-
-Gly-His-Gln-Trp-Tyr-Trp-Ser-Tyr-Glu-Tyr-
-Gly-Tyr-Gln-Trp-Tyr-Trp-Lys-Tyr-Glu-Tyr-
-Gly-His-Gln-Trp-Tyr-Trp-Ser-Tyr-Gln-Tyr-
-Gly-His-Gln-Trp-Tyr-Trp-Ser-Tyr-Glu-Tyr-
-Gly-His-Gln-Trp-Tyr-Trp-Ser-Tyr-Glu-Tyr-

Since this two subunit oxidase , as the mitochondrial oxidases, contains hemes a and a_3, as well as two copper atoms, the location of these prosthetic groups can be restricted to subunits I and II of the mitochondrial oxidases. In addition to the functional relevance of conserved amino acids, the presented data provide a molecular biological contribution to the symbiotic origin of mitochondria (1,11).

References

1   John, P. and Whatley, F.R. (1977) Biochim. Biophys. Acta 463, 129-153.
2   Ludwig, B. and Schatz, G. (1980) Proc. Natl. Acad. Sci. USA 77, 196-200.
3   Solioz, M., Carafoli, E. and Ludwig, B. (1982) J. Biol. Chem. 257, 1579-1582.
4   Anderson, S., de Bruijn, M.H.L., Coulson, A.R., Eperon, I.C., Sanger, F. and Young, I.G. (1982) J. Mol. Biol. 156, 683-717.
5   Steffens, G.J. and Buse, G. (1979) Hoppe-Seyler's Z. Physiol. Chem. 360, 613-619.
6   Macino, G. and Morelli, G. (1983) J. Biol. Chem. 258, 13230-13235.
7   Bonitz, S.G., Coruzzi, G., Thalenfeld, B.E., Tzagoloff, A. and Macino, G. (1980) J. Biol. Chem. 255, 11927-11941.
8   Coruzzi, G. and Tzagoloff, A. (1979) J. Biol. Chem. 254, 9324-9330.
9   Fox, T.D. and Leaver, C.J. (1981) Cell 26, 315-323.
10 Buse, G. and Steffens, G.J. (1976) in Genetics and Biogenesis of Chloroplasts and Mitochondria (Bücher, T. et al., editors) North Holland Publishing Company, Amsterdam and Oxford, pp. 189-194.
11 Steffens, G.C.M., Buse, G., Oppliger, W. and Ludwig, B. (1983) Biochem. Biophys. Res. Comm. (1983) 116, 335-340.

STRUCTURAL PROPERTIES OF MITOCHONDRIAL NICOTINAMIDE NUCLEOTIDE
TRANSHYDROGENASE FROM BEEF HEART

E. Carlenor, B. Persson and J. Rydström
Department of Biochemistry, Arrhenius Laboratory, University of
Stockholm, S-106 91 Stockholm, Sweden

Nicotinamide nucleotide transhydrogenase is localized in the mitochondrial inner membrane and catalyzes the reversible transfer of
hydrogen between NADH and $NADP^+$. The enzyme has recently been shown
to be a proton pump and is regulated by external energy sources
through a proton motive force (1). Transhydrogenase from beef heart
has been purified and reconstituted in liposomes (cf. Persson, B.
and Rydström, J., this volume), which has made it possible to initiate studies on structure-function relationships in the enzyme. At
this stage we have embarked on 4 different approaches in order to
reach an understanding of the proton pump function of transhydrogenase: determination of the amino acid sequence, elucidation of the
pathways for biosynthesis, processing and assembly, proteolytic digestion of the peptide and characterization of the products; and
site-specific modification. The present report concerns mainly the 3
latter points.

Attempts to raise antibodies in rabbits against purified mitochondrial transhydrogenase from beef heart were successful.
Blotting analysis of the specificity of the antibodies using beef
heart mitochondria and beef heart submitochondrial particles showed
a specific binding to the 115.000 molecular weight transhydrogenase
peptide only. A somewhat weaker but still specific interaction occurred against rat liver mitochondria. Cell-free translation experiments indicate that transhydrogenase is coded for by nuclear DNA. A
characterization of polysomal translation products and their processing to native transhydrogenase is presently being pursued.
Preliminary results from amino acid sequencing indicate that the N-
terminal amino acid is blocked.

Transhydrogenase has previously been shown to be a dimer, both in
the intact submitochondrial membrane, as purified active enzyme, and
in reconstituted liposomes (2). Since the pure enzyme on SDS-PAGE
always has appeared as a single peptide, the possibility of the
existence of 2 very similar but different peptides has never been
considered. However, analysis by 2-dimensional PAGE, where one
dimension is isoelectrofocusing, and the other conventional SDS-
electrophoresis, does not support this possibility. If there are 2

separate peptides, they have the same charge and differ by not more than approximately 10 amino acids.

Trypsin treatment of reconstituted and detergent-dispersed trans-hydrogenase, which is known to strongly inhibit the enzyme, showed an interesting difference with respect to product pattern between these 2 preparations. Whereas the fragments 66K, 60K and 37K were formed in the latter case an additional peptide, 68K, was formed in reconstituted liposomes. This may suggest that the 68K fragment contains a minor 2K moiety which is membrane associated and is less prone to be cleaved off in the presence of a membrane. Alfa-chymotrypsin and V-8 (Staphylococcus aureus) were also strongly inhibitory but produced different product patterns.

DCCD is known to inhibit both the catalytic activity and the proton pump activity of transhydrogenase, which has been proposed to involve a modification of a proton channel or carrier in the enzyme (3, 4) similar to that in e.g. the mitochondrial ATPase. Recent results in this laboratory show, however, that under certain conditions DCCD treatment can result in an inhibition of only the proton-pumping capacity essentially without effect on the maximal catalytic activity. This may suggest that the catalytic and proton-pumping activities are not obligatorily linked, or that DCCD separates these 2 functions. In the latter case one would have to assume that some type of "slipping" mechanism would be operative. An important problem in this context is to identify the DCCD-binding moiety and its functional and structural relationship to the 2 catalytic sites. These and related problems are presently being investigated. (Supported by the Swedish Natural Research Science Council).

References
1. Rydström, J. (1977) Biochim. Biophys. Acta 463, 155-184.
2. Andersson, W.M. and Fisher, R.R. (1981) Biochim. Biophys. Acta 635, 194-199.
3. Pennington, R.M. and Fisher, R.R. (1981) J. Biol. Chem. 256, 8963-8969.
4. Phelps, D.C. and Hatefi, Y. (1981) J. Biol. Chem. 256, 8217-8221.

IMMUNOLOGICAL ANALYSIS OF PLANT MITOCHONDRIAL NADH DEHYDROGENASES
I.R. Cottingham°, M.W.J. Cleater" and A.L. Moore°
°Department of Biochemistry, University of Sussex,
Falmer, Brighton BN1 9QG, U.K.
"Department of Biochemistry, University of Southampton,
Southampton SO9 3TU, U.K.

Plant mitochondria are capable of oxidising NADH by a variety of
different pathways (1). The activities associated with the inner
membrane include an external NADH dehydrogenase, which has a substrate
binding site facing the intermembrane space, and one or possibly two
internal dehydrogenases. Unlike mammalian mitochondria, the oxidation
of internal NAD^+-linked substrates by plant mitochondria is only
partially rotenone-sensitive. On the basis of current evidence it is
not possible to distinguish between two internal dehydrogenases, only
one of which is rotenone-insensitive, and a single enzyme, with a
means of bypassing rotenone inhibition. We are developing immunological methods to isolate and characterise the NADH dehydrogenases
of the plant mitochondrial inner membrane. Two different approaches
are being used. The first of these relies on the use of immuno-
electrophoretic techniques, activity staining and precipitate excision
to raise monospecific antibodies to the plant enzymes. The second
approach is based on the observation that antisera raised to purified
beef heart NADH-ubiquinone oxidoreductase (complex I) cross-react
with the equivalent enzyme in plants. Throughout the following work
mung bean (P.aureus) sub-mitochondrial particles (SMP), which are
assumed to be free of outer membrane fragments and matrix enzymes,
were used.

Rabbit antiserum, raised to mung bean SMP, were analysed by crossed
-immunoelectrophoresis (CIE) against the original antigen (2). Under
non-dissociating conditions, 14 different lines containing antibody-
antigen complexes were visualised by coomassie blue staining. When a
similar CIE plate was stained for NADH dehydrogenase activity, using
the NADH-nitro blue tetrazolium reaction, two intersececting bell-
shaped peaks of approximately equal size were observed. This demonstrates the presence of two immunologically different NADH dehydrogenases in plant SMP. Neither of these peaks showed any activity
when NADPH was used instead of NADH, but this does not exclude the
possibility of a minor catalytic activity. Using precipitate
excision methods we are now in the final stages of raising monospecific antibodies to these enzymes.

The cross-reactivity of antibodies raised to beef heart complex I,
with mung bean SMP, were analysed using protein blotting to nitro-

cellulose paper (3). Bound antibodies were then detected using ^{125}I-labelled protein A followed by autoradiography. Antibodies were raised to purified complex I and shown to cross-react with the 75,53,49,30,18 and 15 K molecular weight subunits (as defined in ref. 4) of the purified enzyme. Under identical conditions this antiserum identified four polypeptides in mung bean SMP with approximate molecular weights of 50,47,27 and 13 K. The remarkable degree to which this enzyme is conserved in plant mitochondria was further demonstrated using subunit-specific antibodies raised to the 75,49, 30 and 15 K iron-sulphur subunits of beef heart complex I (Ragan, C.I. and Cleater, M.W.J., unpublished work). Polypeptides from mung bean SMP, of 47 and 27 K molecular weight, cross-reacted strongly with the 49 and 30 K subunit specific antibodies respectively. The cross-reactivity was more variable with the anti-75 K antibodies. A polypeptide of about 60 K molecular weight was usually detected but was completely absent in some preparations. This, and possibly the lower molecular weights of the other two subunits, is most likely due to proteolysis. Experiments are now in progress where protease inhibitors are used in all preparatory stages. Only a faint cross-reaction with the antibody against the smallest subunit was observed. A similar pattern of cross-reactivity with antibodies to the beef heart enzyme was also observed with SMP made from A.maculatum mitochondria. Preliminary studies, using an antiserum raised to one of the dehydrogenases detected in mung bean SMP by CIE, show cross-reactivity with purified beef heart complex I. We are currently proceeding with further characterisation of the two plant NADH dehydrogenases.

1. Palmer, J.M. and Moller, I.M. (1982) Trends Biochem.Sci. 7, 258-261.
2. Owan, P. and Smith, C.J. (1976) In "Immunochemistry of Enzymes and their Antibodies", pp.147-202, (M.R.J. Salton, ed.). Wiley & Sons, New York.
3. Ramarez, P., Bonilla, J.A., Morano, E. and Leon, P. (1983) J.Immunol.Methods 62, 15-22.
4. Ragan, C.I. (1980). In "Sub-cellular Biochemistry", Vol.7, pp.267-307, (D.B. Roodyn, ed.). Plenum Press, New York and London.

INTERACTION OF NITRIC OXIDE WITH NITROUS OXIDE RESPIRATION

K. Frunzke and W. G. Zumft
Lehrstuhl für Mikrobiologie, Universität Karlsruhe, Kaiserstrasse 12,
D-7500 Karlsruhe 1, F.R.G.

Respiration of nitrous oxide (N_2O) is part of an electron transport chain with three to four terminal oxidoreductases which convert nitrate sequentially to dinitrogen. In several Pseudomonads the reaction sequence was found to be truncated at the levels of nitric oxide (NO) and N_2O when the bacteria were grown under oxygen limitation (Table I). This truncation of anaerobic respiration in strains that are known to be complete denitrifiers was due to inhibition of N_2O reductase by NO and a high oxygen sensitivity of the expression of NO reductase. The kinetics of the inhibition of N_2O reduction were measured by use of a dual column gas chromatographic separation system (1) and by dual wavelength spectrophotometry of the cytochrome oxidation levels in suspensions of intact cells (Fig. 1).

TABLE I: Products of nitrite respiration in different Pseudomonads

Organism	Specific Activity (μmol $\cdot$ mg^{-1} $\cdot$ h^{-1})			Products of NO_2^- Reduction (%)		
	NO_2^- Uptake	NO	N_2O	NO	N_2O	N_2
P. perfectomarinus, ATCC 14 405	3.0	1.9	11.1	0	0	100
P. stutzeri, JM 300	2.6	—	10.4	88	11	1
P. fluorescens, PJ 187, Bt B	4.0	—	19.7	85	10	5
P. fluorescens, PJ 188, Bt B	1.7	0.3	0.4	90	2	8
P. fluorescens, DSM 50 111	3.4	—	28.1	100	0	0
P. chlororaphis, B 560, Bt D	0.8	—	0.0 *	88	12	0
P. chlororaphis, B 561, Bt D	3.9	—	0.0 *	10	90	0
P. aureofaciens, ATCC 13 985, Bt E	2.1	0.6	0.0 *	100	0	0

* constitutively without N_2O reduction

Inhibition of N_2O reductase by NO was weak (K_i = 50 mbar) and reversible in cells whose nitrate respiratory system was fully induced (high levels of NO and N_2O reductases). In cells without or only partially expressed NO reductase but fully expressed N_2O reductase the inhibition was strong (K_i = 5 mbar) and was irreversible. Only

partial expression of NO reductase was observed under oxygen limitation, whereas N_2O reductase under the same conditions of reduced oxygen tension was already fully expressed. Both, in fully and partially
induced cell types the inhibition of N_2O reductase by NO was not competitive. Among other inhibitors of N_2O reduction (acetylene, azide,
cyanide, 2,4-dinitrophenol, and oxygen) only azide revealed a similar
differential effect towards both cell types like NO. The dependency
of inhibition of N_2O reduction by NO and azide on the relative activity levels of NO and N_2O reductases suggests some cooperativity between the two enzymes; may be even vicinity of both on the cytoplasmic membrane.

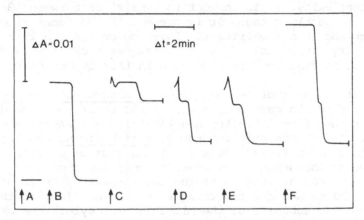

Fig. 1. Effect of NO on the cytochrome oxidation level of intact cells
of _Pseudomonas_ _perfectomarinus_ with active NO reductase during turnover of N_2O. 2 ml cell suspension containing 2.4 mg protein and 150
µmol lactate in 50 mM phosphate buffer, pH 7.0 were incubated under
helium in a gas tight stoppered cuvette. The absorption difference
between 551 nm and 560 nm is being shown. A: Start at lactate reduced
state. B: Addition of 200 µl N_2O. C: Addition of 150 µl NO. D: Second
addition of 150 µl NO. E: Addition of 200 µl N_2O. F: Addition of 500
µl air.

The membrane bound NO reductase was solubilized with non-ionic detergents from membrane fractions of _Pseudomonas_ _perfectomarinus_ and
purified by phase separation (2), ultrafiltration, and gel chromatography. C-type and b-type cytochromes were found to be integral components of the enzyme complex. Both cytochromes were oxidized during
turnover of NO; they were also oxidized by O_2 but not by N_2O.

1. Frunzke, K. and Zumft, W. G., manuscript submitted
2. Bordier, C., J. Biol. Chem. 256, 1604 - 1607

Molecular Characterisation of the NADH Dehydrogenase from
Paracoccus denitrificans

Christina L. George and Stuart J. Ferguson

Department of Biochemistry, University of Birmingham, P.O. Box 363,
Birmingham B15 2TT, England

NADH dehydrogenase [often known as complex 1 in mitochondria] has
proved to be the most complicated enzyme of the mammalian mitochondrial respiratory chain to study. This is due partly to its
polypeptide complexity, it is thought to consist of between 18 to 25
subunits [1] and partly because it is present in the inner mitochondrial membrane at a significantly lower concentration than the
other respiratory complexes. These factors have made it very difficult
to study structure-function relationships in this important enzyme.

The respiratory chain of the bacterium Paracoccus denitrificans is
generally considered to have many similarities to its mitochondrial
counterpart as judged by inhibitor sensitivities and spectroscopic
evidence [2]. Cytochrome oxidase from P. denitrificans has been purified by Ludwig and Schatz [3] and shown to consist of only two subunits compared to the seven found in mitochondria. These two subunits
are of similar molecular weights to the two largest subunits of the
mitochondrial enzyme and are now thought to be equivalent to these
two major catalytic subunits of the mitochondrial cytochrome oxidase.

Following these precedents the NADH dehydrogenase from P.
denitrificans may also be similar to, but simpler than its mitochondrial counterpart and so yield important information about
structure-function relationships in the mitochondrial enzyme. The
P. denitrificans NADH dehydrogenase has been shown to contain iron-
sulphur centres which give similar EPR signals to the mitochondrial
enzyme [4] and also to be rotenone sensitive [2].

Anti P. denitrificans membrane vesicle antiserum was prepared and
used for running crossed immunoelectrophoresis [CIE] with nonidet
P-40 solubilized P. denitrificans membrane vesicles as antigen. This
gave a complex pattern of precipitin arcs but zymogram staining
revealed only a single arc with NADH dehydrogenase activity.

This immunoprecipitation pattern was simplified by partially
purifying the NADH dehydrogenase by a combination of gel filtration
and ion exchange chromatography. The partially purified enzyme preparations gave a less complex immunoprecipitation pattern when analysed by CIE and the NADH dehydrogenase arc became better separated

from the other precipitin arcs. The partially purified enzyme was still catalytically active with ubiquinone analogues as electron acceptors.

The polypeptide composition of the NADH dehydrogenase from $\underline{P}$. $\underline{denitrificans}$ cells grown on $^{35}SO_4^{2-}$ was determined by excision of immunoprecipitates from CIE plates and analysis of the protein eluted from these precipitin arcs by SDS PAGE electrophoresis followed by fluorography [5]. Two polypeptides were consistently seen by this method and were calculated to have molecular weights of 48,000 and 25,000. These two polypeptides were seen whether precipitin arcs were excised from CIE plates run with nonidet P-40 solubilized vesicles or the partially purified NADH dehydrogenase preparation as antigen how-ever only when using the partially purified enzyme was it possible to excise the NADH dehydrogenase precipitin arc free of contamination from other neighbouring precipitin arcs. If the intact NADH dehydro-genase from $\underline{P.\ denitrificans}$ consists of more than two subunits the putative additional subunits must have been lost on solubilising with nonidet P-40.

The complex mitochondrial NADH dehydrogenase can be split with chaotropic agents to give an iron-sulphur fragment which does not contain FMN and a flavoprotein fragment which contains FMN and iron. The flavoprotein fragment is catalytically active with ubiquinone analogues and has been shown [6] to consist of only three subunits, in equal molar proportions, of molecular weights 51,000, 24,000 and 9-10,000. The molecular weights of the subunits of the $\underline{P}$. $\underline{denitrificans}$ NADH dehydrogenase are very similar to those of the two largest subunits of the flavoprotein fragment.

1. Ragan, C.I. (1980) Subcellular Biochemistry 7, 267-307.

2. John, P. and Whatley, F.R. (1977) Advances in Botanical Research, 4, 51-115.

3. Ludwig, B. and Schatz, G. (1980) Proc. Natl. Acad. Sci. 77, 196-200.

4. Albracht, S.P.J., van Verseveld, H.W., Hagen, R. and Kalkman, M.L. (1980) Biochim. Biophys. Acta, 593, 173-186.

5. Owen, P. (1983) in Electroimmunochemical Analysis of Membrane Proteins (Bjerrum, O.J., ed.) pp.347-373. Elsevier, Amsterdam.

6. Galante, Y.M. and Hatefi, Y. (1979) Arch. Biochem. Biophys. 192, 559-569.

SPECIFIC INTERACTIONS BETWEEN TWO BACTERIAL ELECTRON TRANSFER PROTEINS

F. Guerlesquin and M. Bruschi
Laboratoire de Chimie Bactérienne, C.N.R.S.,
B.P. 71, 13277 Marseille Cedex 9, France

Ferredoxin and cytochrome c_3 are specific partners of the sulfate reduction pathway of <u>Desulfovibrio desulfuricans</u> Norway and might be exemplary for electron exchange mechanism studies. Cytochrome c_3 contains four low redox potential haems (- 165 mV, - 305 mV, - 365 mV and - 400 mV) [1] for 13 000 molecular weight. The two haem axial ligands are two histidinyl residues. Amino acid sequence [2] and three dimensional structure [3] have already been published.

Ferredoxin I has been characterized as a dimer of basic unit of M_r 6 000 containing one (4 Fe-4 S) cluster [4,5]. Its amino acid sequence is reported and preliminary crystallographic data have been described [6].

Structural characteristics of cytochrome c_3 and ferredoxin are discussed and should allow a better understanding of the nature of the interaction.

^{1}H-NMR spectra of cytochrome c_3 and ferredoxin I are shown. The modification of the ^{1}H-NMR spectrum of cytochrome c_3 in presence of ferredoxin I is presented.

Preliminary microcalorimetric studies show notable hydrophobic interactions between the two partners.

Rapid kinetic studies by stopped flow spectrophotometry are described. Reduction rate constants of cytochrome c_3 and ferredoxin by dithionite are compared to reduction rate constant of cytochrome c_3 by reduced ferredoxin.

Intramolecular electron exchange between the haems of cytochrome c_3 would be discussed. An intermolecular electron exchange scheme between the haems of cytochrome c_3 and the (Fe-S) cluster of the ferredoxin is proposed.

REFERENCES

(1) Bianco P. and Haladjian J. (1981) Bioelectrochem. Bioenerg. 8, 239-245.
(2) Bruschi M. (1981) Biochim. Biophys. Acta, 671, 219-226.
(3) Pierrot, M., Haser, R., Frey, M., Payan, F. and Astier J.P. (1982) J. Biol. Chem., 257, 14341-14348.
(4) Guerlesquin F., Bruschi M., Bovier-Lapierre G. and Fauque G. (1980) Biochim. Biophys. Acta, 626, 127-135.
(5) Guerlesquin F., Moura J.J.G. and Cammack R. (1982) Biochim. Biophys. Acta, 679, 422-427.
(6) Guerlesquin F., Bruschi M., Astier J.P. and Frey M. (1983) J. Mol. Biol., 168, 203-205.

FLUORESCENCE ENERGY TRANSFER MEASUREMENTS OF THE AVERAGE DISTANCE
BETWEEN CYTOCHROME c AND THE MITOCHONDRIAL INNER MEMBRANE AS A
FUNCTION OF IONIC STRENGTH,

Sharmila S. Gupte and Charles R. Hackenbrock
Laboratories for Cell Biology, Department of Anatomy, School of
Medicine, University of North Carolina, Chapel Hill, NC 27514, USA

We have previously measured the lateral diffusion of cytochrome c
on the mitochondrial inner membrane as well as the kinetics of
electron transfer at various ionic strengths which suggested that
cytochrome c diffuses concomitantly in two and three dimensions (1).
Additional studies indicate that the three dimensional diffusion and
collisional interaction of cytochrome c with its redox partners is an
essential part of the mechanism of electron transfer in whole, intact
mitochondria at physiological ionic strength (2). The purpose of
this communication is to report the measurement of cytochrome c mobility on pure lipid membranes and mitochondrial inner membranes by
determining the average distance of cytochrome c molecules from the
membrane as a function of ionic strength.

Fluorescence energy transfer (FET) from diphenyl hexatriene (DPH)
or tetramethylammonium-DPH (TMA-DPH) probe molecules located in the
membrane bilayer to cytochrome c heme was utilized to measure the
average distance between the probes and c heme (3, eqns. 2 and 5) at
various KCl concentrations. The FET efficiency, hence the average
distance, was measured from the spectral overlap between the emission
spectrum of DPH or TMA-DPH (donor) and absorption spectrum of
cytochrome c (acceptor) (Fig. 1), the quantum yield of the donor (0.8
in asolectin liposomes at 23°C), and the fluorescence intensity of
the donor in the presence and absence of acceptor. The orientation
factor was assumed to be 2/3 for randomly oriented donors and acceptors. When the number of donor and acceptor molecules and their

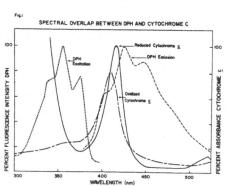

Fig.1
SPECTRAL OVERLAP BETWEEN DPH AND CYTOCHROME C

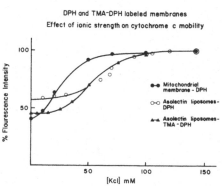

DPH and TMA-DPH labeled membranes

Effect of ionic strength on cytochrome c mobility

orientations remain unchanged, the FET efficiency depends primarily
on the average distance between these molecules.

Labeling of asolectin liposomes or isolated mitochondrial inner
membranes was achieved by incubating the membranes with DPH or TMA-
DPH at probe:lipid ratio of 1:100 for 45 min, at room temperature.
The FET from DPH or TMA-DPH located in the membrane to cytochrome c
heme was determined by monitoring the fluorescence intensity of the
probe at 410 nm (excitation wavelenth 365 nm) in the presence and
absence of 2 μM cytochrome c at various KCl concentrations.

Fluoresence intensity of DPH or TMA-DPH in the presence of 2 μM
cytochrome c was lowest in low ionic strength buffer (no KCl, 2 mM
Hepes, 300 mosm sucrose-mannitol, pH 7.4). The average minimum distance at low ionic strength between DPH located in liposomes and
cytochrome c heme was calculated to be 4.8 nm. The positively charged
TMA-DPH molecule is located closer to the polar headgroups of the
membrane lipids compared to the uncharged DPH molecule and the average minimum distance at low ionic strength between the TMA-DPH and
cytochrome c heme was 4.4 nm. This difference in the average distances between two membrane probes and cytochrome c at low ionic
strength indicates the sensitivity of these FET measurements. The
fluorescence intensity data reveal that the average distance between
the cytochrome c heme and probes in asolectin liposomes and mitochondrial inner membranes is sigmoidal and increases as a function of increased ionic strength up to 100 mM KCl (Fig. 2). Thus the average
time spent by cytochrome c on these membrane surfaces is greater at
low ionic strength and reflects a decrease in its three dimensional
diffusion which increases at higher, more physiological ionic
strengths. Consistent with these findings, the dissociation constants of cytochrome c for asolectin liposomes were 0.85, 0.4 and
2.0 μM in buffers of ~ 0.3, 25 and 60 mM ionic strength, respectively. Previous work showed an increase in duroquinol oxidase activity
of intact mitochondria and isolated inner membranes (2) which parallels the increase in the distance or mobility of cytochrome c from
the mitochondrial inner membrane up to 100 mM KCl. These results
using FET support our hypothesis that electron transfer by cytochrome
c is coupled to its three dimensional diffusion in mitochondria.

References
1. Gupte, S., Wu, E-S., Höchli, L., Höchli, M., Jacobson, K.,
   Sowers, A. and Hackenbrock, C.R. (1984) Proc. Natl. Acad. Sci.
   U.S.A., in press.
2. Gupte, S.S. and Hackenbrock, C.R. (1984) Biophys. J. 45, 297a.
3. Thomas, D.D., Carlsen, W.F., and Stryer, L. (1978) Proc. Natl.
   Acad. Sci. U.S.A. 75, 5746-5750.

DIFFUSION AND COLLISION EFFICIENCY OF OXIDATION-REDUCTION COMPONENTS
IN MITOCHONDRIAL ELECTRON TRANSPORT

Charles R. Hackenbrock and Sharmila S. Gupte
Laboratories for Cell Biology, Department of Anatomy, School of
Medicine, University of North Carolina, Chapel Hill, NC 27514, USA

Studies in this laboratory (1,2) reveal that the redox components
of the mitochondrial inner membrane are independent, free diffusants
in the membrane plane. Such macromolecular random diffusional motion
characterizes the inner membrane as a fluid rather than solid state
structure, and it characterizes the physical arrangement of the redox
components in the membrane as random rather than assembly or chain-
like. It also raises the question of the role of lateral diffusion
of redox components in the mechanism and rate of electron transport.
In this report we present the rates of lateral diffusion (lateral
diffusion coefficients) for the major inner membrane redox components
as determined by fluorescence recovery after photobleaching (Table 1,
column 1). Based on the diffusion coefficients established experimentally, the theoretical diffusion controlled (3) turnover numbers,
which are equal to the theoretical diffusion controlled collision
frequencies (Table 1, column 2) are greater than the experimental
maximum (uncoupled) turnover numbers (Table 1, column 3) for all
redox components. Thus electron transfer between all redox partners
in the mitochondrial inner membrane is diffusion coupled, i.e., in
the native membrane lateral diffusion and collision of redox components precedes electron transfer.

Our findings reconcile the long-standing problem of a physical,
chain-like assembly composed of non-stoichiometric redox components.
Not only ubiquinone, but cytochrome c and the four integral membrane
redox protein complexes behave as common pool, mobile electron
carriers. Their maximum rates of electron transfer can be determined
by their rates of lateral diffusion and their effective redox concentrations, both of which dictate their collision frequencies. Our
study also reveals that ubiquinone and the four integral membrane
redox protein complexes diffuse typically in two dimensions, whereas
the diffusion of cytochrome c occurs in both two and three dimensions. Further, increases in ionic strength result in increases in
electron transfer activity (V_{max}) and the diffusion coefficient of
cytochrome c concomitant with decreases in the affinity of
cytochrome c for the membrane surface.

The ubiquinone-Complex III partners were found to have the lowest
collision efficiency (3.8%), in their interaction with each other,
i.e., one molecule of either of these redox partners collides 26

times with the other to produce one turnover (Table 1, column 4).
The Complex III-cytochrome c partners have the highest collision
efficiency (45%). It was determined that one molecule of cytochrome
c collides 2.2 times with Complex III and 4.6 times with Complex IV
to effect one turnover. The dipole moment of cytochrome c appears to
influence its orientation as it diffuses into the electric fields of
its redox partners (4). Our data tend to support this model since we
found that at artificially low ionic strength (0.3 mM buffer) the
collision efficiencies between cytochrome c and Complex III and IV
were enhanced to become diffusion controlled.

Since experimentally determined electron transport is slower than
the theoretical limit set by the lateral diffusion of all the redox
components, we conclude that ordered chains, assemblies or aggregates
of redox components are not necessary to account for the maximum rate
of electron transport. Rather, mitochondrial electron transport is
diffusion coupled consistent with a "random collision model."

References
1. Hackenbrock, C.R. (1981) Trends in Biol. Sci. 6:151-154.
2. Gupte, S., Wu, E-S., Höchli, L., Höchli, M., Jacobson, K.,
   Sowers, A.E. and Hackenbrock, C.R. (1984) Proc. Natl. Acad.
   Sci. USA, in press.
3. Hardt, S.L. (1979) Biophysical Chem. 10:239-243.
4. Koppenol, W.H. and Margoliash, E. (1982) J. Biol. Chem. 257:
   4426-4437.

TABLE 1

Redox Component	1 Lateral Diffusion Coefficient (cm^2/sec)	2 Theor. Diff. Contr. Collis. Frequency (Collis./s/ Red. Part.)		3 Exper. Turnover Number (Turnov./s/ Red. Part.)	4 Collis. Effic. Col. 2/Col. 3 (Collisions/ Turnover) and %
Complex I	4×10^{-10}	I	36,274	3,360	11/(9.1)
		Q	33	3	
Complex II	4×10^{-10}	II	7,710	1,680	4.6/(22)
		Q	14	3	
Ubiquin.	3×10^{-9}	Q	707	26.7	26.3/(3.8)
		III	1,752	66.7	
Complex III	4.4×10^{-10}	III	1,555	700	2.2/(45)
		C	94	41.9	
Cyto. c	1.9×10^{-9}	C	1,561	339.4	4.6/(21.8)
		IV	27	60	
Complex IV	3.7×10^{-10}	---		---	---

THE HYDROGENASE OF <u>DESULFOVIBRIO</u> <u>GIGAS</u> : PHYSICO-CHEMICAL AND
CATALYTICAL PROPERTIES

E.C. Hatchikian °, R. Aguirre °, V.M. Fernandez" and R. Cammack[Δ]

° Laboratoire de Chimie Bactérienne-CNRS, 31 Chemin J. Aiguier,
  B.P. 71, 13277 Marseille cedex 9 (France)

" Instituto de Catalisis del CSIC, Serrano 119, Madrid 6 (Spain)

[Δ] King's College, Department of Plant Sciences, 68 Half Moon
  Lane, London SE24 9JF (England)

Molecular hydrogen appears to play a significant role in the
energy generating mechanisms of sulfate-reducing bacteria of the
genus Desulfovibrio (1,2). Hydrogenase which catalyzes the reversible
oxidation of hydrogen is the key enzyme of hydrogen metabolism (3).

The periplasmic hydrogenase isolated from Desulfovibrio gigas
contains 12 atoms of iron and labile-sulfide and 1 of nickel in a
molecule of Mr 89,500 comprising two different subunits (4-6). Much
of the iron and sulfide appears to be in ESR-silent [4Fe-4S] clusters
but two types of ESR signals are observed in the enzyme : a narrow
signal centred at g = 2.02, and detected only at low temperature,
which is probably due to an oxidized [3Fe-xS] cluster (7), and a
rhombic signal, usually located at g = 2.32, 2.23, 2.01 which was
assigned to Ni(III) (5,6). The midpoint redox potentials of the
[3Fe-xS] cluster and the ESR-detectable nickel in the enzyme as isolated (-35 mV and -150 mV, respectively, at pH 7) are much higher
than the hydrogen potential (-420 mV) (5).

D. gigas hydrogenase, like many other hydrogenases (3) loses most
of its activity under oxidizing conditions, however, this activity is
gradually restored under reducing conditions (8). This phenomenon is
followed by gradual increase of the catalytic activity and can therefore be considered as a general conversion of the enzyme from a form
which is essentially inert towards hydrogen to an active form. The
data suggest that there are two states of deactivated hydrogenase :
one is rapidly activated under strong reducing conditions (dithionite
+ methyl viologen under argon), the second one required four hours
incubation under hydrogen at 30° C to be activated.

The activity of the reactivated hydrogenase depends on the redox
potential of the reaction media and the experimental data can be
fitted to a Nernst's equation with n = 1. The estimated midpoint
redox potential (-360 mV at pH 7) changes 60 mV per pH unit (8).

The process of reductive activation and oxidative deactivation of hydrogenase has been examined in an attempt to correlate the appearance of enzyme activity with changes in the ESR-detectable centres. The data show that either a modified nickel signal at g = 2.19, 2.14 and 2.01 or an ESR-silent form of the enzyme were observed to appear during the gradual activation that occurred under hydrogen. Reduction of the [3Fe-xS] cluster precedes the reductive activation of D. gigas hydrogenase but does not appear to account for the activation itself. In addition, the results indicate a coupling between the nickel centre in the activated enzyme and the [4Fe-4S] centres of hydrogenase.

The question is still open, as to the physiological significance of the activation/deactivation process of D. gigas hydrogenase as well as the mechanism of reactivation within the cell.

REFERENCES

1. Badziong, W. and Thauer, R.K. (1980) Arch. Microbiol., 125, 167-184.
2. Odom, J.M. and Peck, H.D. (1981) FEMS Lett. 12, 47-50.
3. Adams, M.W.W., Mortenson, L.E. and Chen, J-S. (1980) Biochim. Biophys. Acta 594, 105-176.
4. Hatchikian, E.C., Bruschi, M. and Le Gall, J. (1978) Biochem. Biophys. Res. Commun. 82, 451-461.
5. Cammack, R.C., Patil, D., Aguirre, R. and Hatchikian, E.C. (1982) FEBS Lett., 142, 289-292.
6. Moura, J.J.G., Moura, I., Huynh, B.H., Krüger, H-J., Teixeira, M., Duvarney, R.C., DerVartanian, D.V., Xavier, A.V., Peck, H.D. and Le Gall, J. (1982) Biochem. Biophys. Res. Commun. 108, 1388-1393.
7. Teixeira, M., Moura, I., Xavier, A.V., DerVartanian, D.V., Le Gall, J., Peck, H.D., Huynh, B.H. and Moura, J.J.G. (1983) Eur. J. Biochem. 130, 481-484.
8. Fernandez, V.M., Aguirre, R. and Hatchikian, E.C. (1984) Biochim. Biophys. Acta, in press.

MECHANISM OF THE FIRST STEPS OF SUCCINATE OXIDATION BY THE
MAMMALIAN SUCCINATE DEHYDROGENASE

A.B. Kotlyar, Z.M.Choudry and A.D. Vinogradov

Department of Biochemistry, School of Biology, Moscow State
University, 117234, Moscow, USSR

The following observations relevant to the first steps of succinate
oxidation by the mammalian succinate dehydrogenase (SDH) were taken
into account to suggest the model depicted below:

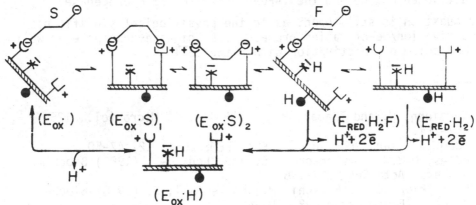

S- Succinate, F- Fumarate, Y^+ - Arginyl, $\bar{X}$ - S^-, ● – FAD, Fe-S.

Both V_{max} and K_i values for the competitive inhibitors are
strongly pH-dependent with apparent pK_a value about 7.0 (1,2). The
reactivity of the enzyme active site sulfhydryl group towards negatively charged 5,5-dithiobis(2-nitrobenzoic acid) (DTNB), neutral 2,2-
dithiopyridine (DTP) or neutral N-ethylmaleimide (NEM) increases over
pH range 6-9, although no simple pH-function is observed for any of
the reagents. At pH 7.0 the enzyme activity of oxaloacetate-free submitochondrial particles can be almost stoichiometrically titrated by
DTNB although SDH sulfhydryl is only about 1 per cent of the total
amount of SH-groups in the particles. In addition to succinate, fumarate and malonate, several inorganic anions: F^-, SO_4^{-2}, HPO_4^{-2} in the
given order of effectiveness completely protect the enzyme against
inactivation by the SH-reagents.

The relative affinities of succinate and oxaloacetate to the oxidized membrane-bound SDH are about 10-times higher than those to the

reduced enzyme; the reverse is true for fumarate. The same K_i for malonate were determined for the membrane-bound or soluble SDH indicating no contribution of the membrane surface charge into free energy of dicarboxylates binding to the enzyme active site (3).

At constant pH an increase of DTNB reactivity to the enzyme is observed when ionic strength is increased from 5 to 20 mM with further decrease at higher ionic strength; only the decrease of SH-reactivity with the increase of ionic strength is seen for DTP. These findings were interpreted as the primary and secondary salt effects for DTNB and as the secondary salt effect for DTP according to Debye-Hückel theory. Thus, electrochemical studies suggest the presence of a positively charged neighbor(s) in a vicinity of the active site sulfhydryl.

Studies on inhibition of the membrane-bound and soluble SDH by the arginine-specific reagents (phenylglyoxal, 2,3-butanedione) revealed the presence of arginine residue(s) at the substrate binding site. The reactivity of an arginyl residue towards phenylglyoxal is decreased (not protected) when the active site SH-group is blocked by mercaptide-forming reagent (4).

On the basis of the findings we suggest the model (see above) for the first steps of succinate oxidation which is in fact the experimentally supported development of the pioneering hypothesis of Van Potter and DuBois (5) and our earlier studies (6).

References

1. Vinogradov, A.D., Gavrikova, E.V. and Zuevsky, V.V. (1976) Europ. J. Biochem. 63, 365-371.
2. Kotlyar, A.B. and Vinogradov, A.D. (1983) Biochim. Biophys. Acta 747, 182-185.
3. Kotlyar, A.B. and Vinogradov, A.D. (1984) Biochim. Biophys. Acta 784, 24-34.
4. Kotlyar, A.B. and Vinogradov, A.D. (1984) Biochem. Int. 8, 545-552.
5. Potter, V.R. and DuBois, K.P. (1943) J. Gen. Physiol. 24, 391-404.
6. Vinogradov, A.D., Zimakova, N.I. and Solntseva, T.I. (1971) Dokl. Akad. Nauk USSR 201, 359-362.

RESPIRATORY PROPERTIES OF THE THERMOACIDOPHILIC ARCHAE-BACTERIUM SULFOLOBUS ACIDOCALDARIUS

G. Schäfer, S. Anemüller and M. Lübben: Institut für Biochemie der Medizinischen Hochschule Lübeck, 2400 Lübeck, Germany

Sulfolobus acidocaldarius (1) was grown heterotrophically on yeast extract under vigorous aeration at 70°C in Brock's medium at pH 2.3. This acidophilic and thermophilic archaebacterium is an obligate aerobic organism of which nothing is known sofar on its bioenergetic systems, as for instance whether it carries out oxidative phosphorylation, or how ΔpH across the plasma membrane is coupled to respiration. Therefore its basic properties with respect to oxygen consumption, response to inhibitors, pH-dependence, temperature dependence, and cytochrome content were investigated. - Freshly grown cells when incubated in glycyl-glycine buffer, pH 3.5, show endogenous respiration of about 13 nmol O_2/mg·min at 60°C. Respiratory activity is strongly temperature dependent vanishing totally between 25-30°C. Towards more alkaline pH the initial respiration increases; however, the cells tolerate higher pH values only for a short period of time. Only succinate (10 mM) caused a pronounced increase of oxygen uptake (1.5-2 fold), other substrates were ineffective including glucose, sucrose, acetate, malate, glutamate, d-lactate, l-lactate and others. Succinate stimulation is observed only at pH values below or above its pK1 (4.2), suggesting that its uptake occurs preferentially in the undissociated form or as a bivalent anion but not as a monovalent anion. Classical uncouplers like FCCP did not increase endogenous or succinate stimulated respiration; only 2.4 DNP (3.8 µM) produced an 1.2-fold increase, restricted to endogenous oxygen uptake; thus a typical uncoupling effect was not measured. Respiration of sulf. ac. is inhibited by cyanide and azide ions, but a cyanide resistant oxygen uptake of about 40 % of maximum respiration persists. Neither K^+/valinomycin, antimycin-A, rotenone, nor oligomycin caused any effect on respiration. Nevertheless, oxygen uptake was totally coupled to the membrane fraction; cytosolic extracts did not take up oxygen significantly (in contrast to Thermoplasma ac. (2). - Spectroscopic analysis of cells and membranes revealed typical spectra of cytochromes as shown in fig. 1. Reduced minus oxidized spectra from native cells or purified membranes (with or without 2 % DOC) show absorption bands at 440, 562 and 587 nm. From pyridine hemochromes a typical a-type cytochrome and a b-type cytochrome could be identified with absorption maxima at 587 and 566 nm, respectively. No soluble cytochrome could be detected. Broken cells and membrane preparations can readily respire with NADH, which like succinate can be used to reduce the cytochromes under anaerobic conditions. The pathway of electrons is not yet known since no inhibitors

of electron transport except cyanide or azide were found. - Since
sulf.ac. has to maintain a large pH gradient across its plasma membrane, it is conceivable that this organism uses its membrane bound
respiratory system mainly as a highly effective proton pump. This is
supported by the finding that a sudden increase of extracellular pH
(from 3.5 to 7.2) causes a burst of oxygen uptake with a transient
8-fold increase of respiratory rate. Isolation of the cytochrome complexes which are assumed to represent efficient proton pumps is in
progress, as well as characterization of membrane associated ATPase
functions (see accompanying report (3)).

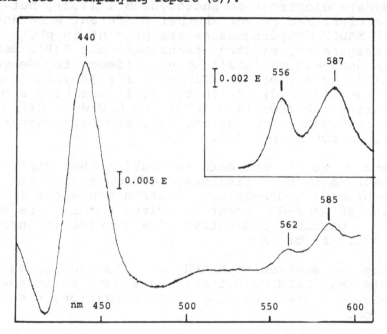

Fig. 1: Reduced minus oxidized difference spectrum of Sulfolobus
acidocaldarius membranes in 10 mM Hepes, pH 7.0; reduction by
dithionite. Inset: pyridin-hemochrome spectrum of sulf.ac.
membranes dissolved in 5 % desoxycholate.

References

(1) Brock, T.D., K.M. Brock, R.T. Belly and R.L. Weiss (1972)
     Arch. Microbiol. 84, 54-68
(2) Searcy, D.G. and F.R. Whatley (1984)
     System Appl. Microbiol. 5, 30-40
(3) Lübben, M. and G. Schäfer (1984)
     this volume, p. 313

DIFFERENT ENZYMES INVOLVED IN NADPH- AND NADH-DEPENDENT RESPIRATION IN THE CYANOBACTERIUM ANABAENA VARIABILIS

S. Scherer, I. Alpes, B. Schrautemeier, and P. Böger
Lehrstuhl für Physiologie und Biochemie der Pflanzen,
Universität Konstanz, D-7750 Konstanz, Germany

Cyanobacteria are photosynthetic prokaryotes with both photosynthetic and respiratory energy conversion. It has been shown recently in the cell-free system that both NADH and NADPH donate electrons to photosystem I (1,2), but it has not been clarified as yet whether different NADH and/ or different NADPH dehydrogenases are present in photosynthetic and respiratory electron transport [e.g.(3)]. Recent results with Anabaena variabilis give evidence for dehydrogenases exhibiting different affinities for NAD(P)H in the dark and in the light (2). Furthermore, it has been shown that ferredoxin-NADP reductase (FNR), an NADPH-specific enzyme in heterocysts of A. variabilis, seemingly donates electrons to photosystem I (4).

In our study, we present data indicating that FNR is linked to NADPH-driven respiratory electron transport. In isolated membranes, NADPH-driven electron transport is twice as high as NADH-dependent activity, but is less sensitive to KCN, possibly indicative of a cyanide-insensitive diaphorase activity of FNR.

By washing the membranes, NADPH oxidation decreases more strongly than NADH oxidation. These data point towards different enzymes, one mediating NADPH oxidation, the other NADH oxidation.

To further elucidate the nature of the NADPH-oxidizing enzyme, different enzyme tests have been performed, some of which are shown in the table. It is known that FNR specifically transfers electrons from NADPH to thio-NADP$^+$ (5). Menadione is an electron acceptor commonly used for assaying NAD(P)H dehydrogenases. Obviously, by washing, FNR activity is released from the membrane in parallel with the menadione-reducing activity (comp. cols. 1 and 2 of the table). In contrast, NADH/menadione activity remains membrane-bound (col.3). No evidence for a membrane-bound NADPH-specific dehydrogenase is found in Anabaena. Whether the membrane-bound NADH dehydrogenase additionally has some NADPH-supported activity is under investigation.

	(1) NADPH → Thio-NADP$^+$	(2) NADPH → menadione	(3) NADH → menadione
First wash:			
supernatant	77	83	28 [O]*)
pellet	22	16	67
Second wash:			
supernatant	11	10	3
pellet	15	7	57
Third wash:			
supernatant	5	4	0
pellet	2	5	52

Isolated membrane material from Anabaena variabilis: NAD(P)H-oxidizing activities in % of total activity after different washing steps after ultracentrifugation*).

Together with the strong NADP$^+$ inhibition of NADPH-driven respiratory oxygen uptake, as reported previously for FNR activity in heterocysts (4), we conclude that NADPH oxidation via respiratory electron transport is mediated by ferredoxin-NADP reductase.

References

(1) Binder, A., Bohler, M., Wolf, M., Muster, P. (1981) In: Selman, B.R., ed., Energy Coupling in Photosynthesis. Elsevier North Holland Inc., Amsterdam, pp. 313-321.
(2) Stürzl, E., Scherer, S., Böger, P. (1984) Physiol. Plant. 60, 557-560.
(3) Houchins, J.P., Hind, G. (1982) Biochim. Biophys. Acta 682, 86-96.
(4) Schrautemeier, B., Böhme, H., Böger P. (1984) Arch. Microbiol. 137, 14-20.
(5) Böger, P. (1971) Z. Naturforsch. 26b, 807-815.

INVESTIGATION OF LATERAL ELECTRON TRANSPORT IN SPINACH THYLAKOIDS

Wolfgang Haehnel, Andreas Spillmann and Garry Newsham*
Biochemie der Pflanzen, Ruhr-Universität Bochum, D-4630 Bochum 1, FRG
*Centre for Mathematical Analysis, Australian National University, Canberra, A.C.T. 2600, Australia

In thylakoid membranes of higher plants many studies have shown an extreme lateral heterogeneity in the distribution of the photosynthetic reaction centers with photosystem II (PSII) and photosystem I (PSI) located preferentially in appressed grana and non-appressed stroma membrane regions, respectively (1). With respect to linear electron transport from PSII to PSI this organization implicates a long-range lateral diffusion of electron carriers. Electron carriers which may be sufficiently mobile are plastoquinone and plastocyanin which functions between PSII and the cytochrome b_6/f complex and the cytochrome b_6/f complex and PSI, respectively. Recently we have summarized our evidence for a limited mobility of plastocyanin and suggested that plastoquinol should function as the fast diffusing molecule (2). With respect to the limited knowledge about the diffusion of plastoquinone and contradicting data on its diffusion coefficient we have investigated the electron transfer reactions which may be associated with the diffusion step of linear electron transport. Millner and Barber (3) have speculated that the rate-limiting step of linear electron transport may be caused by the diffusion of plastoquinol.

Fig. 1 shows the reduction kinetics of $P700^+$ induced by a short flash after oxidation of all electron carriers between the two photosystems by far-red preillumination. The halftime of the $P700^+$ reduction is controlled by the rate-limiting step of linear electron transport and an initial lag of the time course indicates the time for all preceding steps (4). Our approach is based on the possibility to decrease the average distance between PS II and PS I of 200-400 nm in stacked thylakoid membranes of higher plants to about 10-20 nm in unstacked membranes with randomized distribution of the photosystems. This has been realized by incubation of chloroplasts for 1-2 h on ice in the presence of either 5 mM $MgCl_2$ (Fig. 1A) or 10 mM monovalent cations (Fig. 1B) and checked by electron micrographs which showed stacked and unfolded membranes, respectively. The decrease in the average distance between PSII and PSI by more than an order of magnitude does not decrease the halftime of the $P700^+$ reduction. On the contrary the halftime in the presence of $MgCl_2$ was always slightly shorter than that at low salt conditions. Omission of sorbitol or sucrose from the reaction mixture caused an increase of the halftimes. The smaller amplitude of the signal in Fig. 1B as compared to that in Fig. 1A could indicate some P700 not functioning in linear electron transport if the thylakoid membranes are stacked.

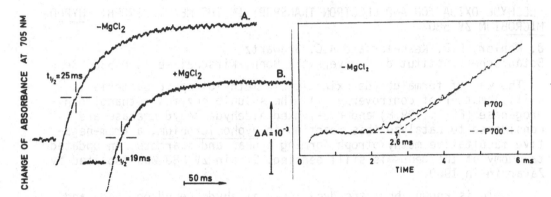

Fig.1 and 2. Kinetics of P700$^+$ reduction at 705 nm induced by a short flash after far-red preillumination at pH 7.5. The reaction mixture contained chloroplasts at a chlorophyll concentration of 20 μM, 10 mM HEPES buffer, 5 mM NaCl, 0.1 M sorbitol and 2 μM gramicidin D. Fig. 1B, NaCl was replaced by 5 mM MgCl$_2$. Fig. 1A and 1B, 50 signals were averaged; Fig. 2, 400 signals were averaged. Repetition rate 0.2 Hz.

It is concluded that the diffusion of a mobile electron carrier from appressed to non-appressed membrane regions is not the rate-limiting step of linear electron transport. A diffusion step of plastoquinol should precede this reaction and would therefore contribute to the initial lag of the P700$^+$ reduction kinetics, shown in Fig. 2. From the small difference in the lag of the P700$^+$ reduction in stacked thylakoids with an average value of 2.7 ms as compared to 2.5 ms (4 preparations) in unstacked thylakoid membranes with randomized distribution of the integral complexes we estimate that the time which could be associated with a diffusion of plastoquinol is ≲ 1 ms . With respect to the distance this is an extremely short time.

We have solved the diffusion equation for a random diffusion between areas of similar dimensions as in stacked thylakoids with a limited boundary. The calculation shows that the diffusion coefficient of plastoquinol would have to be in the range of 10^{-6} cm^2/s to be consistent with our kinetic measurements and the concept of an extreme lateral heterogeneity in the distribution of the photosystems.

REFERENCES
1. Andersson, B. and Anderson, J.M. (1980) Biochim. Biophys. Acta **593**, 427–440
2. Haehnel, W. (1984) Ann. Rev. Plant Physiol. **35**, 659–693
3. Millner, P.A. and Barber, J. (1984) FEBS Lett. **169**, 1–5
4. Haehnel, W. (1976) Biochim. Biophys. Acta **440**, 506–521

ALDEHYDE OXIDATION AND ELECTRON TRANSPORT IN THE METHYLOTROPHIC HYPHO-MICROBIUM ZV 580

J. Köhler, F.P. Kesseler and A.C. Schwartz
Botanisches Institut der Universität Bonn, Kirschallee 1, D-5300 Bonn

The way of formaldehyde oxidation in methylotrophic bacteria is still a matter of controversy (1). The soluble enzymes methanol dehydrogenase (EC 1.1.99.8) and dye-linked aldehyde dehydrogenase are considered to catalyze this reaction in Hyphomicrobium, a gram-negative facultative methylotroph forming hyphae and swarmers. An updated taxonomy of the genus is still pending. Strain ZV 580 was isolated by Zavarzin in 1960.

Little is known about the dye-linked aldehyde dehydrogenase, and the way by which it transfers electrons onto the respiratory chain. The enzyme, estimated with the oxygen electrode or by DCIP reduction, both in the presence of phenazine methosulfate, is present in the genus in much higher activities than previously observed (2,3). In partially purified fractions from ZV 580 grown on methanol, such as an ammonium sulfate precipitate of cell-free extract (45-85 % sat.), the enzyme exists as monomer and dimer of molecular weights of about 40 000 and 80 000, as indicated by chromatography on Sephadex G-100. The dimer oxidizes benzaldehyde with sixfold higher rate than formaldehyde, whereas the monomer oxidizes only formaldehyde. The dimer is stabilized by 500 mM phosphate buffer (pH 7.2). The monomer is predominant in 1 mM buffer. They are interconverted by changing the buffer. The enzyme was purified to homogeneity by the steps of above, and chromatography on DEAE-cellulose (TRIS-HCl pH 7.4) and hydroxylapatite. It was enriched about 300-fold with a yield of 10 %, and had lost most of its activity with benzaldehyde (ratio of rates with benz-/formaldehyde 0.7). The pure enzyme consisted of the dimer only, as shown by electrophoresis in polyacrylamide-gel (PAGE) and activity staining with nitroblue tetrazolium. Estimation of molecular weights with Na-laurylsulfate PAGE gave 76 000 for the dimer, and, after adding 0.4 % thioethanol or 0.2 % dithiothreitol, 44 000 for the monomer. This indicates that during the purification the dimer becomes covalently linked by interchain -S-S-bridges, which cannot be present in the native enzyme. The pure enzyme reduces methylene blue ($E'_0 = 0$ mV) with formaldehyde, but not with benzaldehyde. Of the components of the respiratory chain, only cytochrome c is reduced with benzaldehyde in cell-free extract and intact cells. Benzoate is not degraded.

The four soluble cytochromes c of ZV 580 are present in the ammonium sulfate fraction of 45-95 % sat., and are separated by chromatography on DEAE-cellulose at pH 7.6. They differ in iso-electric points as evident from binding to DEAE-cellulose, and further in mid-point potentials at pH 7.2 (titration after Dutton (4)), amounts and sizes: (I) E_m=333 mV, 10 % of total soluble cyt. c, (II) 218 mV, 50 %, (III) 245 mV, 15 %, (IV) 235 mV, 25 %; molecular weights 12 600 (I-III) and 21 600 (IV), estimated as above.

The membrane-bound cytochromes of ZV 580, as visible in difference spectra of particulate preparations sedimented at 150 000xg (90 min), are b (α 559 nm, β 530 nm), c (α 552 nm, β 523 nm), a/a$_3$ in the early log growth phase (605 nm), and d in the stationary phase (623 nm). Four cytochromes b were distinguished by redox titration (4) with tentative E_m values at pH 7.2 of -140 mV (30 % of total cyt. b), + 20 mV (32 %), 110 mV (27 %) and 250-350 mV (11 %). One of them could be cytochrome o, which was visible in spectra with carbon monoxide. The particles oxidize NADH (1.1 mM), succinate (28 mM) and ascorbate-TMPD (11.1 plus 2.1 mM) with rates of 150, 15, and 2570 nmol·min^{-1} oxygen consumed per mg of protein. Respiration with ascorbate-TMPD proceeds via cytochromes c, and a/a$_3$ or d, and is very sensitive towards cyanide (50 % inhibition at 8 nmol·mg^{-1} protein). When this route of electron transport is completely blocked by cyanide, respiration with NADH is unaffected. It is inhibited by 50 % with 800 nmol cyanide per mg protein. Simultaneous recording of redox changes indicates that only the cytochromes b, including cytochrome o, are active, but not the cytochromes c or others.

The presence of a Rieske iron-sulfur centre in ZV 580 is evident from the inhibition of NADH oxidation with 1,10-phenanthroline, concomitant reduction of cytochromes b as well as oxidation of cytochromes c, and the opening of a shunt with TMPD.

Further, NADH-dependent respiration is effectively inhibited by rotenone, amytal, DBMIB (2,5-dibromo-3-methyl-6-isopropylbenzoquinone), dicoumarol, and NQNO (2-n-nonyl-4-hydroxyquinoline-N-oxide).

Supported by Deutsche Forschungsgemeinschaft.

1. Attwood, M.M. and Quayle, J.R. (1984) in Microbial Growth on C$_1$ Compounds (Crawford, R.L. and Hanson, R.S., eds.), pp. 315 - 323, American Society for Microbiology, Washington

2. Marison, I.W. and Attwood, M.M. (1980) J. Gen. Microbiol. 117, 305 - 313

3. Köhler, J. and Schwartz, A.C. (1982) Can. J. Microbiol. 28, 65 - 72

4. Dutton, P.L. (1978) in Methods in Enzymology, vol. 54 (Fleischer, S. and Packer, L., eds.), pp. 411 - 435, Academic Press, New York San Francisco London

KINETICS OF THE C-CYTOCHROMES IN CHROMATOPHORES FROM <u>RHO-DOPSEUDOMONAS</u> <u>SPHAEROIDES</u> AS A FUNCTION OF THE CONCENTRATION OF CYTOCHROME C_2. INFLUENCE OF THIS CONCENTRATION ON THE OSCILLATION OF THE SECONDARY ACCEPTOR OF THE REACTION CENTERS Q_B.

MARIO SNOZZI[*] AND ANTONY R. CROFTS
University of Illinois, Dept. of Physiology and Biophysics, Urbana, Illinois USA
*present Adress:
Institute of Plant Biology, University of Zurich, CH 8008 Zurich, Switzerland

After the discovery of the membrane bound cytochrome c_1 by Wood [1,2] the role of this cytochrome in the photosynthetic electron transport was studied by Meinhardt and Crofts [3]. They showed, that chromatophores contain about equal amounts of the two cytochromes. Cytochrome c_2 is oxidized rapidly by the reaction centers, whereas cyt c_1 is responsible for the slow oxidation observed earlier [4-6].

The oxidation kinetics of the two cytochromes have been studied in chromatophores with different concentrations of cyt c_2. The changes in the concentration of cyt c_2 were obtained either by selecting preparations of the membranes with different loss of this electron carrier during isolation, or by fusion of chromatophores with liposomes [7]. In both cases, there was a decrease in the initial oxidation rate of cyt c_1 from which it is concluded, that the reaction between the two cytochromes is a collisional second order process, dependent on the concentration of the two reactants. Fusion of the chromatophores with liposomes which lead to an increase of the inner membrane surface by a factor of 1.9 and an increase in the inner volume by a factor of 2.6 resulted in a decrease of the initial oxidation rate by a factor of 1.9 to 2.2 [7,8]. This result does not allow to conclude definitively whether the diffusion of cyt c_2 is through the inner volume or along the inner membrane surface, the latter case is favored, however. The oxidation kinetics of cyt c_2 on the other hand are not affected by the fusion. It is therefore concluded, that this reaction is first order and the reduced cyt c_2 is preferentially bound to the reaction centers.

Isolated reaction centers display an oscillation in the redox state of the secondary acceptor Q_B, with the acceptor being reduced to the semiquinone by the first flash and hence giving rise to an absorption at 446 nm. This absorption disappears after the second flash, which reduces the semiquinone to the fully reduced quinol [9,10]. A similar type of oscillation has been reported by Bowyer et al. [11] for chromatophores; but in this case the oscillations were only present at redox potentials above 300 mV. Investigations with chromatophores with different cyt c_2 content showed that the disappaerence of these oscillations was dependent on the concentration of this cytochrome. Only chromatophores with a high cyt c_2 concentration displayed a complete loss of oscillation below 300 mV. Sphaeroplast derived vesicles, which contain almost no cyt c_2 displayed oscillations at potentials well below 300 mV. Addition of small amounts of cyt c_2 to these membranes enhanced the oscillations at lower potentials, probably due to a better rereduction of the reaction centers between the flashes. Cyt c_2 concentrations above 60 uM resulted in titration curves similar to the ones of chromatophores with high cyt c_2 content. From this we conclude that reduced cyt c_2 binds to the reaction centers and produces cooperating dimers of the latter, so that each flash forms one quinol per two reaction centers and no net production of semiquinone can be observed.

References:

1. Wood, P. (1980) Biochem. J. 189, 385-391

2. Wood, P. (1980) Biochem. J. 192, 761-764

3. Meinhardt, S.W. and Crofts, A.R. (1982) FEBS Lett. 149, 223-227

4. Kihara, T. and Chance, B. (1969) Biochim. Biophys. Acta 189, 116-124

5. Ke, B., Chaney, T.H. and Reed, D.W. (1970) Biochim. Biophys. Acta 216, 373-383

6. Bowyer, J.R. and Crofts, A.R. (1981) Biochim. Bophys. Acta 636, 218-233

7. Snozzi, M. and Crofts, A.R. (1984) Biochim. Biophys. Acta, in Press

8. Snizzi, M. and Crofts, A.R. (1984) in Advances in Photosynthesis Research (Sybesma, C., ed.) I.6.755-758

9. Vermeglio, A. (1977) Biochim. Biophys. Acta 459, 516-524

10. Wraight, C.A. (1977) Biochim. Biophys. Acta 459, 525-531

11. Bowyer, J.R., Tierney, G.V. and Crofts, A.R. (1979) FEBS Lett. 101, 201-206

A METHOD FOR DISTINGUISHING LOCALIZED FROM BULK PHASE PROTON POOLS
USING POST-ILLUMINATION ATP FORMATION AFTER SINGLE-TURNOVER FLASHES.
W.A. Beard and R.A. Dilley
Department of Biol. Sci., Purdue University, W.Lafayette, IN
47907,U.S.A.

A controversy exists within membrane bioenergetics as to whether
the primary route of protons to the ATP synthase complex is through
"bulk phase" domains or "localized" (i.e., membrane-associated)
domains. In chloroplast thylakoids, chemical modification experiments have identified an array of amine buffering groups which are
hypothesized to be sequestered within localized domains associated
with membrane proteins (1) and it has been demonstrated that these
groups must be protonated before the onset of ATP formation can be
observed with single-turnover flashes (2). Further tests of the
"buriedness" of the amine buffering array are needed prior to elucidating what role such arrays play in membrane energy transduction.
One approach is to utilize the phenomenon of post-illumination phosphorylation (PIP) which, in the usual protocol, can directly monitor
the inner bulk phase acid pool (3). Certain permeable buffers,
particularly phenylenediamine (PD), present in the light phase (-ADP)
provide additional bulk phase buffering, greatly increasing both the
observed H^+ uptake in the light and the PIP ATP yield (4). Thus, an
increase in PIP ATP yield due to an added buffer is evidence for bulk
phase acidification sufficient to protonate the buffer. It has been
presumed that the endogenous PIP ATP yield derives from protons
buffered by protein amino acid residues such as $-COO^-$, located in the
bulk phase. We question that presumption on the grounds that our
studies of the amine buffering array suggest a sequestered, intramembrane location of those groups. And by analogy, we can inquire as
to the localization of other, sequestered, lower pK_a groups such as
may be involved as the proton source for the endogenous PIP ATP
yield. The use of single-turnover flash excitation of a cuvette
containing chloroplasts, proper media for ATP formation, and the
luciferin-luciferase (the LKB ATP Monitor Kit was used) system (2)
provides a new way to detect very low amounts of PIP ATP yield. We
can observe PIP ATP yield in the dark after the last flash, either
with ADP present during the flashes (ATP is formed on each flash,
after the lag is over, as well as an amount of ATP formed after the
last flash), or by adding ADP only after the last flash. The latter
is the traditional protocol.

With ADP present during the flash regime, we find: (a) that the
yield of PIP ATP is about 6 nmol per mg chl, (b) the yield is not
increased by the presence of the added buffer phenylenediamine, (c)

the PIP ATP yield requires valinomycin and K^+, even though ATP is formed by every flash (after the lag phase is over) with or without val present, and (d) the PIP yield begins with the same lag as ATP formed during the flash train. The valinomycin and K^+ effect argues against there being a diffusion limitation problem for ATP formed at the CF_1 being the origin of the PIP ATP yield. The same lag ($\simeq$ 15 flashes) for onset of ATP formed during the flash train or for PIP yield suggests that whatever buffering pools are being filled with protons during the 15 flash lag, they are apparently a common, energetic threshold pool. The fact that PD did not increase the PIP yield when ADP was present from the beginning implies that the inner bulk phase is not the "threshold" pool. A critical test of this last point is to note that the PIP yield did increase by 2-fold due to added PD, when ADP was added after the flash train. In that experiment, the PIP yields were 8(-PD) and 16(+PD) nmol ATP per mg chl. It is apparent that when ATP could not be formed in the flash train, protons reached a pool wherein PD could buffer some of them, resulting in the increased PIP yield. That pool must be the bulk inner phase, as explained above. The very different effects of PD in the two protocols for observing PIP ATP yield are consistent with the hypothesis for the primary proton flow pathway being localized, and not including the bulk phase, for the connection between the redox sites of proton release and the CF_0-CF_1. The pathway must include some endogenous buffering groups (at least 20 nmol H^+ per mg chl for a PIP yield of 6 nmol ATP per mg chl) to explain the PIP yield. Moreover, when ADP and P_i are present during the flashes the pK_as of those buffering groups are either considerably above that for PD (6.2), or PD cannot readily reach the domain with the endogenous groups. When ADP is present during the flash train, protons reaching the CF_0-CF_1 must be driven outwards in coupling before equilibrating with PD in the bulk inner phase. When ADP is added after the flash train, protons do interact with the PD in the bulk phase, apparently because with no ADP, the CF_1 does not permit the coupled H^+ efflux and the H^+s equilibrate with the inner phase and protonate the PD present there.

1. Laszlo, J.A., Baker, G.M. and Dilley, R.A. (1984) J. Bioenerg. Biomem. 16, 37-51.
2. Dilley, R.A. and Schreiber, U. (1984) ibid. in press.
3. Vinkler, C., Avron, M. and Boyer, P. (1980) J. Biol. Chem. 254, 10654-10656.
4. Avron, M. (1971) Second Internat. Congress on Photosynthesis, Stresa, Ed. Forti et al., pp. 861-871.

MEMBRANE POTENTIALS AND REACTION RATES

J. Boork and H. Wennerström, Department of Biochemistry, Arrhenius Laboratory, University of Stockholm, S-106 91 Stockholm, Sweden

The kinetic coefficients, whether in a conventional molecular kinetic scheme or in a general non-equilibrium thermodynamic formulation, in general carry a potential dependence. Obviously, the kinetic coefficient can be considered constant over a sufficiently narrow range of potentials, but in many applications it appears that this range is too narrow for the approximation to be useful. This potential dependence has largely been neglected in quantitative models of free energy translocating systems. Based on conventional molecular kinetics we analyze the particular kinetic features of a chemical transformation involving transport of ions across a membrane. We show that membrane potentials change the kinetic coefficients in a way that is dependent on the structural organization of the enzyme system. There is consequently a relation between control behavior and the spatial enzyme organization in the membrane. We also find that under certain conditions the rate is independently dependent on the electrical and dpH components of the proton motive force.

ΔΨ AND ΔpH IN METHANOBACTERIUM THERMOAUTOTROPHICUM

B.M. Butsch and R. Bachofen
Institute for Plant Biology, University of Zürich
Zollikerstrasse 107, CH-8008 Zürich, Switzerland

ΔΨ has been estimated from the distribution of TPP (tetraphenylphosphonium chloride) by an electrode sensitive to the probe in whole cells of Methanobacterium thermoautotrophicum (1). Values in the range of -150 and -200 mV have been obtained. Electron transport was measured by monitoring the methane production rate concomitantly with the concentration of TPP outside the cells. There is a remarkable relation between ΔΨ and electron transport which depends on both H_2 and CO_2 present. The uptake and release of TPP out of the cells is reversible and is regulated by the partial pressure of H_2. In the presence of air, ΔΨ and methanogenesis are irreversibly abolished. TPP itself in concentrations up to 10 μM has only a slight effect upon the apparent ΔΨ ; the rate of methane formation was reduced by TPP by 30% at a concentration of 30 μM.

ΔpH was determined colorimetrically by measuring the pH inside the cells using carboxyfluoresceindiacetate (2) and the pH outside the cells with a pH-electrode. Absorbancy spectra were run at different external pH values under energized conditions in an anaerobic cuvette filled with 2.5 ml of cell suspension on an Aminco DW-2 (American Instrument Company, Silver Spring, Maryland) spectrophotometer. The addition of Triton-X-100 de-energized the cells, and these spectra were used for calibration. As determined so far, the ΔpH contributes only little to the proton motive force at a growth pH of 7.

As an alternative, the pH inside the cells (2 to 3 mg dry weight per ml) was monitored running the photometer in the dual wavelength mode (490-465). Provided that enough electron acceptor (CO_2 or bicarbonate) was present, the cells became alkaline inside rapidly under a gas phase of H_2 (100%) but not of N_2 (Fig. 1A). After about one minute (in accordance with the time expected based on the solubility of H_2 at 60 °C and assuming a methane production rate of 100 ml per h per g dry weight) ΔpH collapsed since all H_2 was consumed; replacing of the hydrogen was simply achieved by shaking the cuvette allowing exchange between gas and liquid phase; this could be repeated many times. When the cells were preincubated for 3 hours at 25 °C and assayed at 60 °C under H_2, the CO_2 support became exhausted; proton pumping from inside to outside could be demonstrated only after potassium bicarbonate (10 mM) had been added (Fig. 1B). The results suggest the splitting of H_2 into electrons and protons by a hydrogenase and the extrusion of the latter into the medium; our data indicate that Δ pH measurements based on distribution methods might be too slow for detecting a Δ pH representative for energized cells. Further experiments will give more quantitative information about ΔpH.

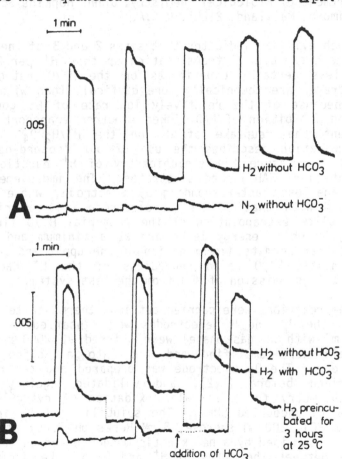

Figure 1: Time course of dual wavelenght absorbancy $(E_{490}-E_{465})$ of cells of Methanobacterium thermoautotrophicum preincubated for 40 min at 60 °C with carboxyfluoresceinediacetate (30 µM). Gas phase: H_2 or N_2, HCO_3^- added as indicated. Energization was achieved by shaking the cuvette, yielding an immediate proton extrusion in the presence of H_2 and HCO_3^-.

1 min

.005

H_2 without HCO_3^-

N_2 without HCO_3^-

A

1 min

.005

H_2 without HCO_3^-

H_2 with HCO_3^-

H_2 preincubated for 3 hours at 25 °C

B

addition of HCO_3^-

1 Butsch, B.M. and Bachofen, R. (1984) Arch. Microbiol. 138, 293-298
2 Thomas, J.A., Kolbeck, P.C. and Langworthy, Th. A. (1982) in Intracellular pH: Its Measurement, Regulation, and Utilization in Cellular Functions (Nuccitelli, R. and Deamer, D.W., eds.), pp. 105-123, Alan R. Liss, Inc., New York

UPPER AND LOWER LIMITS OF THE VECTORIAL H^+ STOICHIOMETRY FOR MALATE + GLUTAMATE OXIDATION IN MITOPLASTS

L. E. Costa, B. Reynafarje, and A. L. Lehninger
Department of Biological Chemistry, Johns Hopkins School of Medicine
Baltimore, Maryland, 21205, U.S.A.

Much evidence indicates that sites 2 and 3 of the respiratory chain are each capable of translocating up to 4 H^+ per $2e^-$. Site 1 data are less certain; experiments on the H^+/O ratio with NAD-linked substrates are technically more difficult than with succinate or cyt c^{2+} because of the relatively low rate of O_2 consumption and the strong inhibition of NAD-linked electron transport by NEM, added to prevent the reuptake of H^+ on the $H^+/H_2PO_4^-$ symporter. This communication describes the use of our "second-generation" methods (1,2) to measure the stoichiometry of H^+ ejection coupled to the oxidation of NAD-linked substrates. The improvements include (i) A membrane-less fast-responding O_2 electrode, whose response time is matched with that of the pH electrode (ii) Experimental conditions that allow extrapolation of the vectorial H^+/O flow ratio to level flow, at which energy leaks are at a minimum and (iii) A rationale (2,3) that permits determination of the upper and lower limits of the mechanistic H^+/O ratio regardless of the H^+ back-decay rate and thus allows omission of NEM from the test system.

The reactions were carried out in a thermostatted closed cell, into which the O_2 and H^+ electrodes were inserted. The cell contents (1.5 ml with no gas phase) were stirred at 2000 rpm. The electrode signals were amplified and fed into a Soltec 330 multichannel recorder. The O_2 electrode was prepared and the response tested as described before (1,2), and validated against the known H^+/O stoichiometry for the scalar oxidation of cyt c^{2+} (2) or NADH (4) by O_2 in uncoupled SMPs. The solubility of O_2 in the test medium (50 mM KCl, 200 mM sucrose, 3 mM Hepes pH 7.05) at 15°C was 560 nmol O/ml, determined by a new kinetic method based on the strict stoichiometry between the rates of H^+ and O_2 uptake during NADH oxidation by uncoupled SMPs (4). Malate + glutamate were employed as electron donors, with rat liver mitoplasts, which are relatively free of endogenous NAD-linked substrates, and with Ca^{2+} as permeant cation. Electron flow and H^+ ejection were initiated by the injection of a small amount of O_2 in the form of air-saturated medium into the anaerobic system containing all other components. The rate of H^+ ejection decreased more rapidly with time than O_2 uptake, owing to the increasing rate of H^+ backflow in response to the increasing ΔpH as the oxidation proceeds. The H^+/O flow ratio will thus be maximal at the beginning of the reaction (i.e. at level flow), where H^+

backflow will be minimal, and decline thereafter. Extrapolation to zero time of the linear semilog plots of O_2 uptake and H^+ ejection rates vs time, carried out by the procedure described in (2), yielded H^+/O ratios for malate-glutamate oxidation in the range 9.6 to 10.4. Since the reactions were run in the absence of NEM, some reuptake of H^+ with $H_2PO_4^-$ probably occurred, yielding values less than the mechanistic H^+/O ratio.

The true H^+/O ratio can be more precisely fixed by a kinetic approach that allows determination of its upper and lower limits, based on a rationale described in (2,3). In brief, the observed rate of H^+ ejection (J_H) under any given set of conditions is $J_H = n\ J_0 - J_L$, in which J_0 is the rate of O_2 consumption, J_L is the rate of all H^+ leaks or slippage and n is the mechanistic stoichiometry. When J_0 and J_H are systematically varied by increasing the concentration of the charge-compensating cation Ca^{2+}, thus increasing the utilization of energy (for Ca^{2+} transport) and decreasing $\Delta\tilde{\mu}_H^+$ (the driving force for H^+ backflow), the slope of a plot of J_H vs J_0 would be higher than n and would set its upper limit. On the other hand, if the rate of production of energy in the form of $\Delta\tilde{\mu}_H^+$ is varied by an inhibitor of electron flow, then the slope of a plot of J_H vs J_0 would set the lower limit of n. When the concentration of Ca^{2+} was varied from 0 to 0.7 mM, in reactions otherwise arranged as described above, the slope of the linear plot of all J_H and J_0 values, which gives the upper limit of the H^+/O ratio, was 12.5. When J_H and J_0 were varied by addition of cyanide from 0 to 15 μM, the lower limit of H^+/O, set by the slope of the linear J_H vs J_0 plot, was 11.1. Similar experiments in which J_H and J_0 were varied by changing the concentration of rotenone from 0 to 0.04 μM or the temperature between 8 and 15°C gave linear plots with slopes of 10.9 and 11.0, respectively for the lower limit of n.

It is concluded that the mechanistic vectorial H^+/O ratio for the NAD-linked oxidation of glutamate + malate is at least 11 and most probably has the integral value 12.

1. Reynafarje, B., Alexandre, A., Davies, P., and Lehninger, A. L. (1982) Proc. Natl. Acad. Sci. USA 79, 7218-7222.
2. Costa, L. E., Reynafarje, B., and Lehninger, A. L. (1984) J. Biol. Chem. 259, 4802-4811.
3. Beavis, A. (1981) Fed. Proc. 40, 1563.
4. Reynafarje, B., Costa, L. E., and Lehninger, A. L. Analytical Biochem. (Submitted).

THE STOICHIOMETRY OF THE BACTERIORHODOPSIN PHOTOCYCLE AND "PROTON CYCLE" IS NOT CONSTANT: IT IS REGULATED BY $\Delta\tilde{\mu}_H^+$.

Zs. Dancshazy[+], G.I. Groma[+], and D. Oesterhelt[++]
[+]Institute of Biophysics, Biological Research Center, Hungarian Academy of Science, 6701 Szeged, Hungary.
[++] Max-Planck-Institute of Biochemistry, D-8033 Martinsried, FRG.

An important parameter of active ion pumping systems such as bacteriorhodopsin (BR) and the electron transport chains in chloroplasts and mitochondria is the stoichiometry S of the number of ions pumped per one turnover of the system. S was extensively studied in many energy coupling systems and evidence is already available that S is not constant for different states of energization of the coupling membrane. For BR it was recently proved that the rate of the photochemical cycle is controlled by $\Delta\tilde{\mu}_H^+$ which exerts a "back pressure" on the molecule (1-3). Since the measured effect of $\Delta\tilde{\mu}_H^+$ on retardation of the photocycle kinetics and on the enhanced occupancy of the M412-intermediate could not quantitatively account for the observed inhibition of proton pumping by $\Delta\tilde{\mu}_H^+$, a branching of the photocycle in a proton pumping and a non-pumping pathway was suggested (4). The ratio of the two pathways would be regulated by $\Delta\tilde{\mu}_H^+$ and analyzed as a change in the stoichiometry S.

In this study we provide experimental evidence for this control of $\Delta\tilde{\mu}_H^+$ on S by showing that the S value of BR decreases with increasing $\Delta\tilde{\mu}_H^+$. For this, the kinetics and the amplitudes of the pH-changes and the M412-intermediate decay were studied in _Halobacterium halobium_ cell envelope vesicles in 4 M NaCl at pH 7. The absorbance changes of BR and a pH-sensitive dye (pyranine, 50 μM) upon laser excitation (Nd-YAG laser, 530 nm) were determined in a flash-photolysis set up (λ of measuring light 450 nm) capable of a time resolution of 1 μsec and a sensitivity of 10^{-5} OD units. The difference in the M-decay curves between two identical samples, but with and without pyranine, was used to determine the ΔpH generated by BR. In control experiments either the uncoupler CCCP (20 μM) was added or isolated purple membranes (PM) examined under identical conditions. Increasing levels of $\Delta\tilde{\mu}_H^+$ were generated by continuous green background light illumination.

Results:

(1) The half time for rise of ΔpH correlates with the kinetics of M-intermediate formation but lags slight behind it.

(2) The decay kinetics of ΔpH, i.e. the passive back flow of protons into the vesicles is much slower that the decay of M.

(3) In PM the M-decay is slightly, but the ΔpH decay is > 10 times faster than in vesicles.

(4) CCCP changes the ΔpH kinetics of the vesicles nearly to that of PM.

(5) At increasing background light more M forms, and at the laser flash less M. Accordingly, the ΔpH generated by flash light is smaller in the presence of background light.

(6) However, the drop in the ΔpH signal is much more pronounced than the decrease in M-formation. Thus, the H^+/M ratios measured (S) at increasing $\Delta\tilde{\mu}_H$+ decrease gradually to half of the original value. The effect of decrease in S reaches saturation at ~50 mW cm^{-2}, with a half saturation value of ~10 mW cm^{-2}.

(7) The decrease in S at increasing background illuminations was already reported on PM and discussed as an effect of light induced surface charge changes (5).

(8) We propose that the drop in S in the case of vesicles is mainly caused by the transmembrane $\Delta\psi$ and/or ΔpH and that increasing $\Delta\tilde{\mu}_H$+ favours a "non-pumping" branch in the photocycle of BR, which may have a regulatory role in halobacterial bioenergetics.

References:

(1) Dancshazy,Zs., Helgerson,S.L. and Stoeckenius,W. (1983) Photobiochem. Photobiophys. 5, 347-357.

(2) Groma, G.I. et al. (1984) Biophys. J., in press.

(3) Westerhoff,H.V. et al. (1983) Biochem. Soc. Trans. 11, 90-91.

(4) Westerhoff,H.V. and Dancshazy,Zs. (1984) TIBS 9, 112-117.

(5) Kuschmitz,D. and Hess,B. (1981) Biochem. 21, 5950-5957.

THE USE OF NEUTRAL RED AS A MONITOR OF PROTON TRANSLOCATION IN PHOTO-
SYSTEM I-ENRICHED SUBCHLOROPLAST VESICLES

F.A. de Wolf, B.H. Groen, and R. Kraayenhof
Biological Laboratory, Vrije Universiteit, de Boelelaan 1087,
1081 HV Amsterdam, The Netherlands

pH-indicating probes have been widely used to monitor proton movements across various types of biological membranes. However, many of
these probes, in combination with such membrane systems, will also show
changes that are not related to pH effects. For instance, redox
changes (1) or changes related to dye binding (2) may interfere with
the pH-monitoring changes of the dye absorbance. Moreover, the location of the probe with respect to the membrane may sometimes be uncertain (see 3, 4). We used neutral red (NR) to monitor possible proton
translocation by photosystem I (PS I)-enriched subchloroplast vesicles
from spinach (5) during cyclic electron flow, mediated by ferredoxin
and NADPH. Although the observed absorbance changes might partly reflect pH changes due to proton translocation, they can also be interpreted in terms of NR binding to the energized membranes and conco-
mittant pKa-shifts of NR.

PS I vesicles were prepared according to (5) and stored under liquid nitrogen. The reaction medium contained 25 mM NaCl, 25 mM KCl,
5 mM $MgCl_2$, 250 mM sorbitol, and 1 mM TES-KOH (pH 7.2), unless stated
otherwise, and was flushed with nitrogen gas before use. Ferredoxin
and NADPH (final conc. 4 µM and 250 µM, resp.) were added just before,
and NR during experiments.

Upon addition of 40 µM (final conc.) NR, the flash-induced fast
carotenoid absorbance changes at 518 nm disappeared (Fig. 1a, c). At
first sight, this effect seems to be due to uncoupling by NR. High valinomycin concentrations (2.5 µM), in the absence of NR, affected
these absorbance changes in a similar way (not shown). However, using
NR-treated vesicles, a flash-induced fast NR absorbance decrease can
be recorded at 545 nm (Fig. 1d), whereas no carotenoid transients can
be detected at 545 nm in control vesicles without NR. Because such an
NR-absorbance decrease will also be present at 518 nm (presumably with
a larger amplitude, according to the NR absorbance spectrum), it will
partly compensate the fast carotenoid absorbance increase at this
wavelength. The fast absorbance decrease of NR presumably represents
fast reduction of NR: it is not affected by buffer in the external
medium (Fig. 1e), nor by valinomycin (2.5 µM) or nigericin (2.5 µM).
NR becomes colourless upon reduction (for instance by dithionite) and
has an E'_m of about -350 mV (1). Thus, it might act as an electron
acceptor for PS I. At 200 µM of NR, or more, the flash-induced absorbance changes at 518 and 545 nm resemble each other very much, in kinetic respect. The slow flash-induced NR absorbance increase (Fig. 1d)

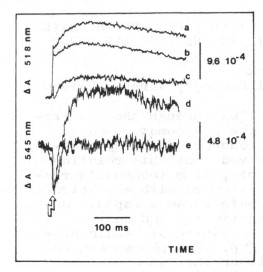

Fig. 1. Flash-induced absorbance changes at 518 and 545 nm in PS I vesicles during cyclic electron flow (cf. (5)) in the absence and presence of NR. Transients, following 16 flashes, fired at 0.125 Hz, were averaged. a: carotenoid absorbance changes at 518 nm without NR; b: as a, in the presence of 10 mM HEPES-KOH (pH 7.2); c: as a, in the presence of 40 μM NR; d: NR (40 μM) absorbance changes at 545 nm, corrected for changes occurring in the absence of NR; e: as d, in the presence of 50 mM HEPES-KOH (pH 7.2).

can be buffered away (Fig. 1e), but is also reduced by valinomycin (not shown; 50% lowering at 80 mM), indicating that it is related to the membrane potential, as is the slow carotenoid change at 518 nm. Like valinomycin, buffers not only affect the slow NR absorbance changes (Fig. 1d,e), but also the slow carotenoid absorbance changes in the absence of NR (Fig. 1a,b), which reflect membrane potential changes.

We explain these results by postulating that NR binds to the (external) membrane surface, which becomes (more) negatively charged after the flash, due to electron flow, that the pKa of NR is thereby shifted to a higher value (cf. 2) and that this causes an additional NR protonation (apparent acidification), which is reflected by the slow NR absorbance increase (Fig. 1d).

REFERENCES

(1) Prince, R.C., Linkletter, S.J.G. and Dutton, P.L. (1981) Biochim. Biophys. Acta 635, 132-148.
(2) Dell'Antone, P., Colonna, R. and Azzone, G.F. (1972) Eur. J. Biochem. 24, 566-576.
(3) Ausländer, W. and Junge, W. (1975) FEBS Lett. 59, 310-315.
(4) de Wolf, F.A., van Houte, L.P.A., Peters F.A.L.J. and Kraayenhof, R. (1984) in Advances in Photosynthesis Research, Vol. 2 (Sybesma, S., ed.), pp. 321-324, Martinus Nijhoff/Dr. W. Junk Publ., The Hague.
(5) Peters, F.A.L.J., van der Pal, R.H.M., Peters, R.L.A., Vredenberg, W.J. and Kraayenhof, R. (1984) Biochim. Biophys. Acta, in press.

RELATIONSHIP BETWEEN THE PROTONMOTIVE FORCE AND THE RATE
OF RESPIRATION OF MITOCHONDRIA IN THE RESTING AND PAR-
TIALLY UNCOUPLED STATES

Jerzy Duszyński and Lech Wojtczak
Nencki Institute of Experimental Biology, Warsaw, Poland

The relationship between the flux through the respiratory chain and the magnitude of the protonmotive force
(Δp) is a crucial point in the concept of energy coupling mechanism. It has been observed that this relationship may not be linear. For example, if mitochondria respiring in the resting state are titrated with a respiratory inhibitor, the respiration rate becomes rapidly decreased, whereas Δp changes slightly [1] and only a
large degree of inhibition of the respiratory chain produces a substantial decrease of Δp. This phenomenon has
been explained by postulating either the non-ohmic character of the proton leak [1] or existence of slips in
the proton pump [2]. According to the latter explanation
slipping of the proton pump(s), together with a passive
leak of protons through the mitochondrial membrane, is
responsible for the resting state respiration.

Both explanations of the non-linearity of the dependence between respiration and Δp are based on assumptions
that are not sufficiently documented experimentally. We
wish to propose an explanation based on a simple and
likely assumption of the heterogeneity of mitochondrial
population. For reason of simplicity let us consider
the case when a mitochondrial population is composed of
a majority of tightly coupled particles, exhibiting a
high respiratory control and high Δp value, and a small
percentage of totally uncoupled mitochondria, whose Δp
is zero but the respiratory chain is fully competent. The
resting state respiration of the tightly coupled particles depends on the proton leak, whereas the respiration
of the uncoupled mitochondria is entirely controlled by
the efficiency of the respiratory chain. In consequence,
when such heterogenous population is titrated with a respiratory inhibitor, the respiration of uncoupled mitochondria becomes strongly inhibited, whereas that of
tightly coupled particles is not affected or affected
only slightly, in accordance with different control sites
[3]. Assuming further a linear relationship between the
respiration rate and Δp in tightly coupled mitochondria
(i.e. ohmic behaviour of the proton leak), one should

expect little, if any, decrease of $\triangle$p by a small amount
of the respiratory inhibitor. Only strong inhibition of
the respiratory chain will affect the protonmotive force
of the tightly coupled portion of the total mitochondrial
population.

When mitochondria were pretreated with a small amount
of a protonophore and subsequently titrated with respiratory inhibitor, $\triangle$p was decreased and became more sensitive to changes of the respiration rate (Fig. 1). This is
interpreted as being due to a partial uncoupling of the
tightly coupled portion of mitochondrial population whose
rate of respiration becomes now less controlled by the
proton leak and more by efficiency of the respiratory
chain.

A simple kinetic model describing the correlation
between $\triangle$p and fluxes in oxidative phosphorylation is
proposed.

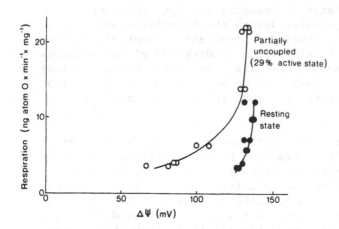

Fig. 1. Depend-
ence between the
membrane poten-
tial ($\Delta \psi$) and
the rate of
respiration.
Rat liver mito-
chondria res-
piring with
succinate were
titrated with
malonate.

References:
1. Nicholls, D.G. (1974) Eur. J. Biochem. 50, 305-315
2. Pietrobon, D., Azzone, G.F. and Walz, D. (1981)
   Eur. J. Biochem. 117, 389-394
3. Tager, J.M., Wanders, R.J.A., Groen, A.K., Kunz, W.,
   Bohnensack, R., Küster, U., Letko, G., Böhme, G., Duszyński, J. and Wojtczak, L. (1983) FEBS Lett. 151,
   1-9

THE MEMBRANE POTENTIAL OF INTACT RHODOSPIRILLUM RUBRUM CELLS IN THE
ABSENCE OF PHOTOSYNTHESIS AND RESPIRATION

C. Fenoll, S. Gómez-Amores and J. M. Ramírez
Instituto de Biología Celular, CSIC, and Facultad de Ciencias, UAM,
Madrid 6, Spain

The reversible nature of the bacterial proton-translocating ATPase
is well documented. In fermentative anaerobes, the enzyme couples ATP
hydrolysis to the outward transport of protons, thus generating the
electrochemical proton gradient required for some types of work.
Alternatively, in photosynthetic and respirative bacteria, membrane-
-linked electron transfer drives the extrusion of protons and,
subsequently, the ATPase couples the movement of protons down their
electrochemical gradient to the phosphorylation of ADP. Physiologically, the enzyme normally works just in one of the two ways.
However, both reactions can be demonstrated in the same organism if
the appropriate experimental conditions are chosen. The data of this
report indicate that intact cells of Rhodospirillum rubrum S1, which
are uncapable of fermentative growth under the culture conditions
used here, maintain a significant level of membrane potential in the
absence of both photosynthetic and respiratory electron flows. Such
residual potential is apparently due to the hydrolytic activity of
the ATPase. Together, the residual membrane potential and the
nonohmic nature of the intrinsic proton conductance of the membrane
explain an unexpected observation: low uncoupler concentrations
enhance the steady state change of membrane potential which goes
along with the aerobic to anaerobic transition of dark suspensions of
R. rubrum cells.

R. rubrum was grown in the dark under low oxygen tensions to
obtain cells that had at the same time high respiratory activities
and high levels of photosynthetic pigments. Changes of membrane potential were estimated from the electrochromic shift of the near
infrared bacteriochlorophyll band, since in that spectral range
changes due to respiratory constituents were not expected to occur.
No attempt was made to calibrate those changes in voltage units.
Routinely, wild type strain S1 was used. However, similar changes
were observed with cells of a reaction centerless mutant, what indicated that they were mostly due to antenna bacteriochlorophyll.

When cells suspended in growth medium were allowed to exhaust
oxygen by respiration in the dark, subsequent addition of phosphorylation uncouplers (CCCP or TCS) elicited a spectral change which
corresponded to a decrease of membrane potential. A similar change in
the carotenoid bands of Rhodopseudomonas capsulata had been observed

(1) and attributed to residual levels of oxygen in the cell suspension. However, that was not the case for R. rubrum because the uncoupler-elicited change was not decreased by gassing of the cell suspension with argon, or by the addition of actively respiring Escherichia coli cells or of an enzymic oxygen-scavenger system. Therefore, it appears that, in the absence of membrane-linked electron flow, R. rubrum cells maintain a residual membrane potential. ATP hydrolysis by the proton-translocating ATPase appears as the likely origin of the residual potential, since the change was significantly reduced by DCCD.

Low concentrations of uncouplers were found to enhance two to threefold the steady state decrease of membrane potential which followed oxygen exhaustion in dark cell suspensions. Such paradoxical observation may be nevertheless understood if it is assumed that the intrinsic ionic conductance of the membrane increases at high membrane potentials, as it has been shown for R. capsulata cells (2). In the presence of added protonophores, voltage independent proton channels appear in the membrane and the membrane potential versus electron flow curve approaches linearity. Given the existence of a significant membrane potential at null rates of electron transport, the mean slope of that curve may be increased by low uncoupler concentrations within a certain range of electron flow. Thus, an enhancement of the potential change which follows the aerobic to anaerobic transition may be caused by uncoupler levels which dissipate a part of the residual potential.

References

1 Clark, A. J. and Jackson, J. B. (1981) Biochem. J. 200, 389-397
2 Clark, A. J., Cotton, N. P. J. and Jackson, J. B. (1983) Eur. J. Biochem. 130, 575-580

PVC-BASED ION SELECTIVE ELECTRODES AS CHEMICAL SENSORS FOR
COMPONENTS OF A PROTON MOTIVE FORCE: AN ELECTRODE TO MEASURE ΔΨ
(INTERIOR POSITIVE) OR ΔpH (INTERIOR ALKALINE)

K.J. Hellingwerf, L.J. Grootjans and P. van Hoorn
Dept. of Microbiology, University of Groningen, Kerklaan 30.
9751 NN HAREN, The Netherlands.

Various methods are available to measure the magnitude of a
proton motive force (pmf). Of these, ion-selective electrodes as
chemical sensors of indicator probes for the magnitude of the
transmembrane electrical potential ($\Delta\psi$) and the pH gradient (ΔpH).
offer considerable advantages. This is particularly true when
simultaneous measurements of several parameters involved in
biological energy transduction are to be measured. Until now
ion-selective electrodes for K^+, TPP^+, TPB^- and NO_3^- (SCN-) have
been described as sensors of $\Delta\psi$. Only when the external medium is
unbuffered and the pH dependence of the internal buffer capacity is
known, in addition, a pH electrode can be used to measure ΔpH.

Here we describe a PVC-based anion selective electrode that can
be used to measure ΔpH (interior alkaline) and $\Delta\psi$ (interior
positive). The construction of the electrode is based on the
procedure for preparation of a TPP^+ electrode, except that the
lipophilic quarternairy ammonium cation
benzyl-dimethyl-hexadecylammonium is used to sensitize the PVC
membrane. In Fig. 1 a calibration curve of such an electrode for the
salicylate anion is given. A linear response is observed at
concentrations above 400 µM. However, a correction for the
non-linearity of the response of the electrode at lower
concentrations allows us to measure changes accurately at salicylate
concentrations ≥ 50 µM. This correction is based on a polynomal fit
of the calibration curve of the electrode and is calculated with a
Basic program, run on an Apple II microcomputer.

The electrode has a similar sensitivity towards perchlorate and
thiocyanate anions (i.e. a Z decreasing from the maximal value of
approximately 55 mV at a concentration between 1 and 0.1 mM).
However, it shows a negligable sensitivity towards phosphate,
sulphate or chloride anions.

This anion selective electrode can be used to measure a ΔpH in
systems that generate an inside alkaline pH gradient, like intact
bacteria or mitochondria. We have demonstrated this via
measurements, with salicylate as the indicator probe, of ΔpH in
Rhodopseudomonas sphaeroides. The outcome of these experiments was
compared with measurements of the same parameter, but with the

Fig. 1. Calibration curve of the anion sensitive electrode with salicylate

Fig. 2. Comparison of electrode measurements (circles) with ^{31}P NMR (squares) to measure ΔpH

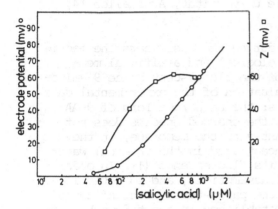

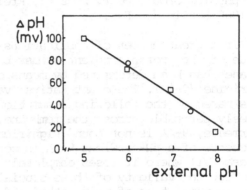

independent technique of 31P NMR (Fig. 2). The salicylate can be used in the physiologically relevant pH range from 5 to 8 and it is clear from this figure that the results of both methods show quantitative agreement. Using chromatophores from this phototrophic bacterium, we have also shown that the electrode can be used to measure Δψ (inside positive). In these experiments thiocyanate was used as a probe. The response time of the electrode towards the salicylate and thiocyanate anions decreases from more than one minute at concentrations below 50 μM to 10 s at concentrations where the Z of the electrode becomes maximal. Full experimental details of these measurements will be published elsewhere (1).

The use of this anion selective electrode allows us to measure, simultaneously with both components of the pmf, several other key parameters in energy transduction. A thermostated incubation vessel (2) with exchangeable sensors allows us to measure in addition to solute transport: redox potential; absorbance and fluorescence (via an optrode); pH; temperature and oxygen and hydrogen concentrations. Further developments can be expected from the construction of enzyme electrodes which may make it possible to construct sensors for methylammonium and amino acids. Such electrodes would make it possible to simultaneously and continuously monitor ΔpH (interior acid) and solute transport.

REFERENCES
1. Hellingwerf, K.J. and van Hoorn, P. (1984)
   J.Biochem.Biophys.Methods, submitted.
2. Lolkema, J.S., Abbing, A., Hellingwerf, K.J., and Konings, W.N.
   (1983) Eur.J.Biochem. 130, 287-292

THE "ΔpH" PROBE 9-AMINOACRIDINE: RESPONSE TIME AND METHODICAL QUERY.

Stefan Grzesiek and Norbert A. Dencher.

Biophysics Group, Dept. Physics, Freie Universität, Arnimallee 14, D-1000 Berlin 33, FRG.

In a great number of publications during the past years the magnitude of the proton gradient across biological and artificial membranes has been determined by means of the fluorescent amine 9-aminoacridine (9-AA). The quantitative evaluation of the experimental data was based on the following assumptions: the uncharged form of 9-AA is freely permeable across the membrane, the charged species does not permeate, 9-AA is not bound significantly to the membrane, in the presence of a pH-gradient 9-AA is concentrated in the internal water phase (Vi) where it loses completely its fluorescence (1). In order to test the validity of these crucial assumptions, and to determine the response time of this method and the parameters which influence it, transmembrane pH gradients were established in about 5 ms by utilizing a rapid-mixing stopped-flow spectrofluorometer, and the induced fluorescence changes of 9-AA were recorded. A variety of pure lipid and reconstituted protein-lipid vesicles were examined. Some of the experimental results obtained challenge the application of 9-AA as accurate monitor of ΔpH-changes, e.g.:

1. in the absence of a pH-gradient 9-AA fluorescence changes occurred upon energization of the system (2);

2. in certain systems, in which a large pH-gradient was present, no fluorescence quenching of 9-AA could be observed;

3. the apparent decay rate of the pH-gradient established was about ten times faster when monitored with 9-AA as compared to other pH-probes applied.

These observations and our important finding that negatively charged membrane constituents are a necessary prerequisite for the energy-dependent 9-AA fluorescence quenching do not agree with the proposed reaction mechanism (1) for this dye. We shall discuss alternative models. Furthermore, data for the dependency of the 9-AA response time on the physical and chemical state of the membrane are presented.

(1) S. Schuldiner, H. Rottenberg and M. Avron, Eur. J. Biochem. 25 (1972) 64.

(2) N.A. Dencher, Biophys. J. 41 (1983) 372a.

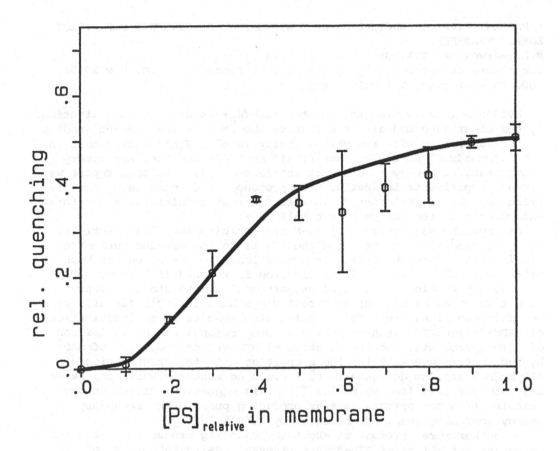

Figure 1: Dependence of the ΔpH-induced fluorescence quenching of
9-aminoacridine on the amount of negative charges in the vesicle
membrane. Vesicles of mixed lipids (diphytanoyl-phosphatidylcholine
and negatively-charged phosphatidylserine) were subjected to a pH-
jump of 2 units (pH_i = 5.7 to pH_o = 7.7) at 25 °C. 9-aminoacridine
concentration:2.5 μM, lipid concentration: 0.5 mg/ml. Mean and standard deviation of 3 - 4 measurements are shown.

DIFFERENT TYPES OF UNCOUPLING MECHANISMS PROVIDE EVIDENCE FOR DIRECT
ENERGY TRANSFER.

M.A. Herweijer, J.A. Berden and A. Kemp
Laboratory of Biochemistry, University of Amsterdam, P.O. Box 20151,
1000 HD Amsterdam, The Netherlands.

Evidence is accumulating that the bulk $\Delta\tilde{\mu}_H+$ is not always a kinetically competent intermediate in oxidative phosphorylation. Recent studies
in our laboratory with the photoaffinity label $8\text{-}N_3ATP$ showed that in
submitochondrial particles (smp's) ATP synthase and the respiratory
chain transfer energy in a direct interaction [1] . In this report we
present experiments indicating that protonophoric uncouplers uncouple
primarily this direct energy transfer, whereas gramicidin, a poreformer,
uncouples via dissipation of the bulk $\Delta\tilde{\mu}_H+$.

As shown by Wagenvoord [2] $8\text{-}N_3ATP$ inhibits the ATPase activity
of F_1 by binding covalently (in the light) to the adenine nucleotide
binding sites. The inhibition is proportional to the amount of bound
nitrenoATP (NATP), and 100% inactivation is reached if 2 moles NATP
are bound per mole F_1. The same proportionality and stoeichiometry
were found upon binding of NATP to ATP synthase in smp's (details will
be published elsewhere). This binding also results in an inactivation
of NADH-driven ATP synthesis which is proportional to the inhibition
of ATP hydrolysis. The inactivation of ATP-driven reduction of NAD^+
by succinate (reversal) is also proportional to the inactivation of
ATP hydrolysis; no decrease in $\Delta\tilde{\mu}_H+$ could be measured while the reversal rate was inhibited up to 60% [1]. This suggests a direct energy
transfer from the primary, energy producing pump to the secundary,
energy consuming pump, not mediated by a bulk $\Delta\tilde{\mu}_H+$.

To gather more information about the coupling process we performed
uncoupler titrations of ATP-driven reversal, using inhibitors of ATP
synthase and NADH dehydrogenase. Both with oligomycin and rotenone the
amount of S13 necessary for full uncoupling was lowered as compared
with the control particles. (Fig. 1). In the case af oligomycin, an
inhibitor of the primary proton pump, this could possibly be explained
by a decrease of $\Delta\tilde{\mu}_H+$. With rotenone however, which inhibits the secundary pump, no decrease in $\Delta\tilde{\mu}_H+$ is expected, according to the Mitchell
theory [3]. Apparently the amount of uncoupler needed is related to
the number of energy transfer reactions that can take place, independent of the kind of inhibitor used. With SF6847, FCCP and valinomycin
+ nigericin the same results are obtained as with S13.

Another indication for a localized action of these uncouplers is
our observation that the uncoupler-induced stimulation of ATP hydrolysis is independent of the level of inactivation of the ATP synthase by
bound NATP. Baum [4] has reported similar results using oligomycin as
ATP synthase inhibitor. If the feed-back pressure of a delocalized $\Delta\tilde{\mu}_H+$
is the force that prevents the maximal ATP hydrolysis turnover, it

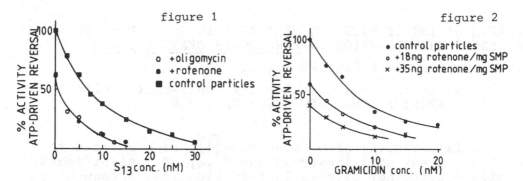

figure 1

figure 2

Figure 1. Titration of ATP-driven reversal in smp's with S13. The effect of inhibition of NADH dehydrogenase and that of inhibition of ATP synthase is shown.
Figure 2. Titration of ATP-driven reversal in smp's with gramicidin. The effect of inhibition of NADH dehydrogenase with various concentrations of rotenone is shown.

should be expected that lowering of the number of active enzyme molecules would release this feed-back inhibition. Clearly this is not the case. In contrast to the direct interactions between the former group of uncouplers and energy-transfer reactions, no change in the amount necessary for uncoupling of reversal is found when gramicidin is used. (Fig. 2). This is at variance with the results of Hitchens and Kell [5] who did not find this difference between gramicidin and other uncouplers in uncoupling photophosphorylation in chromatophores. Gramicidin apparently dissipates the bulk $\Delta\tilde{\mu}_H+$ and can not interact directly with the protons during energy transfer. We suggest that, unlike the other uncouplers, gramicidin has no special affinity for protons, but acts as a passive proton leak in the membrane.

REFERENCES

1. Herweijer, M.A., Berden, J.A. and Kemp, A. (1984) Biochem. Soc. Trans., in the press.
2. Wagenvoord, R.J., Van der Kraan, I. and Kemp, A. (1979) Biochem. Biophys. Acta 548, 85-95
3. Mitchell, P. (1966) Biol. Rev. 14, 455-502
4. Baum, H., Hall, G.S. and Nalder, J. (1971) In: Energy Transduction in Respiration and Photosynthesis (Quagliariello, E. et al., eds.) pp. 747-755, Adriatica Editrice, Bari.
5. Hitchens, G.D. and Kell, D.B. (1983) Biochem. Biophys. Acta 723 308-316

EFFECTS OF THIOCYANATE AND VENTURICIDIN ON RESPIRATION-DRIVEN H+ TRANSLOCATION IN PARACOCCUS DENITRIFICANS

G.Duncan Hitchens & Douglas B.Kell

Department of Botany & Microbiology, University College of Wales, ABERYSTWYTH, Dyfed SY23 3DA, U.K.

1. A fast-responding O_2 electrode[1] has been used to confirm and extend [2] observations [3,4] of a significant kinetic discrepancy between O_2 reduction and consequent H^+ translocation in "O_2-pulse" experiments in intact cells of P.denitrificans. The chaotropic SCN^- ion abolishes this discrepancy, and greatly increases the observable $\rightarrow H^+$/O ratio, to a value approaching its accepted, true, limiting stoichiometry. The observable H^+ decay rates are very slow, particularly in the absence of SCN^-.

2. The submaximal $\rightarrow H^+$/O ratios observed in the absence of SCN^- are essentially independent of the size of the O_2 pulse when this is varied between 4.7 and 47 ng atom, in a manner not easily explained by a delocalised chemiosmotic energy coupling scheme.

3. Osmotically active protoplasts of P. denitrificans do not show a significant kinetic discrepancy between O_2 reduction and H^+ ejection, even in the absence of SCN^-. However, the submaximal $\rightarrow H^+$/O ratios observed in the absence of SCN^- are again essentially independent of the size of the O_2 pulse. As in intact cells, the observable H^+ decay rates are extremely slow.

4. The energy transfer inhibitor venturicidin causes a significant increase in the $\rightarrow H^+$/O ratio observed in P. denitrificans protoplasts in the absence of SCN^-; the decay kinetics are also somewhat modified. Nevertheless, the $\rightarrow H^+$/O ratio observed in the presence of venturicidin is also independent of the size of the O_2 pulse in the above range. This observation militates further against arguments in which (a) a non-ohmic backflow ("leak") of H^+ from the bulk aqueous phase might alone be the cause of the low $\rightarrow H^+$/O ratios observed in the absence of SCN^-, and (b) in which there might be a Δp-dependent change ("redox slip") in the actual $\rightarrow H^+$/O ratio.

5. It is concluded that the observable protonmotive

activity of the respiratory chain of <u>P. denitrificans</u>
in the absence of SCN⁻ is <u>directly</u> influenced by the
state of the H⁺-ATP synthase in the cytoplasmic membrane
of this organism. We are unable to explain the data in
terms of a model in which the putative protonmotive force
may be acting to affect the →H⁺/O ratio.

6. One possibility, which would conveniently serve to
explain these and other [5] data, is that the bulk-to-
bulk phase membrane potential set up in response to protonmotive activity is energetically insignificant. Since
the apparent membrane potential, as judged by steady-state
ion uptake measurements, is insensitive to respiration
rate over a wide range [6], one should predict that the
<u>kinetics</u> of ion uptake (in a chemiosmotic model) would
<u>be</u> similarly insensitive to respiration rate. Such an
experimental test might allow one to distinguish the
veracity of "localised" [7] and "delocalised" energy
coupling models in electron transport phosphorylation [8].

1) Reynafarje,B. et al (1982) PNAS.79,7218-7222
2) Hitchens, G.D. & Kell, D.B. (1984) BBA, submitted
3) Scholes,P. & Mitchell,P.(1970) J.Bioenerg. 1,309-323
4) Kell,D.B. & Hitchens,G.D.(1982) Faraday Disc.Chem.
   Soc. 74, 377-388
5) Kell,D.B. & Hitchens,G.D.(1983) in Coherent Excitations
   in Biological Systems (H.Fröhlich & F.Kremer, eds), pp.
   178-198. Springer, Heidelberg.
6) Kell,D.B. et al (1978) Biochem.Soc.Trans. 6, 1292-1295
7) Westerhoff, H.V. et al (1984) FEBS Lett. 165,1-5
8) Kell,D.B. & Westerhoff,H.V.(1984) in Catalytic Facili-
   tation in Organised Multi-enzyme Systems (Welch,G.R.,
   ed), in press. Academic, New York.

FLASH-INDUCED ATP SYNTHESIS IN PEA CHLOROPLASTS AND OPEN CELLS OF CHLAMYDOMONAS REINHARDII CW-15

C. LEMAIRE, J.M. GALMICHE and G. GIRAULT
SERVICE DE BIOPHYSIQUE, CENTRE D'ETUDES NUCLEAIRES DE SACLAY
91191 GIF-sur-YVETTE Cédex FRANCE

Chloroplasts prepared from preilluminated pea seedlings leaves have an activated ATP synthase and hydrolyse ATP in the dark [1-2]. There is always a pH difference between the inside and the outside of the thylakoids. By giving series of short flashes ($tl_{/2}$ = 2 μs) or flash-groups at a low frequency, 0.1 Hz, we observe an increase of the value of this ΔpH. ATP synthesis is determined by recording the luminescence of the luciferin-luciferase system and proton efflux through the ATP synthase from the kinetics of the dark decay of the absorbance changes at 515 nm, ΔA_{515}. For a low value of the ΔpH, the yield of ATP synthesis is low after the first flash and increases concomitantly with the ΔpH following the number of flashes already fired. Above a certain ΔpH limit, 1.4 - 1.5, the yield remains constant [Fig. 1]. Nevertheless the ratio H^+/ATP is constant whatever is the number of flashes fired or the yield of ATP synthetized, provided that observations are done with the same chloroplast preparation. Indeed the ratio H^+/ATP varies from one to an other chloroplast preparation when they are submitted to the same experimental conditions. Observed values of H^+/ATP range from 7 to 3.4 after single flashes, 5 to 3.7 after groups of 2 flashes, 10 ms apart, 4 to 2 after group of 6 flashes, 3 ms apart. Variations in the ratio H^+/ATP can be explained by the way ATP synthase is operating. The new synthetized ATP molecules remain bound to the enzyme and have variable probabilities to be hydrolyzed or released from the protein, following the level of the membrane energization. A value of 2 is the most probable for the ratio H^+/ATP.

Cells of <u>Chlamydomonas reinhardii</u> CW 15 were grown photohetero-phycally and harvested in late logarithmic phase (approximately 5.10^6 cells/ml). The cells are then opened in hypotonic medium by disrupting at low pressure (5 bars) in a Yeda press. This treatment allows to get good photophospho-rylation activities. Conditions to obtain the more durable an higher activities are determined in continuous light. The ATP synthesis induced by illumination with a group of 6 flashes (3 ms apart, fired at 0.8 Hz) is also followed. Yield of 0.2 ATP synthetized per 1 000 chlorophylls and flash group is observed after a lag time. ΔpH and proton efflux through the ATP synthase are compared with the rate of ATP synthesis.

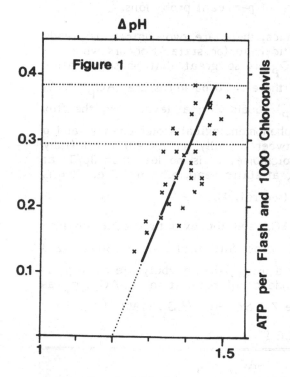

Figure 1

ATP synthetized/flash as a function of the pH across the membrane. Chloroplasts equivalent to 40 µg chorophyll in 2 ml. For ΔpH measurement atebrin (2.5 µM) fluorescence is recorded. Single saturating flashes ($tl_{/2}$=2µs) or groups of 2 flashes (5 to 5 000 ms apart) are fired at 0.1Hz.

1. GIRAULT G, GALMICHE J.M. (1978) Biochim. Biophys.Acta 502, 430-444.
2. GALMICHE J.M., GIRAULT G. (1982) FEBS Lett. 146, 123-128.

ATP STOICHIOMETRIES OF MITOCHONDRIAL OXIDATIVE PHOSPHORYLATION EVALUATED BY NONEQUILIBRIUM THERMODYNAMICS

John J. Lemasters
Laboratories for Cell Biology, Department of Anatomy, University of North Carolina at Chapel Hill, Chapel Hill, NC 27514 USA.

Long established values for the ATP stoichiometries of oxidative phosphorylation have been called into question. Within the last five years, various chemiosmotic schemes have placed ATP/O between $1^1/3$ and 2 for FAD-linked substrates and between 2 and 3 for NAD-linked substrates. ATP/site ratios range from 2/3 to 1 for Site 1, $\frac{1}{2}$ to 1 for Site 2, and 1 to $1\frac{1}{2}$ for Site 3. A goal of this laboratory has been to analyze the overall reaction from unequivocal steady state determinations of products and reactants using the theory of linear nonequilibrium thermodynamics. This approach avoids the difficulties of transient H^+/e^- measurements and the uncertain correction of $\Delta\bar{\mu}_{H^+}$ calculations for nonspecific binding of permeant probe ions.

In linear nonequilibrium thermodynamics, there are two contrasting steady states, static head and level flow. Static head (or state 4) occurs when the back pressure of the output force (ΔG_P) is so great that phosphorylation ceases (Jp=0). Level flow occurs when there is no backpressure (ΔG_P=0). At static head, the force ratio ($-\Delta G_R/\Delta G_P$) equals Z/q; at level flow, the flux ratio (Jp/Jr) equals qZ where Z is the phenomenological stoichiometry and q is the degree of coupling. q varies between 0 (fully uncoupled) and 1 (fully coupled). Since n, the mechanistic stoichiometry, is no less than Jp/Jr at level flow and no more than $-\Delta G_R/\Delta G_P$ at static head, $qZ \leq n \leq Z/q$. If q is close to 1, Z is virtually identical to n (see ref. 1).

At static head, extramitochondrial ΔG_P and the oxidation-reduction free energy changes (ΔG_R) across Sites 1+2, across Site 3, and across Sites 1+2+3 were determined in rat liver mitochondria using 3-hydroxybutyrate as respiratory substrate (2). Based on the dependence of respiration on ΔG_P, q was 0.977. This value was used to calculate Z from $-\Delta G_R/\Delta G_P$ (Table I):

TABLE I

	ΔG_R (kcal/mol)	$-\Delta G_R/\Delta G_P$	Z
Sites 1+2	−26.5	1.80	1.76
Site 3	−22.9	1.56	1.52
Sites 1+2+3	−49.4	3.37	3.28

Z was very close to $1\frac{1}{4}$, $1\frac{1}{2}$ and $3\frac{1}{4}$, respectively, for Sites 1+2, Site 3 and Sites 1+2+3.

For oxidative phosphorylation, level flow is a hypothetical condition, since the standard free energy of ATP synthesis ($\Delta G_p°'$) is so great. ATP/O flux ratios are typically determined from ADP-induced jumps in oxygen uptake (3). In an oxygen jump, it can be shown that ADP/extra O = Z/q. Calibrating ADP and O with the same enzyme assay, ADP/total O and ADP/extra O were determined from the inverse slopes of ADP versus oxygen plots for four different respiratory substrates (4). q was determined independently for each substrate, and Z was calculated (Table II):

TABLE II

Substrate	ADP/Total O	ADP/Extra O	q	Z
Succinate	1.71	2.03	0.981	1.99 $\pm$ .04*
Glutamate/malate	2.71	3.04	0.985	2.99 $\pm$ .07
3-Hydroxybutyrate	2.61	3.23	0.978	3.16 $\pm$ .10
2-Oxoglutarate	3.45	4.15	0.980	4.07 $\pm$ .16

*95% confidence interval

The results from Tables I and II support ideal ATP/O stoichiometries of 2 for succinate, 3 for glutamate/malate, $3\frac{1}{4}$ for 3-hydroxybutyrate, and 4 for 2-oxoglutarate. The lower stoichiometry for glutamate/malate as compared to 3-hydroxybutyrate is accounted for by proton translocation linked to glutamate/aspartate exchange (5). These findings suggest a new thirteen proton scheme for chemiosmotic coupling in which H^+/ATPase is 3, H^+/translocation of ATP, ADP and Pi is 1, H^+/site ratios are 5, 4, and 4, and ATP/site ratios are $1\frac{1}{4}$, $\frac{1}{2}$, and $1\frac{1}{2}$, respectively, for Sites 1, 2 and 3. Supported by GM 28999, AM 30874 and AHA 82-163.

1. Rottenberg, H. (1979) Biochim. Biophys. Acta 549, 225-253.
2. Lemasters, J.J., Grunwald, R., and Emaus, R.K. (1984) J. Biol. Chem. 259, 3058-3063.
3. Chance, B., and Williams, G.R. (1955) J. Biol. Chem. 217, 383-393.
4. Lemasters, J.J. (1984). Submitted for publication.
5. LaNoue, K.F., and Tischler, M.E. (1974) J. Biol. Chem. 249, 7522-7528.

BACKLASH DURING THE FORMATION OF THE ELECTROCHEMICAL PROTON GRADIENT

J.F. Myatt, M.A. Taylor and J.B. Jackson

Department of Biochemistry, University of Birmingham, P.O. Box 363, Birmingham B15 2TT, U.K.

Disagreement has arisen over the role of protons in coupling electron transport to ATP synthesis in energy transducing membranes. The chemiosmotic model predicts that on the time scale of ATP synthesis protons are delocalized within the aqueous bulk phases on either side of the membrane [1]. Alternatively protons may remain localized in a domain distinct from the bulk phases and act as electrochemical intermediates within independent coupling units [2]. This implies a barrier, or resistance, to the free diffusion of protons between independent coupling units and bulk phases of the system.

In chromatophores from Rps. capsulata electron transport, $\Delta\psi$ generation and changes in $[H^+]_{out}$ were measured spectrophotometrically at the onset of illumination. The change in redox state of P870, the development of $\Delta\psi$ and the decrease in $[H^+]_{out}$ followed similar kinetics over the first 100 ms. There was no lag between H^+ disappearance and the changing redox state of P870 suggesting that at least on the outer face of the chromatophore membrane there is no diffusion barrier to protons.

The H^+ disappearance had the form of a burst followed by a much slower uptake. The burst may represent the "backlash" period predicted by the chemiosmotic hypothesis [1]. Three factors might contribute to backlash: (A) non rate-limiting segments of electron transport (e.g. the photosynthetic reaction centre) may proceed rapidly by way of donor and acceptor pools before the steady-state rate of cyclic electron transport is reached. Experiments with antimycin showed that this effect does make a contribution to the burst. (B) During the early turnovers, before the electrical capacitance of the membrane is fully charged, electron transport may be unlimited by back pressure from $\Delta\psi$ and may proceed at a rapid rate. This may not be significant in chromatophores because FCCP or combinations of nigericin/valinomycin/K^+ were found to accelerate the electron flow rate even within a few ms of illumination. (C) As $\Delta\psi$ develops, the rate of H^+ efflux increases and eventually equals the rate of H^+ uptake. Because of the non-ohmic dependence of H^+ efflux on $\Delta\psi$ [3] this effect might be exaggerated. This was supported by a series of experiments in the presence of excess valinomycin and limited concentrations of K^+. The key features in the $\Delta\psi$ kinetics can be explained thus: there was a very rapid rise in $\Delta\psi$ resulting from electrogenic electron transport through the reaction centre. This was followed by a lag period resulting from rapid electrophoretic K^+ efflux and then, as electrochemical equilibrium of K^+ was approached, $\Delta\psi$ generation re-commenced. Significantly, the change in $[H^+]_{out}$ was accelerated, compared to the control, during the period in which $\Delta\psi$ was at a low value. This could have been due to both an increase in the rate of electron transport at low $\Delta\psi$ or a decrease in the passive rate of H^+ efflux. The importance of the latter was demonstrated in experiments carried out under similar conditions but using trains of 4 µs flashes fired at 1.0 Hz. At this frequency, under anaerobic conditions, electron transport is not limiting and so the passive processes dominate. A similar relationship between H^+ uptake and $\Delta\psi$ generation was observed to that in continuous light.

Intact cells of Rps. capsulata in the absence of permeant ions ejected H^+ in response to either oxygenation or illumination. With small oxygen or light pulses, H^+ ejection was faster than the response time of the glass electrode system. Re-entry of H^+ was considerably slower (half time of minutes). The decay of $\Delta\psi$ after a pulse was more rapid than the proton re-entry. In some conditions, in agreement with experiments on other organisms by other workers, the $H^+/2e$ ratio did not increase when the size of the oxygen pulse was decreased [4,5]. Addition of venturicidin, which restricts proton translocation through the F_O component of the ATP synthase, resulted in an increase in the extent of H^+ ejection during an oxygen pulse, which implies that in the absence of such an inhibitor, protons re-enter the cell without being detected by the glass electrode. These results will be discussed with reference to the results obtained with the simpler chromatophore system.

[1] Mitchell, P. (1966) Biol. Rev. 41, 445.

[2] Westerhoff, H.V., Simonetti, A.L.M. and Van Dam, K. (1981) Biochem. J. 200, 193–202.

[3] Clark, A.J., Cotton, N.P.J. and Jackson, J.B. (1983) Biochim. Biophys. Acta 723, 440–453.

[4] Gould, J.M. and Cramer, W.A. (1977) J. Biol. Chem. 5875–5882.

[5] Kell, D.B. and Hitchens, G.D. (1982) Faraday Discuss. Chem. Soc. 74, 377–388.

INTERNAL pH CHANGES AND ION MOVEMENTS IN CYTOCHROME c OXIDASE-
CONTAINING PROTEOLIPOSOMES. P. Nicholls, E. Verghis, and A.P. Singh.
Dept. Biol. Sciences, Brock University, St. Catharines, Ont. L2S 3A1
Canada.

Proteoliposomes were formed by cosonication of phospholipids
(Sigma Type IV lecithin, 'asolectin') and purified beef heart cytochrome c oxidase (1). In some cases, the proteoliposomes were loaded
internally with phenol red by carrying out the sonication in the presence of 1mM indicator, subsequently separating untrapped phenol red
(external) by Sephadex chromatography (2,3). Phenol red-loaded
proteoliposomes have almost the same respiratory control as ordinary
proteoliposomes (RCI > 6) and can be shown to produce protons upon
pulsing with oxygen in an anaerobic system or with ferrocytochrome c
in an aerobic system. At least 90% of the trapped phenol red remains
inside the proteoliposomes for 12 hours at 4°C, as determined by the
dependence on FCCP and valinomycin (or on nigericin) for its titration by externally added alkali.

On addition of ascorbate and cytochrome c to an aerobic suspension
of phenol red-containing vesicles in potassium phosphate buffer, the
indicator is seen to monitor a progressive increase in alkalinity.
This reaches an aerobic steady state ΔpH of between 0.3 and 0.4 pH
units, and collapses slowly upon anaerobiosis. Such internal alkalinization implies the movement of a compensating ion. Potassium
electrode measurements (Fig. 1) identify this ion as potassium,
which enters the proteoliposome during respiration presumably under
the influence of the membrane potential created by electron flow. In
choline or sodium-containing media the internal alkalinization is
considerably smaller. It is also discharged by nigericin addition,
which converts ΔpH into $\Delta\psi$.

In the presence of valinomycin, the rate of internal alkalinization is increased, approximating to the rate of respiration. But
the steady state ΔpH is smaller than in the absence of valinomycin,
and the rate of ΔpH collapse upon anaerobiosis is greater. In the
presence of valinomycin, $\Delta\psi$ is the K^+ Donnan potential and equal to
zero at high potassium levels. Nevertheless, valinomycin addition
does not release the enzyme from steady state respiratory control.
Effective rate control of respiration can thus by achieved by a ΔpH
of no more than 0.3 units (equivalent to a p.m.f. of less than 20mV),
and is thus kinetic rather than thermodynamic in nature.

The system respiring in the absence of valinomycin presents a
number of additional problems of control and function. Firstly, the
higher ΔpH in the absence of valinomycin suggests that the latter
increases the apparent proton permeability; in the absence of valinomycin, protons reenter the liposomal interior more slowly, despite
the presence of a favourable ('driving') $\Delta\psi$. Secondly, in steady

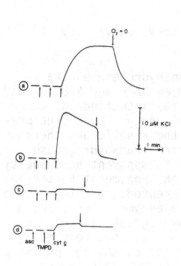

Fig. 1. Energized uptake of K^+ by vesicles in absence (a) and presence (b) of valinomycin. Inhibition by FCCP (c) or by nigericin (d). Proteoliposomes prepared in 50 mM Na HEPES pH 7.4. Medium contained 50mM Na HEPES, 6mM ascorbate, 0.2mM TMPD, 6 μM cyt. c, 40 μM KCl; vesicles with 0.3 μM cyt. aa_3 and 3 mg. ml^{-1} phospholipid, at 30°C. Radiometer (Ruzicka) type K^+ electrode.

state the net entry rate of potassium into the vesicle interior must be zero, despite the permeability indicated by its initial influx and the continued presence of a favourable $\Delta\psi$ across the membrane.

Alternative explanations of these findings will be presented involving either the formation of localized rather than global membrane potentials in the liposomes or the action of cytochrome c oxidase as a cation pump as originally proposed by Fry and Green (4).

References

1. Proteau, G., Wrigglesworth, J.M. & NIcholls, P. (1983) Biochem. J. 210, 199-205.
2. Wrigglesworth, J.M. (1978) in Membrane Proteins, Proc. 11th. FEBS meeting, Vol. 45 (Nicholls, P., Møller, J.V., Jørgensen, P.L. & Moody, A.J., eds.) pp. 95-103, Pergamon Press, Oxford.
3. Wrigglesworth, J.M. & Nicholls, P. (1979) Biochim. Biophys. Acta 547, 36-46.
4. Fry, M. & Green, D.E. (1980) Biochem.Biophys. Res. Comm. 95, 1529-1535.

(Research supported by Canadian NSERC grant A0412).

PROPERTIES OF MITOCHONDRIAL PROTON-TRANSLOCATING NICOTINAMIDE
NUCLEOTIDE TRANSHYDROGENASE FROM BEEF HEART PURIFIED BY FAST PROTEIN
LIQUID CHROMATOGRAPHY

B. Persson and J. Rydström
Department of Biochemistry, Arrhenius Laboratory, University of
Stockholm, S-106 91 Stockholm, Sweden

Mitochondrial nicotinamide nucleotide transhydrogenase catalyzes the reversible transfer of hydrogen between NADH and $NADP^+$ and
the simultaneous transfer of protons across the mitochondrial inner
membrane (1). In the presence of an external energy source, the prevailing proton motive force influences both the kinetic and thermodynamic properties of the enzyme. Mitochondrial transhydrogenase
from beef heart has previously been purified to apparent homogeneity
(2-6). However, during attempts to sequence the enzyme it became apparent that all available methods were too irreproducible to allow a
continous production of large amounts of pure enzyme. Therefore, a
new procedure using fast protein liquid chromatography (FPLC) has
been developed. In principle, an earlier method (2) has been modified to include a high-resolution anion-exchange chromatography step
using a Mono Q HR 10/10 column (Pharmacia Fine Chemicals, Uppsala,
Sweden). The final preparation is close to homogeneous as judged by
silver staining, a major improvement as compared with previous preparations, and is very reproducible. In addition, transhydrogenase
purified by FPLC is reconstitutively active, i.e., when incorporated
in liposomes and in the presence of NAD^+ plus NADPH it catalyzes a
transmembrane proton translocation and the generation of a pH gradient, as monitored by quenching of 9-aminoacridine (9-AA) fluorescence (Fig. 1). It is interesting to note that earlier problems in
demonstrating proton pumping apparently were due to the presence of
residual Triton X-100. An important step in the modified procedure
is therefore to wash the preparation with 0.5% sodium cholate prior
to reconstitution. Washing with 0.05% sodium chelate is not sufficient to achieve fully active vesicles (cf. Fig. 1). In addition,
the pure transhydrogenase isolated by the FPLC procedure is relatively inactive unless it is activated by a cholate-phospholipid
suspension.

The new purification procedure can be used in a somewhat modified form to produce large amounts of transhydrogenase suitable for
sequencing and other protein chemical studies. In this case the same
degree of purity is obtained although the specific activity is lower, and the enzyme is not reconstitutively active.

Fig. 1. Proton-pumping activity
of purified transhydro-
genase washed with 0.05%
(A) and 0.5% (B) sodium
cholate prior to recon-
stitution, assayed by
quenching of 9-AA.

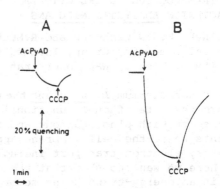

Reconstituted transhydrogenase purified by the present method
also quenches 9-amino-6-chloro-2-methoxy-acridine (ACMA), a compound
that earlier has been assumed to probe a pH gradient. However, in
contrast to the quenching of 9-AA fluorescence, which is markedly
enhanced by valinomycin and the inclusion of potassium sulfate in
the vesicles, ACMA fluorescence is virtually abolished by various
ionophores such as valinomycin and nigericin. These results suggest
that 9-AA probes a pH gradient whereas ACMA more probes a membrane
or surface potential, and that reconstituted transhydrogenase indeed
is a proton pump. Experiments with reconstituted transhydrogenase
carried out under conditions which allow a build-up of a pH gradient
show that transhydrogenase appears to be preferentially regulated by
the internal pH in the liposomes rather than a membrane potential.
(Supported by the Swedish Natural Science Research Council).

References
1. Rydström, J. (1977) Biochim. Biophys. Acta 463, 155-184.
2. Höjeberg, B. and Rydström, J. (1977) Biochem. Biophys. Res.
   Commun. 78, 1183-1190.
3. Andersson, W.M. and Fisher, R.R. (1978) Arch. Biochem. Biophys.
   187, 180-190.
4. Andersson, W.M., Fowler, W.T., Pennington, R.M. and Fisher, R.R.
   (1981) J. Biol. Chem. 256, 1888-1895.
5. Wu, L.N.Y., Pennington, R.M., Everett, T.D. and Fisher, R.R.
   (1982) J. Biol. Chem. 257, 4052-4055.
6. Rydström, J. (1981) in Mitochondria and Microsomes (Lee, C.P.,
   Schatz, G. and Dallner, G., eds.) pp. 317-335, Addison-Wesley
   Publ. Co. Reading, MA, USA.

ENERGY TRANSDUCTION IN THE CYTOPLASMIC MEMBRANE OF INTACT CELLS OF THE CYANOBACTERIUM ANACYSTIS NIDULANS

G.A.Peschek, W.H.Nitschmann and R.Muchl
Biophysical Chemistry Group, Institute of Physical Chemistry,
University of Vienna, Währingerstraße 42, A-1090 Vienna, Austria

In the cyanobacterium A. nidulans the thylakoid (intracytoplasmic) membranes are the site of dual functional respiratory/photosynthetic electron transport with plastoquinone and the cyt f/b-6 complex as common components while the cell membrane appears to be an additional site of respiratory electron transport including a H^+-translocating cyt oxidase (1-3). Here we want to show that the cell membrane of intact Anacystis indeed is an energy-transducing membrane in terms of two fundamental observations:

1. Oxygen pulses to dark anaerobic cells effected H^+ extrusion similar to that observed when intact spheroplasts were pulsed with reduced cyt c (1-3; Fig.1). H^+ extrusion from O_2-pulsed Anabaena had been shown previously but details were not reported (4). H^+-extrusion from our O_2-pulsed Anacystis was only partly DCCD sensitive. Therefore part of the extruded H^+ may stem from an ATP-powered reversible ATPase (DCCD sensitive) while another part (DCCD insensitive; maximum $H^+/O = 2$) originates directly through respiratory electron transport in the cell membrane. The amount of DCCD used (3.5 nmol/µl packed cells) totally inhibited any oxidative phosphorylation. The proton motive force across the cell membrane always was more negative aerobically than anaerobically, with only little influence of DCCD (Fig.2); this was mainly the result of a DCCD insensitive hyperpolarization of $\Delta\psi$ in respiring cells which, however, was greatly diminished by 50 µM CCCP (not shown).

2. In dark anaerobic Anacystis subjected to acid jumps (e.g. pH_o $9 \rightarrow 4.2$) or a valinomycin pulse (10 µM) immediate, and transient, net synthesis of ATP was observed (Fig.3). Pulse-induced ATP formation was abolished by DCCD and CCCP; vanadate was without effect. Thus it was concluded that the cell membrane of Anacystis contains a reversible H^+-ATPase which can give rise for ATP synthesis at the expense of an artificial proton motive force applied to the membrane (5,6).

REFERENCES

1. G.A.Peschek (1983) Subcell.Biochem. 10, 83-189.
2. G.A.Peschek (1983) J.Bacteriol. 153, 539-542.
3. G.A.Peschek (1984) Plant Physiol. (in the press).
4. P.Scholes, P.Mitchell & J.Moyle (1969) Eur.J.Biochem. 8, 450-454.
5. R.Muchl & G.A.Peschek (1983) FEBS Lett. 164, 116-120.
6. R.Muchl, G.Schmetterer, W.H.Nitschmann & G.A.Peschek (1983) Naturwissenschaften 70, 615.

Fig.1 H$^+$ extrusion (left)
and O$_2$ uptake (right) by
whole Anacystis cells with
and without DCCD (3.5 nmol
per µl packed cells). 35°C.
Medium: 40 mM Good buffers,
150 mM KCl, 10 µM valino-
mycin. 50 µM CCCP abolished any H$^+$ extrusion while
1 mM vanadate was without
effect.
(x-x): Difference $\pm$ DCCD.

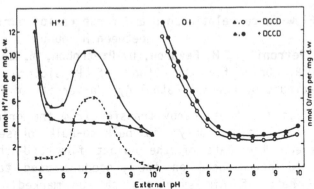

Fig.2 Proton motive force ($\Delta\mu_{H^+} = \Delta\psi -$
$-60.\Delta$pH) at different pH$_o$, aerobically and anaerobically, as determined by
flow dialysis using standard distribution techniques (6). Aerobic pmf's were
only little affected by DCCD at concentrations which abolished any net oxidative phosphorylation (cf.Fig.1) while
anaerobic pmf's were shifted to considerably more positive values. Control cells remained $>$ 90% viable
after 5 hrs incubation at pH 3.2 and
11.0 (35°C; results not shown).

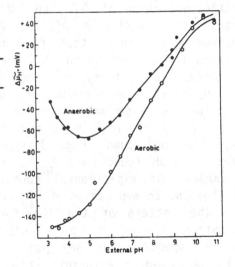

Fig.3 Kinetics of adenylate levels
in dark anaerobic Anacystis subjected
to a 10 µM valinomycin pulse in 40 mM
tris/hepes buffer, pH 9; ATP yields
were lower at lower pH$_o$; external K$^+$
(150 mM) virtually annulled any Val-
induced ATP formation. EDTA-treated
cells (10 mM Na$_2$EDTA, 5 hrs, pH 6.9)
gave the same results as untreated
cells. Net ATP synthesis was also
observed upon acid jumps applied to
intact cells (5,6).

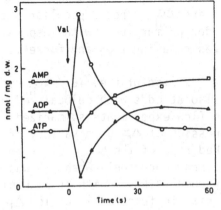

Flow-force relationship and force ratios during energy coupling between H$^+$ pumps

V. Petronilli, M. Favaron, D. Pietrobon, M. Zoratti and G.F. Azzone
C.N.R. Unit for the Study of Physiology of Mitochondria and
Institute of General Pathology, University of Padova, Italy

If the free energy transfer from the primary to the secondary H$^+$ pump occurs only via bulk-to-bulk phase $\Delta\tilde{\mu}_H$, the relation between the rate of the output flow and the magnitude of $\Delta\tilde{\mu}_H$ should not depend on the procedure used to vary $\Delta\tilde{\mu}_H$. However the rate of ATP synthesis can be markedly depressed with very little variation of $\Delta\tilde{\mu}_H$ and $\Delta\tilde{\mu}_H$ can be extensively depressed with little variation of the rate of ATP synthesis (1). The dependence of the output flow of the secondary pump has been reinvestigated by using in various combinations either redox or ATPase as primary H$^+$ pumps and either redox or ATPase or transhydrogenase as secondary H$^+$ pumps.

In the experiment of Fig. 1 the ATPase was the primary (or $\Delta\tilde{\mu}_H$-generating) pump and complex I was the secondary (or $\Delta\tilde{\mu}_H$-consuming) pump. Fig. 1 shows the relationship between the rate of NADH reduction, J_{NADH}, and the dimension of $\Delta\tilde{\mu}_H$ under a variety of experimental conditions, specified in the legend, all leading to depression of $\Delta\tilde{\mu}_H$ and of J_{NADH}.

The pattern of the flow-force relation during the coupled operation of two H$^+$ pumps has the following features: **i)** it is not specific for any particular H$^+$ pump as secondary pump; **ii)** the steep relationship between output flow of the secondary pump and $\Delta\tilde{\mu}_H$ is not a peculiar pattern obtained only when titrating a redox primary pump with respiratory inhibitors; **iii)** very similar curves relating flow and force may be obtained in different $\Delta\tilde{\mu}_H$ ranges.

The rise of the $\Delta G_p/\Delta\tilde{\mu}_H$ ratio as $\Delta\tilde{\mu}_H$ is decreased (see e.g.2), is also at odds with delocalized $\Delta\tilde{\mu}_H$ coupling. In the reversed electron flow experiments summarized in Fig. 2 the output force is ΔG_{ox}. Depression of $\Delta\tilde{\mu}_H$ by a variety of means (see Fig. 2) resulted in a marked rise of the $\Delta G_{ox}/\Delta\tilde{\mu}_H$ ratio as $\Delta\tilde{\mu}_H$ was lowered. The rise was less pronounced only when $\Delta\tilde{\mu}_H$ was depressed by titration of the primary pump (ATPase) with its inhibitor oligomycin. It appears that the rise in force ratio at low $\Delta\tilde{\mu}_H$ is a general characteristic of energy transduction.

References

1) Zoratti, M., Pietrobon, D. and Azzone, G.F. (1982) Eur. J. Biochem. 126, 443-451
2) Azzone, G.F., Pietrobon, D. and Zoratti, M. (1984) Curr. Top. Bioener. 13, 1-77

Fig. 1   **Fig. 2**

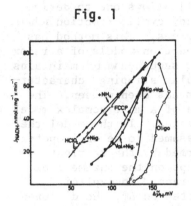 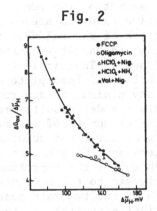

Fig. 1 - The relationship between output flow (rate of NAD reduction) and input force ($\Delta \tilde{\mu}_H$).

The medium contained, 0.1 M sucrose, 0.01 M Tris-MOPS, 6 mM MgAc$_2$, 0.5 mM EDTA, 4.7 mM fumarate, 10 mM succinate, 0.5 μg/mg antimycin, 20 mM phosphocreatine/Na$^+$, 30 μM KSCN, 0.1% BSA, 1 mM NAD, and 0.5 mg/ml submitochondrial particles. The rate of NADH formation was followed at 340 nm. Determination of $\Delta\psi$ was carried out in parallel samples with a SCN$^-$ specific electrode. Under the specified experimental conditions the ΔpH was negligible. pH 7.5. T = 30°C. Variation of $\Delta \tilde{\mu}_H$ were obtained by addition of variable amounts of: a) oligomycin, b) FCCP, c) HClO$_4$ in the presence of 0.1 nmol/mg nigericin (+ 20 mM K$^+$), d) HClO$_4$ in presence of 5 mM (NH$_4$)$_2$ HPO$_4$, e) nigericin in presence of 27 pmol/mg valinomycin and f) valinomycin in presence of 0.1 nmol/mg nigericin.

Fig. 2 - The relationship between output/input force ratio and input force.

Experimental conditions as in Fig. 1 except that NAD was 0.1 mM. The values were obtained on the basis of experiments similar to those reported in Fig. 1 except that the reaction of NADH reduction was allowed to reach steady state.

KINETIC AND THERMODYNAMIC ANALYSIS OF A PROTON PUMP MODEL USING THE
HILL DIAGRAM METHOD

Daniela Pietrobon and S. Roy Caplan
Department of Membrane Research, Weizmann Institute of Science,
Rehovot, Israel, 76100.

A diagram method introduced by T.L. Hill [1] allows one to derive
with relative ease the steady state flows of any cyclic reaction scheme
as a function of the relevant thermodynamic forces. This method has
been applied to one of the possible minimal reaction models of a redox
driven H^+ pump. The simplest ordered reaction sequence which maintains
the possibility of incomplete coupling (molecular slipping) characteristic of any minimal random reaction mechanism has been chosen. The
final 6 state model results from the reduction of a more complex model
with a greater number of enzymatic forms or states. In the model the
electron transfer (from succinate to O_2) is assumed to be an isopoten-
tial process and the electrical work is performed by the reorientation
step which shifts the unprotonated charge groups on the enzyme from
the cytoplasmic to the matrix side of the membrane. An explicit
dependence of the two rate constants of this step on $\Delta\psi$, the difference
in electrical potential, is therefore considered.

The general non-linear relationships between the two coupled steady
state flows, J_e and J_H, and the two relevant thermodynamic forces, A_e,
the affinity of the redox reaction, and $\Delta\tilde{\mu}_H$, have been derived. A
computer-aided analysis of the resulting sigmoidal flow-force curves
has been performed using a set of physically meaningful rate constants.
These were chosen in such a way that the principle of detailed balance
as well as the constraints on the reduced transitions are satisfied,
and that the maximum steady state flows, when $\Delta\tilde{\mu}_H$ is varied keeping A_e
constant at the prevailing experimental value, have values comparable
to the experimental ones. For the regions of approximate linearity
around the inflection points of the sigmoidal flow-force curves, the
phenomenological coefficients L_H and L_{eH} (slopes of J_H vs $\Delta\tilde{\mu}_H$ and J_e
vs $\Delta\tilde{\mu}_H$ plots for a certain constant value of A_e) and L_{He} and L_e (slopes
of J_H vs A_e and J_e vs A_e with $\Delta\tilde{\mu}_H$ constant at the value prevailing at
the inflection point of the flow vs $\Delta\tilde{\mu}_H$ curves), as well as the
additive constants K_e and K_H of the general linear phenomenological
equations, have been derived.

Some of the results of this study are :
(1) Depending on the relative values of the rate constants in the
model, $\Delta\psi$ and ΔpH can be kinetically inequivalent. This kinetic inequiv-
alence determines not only different slopes of the regions of approximate linearity of the flow vs $\Delta\tilde{\mu}_H$ curves but also a shift of the inflection points, depending on whether $\Delta\tilde{\mu}_H$ is varied by varying $\Delta\psi$ or ΔpH.

(2) In general, even for completely coupled H^+ pumps, the phenomenological equations are asymmetric: $L_{eH} \neq L_{He}$, or in other words the two forces A_e and $\Delta\tilde{\mu}_H$ are kinetically inequivalent. A necessary and sufficient condition for reciprocity, i.e. $L_{eH} = L_{He}$, and therefore for the kinetic equivalence of the two forces is that the inflection point of the flow-force curves is a multidimensional inflection point (MIP). The phenomenological stoichiometry $Z = (L_H/L_e)^{\frac{1}{2}}$ is equal to n, the stoichiometry H^+/e^-, only at the MIP.

(3) The relative contribution of the two uncoupled cycles depends on the value of the forces. In the experimental range of forces, the contribution of the cycle corresponding to proton slip is negligible, and the cycle corresponding to redox slip is by far the predominant uncoupled cycle. The contribution of this cycle to the electron flow J_e decreases as $\Delta\tilde{\mu}_H$ decreases in absolute value. In other words the higher the force against which the pumps operate, the higher the probability of slip. More-over, as expected, the number of redox slip cycles completed in unit time decreases as A_e decreases.

(4) Experimentally a linear relationship has been found between J_e, J_H, and $\Delta\tilde{\mu}_H$ when $\Delta\tilde{\mu}_H$ is varied by decreasing $\Delta\psi$ with increasing amounts of K^+ in the presence of valinomycin [2]. This behaviour can be simulated with a set of completely coupled pumps, in the presence of a H^+ leak such that $\Delta\tilde{\mu}_H$ at static head is at the beginning of the region of approximate linearity of the flows vs $\Delta\tilde{\mu}_H$ curves. The simulations also reproduce and explain the finding that, when the input force A_e is decreased, there is a region in which the output force at static head, $\Delta\tilde{\mu}_H^{sh}$, remains constant, so that their ratio decreases. The simple explanation is that the experimental A_e is in the saturated region of the flow vs A_e curves. These simulations predict a linear relationship between the flow of electrons at static head, J_e^{sh}, and $\Delta\tilde{\mu}_H^{sh}$, in contrast with the experimental finding that when A_e is decreased, J_e^{sh} decreases while $\Delta\tilde{\mu}_H$ remains nearly constant (D.Pietrobon et al., unpublished results). However, this result can also be simulated if a slip in the pumps is introduced, or even better, if a certain fraction of the pumps is assumed to be completely uncoupled.

(5) The analysis of the effect of increasing uncoupling of different kinds on the flow-force relationships provides criteria enabling one to distinguish whether a given uncoupling agent acts by increasing a proton leak through the membrane, or by increasing molecular slippage, or by completely uncoupling a certain number of pumps.

References

1 Hill, T.L. (1977) Free Energy Transduction in Biology. Academic Press, N.Y.

2 Pietrobon, D., Zoratti, M., Azzone, G.F., Stucki, J.W. and Walz, D. (1982) Eur. J. Biochem. 127, 483-494.

EFFECTS OF DELIPIDATION ON MITOCHONDRIAL PROTON TRANSLOCATION

M.J.Pringle and M.Taber
Department of Cell Physiology, Boston Biomedical Research Institute,
Boston, Mass., 02114, U.S.A.

Detergent treatment and enzymatic lipid degradation are useful experimental tools for modifying biomembrane composition in order to examine the role of protein/lipid interactions in the function of membrane-bound proteins. Thus, delipidation of sarcoplasmic reticulum vesicles containing Ca^{2+}-ATPase produces a differential effect on ATPase activity and Ca^{2+} transport (1-3).

We have employed detergents and phospholipase A_2 to selectively remove phospholipids from electron tranport particles (ETP_H) prepared from beef heart mitochondria. We have assayed oligomycin-sensitive ATPase activity and ATP (and in some cases NADH)-driven membrane potential, using, for the latter assay, the potential-sensitive dye, oxonol VI (4).

Simple incubation with low levels of Triton X100 ($<$ 0.005 percent) depressed the steady-state membrane potential (oxonol VI absorbance change) as a linear function of detergent concentration. Similar concentrations of octyl glucoside had minimal effect. On the other hand, when ETP_H was incubated at high detergent levels (0.5-2.5 percent), washed and resuspended, the size of the oxonol VI response was inversely proportional to the protein/lipid ratio irrespective of which detergent was used. Adding sonicated asolectin did not restore the ATP-driven oxonol VI response. ATPase activity, although fully oligomycin-sensitive, was impaired at these high detergent levels and, at 2.5 percent Triton X100, was totally abolished.

Since it is difficult to control protein solubilisation with detergents, we have carried out similar experiments with phospholipase A_2. ETP_H was treated with phospholipase for 1 min at room temperature, washed with a low concentration of BSA (1 mg/ml) and resuspended after washing in albumin-free buffer. Under these conditions, there was no change in protein/lipid ratio, ATPase activity, or oligomycin-sensitivity. However, the steady-state membrane potential was reduced by 75 percent with 0.6 units of phospholipase and was totally abolished at higher levels. This suggested that low BSA does not remove the lipase splitting products and that these inhibit the membrane potential. As a test for this, linolenic and linoleic (50 μM) but not stearic acid (150 μM) were shown to discharge the steady-state membrane potential of energised

ETP$_H$ in the manner of oligomycin. EPR data showed a small phospholipase-dependant decrease in lipid mobility in the bilayer interior, consistent with a release of unsaturated fatty acid.

Under different incubation conditions, and where phospholipase splitting products were removed by washing with 25 mg/ml BSA, a range of protein/lipid ratios from 1.8-2.9 were obtained corresponding to a maximum phospholipid depletion of ca. 35 percent. Changes in lipid acyl chain mobility from EPR data reflected the degree of lipid depletion.

At all protein/lipid ratios, oligomycin-sensitive ATPase activity was not statistically different from control values, c.f. Fleischer and Fleischer (5) who showed an initial increase followed by a decrease in ATPase activity of progressively delipidatd mitochondria.

Dye responses to energisation by ATP or NADH were complex. Thus, a 5 percent phospholipid depletion reduced the steady-state oxonol VI response by 60 percent but there was little further reduction up to 33 percent depletion. When the ATP-driven potential was discharged with oligomycin, the discharge rates for all phospholipase-treated samples were equally faster than control values. On the other hand, the size of the steady-state NADH-driven dye responses were enhanced by prior treatment with a low level of oligomycin, and this enhancement was directly proportional to the degree of phospholipid depletion.

The data show that ATPase activity of ETP$_H$ is unaffected by removal of up to 35 percent of membrane phospholipid. The ability of the particles to generate or sustain a membrane potential is highly sensitive to low levels of detergent or phospholipase A$_2$, although the mechanism of these effects is unknown. It would appear that lipid depletion results in a combination of non-specific vesicle leakiness as well as impairment of H$^+$-ATPase complexes.

References

1. Fiehn, W. and Hasselbach, W. (1970) Eur. J. Biochem., 13, 510-18.
2. Meissner, G. and Fleischer, S. (1972) Biochem. Biophys. Acta, 255, 19-33.
3. Martonosi, A., Donley, J.R., Pucell, A.G., and Halpin, R.A. (1971) Arch. Biochem. Biophys., 144, 529-40.
4. Pringle, M.J. and Sanadi, D.R. (1984) Membrane Biochemistry (in press).
5. Fleischer, S. and Fleischer, B. (1967) Methods Enzymol, X, 406-33.

THE LOWER AND UPPER LIMITS OF THE H^+ STOICHIOMETRY OF CYTOCHROME OXIDASE IN MITOPLASTS: EFFECT OF DCCD

B. Reynafarje, A. L. Lehninger, and L. E. Costa
Department of Biological Chemistry, Johns Hopkins School of Medicine
Baltimore, Maryland, 21205, U.S.A.

Although several laboratories have concluded that the vectorial H^+/O ratio for cytochrome oxidase is 4.0 (cf. 1,2), others have reported values of 2, and two groups deny that H^+ translocation is associated with this reaction. Reports on the inhibitory effect of DCCD on H^+ translocation by cytochrome oxidase have also been conflicting. We describe application of a new method for the mechanistic H^+/O ratio of cytochrome c oxidase in rat liver mitoplasts by determination of its lower and upper limits. Also described are experiments on the sensitivity to DCCD of H^+ translocation in rat liver mitoplasts and a possible explanation for some of the reported discrepancies in the H^+/O stoichiometry and effects of DCCD on cytochrome oxidase.

1. H^+/O Translocation Ratio. We have developed a kinetic method to obtain the lower and upper limits of the H^+/O ratio by a rationale (3,4) that in principle yields the ratios at static head and level flow, respectively (see abstract by Costa, Reynafarje, and Lehninger). The lower limit of the H^+/O ratio was obtained from J_H vs J_O plots when energy production by electron flow was systematically varied with cyanide or by changing temperature and yielded values of ~3.80. The upper limit, obtained from titration of energy utilization with valinomycin or K^+, was found to be ~4.12. The mechanistic H^+/O thus is close to the integral value 4.0. Since endogenous electron flow in rotenone-poisoned mitoplasts in the presence or absence of antimycin A, is only ~1% of the total rate of cyt c^{2+} oxidation under the test conditions, the H^+/O ratio of 4 does not include contributions from either site 1 or site 2 electron transport. Moreover, HQNO or myxothiazol had no effect on the observed H^+/O ratios. These data therefore agree with the results of other methods of H^+/O measurements on cytochrome oxidase reported earlier (1) with lower limit-upper limit measurements on succinate oxidation (4) and on NADH oxidation (abstract by Costa et al), and with data of Lemasters et al (cf. 2).

2. Effect of DCCD on Electron Flow and H^+ Translocation. Although all reports indicate DCCD in rather high concentrations blocks H^+ ejection by beef heart cytochrome oxidase in liposomes, it was found to have no effect on H^+ ejection in liposomal cytochrome oxidase from Paracoccus (5). Moreover, DCCD failed to lower H^+/O of

cytochrome oxidase in rat liver mitochondria (6). In all these reports the control H^+/O ratio for cytochrome oxidase was close to 2.

We have found that DCCD up to 80 nmol/mg protein had no effect on electron flow by cytochrome oxidase in rat liver mitoplasts but it inhibited H^+ translocation in an apparently biphasic manner. Fifty percent inhibition of H^+ ejection required no more than 4 nmol DCCD per mg protein, without preincubation. However, inhibition of the remaining H^+ ejection was quite resistant to inhibition by DCCD; significant H^+ ejection still occurred at 80 nmol/mg. In contrast, all of the H^+ ejection was prevented by FCCP. The data strongly suggest that the 4 H^+ ejected per O reduced by cytochrome oxidase under our conditions are translocated by two different pathways, one of which is far more sensitive to DCCD than the other. Further, DCCD at very low concentrations also substantially inhibits H^+ backflow through the mitoplast membrane after reduction of an O_2 pulse in presence of K^+ + valinomycin, but does not inhibit H^+ backflow induced by excess antimycin A. These observations indicate that DCCD closes a specific H^+ channel in both directions.

A possible explanation is offered for some of the reported discrepancies in H^+/O ratios for cytochrome oxidase and DCCD action. The low control H^+/O ratios of 2 for cytochrome oxidase observed in most reports on reconstituted systems and some on intact mitochondria may reflect the action of only that H^+ translocation pathway that is relatively insensitive to DCCD, since they require very high concentrations and long preincubations. The type of H^+ translocation pathway that is highly sensitive to DCCD, as seen in our mitoplast preparations, may not be operative in the systems that yield H^+/O = 2. It is also possible that in liposomal systems the full H^+-translocating activity of native cytochrome oxidase may not have been reconstituted.

1. Reynafarje, B., Alexandre, A., Davies, P., and Lehninger, A. L. (1982) Proc. Natl. Acad. Sci. USA 79, 7218-7222.
2. Lemasters, J. J., Grunwald, R., and Emaus, R. K. (1984) J. Biol. Chem. 259, 3058-3068.
3. Beavis, A. (1981) Fed. Proc. 40, 1563.
4. Costa, L. E., Reynafarje, B., and Lehninger, A. L. (1984) J. Biol. Chem. 259, 4802-4811.
5. Püttner, I., Solioz, M., Carafoli, E., and Ludwig, B. (1983) Eur. J. Biochem. 134, 33-37.
6. Price, B. D., and Brand, M. D. (1982) Biochem. J. 206, 419-421.

MECHANISM OF UNCOUPLING OF OXIDATIVE PHOSPHORYLATION

H. Rottenberg, and K. Hashimoto
Departments of Pathology and Biochemistry,
Hahnemann University
Philadelphia, PA 19102, U.S.A.

It is now recognized that classical uncouplers are protonophores, capable of collapsing the proton electrochemical potential (1) as originally suggested by Mitchell (2). Other ionophores (e.g., valinomycin) that uncouple oxidative phosphorylation collapse the membrane potential, which is the main component of $\Delta\tilde{\mu}_H$ in mitochondria. However, the correlation between uncoupling and the collapse of $\Delta\tilde{\mu}_H$ is not identical. Moreover, there exist a large number of reagents which are neither protonphores, nor ionophores that uncouple oxidative phosphorylation with relatively little effect on $\Delta\tilde{\mu}_H$. These include, general anesthetics (3), fatty acid, detergents (4) and other membrane pertrubing reagents and drugs.

We are studying the quantative relationships between $\Delta\tilde{\mu}_H$, respiration, phosphorylation and the rate of ATP hydrolysis as modulated by the effects of classical uncouplers, ionophores and decouplers. (We use the term decoupler to describe uncouplers which are neither protonophores or ionophores). The results obtained with rat liver mitochondria indicate that each of the three groups show a distinct pattern of respiratory stimulation, ATPase stimulation and inhibition of phosphorylation when expressed as a function of $\Delta\tilde{\mu}_H$. Ionophores which collapse $\Delta\tilde{\mu}_H$, but do not change proton permeability, have relatively small effects on respiration and phosphorylation in relation to the collapse of $\Delta\tilde{\mu}_H$. Protonophores produce larger effects on respiration and phosphorylation by equal collapse of $\Delta\tilde{\mu}_H$. Decouplers uncouple almost completely before a significant reduction of $\Delta\tilde{\mu}_H$ is observed. These patterns are reversed in the effects on the correlation between the rate of ATPase and $\Delta\tilde{\mu}_H$, when generated by ATP hydrolysis. To explain these results, which are incompatible with the chemiosmotic postulate that uncoupling is solely due to the collapse of $\Delta\tilde{\mu}_H$, we employ the parallel coupling model (5) in which both $\Delta\hat{\mu}_H$ and direct proton transfer contribute to the overall efficiency of coupling. Accordingly, the only "pure" uncouplers in the chemiosmotic sense, i.e., uncoupling solely by the collapse of $\Delta\hat{\mu}_H$, are ionophores such as valinomycin. The protonphores uncouple by collapsing $\Delta\hat{\mu}_H$, but also by directly accepting protons in the membrane interior from the F_o- of the ATPase and also, from redox proton pumps. Decouplers uncouple mostly by binding to F_o and the redox proton-pumps thereby interferring with the proton binding capacity of these membrane proteins and their interactions. Studies of fluoresence energy-transfer and fluorescence anisotropy of fatty-acid analogs confirm the strong interaction between integral membrane protein and fatty-acids in mitochondria.

REFERENCES

1. McLaughlin, S.G.A. and Dilger, J.P. (1980) Physiol. Rev. 60, 825-863
2. Mitchell, P. (1966) Biol. Rev. 41, 445-502
3. Rottenberg, H. (1983) Proc. Natl. Acad. Sci U.S. 80, 3313-3317
4. Rottenberg, H. and Hashimoto, K., submitted
5. Rottenberg, H. (1978) in Progress in surface and membrane Science 12, 245-325

THE SODIUM CYCLE: A NOVEL TYPE OF BACTERIAL ENERGETICS
V.P.Skulachev
Belozersky Laboratory of Molecular Biology and Bioorganic
Chemistry, Moscow State University, Moscow 119899, USSR

It is generally accepted that proton operates as a "coupling ion" connecting energy-releasing and energy-consu-
ming processes in biomembranes. As to Na^+, it was shown
to buffer the ΔpH component of $\Delta\bar{\mu}H$ via Na^+/H^+ antiporter.
Such a concept poses difficulties if we deal with alkalo-
philic or alkali-tolerant bacteria. Here $\Delta\Psi$ and ΔpH prove
to be oppositely directed if $\Delta\bar{\mu}H$ generators pump H^+ from
the neutral cytoplasm to the alkaline outer medium.

Recently Tokuda and Unemoto [1] communicated that an
alkali-tolerant marine bacterium Vibrio alginolyticus extruded Na^+ from the cell in a respiratory-chain dependent
electrogenic mode, the process being resistant to protono-
photorous uncouplers. A $\Delta\bar{\mu}Na$-generator seems to be localized between NADH and quinone. In the same group, it was
shown that $\Delta\bar{\mu}Na$ is the driving force in the uptake of 19
amino acids and sucrose into the V. alginolyticus cells.

In our laboratory [2,3], it was shown that the motility
of V.alginolyticus (i) requires Na^+,(ii) can be supported
by artificially-imposed ΔpNa (but not by ΔpH) in a monen-
sin-sensitive fashion and (iii) may be observed in the presence of a high concentration of the uncoupler (1×10^{-4}M
CCCP). A 100-fold lower CCCP concentration completely ar-
rested the motility if the medium was supplemented with
Na^+/H^+-antiporter monensin. Monensin, added without CCCP,
decreased the motility rate only partially. It was concluded that the flagellar motor of V.alginolyticus is driven
by $\Delta\bar{\mu}Na$ rather than by $\Delta\bar{\mu}H$ which is known to be the driving force of motility of neutrophillic bacteria.

The next question is whether $\Delta\bar{\mu}Na$ is employed to perform the chemical work e.g. ATP formation. For that it is
sufficient to have a reversible Na^+-motive ATPase in a
membrane possessing a $\Delta\bar{\mu}Na$-generator. The first indication that it may be the case was obtained in our group when
V.alginolyticus motility was studied. It was found [2]
that $\Delta\bar{\mu}Na$-driven motility can be observed under anaerobic
conditions. Anaerobic motility, in contrast to aerobic,
was sensitive to arsenate which strongly decreases the

level of ATP in the V.alginolyticus cells. Another agent decreasing anaerobic motility proved to be monensin.

In 1980 Dimroth [4] showed that decarboxylation of oxaloacetate to pyruvate by membrane-linked decarboxylase of Klebsiella aerogenes is directly coupled to the uphill Na^+ export from the bacterial cell. Quite recently similar process was described by the same group [5] in Propionigenum modestum, which grows from the fermentation of succinate to propionate. The only biologically useful energy which is gained in this fermentation is $\Delta\bar{\mu}Na$ established upon decarboxylation of methylmalonyl-CoA formed from succinate. This Na^+ gradient drives ATP synthesis via a Na^+-stimulated ATPase which is present in high amounts in the bacterial membrane. This observation directly confirms our hypothesis [2] about a $\Delta\bar{\mu}Na$-driven reversal of Na^+--ATPase as a mechanism of the membrane-linked ATP synthesis, alternative to H^+-ATP-synthase.

Thus in certain bacteria, $\Delta\bar{\mu}Na$, rather than $\Delta\bar{\mu}H$, can perform the role of the membrane-linked convertible energy currency. This conclusion is based on the following observations: (i) There are primary $\Delta\bar{\mu}Na$-generators in bacterial membranes (NADH-quinone reductase and decarboxylases). (ii) Formed $\Delta\bar{\mu}Na$ can support all the three main types of work in the bacterial cell: chemical (ATP synthesis), osmotic (accumulation of solutes) and mechanical (rotation of flagellum).

1. Tokuda, H., Sugasawa, M., and Unemoto, T. (1982) J.Biol. Chem. 257, 788-794
2. Chernyak, B.V., Dibrov, P.A., Glagolev, A.N., Sherman, M.Yu. and Skulachev, V.P. (1983) FEBS Lett. 164, 38-42
3. Dibrov, P.A., Glagolev, A.N., Kostyrko, V.A., Sherman, M.Yu., Skulachev, V.P., Smirnova, I.A. (1984) 16th FEBS Meeting Abstr.(in press)
4. Dimroth, P. (1980) FEBS Lett. 122, 234-236
5. Dimroth, P., Hilpert, B. (1984) 16th FEBS Meeting Abstr. (in press)

FUNCTIONAL CHARACTERIZATION OF BACTERIORHODOPSIN AND ATP SYNTHASE
CONTAINING LIPOSOMES.

R.L. v.d. Bend, J. Petersen, J.A. Berden and K. v. Dam

Laboratory of Biochemistry, B.C.P. Jansen Institute, University of
Amsterdam, Plantage Muidergracht 12, 1018 TV Amsterdam,
the Netherlands.

Bacteriorhodopsin (BRh) and yeast mitochondrial ATP synthase containing liposomes were used to study the relation between $\Delta\tilde{\mu}_H$ and kinetic parameters of the ATP synthesis reaction. BRh and ATP synthase
were co-reconstituted functionally in an optimal way by first sonication of BRh and soy bean phospholipids together, after which ATP synthase was incorporated in the BRh-liposomes by detergent incubation
followed by detergent removal by a quick gelfiltration method (van
der Bend et. al. unpublished results). Light driven ATP synthesis was
measured by trapping formed ATP with hexokinase as glucose-6-phospha-
te, which was enzymatically determined after the light incubation.
Light intensity was varied by using neutral density filters.

The effect of protonophore and light intensity variation on the
$K_m(ADP)$ were found to be contradictory (table 1). Decreasing light
intensity resulted in decreasing $K_m(ADP)$-values, increasing protonophore concentration (carbonyl cyanide-p-trifluoromethoxyphenyl hydrazone (FCCP)) resulted in increasing $K_m(ADP)$-values. Using gramicidin
(which acts as protonophore by forming pores in the membrane) the
same results were found as with FCCP, which makes it less likely that
the effect of FCCP can be explained by a specific interaction with the
ATP synthase. Similar results were found by others (1, 2) with chloroplasts and submitochondrial particles.

In an other experiment it was found that the light driven ATP synthesis activity was independent of protonophore concentration when an
extrapolation was made to infinite light intensity in double reciprocal plots of light intensity versus ATP synthesis activity at different protonophore concentrations (fig. 1). A same result was found
for ΔpH measured in liposomes (3).

Although the experiments in fig. 1 are in agreement with a chemios-
motical way of energy-transduction, the experiments in table 1 can not
be understood from this.

Table 1. Effect of protonophore and light intensity variation on K_m and V_{max} of the light driven ATP synthesis reaction.

		K_m(ADP) (μM)[a)]	V_{max}(ADP)[a)] $(nmol.min^{-1}.mg^{-1})$
light intensity	10%	1.2	21
	25%	4.5	61
	50%	5.9	89
	100%	8.3	109
FCCP	0 nM	6.4	65
concentration	170 nM	9.2	47
	340 nM	13.8	40
gramicidin	0 μM	7.5	169
concentration	3.8 μM	7.9	140
	7.5 μM	11.1	130

[a)] values were calculated from Lineweaver-Burk plots by linear regression. Correlation coeficiets were 0.990 or higher.

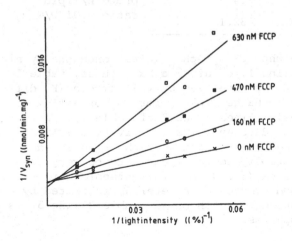

Fig.1
Dependence of the light driven ATP-synthesis reaction on light intensity at different FCCP concentrations.

1. Aflalo, C. and Shavit, N. (1983) FEBS Lett. 154, 175-179
2. Yagi, T., Matsuno-Yagi, A., Vik, S.B. and Hatefi, Y. (1984) Biochemistry 23, 1029-1036
3. Arents, J.C., Hellingwerf, K.J., v. Dam, K. and Westerhoff, H.V. (1981) J. Membrane Biol. 60, 95-104

ARTIFICIAL ENERGIZATION AND ATP SYNTHESIS IN ATPase PROTEOLIPOSOMES
FROM THE THERMOPHILIC CYANOBACTERIUM *SYNECHOCOCCUS* 6716

H.S. van Walraven, M.J.M. Hagendoorn, K.Krab, and R. Kraayenhof
Biological Laboratory, Vrije Universiteit, De Boelelaan 1087,
1081 HV Amsterdam, The Netherlands

In native membrane systems ATP synthesis can be induced by ΔpH or electric field ($\Delta\psi$) pulses (1,2). Our research aims at the resolution of artificial energization by ATP hydrolysis and *vice versa* in reconstituted ATPase proteoliposomes (3). Here we describe ATP synthesis and $\Delta\psi$ and ΔpH calibration in this model system, by ionophore-induced diffusion potentials and ΔpH pulse.

For proteoliposome preparation and experimental conditions, *cf.*(3); the KCl concentrations were as indicated.

Initial K^+ in	out	Perturbation	ATP synthesis rate (nmol.mg prot.$^{-1}$ min^{-1})
100	10	1 µM nig.	16.2
10	100	1 µM nig.	0
100	10	80 nM val.	0.4
10	100	80 nM val.	14.5
100	10	ΔpH pulse =1.5	32.1

Table I. ATP synthesis induced by K^+ and H^+ fluxes. The medium contained in addition 5 mM ADP and 5 mM phosphate; protein/lipid ratio 0.01 w/w.

ATP synthesis was appreciable when $pH_{out} > pH_{in}$ (after exchange by nigericin of K^+_{in} for H^+_{out}, table I line 1, or after a base pulse, table I line 5), or when inward flux of K^+ is catalyzed by valinomycin (K^+ diffusion potential, table I line 4). Exchange of K^+_{out} for H^+_{in} or efflux of K^+_{in} did not result in ATP synthesis (lines 2 and 3 of table I). The rates of synthesis are low, compared to hydrolysis activities (3) but are constant over a long period due to the fact that the K^+ gradients are maintained for hours. The proteoliposomes appeared to be very impermeable for protons and other ions (3). No ATP is synthesized in the presence of the uncoupler S-13 and synthesis is severely inhibited by DCCD. The ATP synthesis rate induced by a ΔpH pulse (table I line 5) is higher than that induced by ionophores.

Membrane energization at different K^+ gradients was followed by the use of different optical probes (Fig. 1). Oxonol VI measures the electric potential gradient induced by valinomycin when K^+_{out} is varied at $K^+_{out} > K^+_{in}$ and appeared to be a suitable probe in our system. It does not react on a reversed gradient. Here, safranine monitors electric potential changes when K^+_{out} is varied at $K^+_{out} < K^+_{in}$ (*cf.*mitochondria, ref. 4), but also responds when $K^+_{out} > K^+_{in}$.

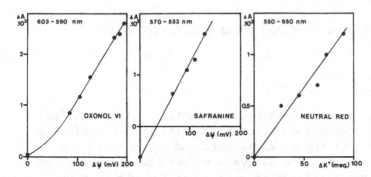

Fig. 1. Calibration of K^+ flux-induced membrane potentials and ΔpH by val. and nig. Electric potentials were monitored by oxonol VI (val., $K^+_{out} > K^+_{in} = 10$ mM), safranine (val., $K^+_{out} < K^+_{in} = 100$ mM) and *internal* neutral red (nig., ΔK^+in both directions). Osmolarity was kept constant with $LiCl_2$ (oxonol VI) or choline chloride (safranine, neutral red). Final lipid concentration was 0.25 mg.ml^{-1}. Concentrations of ionophores as in table 1.

Electric potential changes can also be followed by the native carotenoids in this system. Carotenoids are isolated together with the lipids. In this case the spectral shift is similar in both directions of the applied electric field, from 490 to 510 nm (data not shown). ATP synthesis, given in table I, is achieved at 65 mV (at 50°C). pH changes after nigericin addition have been followed by the pH indicator neutral red (3). The H^+ gradient, inducing ATP synthesis (table I) is estimated from the calibration curve (*cf.* 3) to be about 1 pH unit.

From these results we conclude that the ATPase proteoliposomes are able to synthesize ATP under different artificial conditions due to the fact that the inside medium can be varied during reconstitution. The liposomes can retain a gradient and are stable over a long period and thus are very useful for studying electrical and pH events during energy transduction.

References
(1) Jagendorf, A.T. and Uribe, E. (1966) Proc. Natl. Acad. Sci. USA 50, 170-177.
(2) Gräber, P., Rögner, M., Buchwald, H.E., Samoray, P. and Hauska, G. (1982) FEBS lett. 145, 35-40.
(3) van Walraven, H.S., Lubberding, H.J., Marvin, H.J.P. and Kraayenhof, R. (1983) Eur. J. Biochem. 137, 101-106.
(4) Åkerman, K.E.O. and Wikström, M.K.F. (1976) FEBS lett. 68, 191-197.

THE INNER AND INTERTHYLAKOIDAL ELECTRIC POTENTIAL OF ISOLATED
CHLOROPLASTS IN THE DARK
D. Walz
Biozentrum, University of Basel
CH-4056 Basel, Switzerland

   The two components of the electrochemical potential difference for
H^+-ions, $\Delta\tilde{\mu}_H$, are usually determined from the partitioning of suitable
markers, e.g. methylamine for the chemical part and thiocyanate for
the electric part, respectively. The analysis of such data is based on
the assumption that the concentration of markers in the whole space
outside the thylakoids is constant and equal to that in the suspending
medium [1].

   The surfaces of thylakoid membranes carry charges associated with
the proteins. Hence, diffuse double layers of ions on the membrane
surfaces exist in which the concentration of ions deviates from that
in the bulk phase. This phenomenon is particularly relevant to the
interthylakoidal space in grana stacks since the spacing between thylakoids is of the order of the Debye-Hückel length. Moreover, it becomes the more prominent the lower the concentration of salt. Hence,
when neglecting this phenomenon, one estimates values of $\Delta\tilde{\mu}_H$ for non-
energized chloroplasts at low salt concentrations which deviate significantly from the expected equilibrium value of zero (see Table).

   The extent of curvature of the profiles for electric potential and
concentrations and its effect on average concentrations determined in
partitioning experiments can be assessed from model calculations based
on the Gouy-Chapman theory on diffuse double layers. The thylakoid
membranes are represented by plane sheets of a dielectric with constant
D_m. The charges are considered as smeared over the surface thus yiel-

Chemical and electric part of $\Delta\tilde{\mu}_H$ for dark adapted broken chloroplasts
in a medium with 100mM sucrose and a Good buffer according to 2.5mM K^+.

pH	$\Delta\mu_H$ (kJ/mol)	$F \Delta\Psi$ (kJ/mol)	$\Delta\tilde{\mu}_H$ (kJ/mol)
5.2	2.67 + 0.06	1.10 + 0.05	3.77 + 0.07
7.5	5.68 + 0.06	-2.42 + 0.25	3.26 + 0.26
9.4	7.38 + 0.51	-6.02 + 0.88	1.36 + 1.01

ding a charge density σ. The space in and outside the thylakoids
contains an aqueous solution (dielectric constant D_w) of a 1,1-electro-
lyte [2]. The figure shows an example of the electric potential profile
together with the pertinent concentration c_+ for monovalent cations
with respect to their concentration in the bulk phase c_∞.

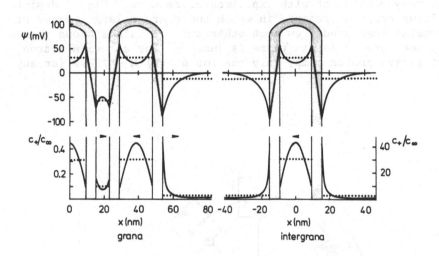

Profiles for the electric potential (top) and the concentration
of positive ions (bottom) in stacked (left) and unstacked (right)
thylakoids. Dotted lines indicate average values and arrows point to
the pertinent scale. Membranes have a charge density σ (C/m^2) of 0.009
on inner surface, -0.01 on outer surface in stacks and -0.02 on surfaces of stacks and unstacked membranes. $D_m = 5$, $D_w = 80$, monovalent
salt 2.5 mM, temperature $20^\circ C$.

The partitioning experiments for markers penetrating into all spaces (e.g. methylamine) are supplemented with that for suitable markers
which move only up to the interthylakoidal space or the surface
layer. The resulting data set can be analysed for the volumes of the
inner and interthylakoidal space as well as for the surface layer
together with the average electric potentials in these spaces.

References
1. Rottenberg, H., Grunwald, T. and Avron, M. (1972) Eur.J.Biochem.
   25, 54-63
2. Walz, D. (1984) in Advances in Photosynthesis Research (Sybesma,
   C., ed.) pp. 257-260, M.Nijhoff/W.Junk Publ., The Hague

STATISTICAL ASPECTS OF PROTONIC ENERGY COUPLING

Hans V. Westerhoff
National Institute of Arthritis, Diabetes, Metabolic and Digestive and
Kidney Diseases, N.I.H., Building 2, Room 318, Bethesda, MD U.S.A.

We recently [1] emphasized that the chemiosmotic coupling hypothesis
may have to be supplemented with an extra postulate to make it
quantitatively consistent with experimental results. Fig. 1 depicts
the resulting coupling scheme, in which there are a large number of
proton domains insulated from each other and the bulk aqueous phases.
For every such proton domain there is just one (or a few) electron
transport linked proton pump, only one (or a few) H^+-ATPase (or any

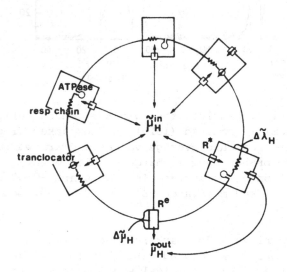

Fig. 1. The 'mosaic protonic coupling' hypothesis. Electron transfer
linked proton pump and H^+-ATPase share a proton domain and consequently
function as coupling unit.

other proton free energy consumer), a leak back across the membrane and a leak towards the bulk aqueous phase adjacent to the proton domain. The number of protons needed to fill such a proton domain up to the level that would make it competent for ATP synthesis is only on the order of 3.

The description of the kinetics of systems in terms of thermodynamic parameters (such as chemical potentials, or concentrations) is only possible if the number of particles in such systems is so large that the fluctuation in that number (which usually amounts to the square root of that number) is insignificant in the relative sense. In the present case discussion of the kinetics of the system in terms of the concentration or potential of the protons in a single proton domain is problematic. An example of the special properties of this system is the fact that once an H^+-ATPase has turned over, it is bound to be inactive until the electron transfer chain that fills its proton domain turns over again. A consequence is that electron-transfer chain and the H^+-ATPase of the same proton domain function as a coupling unit. Implications of this phenomenon for the (control) properties of protonic energy coupling will be discussed using the liposome analogue depicted in Fig. 2.

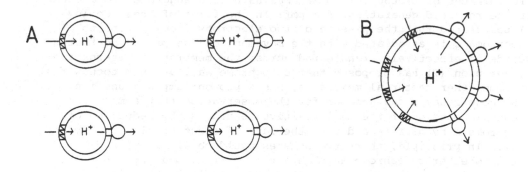

Fig. 2A. Liposome analogue of the protonic coupling hypothesis of Fig. 1. Every liposome contains one electron transfer linked proton pump and one H^+-ATPase. B. The corresponding delocalized chemiosmotic liposome analogue with many electron transfer linked proton pumps and many H^+-ATPases per liposome.
1. Westerhoff, H.V., Melandri, B.A., Venturoli, G., Azzone, G.F. and
    Kell, D.B. (1984) FEBS Lett. 165, 1-5.

THE ENIGMA OF PROTONMOTIVE FORCE GENERATION IN NITROBACTER; IS A PROTON-PUMPING CYTOCHROME OXIDASE ESSENTIAL?

H.G. Wetzstein[1] and S.J. Ferguson[2]

[1]Institut für Mikrobiologie, Universität Göttingen, F.R.G.
[2]Department of Biochemistry, University of Birmingham,
P.O. Box 363, Birmingham B15 2TT, U.K.

Bacteria of the genus Nitrobacter rely upon the oxidation of nitrite by oxygen to provide energy for growth. It is generally accepted that electron flow from nitrite to oxygen is linked to the generation of a protonmotive force (pmf) that in turn drives both ATP synthesis and reversed electron transfer from nitrite to $NAD(P)^+$ in order to provide reducing power for growth. The site of nitrite oxidation is believed to be on the cytoplasmic surface of the plasma membrane from where electrons are passed, possibly via cytochrome a_1, to cytochrome c at the periplasmic surface of the membrane. Cytochrome aa_3 then transfers electrons from cytochrome c to reduce oxygen, presumably at the cytoplasmic surface (1,2). Such a description cannot account for the generation of a pmf because there is neither net inward movement of electrons across the membrane nor outward movement of protons. Two schemes have been suggested to account for the required generation of a pmf: In the first of these Cobley (1) postulated that the passage of two electrons from nitrite to cytochrome c is associated with the movement of one proton so that hydride is effectively transferred across the membrane. Alternatively, Ferguson (2) has proposed that cytochrome aa_3, as in mitochondria and some other bacterial membranes, has a proton-pumping function. Both these schemes can account for the observation that a membrane potential, outside of the cell positive, enhances the reduction of cytochrome c by nitrite and thus the rate of electron flow to oxygen (1,2). In principle, these two schemes could be distinguished by testing whether cytochrome aa_3 has a proton-translocating function. Recently, it has been reported that a two subunit preparation of cytochrome aa_3 isolated from Nitrobacter agilis does not exhibit a detectable proton-translocating activity after incorporation into phospholipid vesicles (3). We have tested for proton-pumping by cytochrome aa_3 in intact cells by using non-physiological substrates that donate electrons to the electron transfer chain at the level of cytochrome c, in combination with the oxygen pulse technique (4). The introduction of approximately 20 to 50 μM O_2 to an anaerobic suspension of cells of Nitrobacter winogradskyi in 1.5 mM glycyl glycine, 50 mM KCl, 100 mM KSCN at pH 8.4 has been observed to be associated with the acidification of the suspension medium when 1 mM potassium ferrocyanide is added several minutes before the oxygen pulse. This acidification is abolished by an uncoupler and corresponds to the translocation of between one and two protons per two electrons flowing. Proton translocation could similarly be detected when ascorbate plus TMPD was the substrate. No proton translocation was observed in the absence of one of the two added substrates. Before it can be concluded that these experiments have demonstrated a proton-translocating role for cytochrome aa₃ it is necessary to consider the following points: First, proton translocation was clearly observed only if cells freshly harvested from a growing culture were used. Second, the rates of both ferrocyanide and ascorbate plus TMPD oxidation were considerably lower than the rates of nitrite oxidation. Finally, we must consider whether the observed proton translocation might be due to electron flow from endogenous substrates via NADH either to ferricyanide formed from oxidation of ferrocyanide or to the oxidized form of TMPD. The evidence that such processes were not significant was that the rate of oxidation of endogenous substrates was not increased upon addition of TMPD and the rate observed was considerably slower than the rate of oxidation of ascorbate plus TMPD. In the case of the experiments with ferrocyanide it was shown that under anaerobic conditions the rate of ferricyanide reduction by endogenous substrates was negligible compared with the rate of ferrocyanide oxidation. Finally, we consider a possible explanation for the recent failure to observe any proton translocation associated with the physiological substrate nitrite (5). It is essential with an oxygen pulse experiment to add reagents that collapse the membrane potential if translocated protons are to be observed. Nitrite oxidation is severely inhibited upon collapse of the membrane potential (1) thus suggesting a reason for this observation (5).

(1) Cobley, J.G. (1976) Biochem. J. **156**, 481-491
(2) Ferguson, S.J. (1982) FEBS Lett. **146**, 239-243
(3) Sone, N., Yanagita, Y., Hon-Nami, K., Fukumori, Y., and Yamanaka, T. (1983) FEBS Lett. **155**, 150-154
(4) Mitchell, P. and Moyle, J. (1969) Eur. J. Biochem. **149**, 471-484
(5) Hollocher, T.C., Kumar, S., and Nicholas, D.J.D. (1982) J. Bacteriol. **149**, 1013-1020

MITOCHONDRIAL STATE 4 ΔG_p IS PROPORTIONAL TO $\Delta \mu_H+$

Henri WOELDERS, Wim VAN DER ZANDE, Annemarie COLEN and Karel VAN DAM

Laboratory of Biochemistry, University of Amsterdam, P.O. box 20151
1000 HD Amsterdam, The Netherlands.

Though in agreement with most experimental observations, Mitchell´s original concept [1] of chemiosmotic coupling cannot adequately explain all available experimental results. One of the aspects of this theory that has been challenged is the assumption that protons freely equilibrate with the bulk aqueous phases at both sides of the energy-transducing membrane. The most prominent anomaly to this "delocalised" coupling concept is that in mitochondrial incubations in State 4 (equilibrium for the ATP-ase) the extramitochondrial ΔG_p is not proportional with $\Delta \mu_H+$ across the membrane when the latter is varied by uncouplers or respiratory inhibitors [2,3,4,5]. This and other anomalies have led Westerhoff et al.[6] to assume that there are many independently operating coupling units, which are functionally separated from each other and from the bulk phase by proton resistances.

We measured ΔG_p and $\Delta \mu_H+$ in mitochondrial incubations in State 4 but we took extra precautions to avoid possible interference with the measurement of ΔG_p and $\Delta \mu_H+$ by factors like intramitochondrial nucleotides, adenylate kinase activity, the quenching method or $\Delta \mu_H+$ dependent matrix volume changes. We found a proportional relationship between ΔG_p and $\Delta \mu_H+$ when $\Delta \mu_H+$ was varied with uncouplers or with malonate (see figure 1). Moreover, we obtained a non-proportional $\Delta G_p/\Delta \mu_H+$ relationship, as found previously [2,3,4,5], when these precautions were omitted (figure 2). Essential modifications of our methods as compared to the methods used previously are:
1. The sample for the determination of ΔG_p is filtrated directly before quenching through Whatman GF/F filters (cf Nicholls [7].
2. An organic quench mixture as described by Slater et al. [8] was used instead of $HClO_4$.
3. Adenine nucleotide concentrations were relatively high (to 11 mM).
4. Matrix volume was measured as ^{14}C Mannitol inaccessible ^3H_2O space for each separate incubation (valinomycin was present and Pi and K^+ distributions were taken as a measure for ΔpH and Δ respectively) K^+ efflux upon lowering $\Delta \mu_H+$ was shown to lead to a significant matrix volume reduction. This implies an overestimation of the uncoupler induced $\Delta \mu_H+$ decrease when a constant matrix volume would be assumed.
5. Adenylate kinase was either blocked by AP5A (and absence of Mg^{2+})

or alternatively the adenylate kinase reaction was allowed to reach equilibrium.

The presently obtained proportional $\Delta G_p/\Delta\mu_H+$ relationship does not support the model of mosaic protonic coupling [6] for the mitochondrial system, as this model implies a non-proportional relationship. Our results do not necessarily conflict with other "localised" coupling schemes. In respect to this , future experiments must show how wide the implications of our present results are.

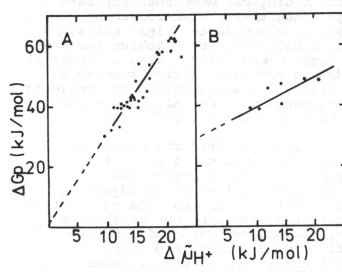

Fig.1 (A) Dependence of ΔG_p on $\Delta\mu_H+$ with succinate as substrate. Summary of four experiments. $\Delta\mu_H+$ was varied using either DNP; DNP + Malonate; Gramicidin d; S13 or FCCP. (B) Experiment using DNP, performed in parallel with one of the DNP experiments from fig.1A using the same mitochondrial preparation, but omitting modifications 1 to 5 listed above.

1. Mitchell, P. (1961) Nature 199, 144-148
2. Wiechman, A.H.C.A., Beem, E.P. and Van Dam, K. (1975) in Electron Transfer Chains and Oxidative Phosphorylation (Quagliariello, E., Papa, S., Palmieri, F.,Slater, E.C. and Siliprandi, N., eds.), pp. 335-342, North-Holland, Amsterdam.
3. Azzone, G.F., Massari, S. and Pozzan, T. (1977) Mol. Cell Biochem. 17, 101-112
4. Van Dam, K., Wiechmann, A.H.C.A., Hellingwerf, K.J., Arents, J.C. and Westerhoff, H.V. (1978) Proc. FEBS Meet. 11th Copenhagen 121-132
5. Westerhoff, H.V., Simonetti, A.L.M. and Van Dam, K. (1981) Biochem. J. 200, 193-202
6. Westerhoff, H.V., Melandri, B.A., Venturoli, G., Azzone, G.F. and Kell, D.B. (1984) FEBS Lett. 165, 1-5
7. Nicholls, D.G. (1974) Eur. J. Biochem. 50, 305-315
8. Slater, E.C., Rosing, J. and Mol, A. (1973) Biochim. Biophys. Acta 292, 534-553

CALCULATION OF THE ELECTRIC POTENTIAL AND THE ION
CONCENTRATION DISTRIBUTION DUE TO CHARGE SEPARATION
IN A CLOSED MEMBRANE VESICLE

L. Zimányi[o] and Gy. Garab"
[o]Institute of Biophysics and "Institute of Plant Physiology
Biological Research Center, Hungarian Academy of Sciences
Szeged, Hungary, H-67ol

The energy transducing biological membranes separate
two compartments of high conductivity electrolyte by
forming closed vesicles. In chloroplasts, algal cells,
photosynthetic bacteria or Halobacterium halobium the conversion and storage of light energy starts with a vectorial
charge separation inside the membrane, giving rise to a
transmembrane electric field. The energization is completed
by proton translocation across the membrane, causing pH
difference between the bulk of the inner and outer
compartments.

In an earlier paper /1/ we calculated the electric
potential in a model where the spherical dielectric
membrane was bounded by metal. Here we report the refine-
ment of our model with the spherical membrane adjoining
electrolyte on both sides. This way the calculated electric
field is not restricted to the membrane, but penetrates
into the aqueous phases, causing variation of ion concentrations /and specifically pH/ across and along the
membrane even before real charge translocation takes
place across the membrane.

The electric field inside the membrane consists of a
limited region of strong local field and of the region
of uniform transmembrane field in the rest of the membrane
/even when the charges are buried deeply/ /fig.1/.
The increment of the intensity of this transmembrane
field when the charge localized close to the donor side
of PS1 enters the inner aqueous phase in the millisecond
range is a possible explanation for the slow rising component /phase b/ of the electrochromic absorbance change
of photosynthetic pigments, which detect the uniform field.

The creation of a dipole may cause depletion of cations
and accumulation of anions in the electrolyte on the side
of the positive charge, and the opposite effect on the
other side. Dissociable groups on the surface of proteins
located close to the site of charge separation may feel
a different pH value than those far from it. This provides different operating conditions for ion pumps

depending on their lateral location. Although the increased transmembrane potential difference and the ion
concentration difference have opposite directions, the
two effects might not compensate completely each other,
yielding also a different driving force for ion pumps
in the vicinity of the reaction centers. Hence the
lateral variation of the electric potential and the
consequent local variation of pH and ion concentrations
along the membrane surface may regulate the functioning
of ion pumps.

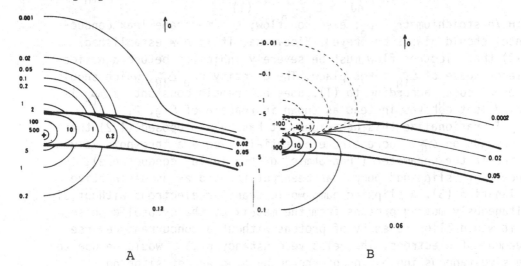

A   B

Figure 1. Equipotential surfaces of a single elementary
charge /A/ or of a dipole /B/ in millivolts. The outer
radius of the sphere is 175 nm, the membrane thickness
is 7 nm, the positive and negative charges are located
at 17o.3 and 172.7 nm, respectively. The relative dielectric constant of the membrane and the aqueous phases
is 2 and 8o, respectively. The Debye length in both
aqueous phases is 3.5 nm, which corresponds to about
1o mM monovalent salt concentration.

Reference
1 Zimányi, L. and Garab, Gy. /1982/ J. Theor. Biol.
 95, 811-821

Mitochondrial Proton Pumps "Slip"

M. Zoratti, M. Favaron, D. Pietrobon and G.F. Azzone
C.N.R. Unit for the Study of Physiology of Mitochondria and Institute of General Pathology, University of Padova, Italy

The chemiosmotic model of energy transduction, along with the postulate that respiration in State 4 mitochondria is due solely to passive proton leaks through the membrane, leads to the prediction that the simple relationship between State 4 respiration and $\Delta\tilde{\mu}_H$:

$$nJ_e = L_H \Delta\tilde{\mu}_H \quad (1)$$

(n:H^+/e stoichiometry; J_e: electron flow; L_H: membrane leak conductance) should always be obeyed. Viceversa, it is now established (1-3) that electron flow must be severely inhibited before a noticeable decrease of $\Delta\tilde{\mu}_H$ takes place. The quantity $nJ_e/\Delta\tilde{\mu}_H$, which should be equal to L_H according to (1), does not remain constant as $\Delta\tilde{\mu}_H$ is varied, but behaves instead as shown in trace A of Fig. 2.

To rationalize this behaviour, it has been proposed (1,2) that L_H depends on $\Delta\tilde{\mu}_H$, increasing exponentially above a threshold value of the latter parameter ("non-ohmic" or "variable" conductance). The concept of "slipping" pumps has been put forward as an alternative explanation (3). A slipping pump would transfer electrons without simultaneously moving protons from the matrix to the cytosolic phase, or it would allow re-entry of protons without a concurrent reverse movement of electrons. The relative constancy of $\Delta\tilde{\mu}_H$ would be due to the simultaneous inhibition of proton pumping and of slipping.

We have now determined the dependence of the basal proton conductivity of the mitochondrial membrane on $\Delta\tilde{\mu}_H$ by measuring the flux of the electrically compensating species, K^+, at varying diffusion potentials (see legend to Fig. 1). Fig. 1 shows the results obtained with State 4 mitochondria as well as with mitochondria uncoupled with 10 μM dinitrophenol. In all cases L_H, calculated as $J_{K^+}/\Delta\tilde{\mu}_H$, remains constant in the range of $\Delta\tilde{\mu}_H$ explored with no indication of an increase at high $\Delta\tilde{\mu}_H$. These results may be compared to those presented in Fig. 2 which shows the behaviour of L_H as calculated from $J_e/\Delta\tilde{\mu}_H$.

We conclude that the "slipping-pump" model provides a more likely explanation than the "variable conductance" model for the experimental results. Furthermore, it appears that in addition to increasing the permeability of the membrane to protons, uncouplers modify the behaviour of the proton pumps, so as to apparently increase slippage.

287

Fig. 1   Fig. 2

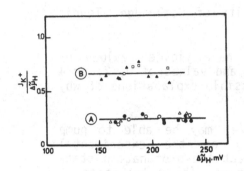
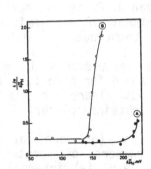

Fig. 1 - **Membrane leak proton conductance (L_H) vs. $\Delta\tilde{\mu}_H$ as determined from K^+ efflux data.**
Mitochondria (2 mg/ml) were incubated in State 4 for 1'. An excess
(50 ng/mg) of Antimycin A was added, followed A) by valinomycin (45
pmol/mg) after 10", or B) by DNP (10 μM) after 5" and valinomycin (90
pmol/mg) after 10". The initial rate of the valinomycin-induced K^+
efflux was used to calculate L_H as $J_{K^+}/\Delta\tilde{\mu}_H$. External K^+ concentration was varied to modulate the K^+ diffusion potential. $\Delta\tilde{\mu}_H$ was calculated as the sum of the K^+ diffusion potential and ΔpH. Medium composition: 0.24 M Sucrose, 1 mM Succinate/Tris, 2 mM P_i/Tris, 0.2 mM
EDTA/Tris, 4 μM Rotenone. pH 7.4.

Fig. 2 - **L_H vs $\Delta\tilde{\mu}_H$ as determined from respiration data.**
Respiration in State 4 (A) or in the presence of 10 μM DNP (B) was titrated with Antimycin. $\Delta\tilde{\mu}_H$ was determined in parallel experiments from
^{14}C-TPMP and ^{14}C-DMO distribution data. Medium composition: 0.15 M Sucrose, 30 mM KCl, 10 mM Choline Cl⁻, 10 mM Succinate/Tris, 10 mM Tris
/Mops, 2 μM Rotenone, 1 μg/mg prot. Oligomycin. pH 7.4. T: 25°C. 2
mg prot/ml.

1) Nicholls, D.G. (1974) Eur. J. Biochem. 50, 305-315
2) Sorgato, M.C. & Ferguson, S.J. (1979) Biochemistry 18, 5737-42
3) Pietrobon, D., Azzone, G.F. & Walz, D. (1981) Eur. J. Biochem.
   117, 389-394

APPARENT VARIABILITY OF THE MITOCHONDRIAL $H^+/2e$ RATIO CAUSED BY N,N'-DICYLOHEXYLCARBODIIMIDE AND Mg^{2+}

Brendan D. Price and Martin D. Brand
Department of Biochemistry, University of Cambridge, Tennis Court Rd., Cambridge, CB2 1QW, U.K.

$H^+/2e$ measurements in mitochondria have yielded values of between 2 and 4 for site I, 4 for site II and values of 0, 2, or 4 for site III. There are a number of possible explanations of why the values obtained differ:

(i) One (or more) of the complexes may be able to pump variable numbers of protons depending on the experimental conditions, e.g. (a) internal pH may affect the protonation state of a proton pump, decreasing its efficiency; (b) redox state of the chain may be important, since some methods evaluate $H^+/2e$ when the chain is oxidised, others when it is reduced; (c) slip reactions may occur, whereby a high PMF thermodynamic back pressure may encourage the respiratory chain to pump fewer protons per electron; (d) other parameters, such as rate of electron transport, the route of electron flow, the ionic composition of the medium, etc., may have important effects.

(ii) Gated leaks. If electron transport is limited by e.g. malonate, a graph of PMF versus electron transport rate shows a near plateau at around 180mV, even though the rate is not yet maximal. One explanation is that there is a PMF gated leak e.g. a rapid H^+ uniport or K^+/H^+ exchange in the presence of a K^+ uniport, leading to K^+ cycling and net proton influx. If such a leak were to operate during measurement of the $H^+/2e$, it could lead to a serious underestimation of the true value.

(iii) Localised chemiosmosis. This explanation says that not all of the translocated protons equilibrate with the bulk phase, causing underestimation of the stoichiometry. The extent of localization of the protons could be dependent on pH, ion concentration, extent of swelling etc.

We have previously shown that DCCD decreases the $H^+/2e$ of b-c1 complex from 3.8+/-0.2 to 2.1+/0.1 (at 100nmol/mg). This effect was not due to uncoupling, inhibition of electron transport, destruction of the membrane or other obvious trivial explanations (1). This led us to conclude that DCCD specifically inhibited the b-c1 proton pump.

Similarly, we now report that Mg^{2+} decreases the $H^+/2e$ for succinate to oxygen from 5.9+/-0.2 to 4.7+/-0.2 but has only a small effect on the $H^+/2e$ of b-c₁ complex (4.0+/-0.2 to 3.6+/-0.1) and none on cytochrome oxidase. In many ways, DCCD and Mg^{2+} are similar in that they both increase state 4 respiration rates, inhibit K^+ uniport, decrease PMF by several mV, and make cytochrome c more reduced. Mg^+ also causes a 30% increase in state 3 rates of respiration.

Are these effects due to change in mechanistic stoichiometry? We have also assessed the $H^+/2e$ for the b-c₁ complex by measuring the force ratio (ratio of the redox drop to the PMF across the b-c₁ complex). Under these conditions, the thermodynamic assessment shows no change in the $H^+/2e$ with DCCD whilst the kinetic method shows a large drop (2). Thus DCCD does not change the $H^+/2e$ of the b-c₁ complex, contrary to our earlier conclusion.

Other experiments with both DCCD and Mg^{2+} do not rule out that these compounds may modulate the gated leak by changing various ion cycling systems, but the results are not clear. In the same way, they may be inducing localised circuits in the mitochondria, but we have been unable to think of any real test for this. We conclude that DCCD does not change the $H^+/2e$ of the b-c₁ complex; whether Mg^{2+} does we have not yet determined.

(1)   Price, D.B. and Brand, M.D., Eur. J. Biochem. (1983), 132, 595-601.
(2)   Brand, M.D. et al. (1984), Biochem. J., submitted

Localization of Adenine Nucleotide Tight Binding Sites on Thylakoid-bound Chloroplast Coupling Factor 1 by the Use of 2-Azido Photo-affinity Analogs

M.S. Abbott[o], J.J. Czarnecki[", K.R. Dunham[o], and B.R. Selman[o]
[o]Department of Biochemistry, University of Wisconsin-Madison
"Institute for Enzyme Research, University of Wisconsin-Madison
Madison, Wisconsin, 53706   USA

The interactions of the 2-azido (photoaffinity) adenine nucleotide analogs with the spinach thylakoid-bound coupling factor 1 (CF_1) have been intensively studied. $2\text{-}N_3\text{-}ADP$ is a substrate for photophosphorylation (K_m ca. 43 µM), binds noncovalently to the ADP tight (exchangeable) binding site (K_{app} ca. 1.0 µM), and inhibits the "light-triggered", Mg^{2+}-dependent CF_1 ATPase activity (K_i ca. 1-2 µM) (1,2). Ultraviolet irradiation of noncovalently bound $2\text{-}N_3\text{-}ADP$, bound to the ADP tight (exchangeable) binding site, results in the covalent incorporation of $2\text{-}N_3\text{-}ADP$ exclusively into thylakoid-bound CF_1 (ca. 0.5 mol/mol CF_1). Covalent modification of CF_1 leads to a loss of both methanol-induced (thylakoid-bound) and octylglucoside-induced CF_1 Mg^{2+}-dependent ATPase activity. The level of binding corresponding to 100% inhibition of the ATPase activity extrapolates to 1.0 mol $2\text{-}N_3\text{-}ADP$ covalently bound/mol CF_1 (3).

The covalent incorporation of $(\beta\text{-}^{32}P)2\text{-}N_3\text{-}ADP$, noncovalently bound to the ADP tight binding site, results in the exclusive modification of the β-subunit of CF_1 (1,2). Fingerprint maps of the modified β-subunit have been constructed using the S. aureus protease and separating the polypeptide fragments by SDS-polyacrylamide gel electrophoresis (4).

Illumination of thylakoid membranes in the presence of $2\text{-}N_3\text{-}ADP$ and (^{32}P)phosphate results in the rapid formation of tightly bound, noncovalent, $(\gamma\text{-}^{32}P)2\text{-}N_3\text{-}ATP$ (ca. 0.1-0.2 mol/mol CF_1). Upon UV irradiation, the $(\gamma\text{-}^{32}P)2\text{-}N_3\text{-}ATP$ covalently binds to the thylakoid-bound CF_1. All of the label in CF_1 has been found to be localized in the β-subunit of the enzyme.

Fig. 1 shows a comparison of the fingerprint maps of $2\text{-}N_3\text{-}ADP$ and $2\text{-}N_3\text{-}ATP$ covalently bound to the β-subunit. It is apparent from these data that the tight binding sites for both analogs, and hence for ADP and ATP, are on the β-subunit of CF_1 and are absolutely identical.

[This research was supported in part by NIH grants GM 31384 (to BRS) and AM 10334 (to Henry A. Lardy).]

References

1. Czarnecki, J.J., Abbott, M.S. and Selman, B.R. (1982) Proc. Natl. Acad. Sci. USA 79, 7744-7748.
2. Czarnecki, J.J., Abbott, M.S. and Selman, B.R. (1983) Eur. J. Biochem. 136, 19-24.
3. Czarnecki, J.J., Dunham, K.R. and Selman, B.R., submitted.
4. Abbott, M.S., Czarnecki, J.J. and Selman, B.R., submitted.

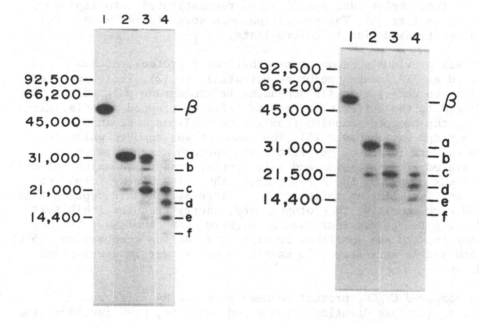

Figure 1: Comparison of the proteolytic cleavage patterns of β-subunits of CF_1 photolabeled with $2-N_3$-adenine nucleotide analogs. β indicates the position of the undigested β-subunits. The letters a-f indicate those proteolytic fragments which contain the majority of the label. Lane 1, autoradiograph of the β-subunit of CF_1 isolated from chloroplast thylakoid membranes photolabeled with $(\gamma-{}^{32}P)2-N_3$-ATP (left) or $(\beta-{}^{32}P)2-N_3$-ADP (right); Lane 2, autoradiograph of photolabeled β-subunit incubated with 15 ng S. aureus protease V8; Lane 3, 100 ng; Lane 4, 300 ng.

ENERGY TRANSDUCTION BY CF_0CF_1 PROTEOLIPOSOMES.

A. Admon and M. Avron
Department of Biochemistry, The Weizmann Institute of Science, Rehovot,
Israel 76100.

The ATP synthase (CF_0CF_1) of chloroplasts is believed to couple
proton flow to ATP synthesis or hydrolysis. The enzyme has been
isolated from thylakoids, purified and reconstituted into liposomes
in an active form (1). The proteoliposomes were studied as a model
for energy transduction in chloroplasts.

It was previously demonstrated that these proteoliposomes
catalysed an ATP induced membrane potential, $\Delta\psi$ (2). Their reported
inability to catalyse an ATP dependent transmembrane pH gradient (ΔpH)
was traced to the buffering effect of leftover trapped ammonia. After
removing the trapped ammonium from the proteoliposomes, an ATP
dependent ΔpH was observed (3). The ammonium was removed while the
liposomes were formed by gel filtration through a sephadex column
equilibrated with potassium and valinomycin. Trapped ammonium buffered
the protons pumped by the ATP synthase, thus, inhibiting formation
of ΔpH and facilitating formation of a large $\Delta\psi$..The ATP dependent $\Delta\psi$
and ΔpH were sensitive to protonophores, energy transfer inhibitors
and detergents. $\Delta\psi$ was increased by nigericin and abolished by
valinomycin. ΔpH was abolished by nigericin. At low temperatures (5°C)
both ΔpH and $\Delta\psi$ were larger in magnitude but slower in development
and decay.

We compared CF_0CF_1 proteoliposomes prepared by different
techniques (cholate dilution, freeze and sonicate, 37°C incorporation
and gel filtration through sephadex with and without valinomycin)
regarding their ability to catalyse acid-base induced phosphorylation.
Proteoliposomes prepared by gel filtration with or without valinomycin
were most effective. Acid-base phosphorylation could be performed by
incubating preformed liposomes in succinate buffer at low pH (around
5) or by forming the liposomes on a sephadex column equilibrated with
succinate or MES buffer at the low pH. The pH of the medium was then
rapidly raised to pH 8-8.5 by injecting Tricine buffer. Both $\Delta\psi$ and
ΔpH were important as energy sources for optimal acid-base phosphorylation. The optimum pH of the acid incubation step was around pH 5,
since the proteoliposomes were apparently unstable at lower values.
MES was effective as an intravesicular buffer only when introduced on
the sephadex column before the liposomes sealed. Succinate, on the
other hand, was effective also when introduced to sealed liposomes.
Ten sec incubation of the proteoliposomes in succinate buffer at low

pH were sufficient for maximal effectiveness. A minimum of about 180-200 mv of transmembrane electrochemical potential were needed for phosphorylation to take place. Low temperatures slowed the kinetics of the phosphorylation but did not affect its magnitude. The decay of the capacity to phosphorylate after the acid-base treatment, was slowed by the presence of low concentrations of ATP or ADP.

It is concluded that in CF_0CF_1 proteoliposomes, as in chloroplasts, transmembrane electrochemical gradients of protons can drive ATP synthesis, and be produced as a result of its hydrolysis.

1. Pick, U. and Racker, E. (1979) J. Biol. Chem. 254, 2793-2799.
2. Shahak, Y., Admon, A. and Avron, M. (1982) FEBS Lett. 150, 27-31.
3. Admon, A., Pick, U. and Avron, M. (1983) Procd. 6th International Congress on Photosynthesis, Brussels. In press.

INVOLVEMENT OF NUCLEOTIDES CONCENTRATION GRADIENTS IN ATP FORMATION BY
THE MEMBRANE BOUND ATP SYNTHETASE

Claude Aflalo and Noun Shavit
Department of Biology, Ben Gurion University of the Negev,
84105 Beer Sheva, Israel.

Several molecular mechanisms proposed for ATP synthetase and its
regulation are based on macroscopic kinetic measurements and studies
of tight binding of nucleotides (1). The data have been interpreted in
terms of conformational changes and interactions between nucleotides
binding sites, implying tacitly a kinetic behavior for ATP synthetase
similar to that of an enzyme in solution. In view of the electrochemical properties of the thylakoid membrane, we expect that the observed
kinetic parameters of the membrane bound enzyme should reflect also
the interactions between the membrane and the substrates, products or
effectors of the enzyme and not only intrinsic properties, which are
relevant to its molecular mechanism. Extensive experimental support
for this is provided by studies conducted in model and biological
systems as well. With enzymes immobilized on solid supports, it has
been shown that impaired distribution or translocation of reactants
between bulk medium and the matrix affects strongly their macroscopic
kinetic behavior (2). Also, the apparent affinity of enzymes bound on
biological membranes to charged substrates seems to depend strongly on
the surface charge of the membrane, whether the turnover of the
enzymes remains unchanged and similar to that obtained after
solubilization (3).

Studies on the accumulation of tightly bound ATP during photophosphorylation in the presence of hexokinase revealed that the ATP accumulated must arise from the rebinding of newly formed ATP to noncatalytic sites (4). Most of the AT^{32}P accumulation is prevented by addition of unlabeled ATP either during or after illumination. Moreover,
the rate of phosphorylation is not affected by the exchange of tightly
bound AT^{32}P during catalysis, indicating that catalytic sites are not
involved in the tight binding process. Addition of an efficient hexokinase trap to the phosphorylation assay fails to prevent the accumulation of tightly bound ATP in a dark period following a pulse of
synthesis. It appears that, under steady state conditions, the newly
released AT^{32}P accumulates in a space near the membrane which is
accessible to external ATP but not to hexokinase; this ATP represents
an intermediate between its formation the membrane on one hand, and
its diffusion to the medium or its rebinding to non catalytic sites of
ATP synthetase on the other hand. The existence of such a nucleotide
concentration gradient should be reflected also in the dependence of
phosphorylation on the bulk nucleotide concentration, and should
depend on factors affecting the microenvironment of the membrane.

Hypotonically treated thylakoid membranes show a marked reduction of the level of rapidly labeled tightly bound ATP (4). Moreover, the apparent affinity of thylakoids for ADP is improved by the hypotonic treatment (higher V_{max}/K_m, lower K_m). A similar effect is observed also when the ionic strengh of the reaction mix is increased; in both cases, the catalytic ability of the system at saturating ADP remains unaffected. The effect is apparent only with "high affinity" substrates (ADP as compared with GDP or P_i). Also, the effect on the apparent affinity tends to vanish upon artificial reduction of the turnover of the enzyme (5,6). These results indicate that the extent of nucleotides gradients could be modulated by the balance between the rate of their utilization and that of their mass transfer to (or from) the bulk medium. The latter process seems to be affected by electrostatic interactions between charges on the membrane and those of the nucleotides. To test this, we altered directly the surface charge of the membrane using sub-solubilizing concentrations of charged detergents or lipophilic ions. However, all compouds tested inhibit the catalytic ability (V_{max}) of phosphorylation and affect equivocally the apparent affinity for ADP. While charged detergents lowered the pseudo first order rate constant for ADP utilization (V_{max}/K_m) without changing the apparent K_m, the lipophilic ions acted differently (constant V_{max}/K_m, lower K_m). The effect on the affinity is independent of the nature of the charge added to the membrane. The charged surfactants appear to act primarily on electron transfer. However the phosphorylation induced by an acid-base transition is also inhibited independently of the charge on the membrane.

Thus, the approach used to test the involvement of membrane charge on heterogenous distribution of reactants, does not provide the answers expected. Nevertheless, the existence of nucleotides gradients between the membrane and the bulk phase still represents a good alternative to the various models proposed for ATP synthesis based on affinity binding changes, "intermediate" or "medium" exchange reactions and inaccessibility of "bound" products to enzymes added in the medium (1).

1. Cross, R. L. (1981) Ann. Rev. Biochem. 50, 681-714.
2. Engasser, J-M. & Horvath, C. (1976) Appl. Biochem. Bioeng. 1, 127-220.
3. Wojtczak, L. & Nalecz, M.J. (1979) Eur. J. Biochem. 94, 99-107.
4. Aflalo, C. & Shavit, N. (1982) Eur. J. Biochem. 126, 61-68.
5. Aflalo, C. & Shavit, N. (1983) FEBS Lett. 154, 175-179.
6. Aflalo, C. & Shavit, N. (1984) Current Topics in Cellular Regulation, 24, 435-445.

INTERACTION OF NITROPHENOLS WITH THE MITOCHONDRIAL ATPase

O. Akinpelu, R. Kiehl, and W.G. Hanstein

Institut für Physiologische Chemie, Ruhr-Universität Bochum,
D-4630 Bochum, West-Germany

It has long been known that uncouplers stimulate the mitochondrial ATPase in the membrane-bound and soluble form(F_1).The present study focusses on the interaction of F_1 with nitrophenols, nucleotides and the mitochondrial inhibitor protein (IF_1).

2-Azido-4-nitrophenol (NPA) stimulates the ATPase activity of F_1 by 30%, and in submitochondrial particles by a factor of up to 2.5, depending on the method of particle preparation. In S-particles (where IF_1 has been removed by Sephadex treatment), no such stimulation occurs. In these particles, incorporation of NPA into the β subunits by photoaffinity labeling is up to two times higher than in untreated particles. Similarly, addition of IF_1 to F_1 in the presence of Mg-ATP decreases NPA-labeling in β, but not α . These data indicate that IF_1 inhibits the binding of nitrophenols, and suggest that uncouplers stimulate the ATPase by weakening the interaction between IF_1 and F_1. This is in agreement with findings of others that i) high uncoupler concentrations solubilize the inhibitor protein [1], ii) the β subunit interacts with the inhibitor protein [2], and iii) in intact mito-chondria, the α, but not the β subunit is labeled by NPA [3].

With repeated cycles of photoaffinity labeling of F_1 with NPA (in-cluding removal of non-covalently bound photo-products by the centri-fuged column procedure [4]),it is possible to saturate NPA binding (Fig. 1). The number of binding sites is 5-6, the same as for nucleo-tides [4]. During this treatment, the ATPase activity does not be-come inhibited as compared to the control. The number of tightly bound nucleotides, however, decreases. Conversely, addition of Mg-ATP decreases the labeling of subunit β, while the incorporation of NPA into subunit α remains unchanged.

These and the above data suggest interactions between the binding sites of nucleotides, nitrophenols and IF_1 on the β subunit of F_1.

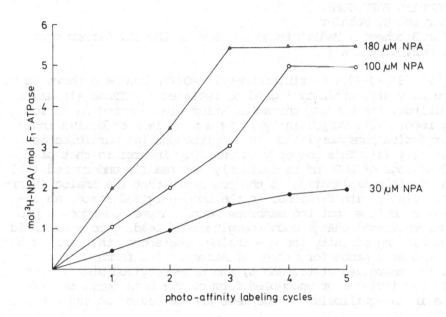

<u>Fig. 1:</u> Determination of the number of nitrophenol binding sites in
F₁. Repetitive photoaffinity labeling with the NPA anion.
Each cycle includes photolabeling, removal of remaining NPA
and its non-covalently bound photo products, and addition of
the indicated concentrations of fresh NPA.

[1] Yamada, E.W., and Huzel, N.J. (1982) Can. J. Biochem. Cell Biol.
<u>61</u>, 1006-1011

[2] Klein, G., Satre, M., Dianoux, A.-C., and Vignais, P.V. (1981)
Biochemistry <u>20</u>, 1339-1344

[3] Hanstein, W.G. (1979) Meth. Enz. <u>56</u>, 653-683

[4] Cross, R.L., and Nalin, C.M. (1982) J. Biol. Chem. <u>257</u>, 2874-2881

This work was supported by DFG grant Ha 1124.

INFLUENCE OF TNP-ATP ON THE STEADY-STATE PHOSPHORYLATION POTENTIAL OF SUBMITOCHONDRIAL PARTICLES

S. Anemüller and G. Schäfer
Institut für Biochemie, Medizinische Hochschule Lübeck, Ratzeburger
Allee 160, D-2400 Lübeck 1

The analog 2'-(3')-0-(2,4,6-trinitrophenyl)-AD(T)P has been shown to be
an extraordinary high affinity ligand to isolated F_1-ATPase (1) and a
strong inhibitor. Its binding characteristics are reported in an accompanying report (2). Surprisingly it is a much less effective inhibitor of oxydative phosphorylation in submitochondrial particles as
shown previously (3). This property is entirely inverse to that of
other 3'-0-analogs of ADP and is certainly not readily understood (3).
However, a possible explanation is the postulate that the analog interacts preferentially with F_1-ATPase in the non energized state, an assumption which implies that the membrane bound enzyme undergoes a pronounced conformational change upon energization, leading to a dramatic
decrease of binding affinity for the analog. Therefore, this analog has
been suggested as a probe for differentiation of two forms of the
enzyme E*, the energized state, and E, the non-energized state, corresponding to the isolated or uncoupled form of F_1; both forms are assumed to be in an equilibrium on the membrane depending on the overall
ΔpH or $\Delta \Psi$.
Former studies in support of this hypothesis were performed with a lu-
ziferase coupled direct assay of the steady-state phosphorylation potential in suspensions of respiring submitochondiral particles. This
assay is unsuitable, however, for quantitative measurements in SMP suspensions because the actual ATP-concentrations are far above the range
of linear response.
Therefore, the measurements have been repeated by determination of the
phosphorylation potential ΔGp via ^{32}P-incorporation. Our former results
could be confirmed and extended, demonstrating that addition of TNP-ATP
to respiring SMPs in a phosphorylating steady state increases the phosphorylation potential in a concentration dependent manner. The effect
was titrated as shown in fig. 1. Presumably ATPase due to a fraction of
uncoupled (leaky) SMPs is inhibited causing an increase of the overall
ΔGp in the mixture. A decrease of ΔGp by TNP-ATP was never observed.
It is concluded that the ratio of E*/E is increased by TNP-ATP in respiring SMPs.
Whereas a sudden stop of respiration by antimycin/rotenone addition
leads to a rapid decrease of [ATP] and the ΔGp in respiring SMPs, in
presence of TNP-ATP the drop of [ATP] and ΔGp is strongly decreased
or totally abolished depending on the concentration of the analog. Obviously immediately on transition from E* to E the latter is chased by
TNP-ATP and converted into an inactive inhibitor-enzyme complex.

Furthermore, the ratio of E*/E can be artificially modified by addition of soluble F_1-ATPase to respiring SMPs; the immediate decrease of the overall ΔGp is readily reversed by TNP-ATP and ΔGp is recovered. Thus, the isotope studies support our former conclusion that E and E* expose significantly different structures of the substrate binding domains and that F_1 indeed is a 'dual state enzyme', the ratio of E*/E being regulated by the proton motive force across the membrane.

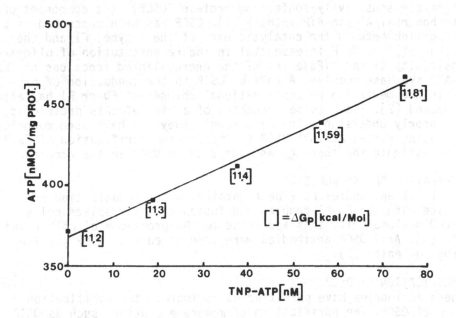

Fig. 1: Dependency of steady state [ATP] and ΔGp on concentration of [TNP-ATP] added during steady state phosphorylation to SMPs. ΔGp given in paranthesis.
(SMP = 123 ug/ml; AP$_5$A = 40 uM; [ADP]$_O$ = 67 uM; T = 25°C; NADH-regenerating system; standard phosphorylation buffer (3)

References
(1) Grubmeyer, C., and Penefsky, H. S. (1981), J. Biol. Chem. 256, 3718-3727
(2) Tiedge, H., and Schäfer, G. (1984), this volume p. 367
(3) Schäfer, G. (1982), FEBS-Letters 139, 271-275

300

USE OF MONOCLONAL ANTIBODIES IN THE STUDY OF THE OLIGOMYCIN SENSITIVITY
CONFERRING PROTEIN

Ph. Archinard, F. Penin and C. Godinot
Laboratoire de Biologie et Technologie des Membranes du CNRS,
Université Claude Bernard , F 69622 Villeurbanne, FRANCE

Oligomycin sensitivity-conferring protein (OSCP) is a component of
the mitochondrial ATPase-ATPsynthase [1]. OSCP has been described as a
connecting link between the catalytic part of the enzyme, F1 and the
proton channel, FO. OSCP is essential in the reconstitution of oligomycin sensitivity of the ATPase and of the energy-linked functions of the
ATPase-ATP synthase complex. A role of OSCP in the conduction of protons or in the mediation of conformational changes of FO or F1 has also
been proposed [2]. The precise mechanism of action of this protein is,
however, poorly understood. In the present study, we have used monoclonal antibodies raised against OSCP to improve the purification of OSCP
and to investigate the topology and function of OSCP in the complex.

PREPARATION OF ANTIBODIES
Monoclonal antibodies have been obtained after immunization of
Balb/c mice with the FO-F1 complex and fusion of their spleen cells
with SP2-0 myeloma cell lines according to the procedure of Kölher and
Milstein [3]. Anti-OSCP antibodies were identified with OSCP purified
according to Senior [2].

PURIFICATION OF OSCP
These antibodies have permitted us to improve the purification
procedure of OSCP. The purification of membrane peptides such as OSCP
is difficult since it is devoided of intrinsic enzymatic activity and
only active in reconstitution experiments. Besides, in SDS-PAGE, OSCP
is poorly separated from two other neighbouring peptides. Monoclonal
antibodies can serve as probes to follow the purification of OSCP. At
each step of the purification procedure, samples were analyzed by SDS-
PAGE followed by electroblotting onto nitrocellulose sheets [4] and
the peptide was visualized by immunodecoration with its specific
antibodies. This technique has been applied to test the efficiency of
the purification procedure of OSCP. The first step of our new method
(alkaline extraction of mitochondria) was the same as that used by
Horstman and Racker [5] to purify IF1 (inhibitory peptide of F1). The
membranes were then extensively washed to eliminate some peptides which
would otherwise be difficult to separate from OSCP. OSCP was finally
solubilized by ammonia as described by Senior [2] and purified by an
improved ion-exchange chromatography. This method is simple and can be
performed within one day.

ACCESSIBILTY OF OSCP TO THE ANTIBODIES

Various experiments were conducted to study the accessibility of the epitopes recognized by the antibodies when OSCP was reconstituted with F1 or in electron transport particles (ETP). In solid phase radio-immunoassays (SPRIA), F1 did not decrease the binding of the antibody to OSCP when purified F1 was reassociated with purified OSCP. In addition, OSCP was accessible to the antibodies in ETP whether the accessibility was tested by direct or competitive SPRIA or by electron microscopy with the immunogold method [6]. (Thanks are due to Dr. J. Comte for performing the electron microscopy experiments.)

EFFECTS OF THE ANTIBODIES ON ATP SYNTHESIS

The antibodies had no effect on the succinate-dependent net ATP synthesis of ETP or of ETP depleted from F1 by urea treatment and reconstituted with F1 and OSCP except if the antibodies were preincubated with OSCP before the reconstitution. This means that the antibodies bind to OSCP integrated in the membrane without interfering with ATP synthesis or proton translocation. The OSCP-antibody complex cannot however be used to reconstitute F1 with F1-depleted ETP.

PEPTIDES CLOSE TO OSCP IN THE COMPLEX

Cross-linking experiments made either by using the antibodies to identify the cross-linked products of OSCP or by direct photolabeling of the cross-linked products of OSCP revealed that major neighbours of OSCP were the alpha and gamma subunits of F1 and at least two other peptides of about 30 and 10 kdalton, not yet identified.

All these results are not easily explained by the model generally proposed where OSCP together with F6 constitutes the stalk of the ATPase-ATP synthase complex. Our results are in agreement with the role of OSCP in the binding of F1 to the membrane but the exact location of OSCP must rather be peripheral, on a side or surrounding the complex to explain its easy accessibility to the antibodies. OSCP must also be in close interaction with the alpha and gamma subunits of F1. Further studies are in progress to determine the sequences recognized by the antibodies and their location on the complex.

1. Mc Lennan, D.M. and Tzagoloff, A. (1968) Biochemistry 7, 1603-1610.
2. Senior, A.E. (1979) Methods Enzymol. 55, 391-397.
3. Köhler, G. and Milstein, C. (1975) Nature (London) 256, 495-497.
4. Towbin, H. Staehlin, T. and Gordon, J. (1979) Proc. Nat. Acad. Sci.
5. Horstman, L.L. and Racker, E. (1970) J. Biol. Chem. 245, 1336-1344.
6. Faulk, W.P. and Taylor G.N. (1971) Immunochemistry, 8, 1081-1083.

MODULATION BY PHOSPHATE AND ACIDIC PH OF THE ADP-INDUCED INHIBITION
OF MITOCHONDRIAL F_1-ATPASE.

A. Di Pietro and G. Fellous.
Laboratoire de Biologie et Technologie des Membranes du CNRS,
Université Claude Bernard de Lyon, 69622 Villeurbanne cedex, France.

It was previously shown that preincubation of pig heart mitochondrial F_1-ATPase with ADP in the presence of Mg^{2+} induced ADP binding at regulatory site (1) specific of adenine nucleotides (2, 3). This ADP binding occured on a β-subunit (4) and induced a conformational change leading to a more compact structure of the enzyme as monitored by protection against the subunits proteolysis by trypsin and the concomitant inactivation of the enzyme (5). Upon dilution into an ATPase assay medium, the bound ADP induced a hysteretic inhibition which progressively increased during MgATP hydrolysis to finally lead to stable low rate of activity (1).

The present data show that preincubation with Pi in addition to ADP suppresses the lag period preceding the stable inhibited-rate of activity. The figure indicates that although Pi has no significant effect when preincubated alone with F_1, it is however able to induce an immediate inhibited-rate of ATPase activity when preincubated together with ADP whereas the final level of the ADP-induced hysteretic inhibition is reached after a lag of about 1 min. The Pi effect is half maximal for a concentration of 35 µM at pH 6.6. On the contrary, only a very limited effect is observed at pH 8.0 even for high Pi concentrations. This pH dependence indicates that the monovalent $H_2PO_4^-$ form is involved, although a direct effect of protons cannot be excluded. The Pi effect requires a 30 min-preincubation and is directly correlated with the binding of one mole ^{32}Pi, (K_D = 33 µM) per mol of enzyme.

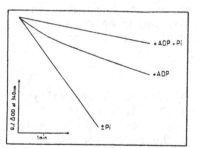

EFFECTS OF INORGANIC PHOSPHATE ON THE TIME
DEPENDENCE OF HYSTERETIC INHIBITION.

If F_1-ATPase with variable activity is preincubated with ADP in the absence of Pi, the stable low-rate of hysteretic inhibition is always obtained for a constant amount of ATP hydrolyzed and therefore of Pi produced. All these results allow to interpret the hysteretic inhibition as due to binding of Pi generated by MgATP hydrolysis to the ADP-preincubated enzyme; however the setting up of inhibition do provided Pi is already present with ADP during preincubation.

Kinetic experiments indicate that the inhibition induced by ADP plus Pi is uncompetitive, since both Vm and Km are decreased. Such a change in affinity for MgATP, together with the slowness of Pi effect during preincubation, favour a conformational change, consistent with that previously monitored by decreased sensitivity of the activity to anion activation (6) and by increased α-helix content of the circular dichroism spectrum(7).

The high-affinity Pi binding site(8) has been localized on a β-subunit(9). It is not known whether the ADP regulatory site is located on the same β-subunit(4) and in the vicinity of the Pi site nor whether this binding of ADP and Pi plays a regulatory or a catalytic role in ATP synthesis by the membrane-bound enzyme. The inhibition induced by ADP plus Pi and that induced by the natural ATPase inhibition IF$_1$ show a number of common features and preliminary results suggest that the enzyme conformation is similar in both cases. Experiments are in progress to better characterize these points.

1 - Di Pietro, A., Penin, F., Godinot, C. and Gautheron, D.C. (1980) Biochemistry 19, 5671-5678.
2 - Baubichon H., Godinot, C., Di Pietro, A. and Gautheron, D.C. (1981) Biochem. Biophys. Res. Commun. 100, 1032-1038.
3 - Di Pietro, A., Godinot, C. and Gautheron, D.C. (1981) Biochemistry 20, 6312-6318.
4 - Fellous, G., Godinot, C., Baubichon, H., Di Pietro, A. and Gautheron D.C., submitted.
5 - Di Pietro, A., Godinot, C. and Gautheron, D.C. (1983) Biochemistry 22, 785-792.
6 - Baubichon, H., Di Pietro, A., Godinot, C. and Gautheron, D.C. 1982 FEBS Lett. 137, 261-264.
7 - Roux, B., Fellous, G. and Godinot, C. (1984) Biochemistry 23, 534-537.
8 - Penefsky, H.S. (1977) J. Biol. Chem. 252, 2891-2899.
9 - Lauquin, G.J.M., Pougeois, R. and Vignais, P.V. (1980) Biochemistry 19, 4620-4626.

STUDY OF THE INTERACTIONS BETWEEN THE OLIGOMYCIN SENSITIVITY CONFER-RING PROTEIN (OSCP) AND MITOCHONDRIAL F_1 SUBUNITS BY CROSS-LINKING AND IMMUNOCHARACTERIZATION OF THE CROSS-LINKED PRODUCTS

A. DUPUIS and P.V. VIGNAIS
DRF/Biochimie (CNRS/ERA 903 et INSERM U.191), Centre d'Etudes Nucléaires, 85X, 38041 Grenoble cedex, France.

The oligomycin sensitivity conferring protein (OSCP) (1) is a small basic protein of Mr 20967 (2) which belongs to the mitochondrial H^+-ATPase complex and may be the stalk which links the F_0 and F_1 sectors of the complex. It has an elongated shape with an axial ratio of 1/3 (3). OSCP is homologous with the δ subunit of the *E. coli* F_1-ATPase (4). The nature of the subunits which in F_1 interact with OSCP has been investigated so far by indirect approaches. By analogy with *E. coli* F_1-ATPase where limited proteolysis of the α subunit results in loss of the ability of F_1 to bind δ (5), beef heart F_1 modified in its α subunit by mild trypsin treatment fails to bind OSCP (6). This suggested direct interaction between α of the mitochondrial F_1 and OSCP. However, as recently reported, the β subunit is also partially digested by trypsin (7).

The present work concerns the direct identification of beef heart mitochondrial F_1 subunits that interact with OSCP by means of cross-linking reagents and immunochemical characterization of the cross-linked products. Two zero-length cross-linkers were used, namely the 1-(3-dimethylaminopropyl)-3-ethylcarbodiimide (EDAC) and the 1-ethoxy-carbonyl-2-ethoxy-1-2-dihydroquinoline (EEDQ). The cross-linked products were separated by Na Dod SO$_4$ polyacrylamide gel electro-phoresis. Compared to control F_1 and OSCP that were cross-linked separately, cross-linking of the mixture of F_1 and OSCP resulted in two supplementary products of Mr 75000-78000. The specificity of such crosslinks was ascertained by limited proteolysis and OSCP titration assays. The subunits in the two cross-linked products were identified after electrophoretic transfer from the polyacrylamide gel to nitro-cellulose sheets by reaction with specific antibodies directed against OSCP, α and β, followed by revelation of the immune-complexes by a second antibody conjugated to peroxidase and directed against the first one. The two cross-linked products of Mr 78000 and 75000 were identified as OSCP-α and OSCP-β, showing unambiguously that OSCP interacts with both α and β subunits. These results, however, did not indicate which of the two subunits α or β play a strategic function in the transfer of information by OSCP from the F_0 sector, where the binding site of oligomycin is located, to the F_1 sector which contains the catalytic site.

References

(1) Mac Lennan, D.H. and Tzagoloff, A. (1968) Biochemistry $\underline{7}$, 1603-1610.

(2) Ovchinnikov, Y.A., Modyanov, N.N., Grinkevich, V.A., Aldanova, N.A., Trubetskaya, O.E., Nazimov, I.V., Hundal, T. and Ernster, L. (1984) FEBS Lett. $\underline{166}$, 19-22.

(3) Dupuis, A., Zaccaï, G. and Satre, M. (1983) Biochemistry $\underline{22}$, 5951-5956.

(4) Walker, J.E., Runswick, M.J. and Saraste, M. (1982) FEBS Lett. $\underline{146}$, 393-396.

(5) Dunn, S.O., Heppel, L.A. and Fullmer, C.S. (1980) J. Biol. Chem. $\underline{255}$, 6891-6896.

(6) Hundal, T., Norling, B. and Ernster, L. (1983) FEBS Lett. $\underline{162}$, 5-10.

(7) Hollemans, M., Runswick, M.J., Fearnley, I.M. and Walker, J.E. (1983) J. Biol. Chem. $\underline{258}$, 9307-9313.

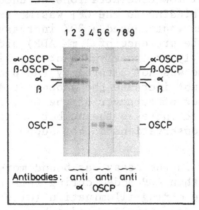

Figure : Immunochemical detection of the OSCP-F_1 cross-linked products obtained by means of EDAC reticulation, after electrophoretic blotting. The anti α, anti β and OSCP antibodies were raised in rabbit. The second antibody was an antirabbit antibody raised in goat and conjugated to peroxidase.
Lane 1 to 3 : Reaction with anti α antibody. Lane 1 : Control F_1.
Lane 2 : Control F_1 cross-linked by EDAC. Lane 3 : F_1-OSCP complex cross-linked by EDAC.
Lane 4 to 6 : Reaction with anti OSCP antibody. Lane 4 : F_1-OSCP complex cross-linked by EDAC. Lane 5 : Control OSCP cross-linked by EDAC. Lane 6 : Control OSCP.
Lane 7 to 8 : Reaction with anti β antibody. Lane 7 : Control F_1.
Lane 8 : Control F_1 cross-linked by EDAC. Lane 9 : F_1-OSCP complex cross-linked by EDAC.

ON THE MECHANISM OF UNCOUPLER-INDUCED STIMULATION OF PHOTOPHOSPHORY-
LATION IN CHLOROPLASTS

Ch. Giersch and J. Schumann
Botanisches Institut der Universität Düsseldorf
Universitätsstraße 1, D-4000 Düsseldorf, F.R.G.

The rate of photophosphorylation (H_2O → ferricyanide or methylviologen) by thylakoids freshly released from isolated intact chloroplasts
by osmotic shock is increased by the addition of uncouplers like nigericin, monensin or ammonium chloride at low concentrations (1). Preliminary
results suggested that this stimulation might be induced by an uncoupler-
mediated transient increase of the transmembrane electrical potential;
the latter is known to contribute to activation of the chloroplast coupling factor CF_1 (2).

As adenylate binding to the CF_1 is known to monitor activation of this
protein (3) the effect of low concentrations of uncouplers on the kinetics
and extent of adenylate binding to the CF_1 was measured. Addition of nigericin at concentrations that caused a 30% increase of the phosphorylation rate (measured in the presence of 1 mM ADP) did not affect the extent of nucleotides tightly or loosely bound; the kinetics of light-
induced adenylate release was not altered by the uncoupler. Moreover,
uncoupler treatment did not increase the amount of ATP produced in the
course of a limited number of turnover of the CF_1, suggesting that addition of uncouplers did not increase the number of activated coupling
factor molecules. These observations do not support the suggestion given
above.

ADP present initially in the assay was bound more rapidly during deenergization to the CF_1 than ADP added during this procedure. Addition
of glucose and hexokinase during illumination (either in the absence or
presence of stimulating uncouplers) caused enhancement of the steady-
state phosphorylation rate even at saturating ADP concentrations. These
findings are understood not to indicate an effect of uncouplers on
nucleotide binding; they rather suggest that adenylates in the neighbourhood of the CF_1 do not rapidly exchange with those present in the medium,
and that the CF_1-associated pool is small even in the presence of saturating ADP concentrations.

As the CF_1 is likely not to be responsible for uncoupler-induced
stimulation the site of action was further characterized by studying the
effect of different electron transport systems on phosphorylation. No
stimulation was observed with cyclic (PMS) or linear (DAD → methylviologen)
PSI electron transport. In the H_2O → methylviologen system, stimulation
was observed also under conditions where the phosphorylation rate was

decreased by DCMU. These data indicate that the site of uncoupler action is located in the electron transport chain between the DCMU binding site and the oxidizing side of PSI. However, though the CF_0 is largely considered as a non-regulated protein, from these data an involvement of this portion of the coupling factor in uncoupler-induced stimulation of phosphorylation cannot be excluded.

References

(1) Giersch, Ch. (1983) Biochim. Biophys. Acta 725, 309-319
(2) Strotmann, H. and Bickel-Sandkötter, S. (1984) Ann. Rev. Plant Physiol. 35, 97-120
(3) Harris, D.A. and Crofts, A.R. (1978) Biochim. Biophys. Acta 502, 87-102

K_M-VALUES OF P_i FOR ACID-BASE INDUCED ATP-SYNTHESIS IN
CHLOROPLASTS.

Peter Gräber and Gerlinda Thulke

Max-Volmer-Institut für physikalische und biophysikalische
Chemie, Technische Universität Berlin,
D-1000 Berlin 12, Germany

Recent investigations have shown, that the apparent K_M-
values of ADP in light-driven phosphorylation change with
energization (1,2) : when the ΔpH is decreased by decreasing light intensity or by addition of electron-transport
inhibitors, the K_M-values decrease ; when ΔpH is decreased
by addition of uncouplers, the K_M-values increase (2).

We have investigated the rate of ATP-synthesis by generating artificially a transmembrane ΔpH (by an acid-base-
transition) and a $\Delta\Psi$ (by a K^+/valinomycin diffusion potential). A rapid mixing quenched flow apparatus was used
to limit the reaction time to about 200 ms. In this range
the initial ΔpH and $\Delta\Psi$ remain constant (3).
At pH_{out}= 8.2 the rate of ATP-synthesis was measured as a
function of the P_i-concentration at different ΔpH and $\Delta\Psi$
values. The results are listed in Table I :

| Reaction conditions | | Rate | P_i-concentration at halfmax. rate |
pH ; $\Delta\Psi$/mV		V_{max} / $\dfrac{mM\ ATP}{M\ Chl\ s}$	P_i/M
2.8	0	29	$3 \cdot 10^{-4}$
2.8	85	170	$1 \cdot 10^{-4}$
3.2	0	75	$2.5 \cdot 10^{-4}$
3.2	85	275	$5 \cdot 10^{-5}$

It is obvious, that with increasing membrane energization the P_i-concentration for the half maximal rate decreases. The data obtained at low P_i-concentrations might indicate that there exist two binding sites for P_i; the first site is saturated at a P_i-concentration of about $2\ 10^{-5}$ M, the second site is saturated at about 10^{-3}M P_i.

(1) Bickel-Sandkötter, S., and Strotmann, H. (1981) FEBS Lett. 125, 188 - 192

(2) Vinkler, C. (1981) Biochem. Biophys. Res. Comm. 99, 1095 - 1100

(3) Gräber, P., Junesch, U., and Schatz, G.H. (1984) Ber. Bunsenges. Phys. Chem. 88, 599 - 608

SEQUENCE HOMOLOGY OF THE OLIGOMYCIN SENSITIVITY CONFERRING PROTEIN
(OSCP) AND FACTOR F_6 WITH OTHER SUBUNITS OF H^+-ATPase.

V.A. Grinkevich*, N.A. Aldanova*, P.V. Kostetsky*, N.N. Modyanov*,
T. Hundal", Yu.A. Ovchinnikov*, L. Ernster"
*Shemyakin Institute of Bioorganic Chemistry, USSR Academy of
Sciences, Moscow, USSR; "Department of Biochemistry, Arrhenius
Laboratory, University of Stockholm, Stockholm, Sweden.

"Oligomycin sensitivity conferring protein" (OSCP) and coupling
factor 6 (F_6) are components of the mitochondrial H^+-ATPase, participating in the interaction between the catalytic (F_1) and the proton-
translocating (F_0) moieties of the enzyme.

Determination of the complete amino acid sequence of OSCP (1) confirms the notion that this protein is a counterpart of the δ subunit
of E.coli F_1 (2,3). The N-terminal region of OSCP is more hydrophobic
(75% hydrophobicity) than the corresponding segment of the δ subunit
(60%):   OSCP /7-26/ P P V Q I Y G I E G R Y A T A L Y S A A
           δ /1-18/   M S E I T V A R P Y A K A A F D F A
and may serve to anchor OSCP in the membrane (4). This may explain
why OSCP remains bound to the membrane sector while the δ subunit in
E.coli accompanies F_1 when the latter is removed from the membrane.

Comparative analysis of the OSCP sequence with itself reveals a
striking pattern of a homologous repeat (4):
     1- 52 FAKLVRPPVQIYGIEGRYATALYSAASKQNKLEQVEKELLRVGQILK*EPKMA
   105-155 FSTMMSVHRGEVPCTVTTASALNEATLTELK**TVLKSFLKKGQVLKLEVKID

    53-104 ASLLNPYVKRSVKVKSLSDMTAKEKFSPLTSNLINLLAENGRLTNTPAVISA
   156-190 PSIMGGMIVRIGE*KYV*DMSAKTKIQKLSRAMRQIL

Comparison of the amino acid sequences of OSCP and the b subunit
of E.coli F_0F_1-ATPase (5-7) also shows a homology between the repeating portion of OSCP and the central part of the b subunit (4):
     OSCP / 1- 48/ FAKLVRPPVQIYGIEGRYATALYSAASKQNKLEQVEKELLRVGQILKE
       b /21- 67/ CMKYVWPPL*MAAIEKRQKEIADGLASAERAHKDLDLAKASATDQLKK

     OSCP /49- 94/ PK*MAASLLNPYVKRSVKVKSLSDMTAK*EKFSPLTSNLINLLAENGR
       b /68-115/ AKAEAQVIIEQANKRRSQILDEAKAEAEQERTKIVAQAQAEIEAERKR
Thus mitochondrial OSCP appears to contain structural elements of both
the δ and the b subunits of E.coli F_0F_1-ATPase.

A considerable sequence homology was unexpectedly found between the
repeating parts of OSCP/1-89/ and the ADP/ATP carrier/15-105/ (8,9).

The amino acid sequence of beef heart F_6 has recently been completed (10); it differs from that obtained by Bradshaw (11) in position
62: Thr instead of Phe. The polypeptide chain of F_6 as determined in
the present study (10) consists of 76 residues and contains an internal repeat: 1-35 NKELD*PVQKLFVDKIR*EYRTKRQ*TSGGPVDAGPEY
        36-76 QQDLDRELFKLKQMYGKADMNTFPNFTFEDPKFEVVEKPQS

The internal homology is not as pronounced as in case of OSCP; however, both of these proteins appear to have evolved by a process of gene duplication. In addition, OSCP and F_6 have homologous structural elements, mostly manifested when comparing the following regions:

OSCP / 1-40/ FAKLVRPPVQIYGIEGRYATALYSAASKQNKLEQVEKELL

F_6 /39-76/ LDRELFKLKQMYG*KADMNT*FPNFTFEDPKFEVVEKPQS

The N-terminal part of F_6 is homologous with the C-terminal region of the E.coli subunit c(12), including the DCCD-binding site /Asp-61/:

F_6 /1-39/ NKELDPVQKLFVDKIREYRTKRQTSGGPVDAGPEYQQDL

c /33-70/ GKFLEGAARQP*DLIPLLRTQFFIVMGLVDAIPMIAVGL

A certain homology is also found between F_6 and other H^+-ATPase subunits such as the mitochondrial ATPase inhibitor (13), the E.coli ε subunit (14) and its mitochondrial counterpart δ (2), as well as the amino acid sequence that can be deduced from URF A6L of mitochondrial DNA (15). Further investigations will have to establish whether these structural homologies are of functional significance.

References:
1. Ovchinnikov,Yu.A.,Modyanov,N.N.,Grinkevich,V.A., Aldanova, N.A., Trubetskaya,O.E.,Nazimov,I.V.,Hundal, T.,Ernster, L.(1984) FEBS lett. 166, 19-22
2. Walker,J.E.,Runswick,M.J.,Saraste,M.(1982) FEBS Lett.146,393-396
3. Grinkevich,V.A.,Modyanov,N.N.,Ovchinnikov,Yu.A.,Hundal, T.,Ernster,L.(1982) EBEC Reports 2, 83-84
4. Grinkevich,V.A.,Aldanova,N.A.,Kostetsky,P.V.,Trubetskaya,O.E., Modyanov,N.N.,Hundal,T.,Ernster,L. (1984) in "H^+-ATP Synthase: Structure, Function, Biogenesis" (S. Papa et al., eds.) ICSU Press and Adriatica Editrice, in press
5. Gay,N.J.,Walker,J.E.(1981) Nucleic Acids Res. 9, 3919-3926
6. Kanazawa,H.,Mabuchi,K.,Kayano,T.,Noumi,T.,Sekiya,T. and Futai, M. (1981) Biochem. Biophys. Res. Commun. 103, 613-620
7. Nielsen,J.,Hansen,F.G.,Hoppe,J.,Friedl,P.,von Meyenburg, K.(1981) Mol. Gen. Genet. 184, 33-39
8. Aquila,H.,Misra,D.,Eulitz,M.,Klingenberg,M.(1982) Hoppe-Seyler's Z. Physiol. Chem. 363, 345-349
9. Saraste,M.,Walker,J.E.(1982) FEBS Lett. 144, 250-254
10.Grinkevich,V.A.,Modyanov,N.N.,Ovchinnikov,Yu.A.,Hundal, T.,Ernster, L. manuscript in preparation.
11.Bradshaw,R.A. personal communication by E.Racker
12.Sebald,W.,Hoppe,J.(1981) Curr.Top.Bioenerg.12, 1-64
13.Frangione,B.,Rosenwasser,E.,Penefsky,H.S.,Pullman, M.E.(1981) Proc. Natl. Acad. Sci. USA 78, 7403-7407
14.Saraste,M.,Gay,N.J.,Eberle,A.,Runswick,M.J.,Walker, J.E.(1981) Nucleic Acids Res. 9, 5287-5296
15.Anderson,S.,de Bruijn,M.H.L.,Coulson,A.R.,Eperon, I.C.,Sanger,F., Young, I.G. (1982) J. Mol. Biol. 156, 683-717.

MECHANISM OF PROTON CONDUCTION BY MITOCHONDRIAL H^+-ATPase. ROLE OF THIOL GROUPS

F. Guerrieri[°], F. Zanotti[*], R. Scarfò[°] and S. Papa[°]
°Institute of Biological Chemistry, Faculty of Medicine and *Centre for the Study of Mitochondria and Energy Metabolism, University of Bari, Bari, Italy

The mitochondrial ATPase complex (H^+-ATPase synthase) (1) consists of two main portions: F_1, which is responsible for the catalytic events, and F_0 that is a transmembrane complex mediating H^+-translocation across the membrane (1). In the mitochondrial enzyme some other polypeptides are present (OSCP, F_6, F_B), apparently necessary for binding of F_1 to F_0 and/or for functional connection between H^+-conduction in F_0 and catalytic events in F_1 (1). Chemical modification by DCCD and genetic aminoacid substitution indicate a direct involvement in H^+-conduction by F_0 of a glutamic residue located in hydrophobic strech of a 8 kDa proteolipid of F_0 (2). Studies with specific aminoacid modifiers have shown the involvement of other aminoacid residues in H^+-conduction i.e. arginine (3), tyrosine (3,4) and cysteine (5).

In this paper the role of thiol groups in H^+-conduction by F_0 has been studied by analysis of release of a proton gradient set up by respiration in sonic submitochondrial particles (3-5) or proton release promoted by a diffusion potential (valinomycin-induced K^+-accumulation).

Passive H^+-conduction in F_0 is inhibited by oligomycin (Table I; see also ref. 6) and by DCCD (6,7). N-ethylmaleimide (NEM) and dithiopiridine (DTP), permeant thiol blocking reagents, also inhibit the proton conductivity by F_0 (Table I) and their inhibition is additive with that exerted by oligomycin.

On the contrary the reagent for vicinal dithiols, diamide, enhances the proton conduction. This stimulation of H^+-conduction is prevented by pretreatment of particles with oligomycin, indicating that the effect of diamide is not due to an unspecific effect on the membrane but to a direct action on thiol groups in the H^+-ATPase complex.

The data reported in the table and other observations (8) indicate that NEM or DTP can modify residues different from that modified by diamide.

In line with this is the observation that the effect of diamide is lost in ammonia particles where the effect of NEM and DTP is still present (8). Evidence is presented showing that the thiol-residue(s) modi-

TABLE I. Comparison of the effect of SH reagents on passive proton release from EDTA submitochondrial particles in the presence
and in absence of oligomycin. For experimental conditions see
ref.s 6 and 8.

Additions	Control		+ Oligomycin 1.5 $\mu g \cdot mg$ prot^{-1}	
	$1/t_{\frac{1}{2}}$ (s^{-1})	%	$1/t_{\frac{1}{2}}$ ($s-1$)	%
-	0.97		0.31	
NEM 3 mM	0.56	-42	0.2	-36
DTP 3 mM	0.37	-62	0.13	-58
Diamide 3 mM	7.16	+638	0.24	-22

fied by NEM and DTP is (are)closely associated to the critical glutamic
residue in the 8 kDa polypeptide. It is possible that the thiol groups
modified by diamide (oxidation followed by formation of dithiol bridge)
belong to polypeptides involved in the gating function of the H^+-ATPase.

References

1. Papa, S., Altendorf, K., Ernster, L. and Packer, L., Editors "H^+-
   ATPase-Synthase: Structure, Function, Regulation", 1984, ICSU Press-
   Adriatica Editrice, in press.
2. Papa, S., Guerrieri, F., Zanotti, F. and Scarfò, R. in "Biological
   Membranes: Information and Energy Transduction in Biological Membranes" (E. Helmreich, ed.), 1984, Alan R. Liss Inc., in press.
3. Guerrieri, F. and Papa, S., 1981, J. Bioenerg. Biomembr. 13, 393.
4. Guerrieri, F., Yagi, A., Yagi, T. and Papa, S., 1984, J. Bioenerg.
   Biomembr., in press.
5. Guerrieri, F. and Papa, S., 1982, Eur. J. Biochem. 128, 9.
6. Pansini, A., Guerrieri, F. and Papa, S., 1978, Eur.J.Biochem. 92,545.
7. Kopecky, J., Guerrieri, F. and Papa, S., 1983, Eur.J.Biochem. 131,17.
8. Papa, S., Guerrieri, F., Zanotti, F. and Scarfò, R. in "H^+-ATPase-
   Synthase: Structure, Function, Regulation" (S. Papa et al.,eds.),
   1984, ICSU Press-Adriatica Editrice, in press.

MODULATION OF ATP SYNTHESIS BY THE ATPase INHIBITOR PROTEIN

D. Harris and I. Husain
Department of Biochemistry,
University of Leeds,
Leeds LS2 9JT, U.K.

A small protein (mol. wt. 10,000) inhibits ATP hydrolysis of the F_1-ATPase. When phosphorylation is induced by an acid base transition (period of phosphorylation ≈ 3 s), ATP synthesis rates by submitochondrial particles are proportional to the number of F_1 molecules free of the inhibitor protein [1].

Steady state phosphorylation rates, however, are little (<20%) affected by the number of F_1-ATPase molecules initially active. This is apparently because the inhibitor is displaced from its inhibitory site into free solution, under the influence of an energised membrane [2]. The time course of this displacement is biphasic, with the rate of the fast phase (t ≈ 23 s) comparable with the rate of the onset of phosphorylation (t ≈ 11 s), within experimental error. The second phase, although comparable in amplitude, was an order of magnitude slower, and did not correlate with changes in phosphorylation rate, as observed by Schwerzmann & Pedersen [3].

Since only about 30% of F_1 molecules appear to be engaged in steady state phosphorylation in submitochondrial particles at any one time [4], we conclude that only the kinetics of the fast phase is relevant to the initiation of phosphorylation. Thus during the attainment of steady state conditions, as in transient phosphorylation, inhibitor protein release is probably the major factor limiting ATP synthesis. This has implications when developing a 'thermodynamic' model for control of phosphorylation - under these conditions at least, phosphorylation rates depend on the history of the system under investigation rather than the energy available in the electrochemical gradient.

(1)   Husain, I. and Harris, D.A. (1983) FEBS Lett. 160, 110-114

(2)   Power, J., Cross, R.L. and Harris, D.A. (1983) Biochim.
         Biophys. Acta. 724, 128-141.

(3)   Schwerzmann, K. and Pedersen, P.L. (1981) Biochemistry
         20, 6305-6311.

(4)   Harris, D.A., von Tscharner, V. and Radda, G.K. (1979)
         Biochim. Biophys. Acta 548, 72-84.

ANTIBODIES AS PROBE OF F_1-ATPASE BINDING TO ITS NATURALLY OCCURRING
INHIBITOR PROTEIN

I. Husain, P.J. Jackson and D.A. Harris
Department of Biochemistry,
University of Leeds,
Leeds LS2 9JT,
U.K.

The mitochondrial ATPase inhibitor protein is a small protein
(MW 10,000) which inhibits both ATP hydrolysis and synthesis by the
ATP synthase complex. An antibody to the ox heart mitochondrial
inhibitor was prepared in rabbits. The antibody was specific for
the inhibitor protein, as shown by immunoblotting [1] and competition studies. Rocket immunoelectrophoresis of inhibitor protein
into this antibody yields no precipitin line, leading to the conclusion that there is probably only one antigenic determinant per
inhibitor molecule. Labelling experiments suggest this site is
near the C terminus of the inhibitor.

The antibody was used to probe the interaction between the F_1-
ATPase and its inhibitor. Unlike that described by Dreyfus et al.
[1], this antibody did not affect inhibition of F_1-ATPase by the
inhibitor protein nor phosphorylation, although it did slow down the
response of the inhibitor protein to membrane energisation. More
particularly, titration of Mg-ATP (untreated) and of state III
(phosphorylating) submitochondrial particles with antibody showed
that on energisation inhibitor was lost from the membranes into
solution.

(1) Jackson, P.J. and Harris, D.A. (1983) Bioscience Reports
         3, 921-926.

(2) Dreyfus, G., Gomez-Puyou, A. and Tuena de Gomez-Puyou, M.
         Biochem. Biophys. Res. Commun. 100, 400-406.

REGULATION OF COUPLING FACTOR ACTIVITIES BY SOME COMPONENT OF
CHLOROPLAST ELECTRON TRANSPORT SYSTEM

Bernhard Huchzermeyer and Indra Willms
Botanisches Institut der Tierärztlichen Hochschule Hannover
Bünteweg 17d, D-3000 Hannover 71, W.-Germany (RFA)

A correlation between non-exchangeable ADP binding, enzymatical
activity of CF_1, and energy supply to the thylakoids has been found
(1, 2). Tightly bound ADP can be understood as indicating enzymatical
inactive CF_1 when investigating photosystem I dependent reactions.
An energy dependent conformational change in the CF_1 protein transforming tightly bound ADP to exchangeable bound ADP precedes PS I
driven ATP synthesis (3).

When investigating effects of different sections of electron transport chain, no nucleotide exchange on high affinity binding sites was
induced by PS II. With PS II electron transport from water to DAD,
PdA, and DBMIB, respectively, phosphorylation took place without previous nucleotide exchange on high affinity sites. With PS I electron
transport employing PMS or DCPIP/ascorbate-MV as well as with complete
electron transport chain employing Fecy or MV high affinity nucleotide
exchange took place.

We determined apparent K_m values for ADP in phosphorylation
employing the above electron transport systems. We found constant K_m
values with PS II electron transport. But when employing PS I electron
transport apparent K_m values above 10µM varying with different light
dependent electron transport rates were observed.

When observing ADP effects on transmembrane proton gradient, Löhr
(4) did not find any ADP effect when pH of incubation medium was below
pH 7.2. We measured apparent K_m values for ADP in phosphorylation
using an incubation medium adjusted to pH 7.0. In contrast to experiments performed at pH 8.0 we did not observe any effect of different
light intensities on PS I driven phosphorylation when employing this
low pH. In a parallel test we did not observe nucleotide exchange
at high affinity sites when using an incubation medium adjusted to
pH 7.0.

Our results indicate a regulation of CF_1 dependent reactions by
some component of the PS I electron transport chain. This regulation
is paralleled or even mediated by binding of nucleotides to the high
affinity binding sites.

Legends
1) J. Schumann and H. Strotmann (1980)
   Proc. V. Int. Congr. Photosynth. 881 - 892
2) H. Strotmann, S. Bickel-Sandkötter, U. Franek, and V. Gerke (1981)
   Energy Coupling in Photosynth. 20, 187 - 196
3) D.J. Smith and P.D. Boyer (1976)
   Proc. Nat. Acad. Sci. USA 73, 4314 - 4318
4) A. Löhr (1983)
   Proc. VI. Int. Congr. Photosynth., in press

INTERACTION OF MITOCHONDRIAL F_1-ATPase WITH THE OLIGOMYCIN
SENSITIVITY CONFERRING PROTEIN (OSCP) OF THE MITOCHONDRIAL
F_0F_1-ATPase COMPLEX

T. Hundal, B. Norling and L. Ernster
Department of Biochemistry, Arrhenius Laboratory, University of
Stockholm, S-106 91 Stockholm, Sweden

When beef heart submitochondrial particles (1) are treated with
suitable reagents, e.g. urea, ammonia and silicotungstic acid, coupling factors F_1 (2), OSCP (3) and F_6 (4), in that order, are removed
from the membrane, giving rise to particles incapable of ATP hydrolysis and ATP synthesis. By the addition of the same purified proteins
it is possible to reconstitute these activities. These preparations
are suited for studies of the protein-protein interactions necessary
for the functional integrity of the complex.

We have investigated the functional reconstitution of the ATPase
complex with emphasis on the interaction of OSCP with F_1, in particular the effects of modifications of F_1 (by trypsin digestion and
cold exposure) on its ability to aquire oligomycin sensitivity as mediated by OSCP.

In an earlier report we showed a stable complex formation between
F_1 and OSCP (5). OSCP renders F_1 appr. 50% cold-stable. Free F_1 under
the same conditions loses more than 85% of its activity. Free F_1 and
the $F_1 \cdot$OSCP complex show the same dissociation pattern and are converted into a form which contains a diminished number of β subunits.
The cold-exposed F_1.OSCP complex is capable, just as the native complex, of rebinding to F_1- and OSCP-depleted particles, giving rise to
an oligomycin sensitive ATPase. A stoichiometry of 3 moles OSCP per
mole F_1 was found to be needed for maximal cold-stabilization and
for conferral of full oligomycin sensitivity in reconstitution
experiments with particles devoid of F_1 and OSCP (Fig. 1) (6).

Mild trypsin treatment of F_1 has been reported to result in a
partial digestion of the α subunit (7-9). The proteolytic cleavage
causes the loss of a small segment of the peptide and results in a
lack of ability of F_1 to bind OSCP (10). This would indicate that
OSCP needs an intact α subunit for binding. When OSCP is bound to
F_1 before proteolytic digestion, no cleavage of the α subunit can be
seen on a polyacrylamide gel (Fig. 2). This indicates that OSCP binds
to the α subunit and, in doing so, protects it from proteolysis (6).
From the same experiment it can be deduced that F_1 protects OSCP from
total digestion, since the trypsin-treated F_1.OSCP complex contains
OSCP with a reduced molecular weight. Free OSCP under the same conditions is completely cleaved into small peptides not detectable on the
gel. It seems likely that OSCP protects all three α subunits which

again would indicate a 3:1 stoichiometry of OSCP to F_1.

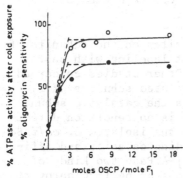
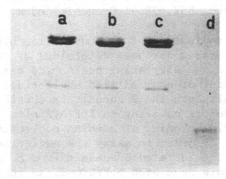

Fig. 1
The effect of increasing amounts
of OSCP on the ATPase activity
of soluble F_1 after cold exposure (●-●), and on the conferral of oligomycin sensitivity
on F_1 in reconstitution experiments with F_1- and OSCP-depleted
particles (o-o).

Fig. 2
Protection by OSCP of the α subunit of F_1 against trypsin digestion. a: control F_1, b: trypsin-
treated F_1, c: trypsin-treated
$F_1 \cdot$OSCP complex, d: control OSCP.

(Work supported by the Swedish Natural Science Research Council).

References
1. Lee, C.P. and Ernster, L. (1967) Methods Enzymol. 10, 543-548.
2. Racker, E. and Horstman, L.L. (1967) J. Biol. Chem. 242,
   2541-2551.
3. MacLennan, D.H. and Tzagoloff, A. (1968) Biochemistry 7,
   1603-1610.
4. Fessenden-Raden, J.M. (1972) J. Biol. Chem. 247, 2351-2357.
5. Hundal, T. and Ernster, L. (1979) in: Membrane Bioenergetics
   (Lee, C.P. et al., eds.) pp. 429-445, Addison-Wesley, Reading,
   MA., USA.
6. Hundal, T., Norling, B. and Ernster, L. (1984) submitted.
7. Hundal, T. and Ernster, L. (1981) FEBS Lett. 133, 115-118.
8. Skerret, K.J., Wise, J.G., Richardson Latchey, L. and Senior,
   A.E. (1981) Biochim. Biophys. Acta 638, 120-124.
9. DiPietro, A., Godinot, C. and Gautheron, D.C. (1983) Biochemistry
   22, 785-792.
10. Hundal, T., Norling, B. and Ernster, L. (1983) FEBS Lett. 162,
    5-10.

NUCLEOTIDE BINDING SITE ON ISOLATED β SUBUNIT FROM *E. COLI* F_1-ATPase

J.P. ISSARTEL and P.V. VIGNAIS
DRF/Biochimie (CNRS/ERA 903 et INSERM U.191), Centre d'Etudes
Nucléaires, 85X, 38041 Grenoble cedex, France.

The localisation of the nucleotide binding sites on the subunits of
the H^+-ATPases has been determined by covalent labeling with photo-
activable and alkylating nucleotide analogs. Other studies concerned
the binding of non modified nucleotides to isolated subunits of F_1
(1-3). Although the β subunit is considered as the catalytic subunit
and therefore contains an ADP/ATP site, there is no report on direct
equilibrium binding of adenine nucleotides to the isolated *E. coli* β
subunit. In order to assay the nucleotide binding capacity and affinity of the isolated β subunit from *E. coli* F_1-ATPase two kinds of
experiments have been performed, an indirect one using (^{14}C)aurovertin
D and a direct one with radiolabeled nucleotides.

ADP binding to isolated β subunit revealed through (^{14}C)aurovertin D
binding :

By means of (^{14}C)aurovertin D we have demonstrated that the isolated
β subunit binds aurovertin with a stoichiometry of 1 and a Kd value of
about 6 μM or 1.4 μM in the absence or presence of ADP respectively
(4). We took advantage of this ADP-dependent effect and used (^{14}C)-
aurovertin D as a specific probe to investigate the ADP binding parameters of isolated β. Measurements of (^{14}C)aurovertin D binding to β
with increasing amounts of ADP revealed that ADP can bind to β with a
Kd value of about 28 μM. The binding affinity of ADP to the β-auro-
vertin complex was increased (Kd = 4.7 μM). A nearly similar experiment was carried out by Verschoor (5) who measured the aurovertin
fluorescence intensity in the presence of β isolated from beef heart
mitochondria at different concentrations of ADP. The Kd value for ADP
binding found by Verschoor was much higher (420 μM) than that found in
our present work.

Binding of radiolabeled ADP to isolated β subunit :

Direct binding of radiolabeled ADP to β was performed by an equilibrium dialysis technique. In EDTA medium, each β subunit was found to
bind 1 mol of ADP with a Kd value of 25 μM (Fig. 1). This result
clearly shows that isolated β possesses one adenine nucleotide binding
site. Furthermore, it was confirmed that no heterogeneity exist in the
isolated β subunit molecules (4). In the presence of aurovertin, the
Kd value for ADP binding was lowered to 4.8 μM without any change in
the binding stoichiometry (Fig. 1). Hence, it appears that aurovertin
is an activator for the binding of nucleotide to the isolated β
subunit and vice versa.

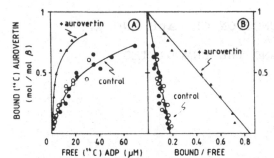

Fig. 1: (^{14}C)ADP binding to isolated β subunit.
(A) Direct binding curves.
(B) Scatchard plots.
(▲—▲) aurovertin added (26 μM).

We have previously demonstrated the existence of 3 aurovertin binding
sites of different affinities per *E. coli* F_1-ATPase (4). Interestingly,
the high affinity aurovertin binding site, revealed in the presence of
ADP, was not apparent in the absence of nucleotide (Fig. 2). Binding
of aurovertin to the high affinity site is responsible for the inhibition of F_1 by aurovertin.

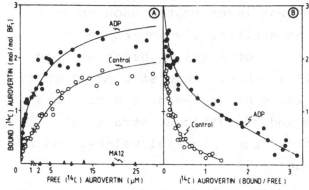

Fig. 2: Binding of (^{14}C)-
aurovertin D to *E. coli*
F_1-ATPase (from wild type
strain or aurovertin
resistant mutant: MA 12).
(A) Direct binding curves.
(B) Scatchard plots.

The enhancement of the binding affinity of F_1 for aurovertin that is
caused by ADP is possibly due to a conformational change of β upon
binding of ADP. As F_1-ATPase contains 3 β subunits it can be concluded
that only one β subunit out of the 3 β of F_1 is subject to ADP-
dependent conformational changes.

References
(1) Ohta, S., Tsuboi, M., Oshima, T., Yoshida, M. and Kagawa, Y.
    (1980) J. Biochem. (Tokyo) 87, 1609-1617.
(2) Dunn, S.D. and Futai, M. (1980) J. Biol. Chem. 255, 113-118.
(3) Gromet-Elhanan, Z. and Khananshvili, D. (1984) Biochemistry 23,
    1022-1028.
(4) Issartel, J.-P., Klein, G., Satre, M. and Vignais, P.V. (1983)
    Biochemistry 22, 3485-3492.
(5) Verschoor, G.J. (1982) Thesis, Netherlands.

KINETICS OF ATP-SYNTHESIS AS FUNCTION OF ΔpH AND ΔΨ CATALYZED BY THE DTT-MODIFIED AND THE UNMODIFIED CHLOROPLAST ATPase

Ulrike Junesch und Peter Gräber

Max-Volmer-Institut für physikalische und biophysikalische Chemie, Technische Universität Berlin, D-1000 Berlin 12, Germany

The kinetics of proton transport coupled ATP synthesis at the chloroplast membrane was investigated upon energization of the membrane by an artificially generated ΔpH and a K^+/Valinomycin diffusion potential, $\Delta\Psi$. Using a rapid mixing system, rates of ATP synthesis were studied at short reaction times ($<$ 150 ms) where all relevant kinetic parameters (ΔpH, $\Delta\Psi$, substrate and product concentrations) remain practically constant at their initial values. Under these conditions it was found (1):

I Unmodified ATPase:

a) The maximal rate of ATP synthesis obtained under these artificial conditions is the same as that obtained by light-induced ATP synthesis.

b) The turnover number of the ATPase is 410 s^{-1}.

c) The rate of ATP synthesis depends in a sigmoidal way on the transmembrane electrochemical potential difference of protons, $\Delta\tilde{\mu}_{H}+$, regardless of the relative contributions by ΔpH and $\Delta\Psi$.

d) The functional dependence of the rate of ATP synthesis
on $\Delta\tilde{\mu}_H{+}$ does not reflect the kinetics of the catalytic
events at the ATPase but the transformation of the ATPase
from an inactive to a catalytically active state.
e) The binding of two protons to the ATPase from the inside
of the membrane is necessary for activation.

II DTT-modified ATPase (2)

a) The same maximal rate of ATP synthesis is found with or
without DTT-modification.
b) The functional dependence of the rate of ATP synthesis
is different from that catalyzed by the unmodified ATPase.
c) Near the equilibrium exists a linear relation between
the rate of ATP synthesis and driving force which extrapolates to zero for n=3, as expected from non-equilibrium
thermodynamics.
d) Three protons are involved in the catalytic reaction.
e) The functional dependence of the rate of ATP-synthesis
   on Δ pH is shifted parallel to lower ΔpH-values if an
   additional diffusion potential is applied, i.e. the rate
   depends only on $\Delta\tilde{\mu}_H{+}$.

1) Gräber,P., Junesch,U. and Schatz,G.H. (1984) Ber. Bunsenges.
   Physik. Chem. 88, 599-6o8
2) Junesch,U. and Gräber,P. (1983) in: Advances in Photosynthesis Research, (C.Sybesma,ed.) pp.II,5.431-436, M.
   Nijhoff/Dr. W. Junk Publ., The Hague

DEMONSTRATION OF TWO BINDING SITES FOR ATP AND ADP AND ONE FOR
INORGANIC PHOSPHATE ON THE ISOLATED β-SUBUNIT OF THE Fo·F1-ATP
SYNTHASE

D. Khananshvili and Z. Gromet-Elhanan
Department of Biochemistry, Weizmann Institute of Science, Rehovot,
Israel.

The reconstitutively active β-subunit, that has been removed from
the *Rhodospirillum rubrum* membrane-bound Fo·F1-ATP synthase (1) and
purified to homogeneity (2) has recently been shown to contain two
nucleotide binding sites (3). One of them is a high affinity binding
site, that binds 1 mol of either ATP or ADP with a Kd of 4-7 µM
and a $t\frac{1}{2}$ of 3 min, and is not affected by $MgCl_2$. The second is
a low affinity site that is revealed only in the presence
of $MgCl_2$. So, under these conditions 2 mols of ATP or ADP are bound
per mol of isolated β-subunit: one molecule of the nucleotide is bound
to the β with a Kd and $t\frac{1}{2}$ identical to those recorded in the absence
of $MgCl_2$, whereas the second molecule of either ATP or ADP, binds to
the β-subunit with a Kd of 200 µM and 80 µM, respectively, and the $t\frac{1}{2}$
for both nucleotides is about 20 min. The Kd of this Mg-dependent
nucleotide binding site is similar to the Km for ATP hydrolysis by
R. rubrum chromatophores.

The isolated β-subunit is also able to bind 1 mol of Pi with a Kd
of 280 µM and a $t\frac{1}{2}$ of 15 min. This binding is absolutely dependent on
the presence of $MgCl_2$ and is inhibited in a noncompetitive manner by
low concentrations (5-50 µM) of ADP or ATP. At higher nucleotide
concentrations (0.1-10 mM) a competitive inhibition of Pi binding to
β is obtained with ATP being about a 30 fold more effective inhibitor
(Ki=0.35 mM) than ADP (Ki=10 mM). These results indicate that: a. Pi
is bound to the Mg-dependent low affinity nucleotide binding site on the
isolated β-subunit of the RrFo·F1-ATP synthase complex, and b. that
the affinity of Pi binding to this site is decreased when the high
affinity nucleotide binding site becomes occupied.

Further proof for the identification of the Pi binding site with
the Mg-dependent low affinity nucleotide binding site and for the
specific role of this site in the catalytic activity of the RrFo·F1
complex has been obtained from studies with a chemically modified
β-subunit. We have found that modification of the isolated β with
either Woodword's reagent K or diethyl pyrocarbonate resulted in
complete inhibition of the binding of Pi as well as in inhibition of
the binding of ATP to the Mg-dependent low affinity site, but not to
the Mg-independent high affinity site. Moreover, the β-subunit, that

has been modified with either one of these two reagents and lost its Mg-dependent low affinity substrate binding site, also lost its capacity to restore photophosphorylation to β-less chromatophores, although it retained its capacity to bind to these chromatophores. It is therefore concluded that the low affinity binding site identified on the isolated β-subunit of the RrFo·F1 is located at the catalytic site of this complex. This conclusion is also corroborated by the above stated similarity between the Kd of ATP binding and the Km of its hydrolysis.

References:

1. Philosoph, S., Binder, A. and Gromet-Elhanan, Z. (1977) J. Biol. Chem. 252, 8747-8752.
2. Khananshvili, D. and Gromet-Elhanan, Z. (1982) J. Biol. Chem. 257, 11377-11383.
3. Gromet-Elhanan, Z. and Khananshvili, D. (1984) Biochemistry 23, 1022-1028.

CONFORMATIONAL STATES OF BHM-ATP SYNTHETASE AS PROBED BY THE POTENTIAL SENSITIVE DYE OXONOL VI

R. Kiehl, W.G. Hanstein

Institut für Physiologische Chemie der Ruhr-Universität Bochum, 4630 Bochum.

Complex V has been isolated as the segment of the mitochondrial oxidative phosphorylation system which is concerned with ATP synthesis and hydrolysis. In the isolated state, complex V catalyzes, without reconstitution procedures, oligomycin-sensitive hydrolysis of ATP, oligomycin and uncoupler-sensitive ATP-P_i-exchange [1], shows oligomycin and uncoupler-sensitive energy dependent oxonol VI response but no energy dependent ACMA-quenching [2] even after removal of $AmSO_4$ from the ATPase complex. The preparation is reported to be essentially free of vesicularized structures in electron microscopic studies [1]. KSCN inhibits the dye response by about 10-27%, depending on the preparation, which may be an estimate of the amount of vesicularized structure present. The extent of the oxonol response depends on the ATP/ADP ratio rather than the phosphorylation potential. Studies on complex V activities suggest that oxonol VI reflects, at least in part, a more local, ATP-dependent and energy-related process [3]. The oxonol VI response may be used, therefore, as a probe for conformational states of the ATP synthetase, as demonstrated by the following results.

The nucleotides ITP, GTP and UTP are hydrolysed by the complex, but considerable NTP-P_i-exchange is observed only with ATP. ITP and GTP give rise to 50%, and UTP only to 25% dye response. An ATP concentration far below the K_m ($\sim$160 µM) for ATPase elicits also partial (about 50%) and declining dye response, which suggests that energy is required to maintain the response. 50 µM AMP-PNP prior to addition of 5 mM ATP inhibits ATPase activity to 90%, increases ATP-P_i-exchange and limits the dye response to about 50%. The dye response can be titrated with ATP in the presence of an ATP regenerating system and 10 mM Mg^{2+}. There are four steps in the titration curve. Breaks appear at about 25,50 and 75% dye response and at about 8,45 and 140 µM ATP. 100% dye response is reached at about 300 µM ATP. This may be explained by negative cooperativity in four discrete ATP binding steps which induce progressive dye response. The response at low steady state ATP can be modulated by K^+, Mg^{2+} or Ca^{2+}-ions. K^+ enhances the negative cooperativity and limits the dye response to 50% at 290 µM ATP (10 mM Mg^{2+} present). Low steady state ATP (100 µM) induces, in the absence of added Mg^{2+}, partial dye response. Subsequent addition of Mg^{2+} (1 mM) results then in almost 100% response.

The difference between the response at low and saturating Mg^{2+} concentration is about 50%. Ca^{2+} in the presence of saturating Mg^{2+} also lowers dye response.

The temperature dependence of the ATP induced oxonol response reveals a break at about 50%: at 10°C, there is rapid response to about 50% and much slower response to 100%.

Aurovertin, which has been used to study conformational changes in F_1, inhibits ATPase to 90%, inhibits ATP-P_i-exchange to 97%, but inhibits dye response only to 50%. Addition of limited amounts of the nitrophenol compounds, HE-DNP, DNP, AE-DNP, picrate and NPA (causing 2 to 10% abolishment of dye response) results in cumulative abolishment of the response. No such effects are observed in the presence of oligomycin and uncouplers. The enhanced effects of the compounds may be explained with an aurovertin promoted conformation and an enhanced binding of nitrophenols.

The partial dye responses as described above may be explained assuming different conformational states of the complex. The progressive conformational changes induced by ATP are similar to the changes induced by NAD-binding to GPDH [4]. The multisubunit ATP synthetase may be a case in which the ligand induced changes show predominantly negative cooperativity, and where the half-of-the sites reactivity [5] may only be one extreme form.

[1]   Stiggall, D.L., Galante, Y.M., and Hatefi, Y. (1978) J.Biol.Chem. 253, 956-964.

[2]   Kiehl, R., Hanstein, W.G. (1981) Developments in Bioenerg. and Biomembr. 5, 217-222.

[3]   Kiehl, R., Hanstein, W.G. (1984) Biochim.Biophys. Acta, in press.

[4]   Levitzki, A., Koshland, D.E. (1976) Current Topics in Cell Reg. 10, 1-40.

[5]   Choate, G.L., Hutton, R.L. and Boyer, P.D. (1979) J.Biol.Chem. 254, 286-290.

This work was supported by DFG grant Ha 1124.

LATERAL DISTRIBUTION OF CF_1 AND CF_0 IN DIFFERENT THYLAKOID REGIONS

L. Klein-Hitpaß and R.J. Berzborn
Biochemie der Pflanzen, Ruhr-Universität Bochum, D-4630 Bochum, FRG

INTRODUCTION AND PROBLEM

As could be demonstrated by inhibition experiments (1), electron microscopy (2) and recently by quantitative immunological determination (3)
the coupling factor CF_1 is absent from stacked grana partitions. The
distribution of the membrane part (CF_0) of the ATP synthase complex
(CF_1-CF_0), which serves besides its function in proton translocation
as the binding site for the peripheric CF_1, has not yet been investigated since a specific detection method for CF_0 was lacking. We produced special antisera against the CF_0 complex and elaborated a specific and sensitive quantitative immunological assay, which enabled us
to determine the relative distribution of CF_0 in thylakoid membrane
subfractions; now the CF_0 distribution can be compared to the CF_1 distribution.

RESULTS

By immunization of rabbits with CF_1-CF_0 complex purified from spinach
thylakoids by Triton X-100 extraction, anion-exchange chromatography
and sucrose density centrifugation (Klein-Hitpaß, L. and Berzborn, R.J.
to be published) and subsequent absorption of these CF_1-CF_0 antisera by
purified CF_1 we produced monospecific precipitating antisera against
the CF_0. Electroimmunodiffusion (EID) as described (4) for the determination of CF_1 in membrane fractions after solubilization with 3% Triton X-100 also was possible for quantitation of CF_0.
For our measurements it was essential that the CF_0 rocket lengths of
free CF_0 and CF_0 as constituent of CF_1-CF_0 complex are identical. To
varify this EDTA-treatments of thylakoids under different sodium chloride concentrations were performed to yield membranes with different
CF_1 and free CF_0 content and assayed for CF_0 by EID. Independent of the
CF_1 content the dissolved membranes gave identical CF_0 lengths;EDTA supernatants never contained any CF_0.
The various vesicle fractions obtained by the procedure described by
Andersson et al. (5) were analyzed in parallel for their CF_0 and CF_1
content by EID. The results (table I) show that on a chlorophyll basis
the stroma lamellae fraction (100K) contains much more CF_1 than the
fraction (B3) enriched in inside-out vesicles originating from stacked
grana partitions. The ratios of CF_0 and CF_1 rockets in the different
membrane fractions show no significant deviation.

Table I: CF_0 and CF_1 "rockets" in the various fractions of an "inside-
out" preparation (high salt). For preparation and nomenclature see ref.
5. Three dilutions of each vesicle fraction dissolved in 3% Triton X-
100 were analyzed. The rocket lengths of these three samples were plot-
ted against the chl content. From the slope of the linear curve the

fraction	chl a/b	CF_0 (mm r./3 μg chl)	CF_1	CF_0/CF_1 (mm/mm)
CII	3.05	29.0	38.0	0.76
YPH	3.08	31.0	40.0	0.78
40K	2.95	25.0	33.5	0.75
100K	7.16	78.0	104.0	0.75
T2	2.95	28.5	37.0	0.77
B3	2.33	16.5	21.0	0.79

rocket length per 3 μg chl was calculated. Electrophoresis: 3.5 V/cm 15°C, 18h; antiserum against CF_0 (CF_1), 169-4A (150-4), 150 (30) μl/10 ml gel.

DISCUSSION

In accordance with earlier publications (1-3) this study presents evidence for an extreme heterogeneous distribution of CF_1 in the thylakoid system. For the first time it could be shown that also CF_0 is highly enriched in exposed areas. Furthermore the identity of CF_1 and CF_0 distribution is obvious (table I). In control thylakoids no CF_0 without bound CF_1 is present, since after solubilization of the membranes all CF_0 was precipitated by an antiserum against CF_1 (7). Different CF_0 and CF_1 distributions could have been observed if a significant amount of (assembled) CF_0 without bound CF_1 would be present in the thylakoids and if CF_0 - in contrast to CF_1-CF_0 - would not be restricted to exposed membrane areas. Under artificial conditions (restacked EDTA treated membranes) CF_0 without bound CF_1 indeed was randomly distributed (7) i.e. also close to PS 2 in stacked membrane regions (6). The relation of our results to the photosystem 2-specific light-dependent modulation of acetic anhydride labeling of subunit III (proteolipid) of the CF_0 complex (8) remains to be elucidated.

1 Berzborn, R.J., Menke, W., Trebst, A. and Pistorius, E. (1966) Z. Naturforsch. 21b, 1057-1059
2 Miller, K.R. and Staehelin, L.A. (1976) J. Cell Biol. 68, 30-47
3 Berzborn, R.J., Müller, D., Roos, P. and Andersson, B. (1981) in: Photosynthesis (Akoyunoglou, G., ed.), Vol. 3, 107-120. Balaban Int. Sci. Services, Philadelphia, PA
4 Roos, P. and Berzborn, R.J. (1983) Z. Naturforsch. 38c, 799-805
5 Andersson, B., Sundby, C. and Albertsson, P.A. (1980) BBA 599, 391-402
6 Andersson, B. and Haehnel, W. (1982) FEBS Lett. 146, 13-17
7 Klein-Hitpaß, L. (1983) Thesis, Ruhr-Universität Bochum
8 Tandy, N.E., Dilley, R.A., Hermodson, M.A. and Bhatnagar, D. (1982) JBC 257, 4301-4307

REGULATION OF PHOSPHORYLATION EFFICIENCY BY SOME COMPONENT OF CHLOROPLAST ELECTRON TRANSPORT CHAIN

Andreas Löhr and Bernhard Huchzermeyer

Botanisches Institut der Tierärztlichen Hochschule Hannover

Bünteweg 17d, D-3000 Hannover 71, W.-Germany (RFA)

Effects of ADP on basal electron transport rate or the size of steady state proton gradient are understood as indicating ADP-CF_1 interactions (1). We determined the ADP concentration resulting 50% of maximal nucleotide effect. We compared these concentrations to the binding constants of high and low affinity binding sites as well as the apparent K_m value for ADP in phosphorylation (2).

When reducing light intensity or adding DCMU type electron transport inhibitors, we found reduced $c_{50\%}$- and K-values with reducing light mediated electron transport rate. When affecting transmembrane proton gradient by addition of uncouplers the measured values increased with increasing electron transport rate, corresponding to results published by Vinkler (3). In contrast the values were lowered with increasing electron transport rate when phosphorylation was inhibited by addition of gramicidin.

In all tests we observed strong correlation between $c_{50\%}$- and K-values. Two types of nucleotide affinities were found: High affinity binding sites and the ADP effects on steady state proton gradient and basal electron transport shew values below 10µM; apparent K_m in phosphorylation and low affinity binding sites resulted in values above 10µM.

When measuring P/e_2-values employing the above experimental conditions we observed increasing P/e_2 ratios with decreasing $c_{50\%}$- and K-values. As P/e_2-values were determined in presence of 5mM P_i and 1mM ADP changing P/e_2-ratios are not due to restricted substrate transfer to the CF_1 reaction center.

When employing an incubation medium adjusted to pH 7.0 neither ADP effects nor changing P/e_2 ratios could be measured with varying light intensities. Even more, after inhibition of PS I constant P/e_2 ratios of PS II driven phosphorylation were observed with varying light intensities. This effect was paralleled by lacking ADP exchange on tight nucleotide binding sites.

We understand these results as indicating a regulatory effect of high affinity ADP binding on the CF_0CF_1-complex. This effect seems to be mediated by some component of the PS I system.

Legends
1) R.E. McCarty (1979)
   Trends Biol. Sci. 4, 28 – 30
2) B. Huchzermeyer and A. Löhr (1983)
   Hoppe—Seyler's Z. Phys. Chem. 364, 1148 – 1149
3) C. Vinkler (1979)
   Biochem. Biophys. Res. Comm. 99, 1095 – 1100

F_1-F_0 ATPASE OF E. COLI : INHIBITION OF THE PROTON CHANNEL BY A WATERSOLUBLE CARBODIIMIDE

H.R. Lötscher* and R.A. Capaldi

Institute of Molecular Biology, University of Oregon,
Eugene Oregon 97403 USA
*present address: c/o F. Hoffmann-La Roche & CO., Central Research
Unit, CH-4002 Basle, Switzerland

1-ethyl-3 [3-(dimethylamino) propyl] carbodiimide (EDC), a water-soluble carbodiimide, inhibits ECF_1-F_0 ATPase activity and proton-translocation through F_0 when reacted with E. coli membrane vesicles. The site of modification is clearly different from that of the hydrophobic carbodiimide DCCD, although both reagents have the same effect on the functioning of the enzyme. [^{14}C]DCCD has been found to modify Asp 61 of subunit c of the E. coli ATP synthase with incorporation of radioactivity presumably through formation of an N-acylisourea derivative of the carboxyl group (1). We find no incorporation of [^{14}C] ETC into the F_0 portion of ECF_1-F_0 in membranes under reaction conditions in which there is 95% loss of ATPase activity and complete blocking of the proton channel through F_0.

Fragmentation of subunit c with cynogen bromide followed by high pressure liquid chromatography and thin layer chromatography indicates that reaction of ECF_1-F_0 with EDC does lead to modification of polypeptide c. Unambiguous evidence has been obtained that the C terminal four amino acid strech of this subunit is modified. The modification involves addition of material containing phosphorus and the altered fragment is sensitive to mild alkaline alcoholysis. Subunit c, isolated from E. coli membranes containing [^{14}C] phosphatidylethanolamine (PE), contains this radioactive lipid after EDC reaction but no [^{14}C] PE is bound in the control, i.e. unreacted membranes.

These findings strongly suggest that EDC catalyzes the crosslinking of the C terminal carboxyl of subunit c to PE, presumably through the amino group of the phospholipid. We propose that it is this alteration of subunit c which inhibits ATPase activity of the ECF_1-F_0 complex and blocks proton movements through F_0.

Reference :
(1) Wachter, E., Schmid, R., Decker, G. and Altendorf, K. (1980)
FEBS Lett. 113, 265-270.

A PLASMA MEMBRANE ASSOCIATED ADENOSINE TRIPHOSPHATASE OF HETEROTROPHICALLY GROWN SULFOLOBUS ACIDOCALDARIUS

Mathias Lübben and Günter Schäfer
Institut für Biochemie, Medizinische Hochschule Lübeck, Ratzeburger Allee 160, D-2400 Lübeck 1

The thermoacidophilic archaebacterium Sulfolobus acidocaldarius is growing optimally in media with a pH of 2.5 - 3.5 at temperatures between 70 and 80°C (1). Life in this extreme environment requires effective mechanisms for balancing the intracellular ionic composition and maintaining energy conservation. By analysis of cell components, which might have a bioenergetic function, a plasma membrane bound ATPase has been detected.

A membrane fraction of heterotrophically grown Sulfolobus acidocaldarius (DSM 639) was obtained by sonic oscillation of freshly harvested cells, followed by a differential centrifugation and repeated washing prodecure.

As measured after sucrose density gradient centrifugation and fractionation, the particles with a buoyant density of 1.25 - 1.26 g/cm^3 concentrate in a single band, which contains the ATP hydrolytic activity, as determined by phosphate analysis after incubation at 70 C.

For comparison, a cell wall preparation after the method described in (2), is found to have a density of 1.34 g/cm^3. Thus the relatively high density of membrane particles could be explained either by assuming cell wall contamination or possibly by a high protein content and the specific constitution of the archaebacterial lipids.

The membrane associated ATPase has a broad pH-optimum ranging from 5.0 - 6.5. It requires divalent cations for activation, as shown in table I. By using MgCl$_2$ as activator, the optimal Mg^{++}/ATP ratio is 0.5. Additional presence of 20 mM NaCl, 20 mM KCl or 20 mM of each NaCl and KCl does not yield further stimulation, 10 mM NaHCO$_3$ or 10 mM Na$_2$SO$_3$ (which is spontaneously oxidized to Na$_2$SO$_4$ at 70°C) are also ineffective. As can be seen in table I, MgSO$_4$ does not affect the ATPase markedly, which is in contrast to the strongly MgSO$_4$ - dependent membrane bound ATPase of Thermoplasma acidophilum (3). The Sulfolobus ATPase is not susceptible to uncouplers in the presence or absence of KSCN.

Among the ATPase inhibitors tested, 0.5 mM Ouabain, 1 mM KCN, 1 mM NaN$_3$ and 0.2 mg/ml Oligomycin do not impair the activity significantly, while others like sodium orthovanadate (1 mM, 24 % inhibition) and DCCD (1 mM, 2 h preincubation at 37°C, 50 % inhibition) act only slightly at high concentrations. Incubation with 200 µM Quercetin, however, results in 40 % inhibition.

The extreme thermostability seems to be an interesting property of the membrane bound ATPase. This becomes obvious, because the activity does not reach its maximal value, when measured at 95°C.

As listed in table II, the specificity of the ATPase is tested for various nucleosidtriphosphates and other energy - rich compounds. The enzyme prefers purine nucleotides and has little ADPase and no AMPase activity. A weak phosphatase activity is detectable, as reflected by the slow hydrolysis of substrates like para-nitrophylphosphate or phosphoenolpyruvate. The K_m value for ATP is in the order of 2.5 - 4.5 mM, depending on the Mg^{++}/ATP ratio.

table I

salt	nMol P_i/mg x min	%
Control	32	21
+ $MgCl_2$	149	100
+ $MgSO_4$	138	93
+ $MnCl_2$	183	123
+ $CaCl_2$	51	34

P_i - release is tested at 70°C in 50 mM Na-Maleat buffer, pH 6.0.
table I: 10 mM disodium ATP and 10 mM salt present.
table II: 10 mM $MgCl_2$ and 10 mM substrate present; 100 % refers to 149 nMol P_i/mg x min

table II

substrate	% activity
ATP	100
dATP	88
GTP	84
ITP	84
CTP	63
UTP	45
ADP	9
AMP	<1
pyrophosphate	<1
nitrophenylphosphate	7
phosphoenolpyruvate	7

The relation of the plasma membrane bound ATPase to the bioenergetic apparatus of Sulfolobus acidocaldarius remains to be elucidated. A possible involvement of the ATPase in energy linked functions will be further investigated.

References
(1) Brock, T. D., K. M. Brock, R. T. Belly and R. L. Weiss (1972), Arch. Mikrobiol. 84, 54-68
(2) Michel, H., D.-Ch. Neugebauer and D. Oesterhelt (1980), in "Electron Microscopy at Molecular Dimensions" (Baumeister, W. and Vogell, W., eds.), pp. 27-35, Berlin, Springer-Verlag
(3) Searcy, D. G. and F. R. Whatley (1982), Zbl. Bakt. Hyg., I. Abt. Orig. C3, 245-257

NUCLEOTIDE BINDING MODELS OF MITOCHONDRIAL F₁-ATPase

U. Lücken*, F. Peters° and G. Schäfer*
* Institut für Biochemie, Medizinische Hochschule Lübeck,
  Ratzeburger Allee 160, D-2400 Lübeck 1
° Biophysikalische Meßgeräteabteilung, Medizinische Hochschule
  Hannover, Konstanty-Gutschow-Str. 1, D-3000 Hannover

The knowledge of nucleotide binding behavior is a prerequisite for understanding of the mechanism of ATPase and ATP-synthetase. Recent data obtained with nucleotide analogs strongly suggest anticooperativity of binding to the high affinity sites (1,2) and cooperativity in the catalytic cycle (3,4). Though thermodynamic binding data are frequently evaluated by Scatchard-analysis, this is not necessarily the most suitable method for analysing complex binding systems (5). Multiple site interactions of small molecules with proteins can be described on the basis of the Adair equation (6) for a given model by site binding constants as described in (7). We have developed a non-linear least squares procedure for calculation of macroscopic as well as site binding constants for several binding models. This approach has been applied to nucleotide depleted mitochondrial F₁-ATPase, which normally exhibits three reversible high affinity nucleotide binding sites. ADP and ANA-ADP* were used as ligands (the latter is a photo-reactive analog (8)).

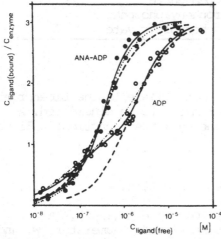

*) ANA-ADP = 3'-O-((5)-azi-donaphtoyl(1))-ADP

Fig. 1

Typical titration data obtained for ^{14}C-ADP (o) by centrifuged column technique and ultrafiltration method with .05, .27 and 1 μM F₁; for ANA-^{14}C-ADP (●) by centrifuged column technique with .1 and .4 μM F₁. For details see (8).

Titration data of the trivalent binding system (see Fig. 1) were analysed on the basis of the following models:
1. Three identical independent binding sites defined by one microscopic binding constant (dashed lines).
2. Three different and independent binding sites; three microscopic constants (dashed/dotted line for ADP; for ANA-ADP identical to dashed line).

3. All sites a priori equal; second and third site dependent on the first site in the same manner; 2 microscopic constants (dotted line).

4. All sites a priori equal and dependent; 3 microscopic constants (solid lines).

In all cases the fit of macroscopic binding constants yields the best results. For none of the ligands the curves based on model 1 fit the data (dashed lines). The flattened titration curve of ADP is optimally fitted by model 4 (solid line) and shows a significant negative cooperativity; whereas the approximation with two independent classes of sites (model 2; dashed/dotted) is insufficient. Binding data of ANA-ADP show a steeper slope than the curve of model 2 (independent unequal; dashed line), which points to a modest positive cooperative interaction. The curve based on model 4 agrees well with the optimal fit of the macroscopic affinity constants, whereas model 3 (dotted line) shows a slightly different behavior and cannot be excluded.

Exchange experiments suggest binding of both ligands to the same sites on F_1-ATPase. ADP prevents binding of ANA-^{14}C-ADP and vice versa. The dissociation kinetics of a saturated enzyme measured by an isotope chase experiment are biphasic. Two sites exchange within mixing time, whereas the dissociation of one ligand is much slower. If asymmetry of binding for ANA-ADP would be caused by independent sites (model 2) the equilibrium binding curve would be expected to be biphasic; this was not the case. A probable explanation is that the analog overcomes negative cooperativity or induces other types of site-site interaction. Further information of site-site interaction will be obtained by comparing the dissociation velocity of an 1:1 complex in a dilution experiment and in an isotope chase.

References
(1) Grubmeyer, Ch. and Penefsky, H.S.(1981) J.Biol.Chem.256, 3728-3734
(2) Tiedge, H., Lücken, U., Weber, J. and Schäfer, G. (1982)
    Eur. J. Biochem. 127, 291-299
(3) Cross, R. L., Grubmeyer, Ch. and Penefsky, H. S. (1982)
    J. Biol. Chem. 257, 12101-12106
(4) Gresser, M. J., Myers, J. A. and Boyer, P. D. (1982)
    J. Biol. Chem. 257, 12030-12038
(5) Peters, F. and Pingoud, A. (1979)
    Int. J. Bio-Medical Comput. 10, 401-415
(6) Adair, G. S. (1925) J. Biol. Chem. 63, 529-545
(7) Klotz, I. M. and Hunston, D. L. (1979)
    Arch. Biochem. Biophys. 187, 132-137
(8) Lübben, M., Lücken, U., Weber, J. and Schäfer, G. (1984)
    Eur. J. Biochem., in press

KINETIC EFFECT OF CHEMICAL MODIFICATION OF F_1-ATPase ON ATP SYNTHESIS AND HYDROLYSIS IN RECONSTITUTED SYSTEMS
A. Matsuno-Yagi and Y. Hatefi
Scripps Clinic and Research Foundation, La Jolla, California 92037, U.S.A.

The purified, soluble F_1-ATPase was modified by several covalently reacting inhibitors, either known or considered to bind to the active site-bearing β subunit, to cause partial inhibition up to 99%. The modified enzyme was then reconstituted in the presence of OSCP (oligomycin sensitivity-conferring protein) with submitochondrial particles (SMP) almost completely (> 99%) denuded of active F_1-ATPase, and assayed for oligomycin-sensitive ATPase and oxidative phosphorylation activities. The inhibitors used were 1-fluoro-2,4-dinitrobenzene (FDNB), N-ethoxycarbonyl-2-ethoxy-1,2-dihydroquinoline (EEDQ), 1-cyclohexyl-3-(2-morpholino-4-ethyl)-carbodiimide metho-p-toluene-sulfonate (CMCD), quinacrine mustard (QM), 5-dimethylaminonaphthalene-1-sulfonyl chloride (dansyl-Cl), p-fluorosulfonylbenzoyl-5'-adenosine (FSBA), and N,N'-dicyclohexylcarbodiimide (DCCD). The SMP reconstituted with unmodified F_1 exhibited oxidative phosphorylation and oligomycin-sensitive ATPase (in the presence of uncouplers) activities as high as 500 nmol/min/mg and 8 μmol/min/mg, respectively. As shown in Fig. 1, the systems reconstituted with F_1 modified to cause various degrees of inhibition with FDNB, EEDQ, CMCD, QM and dansyl-Cl exhibited the same degree of inhibition of oxidative phosphorylation and oligomycin-sensitive ATPase activities as the inhibition of the ATPase activity of the modified F_1 before reconstitution. The systems reconstituted with FSBA-modified F_1 showed the following relative degrees of inhibition: oxidative phosphorylation > oligomycin-sensitive ATPase of particles > ATPase of soluble F_1. In contrast, the systems reconstituted with DCCD-modified F_1 showed much greater inhibition of oligomycin-sensitive ATPase than of oxidative phosphorylation activity. This DCCD effect agrees with the results of Steinmeier and Wang (1) and Kohlbrenner and Boyer (2) on the ATP synthetic activity of systems reconstituted with Nbf-modified F_1. In the case of DCCD, experiments with [^{14}C]DCCD showed the stable binding of ≥ 1 mol [^{14}C]DCCD/mol F_1 exclusively on the β subunit(s) with a concomitant inhibition of ATP hydrolysis by about 95%. However, the F_1 thus modified was still capable of ATP synthesis at about 50% of the control rate after reconstitution. These results are consistent with the findings of Grubmeyer et al (3) and Cross et al (4) who showed that F_1-ATPase can function without the participation of all three active sites in catalysis.
Regardless of the nature of the inhibitor used (e.g., EEDQ, FSBA or DCCD), all the systems reconstituted with F_1 modified to

cause various degrees of inhibition exhibited the same apparent K_m^{ATP} in oligomycin-sensitive ATPase assays and the same apparent K_m^{ADP} and K_m^{Pi} in oxidative phosphorylation assays. The apparent K_m^{ADP} of the reconstituted systems (including the control SMP reconstituted with unmodified F_1) were 4-5 times that of "intact" SMP, and the apparent K_m^{Pi} of reconstituted systems containing modified F_1 were twice that of systems containing unmodified F_1.

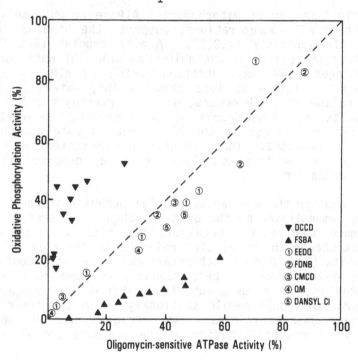

Figure 1. Correlation between the oxidative phosphorylation and the oligomycin-sensitive ATPase activities (% relative to controls) of particles reconstituted with F_1 that had been inhibited to various extents by modification with the reagents shown.
(A. Matsuno-Yagi and Y. Hatefi, Biochemistry (1984) in press.)

References
1. Steinmeier, R.C. and Wang, J.H. (1979) Biochemistry 18, 11-18.
2. Kohlbrenner, W.E. and Boyer, P.D. (1982) J. Biol. Chem. 257, 3441-3446.
3. Grubmeyer, C., Cross, R.L., and Penefsky, H.S. (1982) J. Biol. Chem. 257, 12092-12100.
4. Cross, R.L., Grubmeyer, C., and Penefsky, H.S. (1982) J. Biol. Chem. 257, 12101-12105.

PROBES OF THREE-SITE PARTICIPATION AND ROTATIONAL CATALYSIS WITH ATP
SYNTHASE

T. Melese, R.P. Kandpal, S.D. Stroop and P.D. Boyer
The Molecular Biology Institute and the Chemistry
and Biochemistry Department, University of California
Los Angeles, California 90024 U.S.A.

The catalytic behavior of mitochondrial ATPase at extremely low
compared with high ATP concentrations, supports the binding change
mechanism for ATP synthesis (1,2,3). A more crucial test is to
determine whether there is a predictable modulation of water-oxygen
exchange by changes in ADP concentration during net ATP synthesis.
During photophosphorylation at less than μM ADP, extensive water
oxygen incorporation into ATP occurs, while increasing the ADP concentration results in a sharp drop in oxygen exchange. In addition,
there is a biphasic increase in the ATP synthesis rate as the ADP
concentration is increased. These results are consistent with the
participation of two or three catalytic sites as proposed in the
binding change mechanism.

In order to explain the similar sequential behavior of each catalytic site, as demonstrated by the oxygen exchange patterns, a suggestion was made that during catalysis the beta subunits might
rotate and thus change their position relative to the core of noncatalytic subunits (1). During such rotational catalysis, conformational changes could cause the beta subunits to move relative to
other F_0-F_1 subunits, or the beta subunits combined with clusters of
DCCD-binding proteins could remain stationary while the inner core
of non-catalytic subunits rotates. An intriguing analogy can be
drawn between this possible behavior of the synthase and flagellar
rotation.

If rotational catalysis occurs with an equal participation of all
subunits in the catalytic cycle, then at any one time during catalysis all the subunits will not be conformationally or functionally
identical, but will show time-averaged symmetry. In the absence of
catalysis, definitive assymetry of the beta subunits should be present as illustrated by the tight binding of nucleotide to only one
catalytic site. To probe these possibilities, measurements have
been made of the reactivity of the beta subunits with succinic anhydride and iodoacetate under conditions where the catalytic capacity
of the enzyme is retained. Subunit separation by isoelectric focusing allows quantitation of the extent of reaction of these reagents
with a chloroplast ATPase that has an initial near isoelectric
homogeneity of beta subunits. The patterns observed display

heterogeneity of reaction for inactive enzyme and a more homogeneous pattern for enzyme derivatized during catalysis. Also, derivatization of inactive enzyme followed by transient catalysis leads to an increase in reactivity of some subunits. These results warrant additional exploration of possible rotational catalysis.

Although three-site catalysis is likely under favorable conditions, relatively rapid two-site catalysis can occur. Two-site catalysis may also occur when one subunit is absent (4) or covalently derivatized (5,6,7). We have found, however, that the DCCD- or NBD-Cl-treated MF_1 and CF_1 retain a capacity for slow catalysis and show retarded binding changes after the addition of ATP. Ascertaining if the same subunits are derivatized by DCCD and by active-site directed reagents may help discern between the participation of 2 normal and 1 crippled subunit or only 2 subunits.

A final suggestion to be explored arises from the evidence of various laboratories for a localized energy transmission from redox enzymes to the ATP synthase. One proposal is that charged group migration at the interface between F_0 and the membrane can be driven by, and result in, transmembrane proton movement, or can be driven by direct interaction with redox enzymes (8).

1. Gresser, M.J., Myers, J.A. and Boyer, P.D. (1982) J. Biol. Chem. 257, 12030-12038.
2. Cross, R.L., Grubmeyer, C., and Penefsky, H.S. (1982) J. Biol. Chem. 257, 12101-12105.
3. O'Neal, C.C. and Boyer, P.D. (1984) J. Biol. Chem. in press.
4. Hundal, T. and Ernster, L. (1979), in Membrane Bioenergetics, C.P. Lee et al., Eds., Addison-Wesley, 429-455.
5. Soong, K.S. and Wang, J.H. (1984) Biochemistry, 23, 136-141.
6. Matsuno-Yagi, A. and Hatefi, Y. (1984) Biochemistry, in press.
7. Schafer, G. (1984) Report at conference on H^+-ATP Synthase, Italy, April 1984.
8. Boyer, P.D. (1984) Ibid.

DIFFERENCE IN INHIBITORY EFFECTS OF ANTI-F_1 β SUBUNIT MONOCLONAL
ANTIBODIES ON ATPASE AND ATPSYNTHASE ACTIVITY

M. Moradi-Améli and D.C. Gautheron
LBTM-CNRS, Univ.Cl.Bernard de Lyon, 69622 Villeurbanne cedex,France

Since monoclonal antibodies react with only one epitope, they
can be precious tools to identify differences in conformation of
proteins. Two monoclonal antibodies (5G11, 19D3), which recognized
the β subunit and were specific of a native conformation of heart
mitochondrial F_1(1), are shown to behave differently on ATPase and
ATPsynthase activities.

As shown in Table I, the antibody 5G11 significantly inhibited
the ATP synthesis of MgATP submitochondrial particles (ETP) while
the ATP hydrolysis activity of these particles as well as F_1,
remained unchanged. Since this antibody recognized an active conformation of F_1, this result suggests that it distinguished a conformation competent for ATP synthesis which is different from that of
ATP hydrolysis. It should be noted that the number of moles of 5G11
bound per mole of F_1 was estimated to be 1.4 (1), indicating that
5G11 binds to a maximum of two β subunits per mole of F_1.

Table I
Inhibitory effect of 5G11 anti-β subunit antibody on ATP synthesis

	Net ATP synthesis *		ATP-^{32}Pi Exchange **		[γ^{32}P]-ATP Hydrolysis**	
	Activity	% I	Activity	% I	Activity	% I
ETP**	0.34		0.247		5.09	
ETP+5G11	0.151	56	0.173	30	5.73	0

Activity expressed as μ moles of ATP synthesized or
hydrolyzed/min/mg protein.
* Activity measured as in (2)
** ETP prepared and activity measured as in (3). I = Inhibition

On the contrary the antibody 19D3 had no effect on ATP synthesis
but showed an inhibitory effect on ATP hydrolysis under appropriate
conditions (TableII). The presence of Pi at pH 6.6 (monovalent Pi)
seemed to be compulsory for promoting the inhibitory effect of the

Table II
Inhibitory effect of 19D3 anti-β subunit antibody
on ATP hydrolysis in the presence of Pi at pH 6.6

Particles	Conditions	ATP hydrolysis % no addition	ATP synthesis % no addition
F_1, ASparticles or ETP alone	various	(100)	
F_1 + 19D3	+Pi pH 7.5	90	
" "	-Pi pH 6.6	113	
" "	+Pi pH 6.6	58	
F_1 + 14D5	" "	106	
ASparticles+19D3	" "	50	
ETP+19D3	" "	100	108
" "	+Pi pH 7.5	97	87
ETP+14D5	" "		88

antibody on the ATPase activity of F_1 and of AS particles (EDTA particles). In fact, no inhibitory effect can be seen either in the presence of Pi at pH 7.5 or by lowering the pH at 6.6 in the absence of Pi. The inhibitory effect may be due to a conformational change, induced by 19D3 in the presence of monovalent Pi on F_1 or AS particles both depleted from the protein inhibitor. However, under the same conditions the ATPase activity of the ETP (MgATP particles) was not affected by this antibody. One explanation could be that the inhibited conformation is already reached in these particles. Another explanation could be that the monovalent Pi could not affect the conformation of the membrane integrated system associated to the protein inhibitor.

Since under the inhibitory condition, another anti-β subunit antibody, 14D5, inhibited neither the ATPase activity of the F_1 nor the ATPsynthesis of ETP (Table II), the above effects cannot be attributed to a mere steric hindrance due to the IgG molecules 19D3 and 5G11.

1 - Moradi-Améli, M. and Godinot, C. (1983) Proc. Natl. Acad. Sci. U.S.A 80, 6167-6171
2 - Deléage, G., Penin, F., Godinot, C. and Gautheron, D.C. (1983) Biochim. Biophys. Acta 725, 464-471
3 - Penin, F., Godinot, C., Comte, J. and Gautheron, D.C. (1982) Biochim. Biophys. Acta 679, 198-209

TEMPERATURE DEPENDENCE OF KINETIC PROPERTIES OF MITOCHONDRIAL ATPASE

G.Parenti Castelli, A.Baracca, G.Solaini, A.Rabbi and G.Lenaz
Istituto di Chimica Biologica and Istituto Botanico,University of Bologna,40126 Bologna,Italy.

A break in the Arrhenius plot of oligomycin sensitive ATPase activity in mitochondrial membranes was previously found (1). The isolated ATPase (Complex V) exhibits breaks in both the Arrhenius and Van't Hoff plots, with increase of activation energy and decrease of both k_{cat} and k_m for ATP below 20°C. In accordance with the presence of an isokinetic point at the break, $\Delta G*$ of formation of the transition state are unchanged in the temperature span 10-37°; since $\Delta H*$ strongly increases below the break, a compensating increase of $\Delta S*$ is calculated (Table 1).

Table 1. Activation parameters of ATPase in Complex V.

Parameter	at 10°C	at 30°C
V_m (μmol.min^{-1}.mg^{-1})	0.057	1.82
k_{cat} (s^{-1})	0.48	15.2
k_m for ATP (mM)	0.09	0.38
k_{app} 2nd order (k_{cat}/k_m) (M^{-1}s^{-1})	5.4 x 10^3	39.4 x 10^3
E_a (kcal.mol^{-1})	39.7	21.0
$\Delta H*$ (kcal.mol^{-1})	39.2	20.4
$\Delta G*$ (kcal.mol^{-1})	16.9	16.1
$\Delta S*$ (cal.mol^{-1}.K^{-1})	78.6	14.3

A discontinuity in the Arrhenius plot of an enzymic activity could be the result of (a) a different temperature dependence of individual steps of catalysis, shifting the rate-limiting step of the overall k_{cat} to a different one below a critical temperature; (b) a structural change in the solvent surrounding the enzyme; (c) a temperature dependent conformational change in the enzyme, with a conformation having higher E_a stabilized below the break. In the mitochondrial membrane(2) and in isolated OS-ATPase (3) temperature plots of spin labels motion parameters have revealed breaks, which can be related to the presence of protein (2). At the same time the temperature dependence of intrinsic tryptophan fluorescence of isolated Complex V exhibits a plateau near 25° (4) and the CD spectra show a decrease of α-helix below 20-25°C (5).

Studies with isolated F_1 have revealed no break (3) or breaks at 18° (6) or 30° (7); on the other hand the H+ conduction by bacterial TF_0 in liposomes exhibits a break which depends on lipid composition (8). Since ATPase activity in F_1 is several-fold higher than in Complex V, where it is limited by H+ conduction, the breaks in the overall activity are likely to represent a property of the H+-translocating membrane sector of the enzyme. Reconstitution of ATPase with dioleyl lecithin (DOL) having a gel-liquid phase transition at -21°, is shill accompanied by a break in V_m but no change in K_m, and little change in α-helix content, whilst reconstitution with dimyristoyl lecithin (DML) (transition at 23°) is accompanied by a break in both V_m and K_m (Table 2).

Table 2. Effect of lipid substitution on ATPase activity of Complex V

	+ DOL	+ DML
V_m at 32° (μmol.min^{-1}.mg^{-1})	0.31	0.11
Break in V_m (°C)	18	24-28
ΔH^* (kcal.mol^{-1} above/below break)	12.6/27.3	12.5/ 40,9
ΔG^* (kcal.mol^{-1} above/below break)	17.3/17.3	17.9/18.5
ΔS^* (cal.mol^{-1}.K^{-1} above/below break)	-15/35	-18/77

The results suggest that the breaks in the Arrhenius plots are related to conformational changes of ATPase in its membrane-bound form, as a result of lipid-protein interactions in the F_0 sector, and possibly due to the viscosity and/or the surface properties of the membrane.

1. Parenti Castelli G., Sechi A.M., Landi L., Cabrini L., Mascarello S. and Lenaz G. (1979) Biochim. Biophys. Acta 547, 161-169.
2. Lenaz G., Curatola G., Mazzanti L., Zolese G. and Ferretti G. (1983) Arch. Biochem. Biophys. 223, 369-380.
3. Solaini G. and Bertoli E. (1981) FEBS Lett. 132, 127-128.
4. Parenti Castelli G., Baracca A., Fato R. and Rabbi A. (1983) Biochem. Biophys. Res. Commun. 111, 366-372.
5. Curatola G., Fiorini R.M., Solaini G., Baracca A., Parenti Castelli G. and Lenaz G. (1983) FEBS Lett. 155, 131-134.
6. Harris D.A., Dall-Larsen T. and Klungsyor L. (1981) Biochim. Biophys. Acta 635, 412-428.
7. Gomez-Puyou M.T., Gomez-Puyou A. and Cerbon J. (1978) Arch. Biochem. Biophys. 187, 72-77.
8. Okamoto H., Sone N., Hirata H., Yoshida M. and Kagawa Y. (1977) J. Biol. Chem. 252, 6125-6131.

A CRITICAL EVALUATION OF VARIOUS PROCEDURES USED TO RECONSTITUTE CHLOROPLAST ATP SYNTHETASE INTO LIPOSOMES

M. Paternostre, J.M. Galmiche, G. Girault, J.L. Rigaud
Service de Biophysique, Département de Biologie
CEN Saclay, 91191 Gif-Sur-Yvette cédex, France

Various procedures have been used to reconstitute purified chloroplast ATP synthetase (CF_1-CF_0) into liposomes. These include cholate dialysis (1) and direct incorporation of detergent solubilized protein into small lipid vesicles either by freeze-thawing (2) or dilution (3). However, few attepts have been made to characterize in details the morphology of the reconstituted systems.

We have carried out a critical comparison of various reconstitution procedures of the ATP synthetase. Our ultimate goal is to obtain reconstituted proteoliposomes of large internal volume suitable to study electrochemical proton gradient mediated energy coupling. ATP synthetase was purified in Triton X_{100} from spinach chloroplasts (4). Four different incorporation methods were tested : cholate dialysis, freeze thawing with small sonicates lipid vesicles, reverse phase evaporation (a procedure previously used by us with bacteriorhodopsin (5)) and direct incorporation into large (0.4μm) unilamellar pure lipid vesicles by dilution. The proteoliposomes were characterized with regard to their activity (ATP-Pi exchange and ATP driven proton fluxes), protein orientation and size (freeze-fracture electron microscopy).

Protein incorporation into preformed large liposomes by dilution was found to yield the more satisfactory results with regards to the activity of the ATP synthetase, in particular, ATP-[32]Pi exchange activities of 700-800 nmoles/mg/min could be obtained using this procedure. In comparison, other reconstitution methods gave values which were 5 to 10 times lower. Furthermore, it was found that subsequent dialysis of proteoliposomes formed, using the dilution increased exchange activities up to 2-3000 nmoles/mg/min. This was shown to be due to the removal of residual Triton X_{100}.

The efficiency of protein incorporation in the liposomes was tested by electron microscopy and flotation on ficoll gradient and it was found to be critically dependent upon the protein to detergent weight ratio. Yields of incorporation near 100 % could be obtained at ratios above 5. Freeze fracture micrographs showed that the proteoliposomes thus formed were 0.3-0.4 μm in diameter and that the protein was asymetrically incorporated with the CF_1 moiety oriented towards the exterior. The latter result was confirmed by ATPase hydrolysis activity in the presence of uncouplers.

Incorporation of the ATP synthetase by dilution could also be performed in large liposomes containing bacteriorhodopsin (prepared

by reverse phase evaporation (5)). This results in large proteoliposomes in which the two proteins were properly oriented for the study
of proton gradient mediated energy coupling.

The influence of lipid to protein ratio, lipid composition and
liposome size upon the efficiency of the dilution incorporation procedure will also be discussed.

References
1. Sone Y., Yoshida M., Hirata H., Kagawa Y. (1977) J. Biochem. 81,
   519-528
2. Kasahara N., Hinkle P. (1977) J. Biol. Chem. 252, 7384-7390
3. Shahak Y., Pick U. (1983) Arch. of Biochem. and Biophys. 223,
   393-406
4. Pick U., Racker E. (1979) J. Biol. CHem. 254, 2793-2799
5. Rigaud J.L., Bluzat A., Buschlen S. (1983) Biochem. Biophys. Res
   Comm. 111, 373-382

ROLE OF SOME COMPONENTS OF THE PIG HEART MITOCHONDRIAL ATPase-ATP
synthase ON PROTON FLUX, ATP HYDROLYSIS AND ATP SYNTHESIS IN THE
RECONSTITUTED COMPLEX

F. Penin, G. Deléage and C. Godinot.
Laboratoire de Biologie et Technologie des Membranes du CNRS,
Université Claude Bernard. F69622 Villeurbanne cedex, France.

The coupling between proton flux and ATP synthesis catalyzed by
the ATPase-ATPsynthase complex in energy transducing membranes is
poorly understood. Recent studies from our laboratory have permitted
to correlate either ATP hydrolysis with the generation of a proton
gradient accross the mitochondrial membrane or ATP synthesis with
the utilization of the proton gradient [1]. The role of peptides involved in the attachment of F1 such as the Oligomycin Sensitivity
Conferring Protein (OSCP) or in its activity such as the protein inhibitor IF1 has been investigated on proton flux, ATP hydrolysis and
ATP synthesis.

Proton flux was monitored by using fluorescence quenching of
ACMA (9-amino-6-chloro-2-methoxyacridine,[2]) induced by either succinate or ATP with F1-depleted submitochondrial particles (ETPu) reconstituted or not with F1, IF1 and OSCP. In such reconstituted systems, the ATP-dependent fluorescence quenching only reflects the
ATPase activity of the F1 efficiently associated to the membrane.
Indeed, free F1 or F1 adsorbed on the membrane does not induce any
ATP-dependent fluorescence quenching. Besides, the succinate-dependent fluorescence quenching of ACMA was used as an estimation of the
passive H^+ leak through Fo. The measurements of net ATP synthesis
induced by succinate oxidation were performed by phosphorylation of
ADP with ^{32}Pi.

Submitochondrial particles depleted of F1 by urea treatment were
devoided of net ATP synthesis activity and ATP-dependent proton flux
even in the presence of nucleotides in the reconstitution medium.
Besides, the succinate-dependent fluorescence quenching was low but
could be highly increased by oligomycin which blocks the large
passive leakage of protons through Fo occurring in this type of
depleted particles. The binding of F1 to the membrane also decreased
this leakage as shown by the stimulation of the succinate-dependent
fluorescence quenching; the presence of Mg in the reconstitution
medium greatly increased this succinate-dependent fluorescence quenching indicating that Mg is involved in the binding of F1 to the
membrane. The reconstitution of an ATPase-ATPsynthase complex

competent for ATP synthesis and ATP-dependent proton flux requires the preincubation of the ETPu with F1 in the presence of Mg or/and nucleotides which modulate the efficiency of the reconstitution.

In all reconstitution conditions both ATP synthesis and ATP-dependent proton translocation were completely inhibited by oligomycin. As demonstrated in our laboratory, the incubation of isolated F1 with nucleotides, Pi and Mg induces a conformational change [3] giving a more compact structure of the enzyme [4]. Under these conditions, i.e. reconstitution in the presence of ATP and Mg, the maximal rates of ATP synthesis, ATP- and succinate-dependent proton fluxes were observed, which can be interpreted by the binding of F1 in a conformation efficient for ATP synthesis.

IF1, incubated with F1 (under conditions giving maximal inhibition of ATPase activity, i.e. 10 minutes at pH 6.7 with ATP and Mg, [5]) before reconstitution with ETPu in the presence of Mg and ATP did not modify the succinate-dependent fluorescence quenching indicating that IF1 does not seem to be directly involved in proton translocation. However IF1 inhibited the ATP-dependent proton flux as expected by its inhibitory effect on the ATPase activity. Moreover, IF1 had no effect on the rate of net ATP synthesis.

When OSCP was reassociated with ETPu and F1 in the presence of Mg and ATP, a three-fold stimulation of the net ATP synthesis activity was obtained. However, under all conditions tested up to now, OSCP did not seem to modify the maximal ATP- or succinate-induced fluorescence quenching of ACMA. Other experiments are in progress to check whether OSCP may play any role in the proton translocation through the ATPase-ATPsynthase complex or if OSCP is only involved in the efficient association of F1 to Fo.

1 - Deléage, G., Penin, F., Godinot, C. and Gautheron, D.C. (1983) Biochim. Biophys. Acta 725, 464-471.
2 - Kraayenhof, R. and Fiolet, J.W.T. (1974) in Dynamics of Energy Transducing Membranes (Ernster, L., Estabrook, R.W. and Slater, E. C. eds) pp. 355-364, Elsevier, Amsterdam.
3 - Di Pietro, A., Penin, F., Godinot, C. and Gautheron, D.C. (1980) Biochemistry 19, 5671-5678.
4 - Roux, B., Fellous, G. and Godinot, C. (1984) Biochemistry 23, 534-537.
5 - Pullman, H.E. and Monroy, C.C. (1963) J. Biol. Chem. 238, 3762-3769.

352

ISOLATION AND PROPERTIES OF THE ε-SUBUNIT FROM YEAST F₁-ATPase

ISOLATION AND PROPERTIES OF THE ε-SUBUNIT FROM YEAST
F_1-ATPase

B. Pevec[o], E. Fassold[''], N. Uwira[o] and B. Hess[o]
[o]Max-Planck-Institut für Ernährungsphysiologie,
 Rheinlanddamm 201, 4600 Dortmund 1, FRG
['']Max-Planck-Institut für medizinische Forschung,
 Jahnstraße 29, 6900 Heidelberg 1, FRG

Mitochondrial F_1-ATPase was prepared in highly pure
form from baker's yeast (ATCC 24967) grown aerobically on
2% ethanol or 0.8% glucose. These preparations are free
of any contaminating proteins, e.g. those of 63 kD and 42
kD, which were present in preparations of F_1-ATPase from
commercial baker's yeast in nearly stoichiometric amounts,
and the 51 kD protein which accompanied the α-subunit in
every purification procedure.

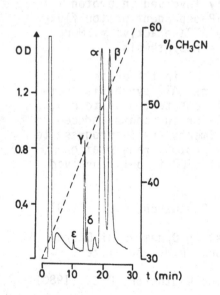

So far, the purification and
the analysis of the subunits
was complicated by the tendency of the proteins to aggregate into complexes with
$> 10^3$ kD. However, acqueous
CH_3CN and/or 10 mM CF_3COOH
seem to favour the monomeric
state, which made it possible
to preparatively separate all
five subunits by reversed
phase HPLC with a gradient
from 30% to 60% CH_3CN in 10mM
CF_3COOH. The isolated and
salt free subunits are homogeneous with regard to SDS-
electrophoresis. They are readily soluble in acqueous ace-
tonitrile, a prerequisite for
sequencing and reconstitution
experiments.

Fig. 1: Separation of F_1-ATPase by reversed phase HPLC
        Solvent A: 10 mM CF_3COOH in H_2O
        Solvent B: 10 mM CF_3COOH in CH_3CN
        Flow rate, 1 ml/min, room temperature

The amino acid composition of the ε-subunit differs markedly from all other subunits. With a molecular weight of 8 kD, it is the smallest of all the F_1-ATPase subunits. It does not contain cys and met, only one trp and two tyr, but 16 mole % basic amino acids in contrast to 8-13 mole % in all the other subunits. This accumulation of basic amino acids leads to a pk > 10. The high content of lys, arg and his of the ε-subunit was also observed in other F_1-ATPases (1,2). In contrast the pK of the other subunits varies from 7(γ) to 5 (ß).

The N-terminus of ε was determined to be Ser- Ala-, which is different from the corresponding N-terminus of the ATPases of E. coli and spinach chloroplasts (3,4).

The δ-subunit proved to have an unusual N-Terminus with a high accumulation of Ala: Ala - Glu - Ala - Ala - Ala - Ala. Cys was only found in α, γ and δ. The ε- and, in contrast to published DNA-data (5), the ß-subunit do not contain cys.

1. Yoshida, M., Sone, N., Hirata, H., Kagawa, Y. and Ui, N. (1979) J. Biol. Chem. 254, 9525-9533.
2. Knowles, A.F. and Penefsky, H.S. (1972) J. Biol. Chem. 247, 6624-6630.
3. Saraste, M., Gay, N.J., Eberle, A., Runswick, M.J. and Walker, J.E. (1981) Nucl. Acid. Res. 9, 5287-5296.
4. Zurawski, G., Bottomley, W. and Whitfield, P.R. (1982) Proc. Natl. Acad. Sci. USA 79, 6260-6264.
5. Saltzgraber-Muller, J., Kunapuli, S.P. and Douglas, M.G. (1983) J. Biol. Chem. 258, 11465-11470.

354

A $\Delta\mu_H^+$-DEPENDENT DCCD SENSITIVE Ca-ATPase ACTIVITY IN CHLOROPLASTS NOT COUPLED TO PROTON TRANSLOCATION.

U. Pick
Biochemistry Department, Weizmann Institute of Science, Rehovot, Israel

Illuminated chloroplasts catalyze ATP synthesis in the presence of Mg but not in the presence of Ca. Similarly, chloroplasts catalyse also the reverse reaction, namely, a Mg-specific ATP hydrolysis coupled to proton translocation in the dark following preillumination with dithiol reducing agents (light-triggered Mg-ATPase). In contrast, Avron described a light-dependent Ca-stimulated ATPase activity in chloroplasts whose relation to the mechanism of ATP formation is not clear (1).

It is demonstrated that this light-dependent Ca-ATPase activity is catalysed by CF_o-CF_1, is not coupled to proton translocation and is $\Delta\mu_H^+$-dependent.

The dependence on $\Delta\mu_H^+$ is indicated by: (a) An absolute dependence on illumination. (b) Requirement for an electron transport carrier (PMS, MV or FeCN). (c) Sensitivity to uncouplers. Ca-ATPase activity is unaffected by thiol modulation and does not persist in the dark after preillumination in the presence of dithiolthreitol. A comparison of the energetic requirements for activation of Ca-ATPase and of ATP formation shows that a higher $\Delta\mu_H^+$ is needed to activate Ca-ATPase

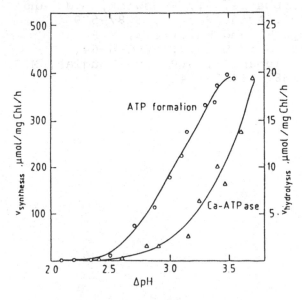

ATP-synthesis and hydrolysis was measured by incorporation or release of ^{32}P into/from $\gamma^{32}P$-ATP in a medium containing: 30 mM Na-tricine (pH 8), 30 mM KCl, 50 µM pyocyanine, 2 mM P_i, 5 mM DTT, chloroplasts, equivalent to 30 µg/ml, 2 µM 9-aminoacridine and either 5 mM $MgCl_2$, 0.5 mM ADP and $^{32}P_i$ (ATP synthesis) or 5 mM $CaCl_2$ and 0.5 mM $\gamma^{32}P$-ATP (ATP hydrolysis). ΔpH was measured simultaneously by 9-AA fluorescence.

Ca-ATPase activity in illuminated chloroplasts is not coupled to proton translocation since: (a) The addition of Ca-ATP to chloroplasts in the dark or in the light does not induce formation of ΔpH or $\Delta\Psi$ and turning off the light leads to a complete dissipation of $\Delta\mu_H^+$. In contranst, addition of Mg-ATP to illuminated chloroplasts leads to an additional buildup of ΔpH and $\Delta\Psi$ which only partially decays in the dark. (b) Uncouplers, which stimulate Mg-ATPase in illuminated chloroplasts as expected from a proton ATPase, inhibit Ca-ATPase activity.

Light-dependent Ca-ATPase activity of chloroplasts is sensitive to DCCD indicating that activation of Ca-ATPase does not result from a light-induced dissociation of CF_1 from the thylakoid membrane. Light dependent Ca-ATPase is also inhibited by tentoxin, quercetin and anti-CF_1 serum.

These results clearly demonstrate that light-dependent Ca-ATPase in chloroplast is catalysed by CF_0-CF_1 by a mechanism different from ATP formation and of light-triggered ATP hydrolysis. The possibility that it reflects a partial reaction of ATP synthesis in the reverse direction seems unlikely also in view of the different energetic requirements for activation of ATP synthesis and of Ca-ATPase which indicate different mechanisms.

An alternative mechanism is that high energization levels induce a functional uncoupling in a small fraction of the CF_0-CF_1 units which is associated with Ca-ATPase activation and possibly also with a partial exposure of a proton leak through the enzyme. The latter hypothesis is supported by: (a) The close correlation between the energetic requirements for Ca-ATPase activation and for the increase in proton-conductance through CF_0-CF_1 in the thylakoids under non-phosphorylating conditions (2). (b) The recent observation that binding of an ε-deficient CF_1, which has high Ca-ATPase activity (3, 4),does not seal the proton leak through CF_0 (4).

Such a controlled regulation of proton conductance through CF_0-CF_1 may have important physiological implications in preventing extreme acidification of the intrathylakoid space or in avoiding over inhibition of electron flow by super optimal ΔpH at very high energization levels.

References
1. Avron, M. (1962) J. Biol. Chem. 237, 2011-2017
2. Shönfeld, M. and Neumann, J. (1977) FEBS Lett. 73, 51-54
3. Pick, U. (1983) FEBS Lett. 152, 119-124.
4. Richter, M.L., Patrie, W.J. and MacCarty, R.E. (1984) J. Biol. Chem., in press.

LIGAND BINDING TO ISOLATED SUBUNITS AND THE NATIVE
H^+-ATPASE OF THE THERMOPHILIC BACTERIUM PS3

M.Rögner[o], U.Lücken["], H.Tiedge["], J.Weber["], and P.Gräber[o]

[o]Max-Volmer-Institut für Biophysikalische Chemie,TU Berlin
1000 Berlin 12, West-Germany
["]Institut für Biochemie,Medizinische Hochschule Lübeck,
2400 Lübeck, West-Germany

The isolated α- and β-subunits and the reconstituted
ATPase (TF_1 and TF_oF_1-vesicles) of the thermophilic bacterium PS3 (1) have been tested for their ligand binding
characteristics. Ligands tested consist of the natural
substrates (ADP, ATP, P_i) and the fluorescent ATP-analo-
gue naphthoyl-ATP (N-ATP) (2).
Binding characteristics were examined by measuring the
following parameters :
a) change of intrinsic protein fluorescence upon ligand
   binding
b) change of anisotropy of the fluorescent N-ATP upon
   ligand binding and release (3)
c) change of ATP-hydrolysis activity of TF_1 and TF_oF_1-
   vesicles upon addition of different inhibitors (ADP,
   P_i, N-ATP)

The following results have been obtained :
1) Binding sites : Both isolated α and β has 1 binding
   site for N-ATP, which can be occupied by ADP and ATP.
   Surprisingly, TF_1 ($\alpha_3\beta_3\gamma\delta\epsilon$) shows only two binding sites for N-ATP (below 5 μM).
2) Dissociation constants : K_D(TF_1-N-ATP) $\ll$ K_D(α-,β-N-ATP)
   While the dissociation constants K_D(N-ATP), K_D(ATP) and
   K_D(ADP) for both isolated α- and β-subunits are of similar magnitude (20-15 μM), K_D(N-ATP) of TF_1 is much
   smaller than its K_D for ADP and ATP.
3) Binding kinetics : Ligand binding and ligand release
   is much faster with isolated α- and β-subunit than
   with TF_1. Addition of excess ATP to the $TF_1 \cdot$2N-ATP-com-
   plex results in a biphasic release of N-ATP, while the
   binding of 2 N-ATP is a monophasic process (without
   AT(D)P)
4) Inhibition constants : K_i-values differ remarkably with
   both N-ATP and ADP as inhibitors of ATP-hydrolysis activity.

These results   indicate :
a) Subunit-subunit-interactions in the reconstituted
   ATPase change the properties of the isolated subunits.
b) Specific interactions of the natural substrates (AT(D)P)
   with the ATPase induce a negative cooperativity of binding sites. The binding of N-ATP does not cause such an
   effect.

1) Kagawa,Y., Sone,N., Hirata,H., and Yoshida,M. (1979)
     J.Bioenerg.Biomembr. 11, 39-78

2) Onur,G., Schäfer,G., and Strotmann,H. (1983) Z.Natur-
     forsch.38c, 49-59

3) Tiedge,H., Lücken,U., Weber,J., and Schäfer,G. (1982)
     Eur.J.Biochem. 127, 291-299

^{1}H NMR STUDY AT 500 MHz OF CF1 FROM SPINACH CHLOROPLASTS

Madeleine Roux-Fromy, Guy Girault and Jean-Michel Galmiche
Service de Biophysique, Département de Biologie
CEN Saclay, 91191 Gif-sur-Yvette cédex, France.

^{1}H NMR spectra of coupling factor 1 from chloroplasts (CF1) usually consist of a few narrow resonances embedded in a broad structureless band. To get more tractable NMR results the broader components of the spectra are eliminated using spin-echo (90°-τ -180°- τ) sequences with τ values ranging from 2 to 20 ms. For τ ⩾ 10 ms a straight baseline is obtained and on the basis of the chemical shifts of the remaining resonances the more mobile residues are identified. For the latent ATPase (one residual ADP per molecule) residues with T_2 ⩾ 15 ms comprise about 8 Leu, Ile or Val, 5 Thr, 4 Ala, 5 Met and 10 Lys (and/or Asn) per molecule. For DTT activated CF1, the salient feature of the spectra is the appearance of very mobile Thr residues.

The paramagnetic broadening induced by Mn^{II} affects more or less all the above residues, though the effect on Thr and Ala is more specific for activated CF1.

Changes induced by binding of Mg-ADP on the second nucleotide binding site have also been investigated and are indicative of immobilisation of some Leu, Ile or Val residues.

CATION-DEPENDENT BINDING OF F_1 TO F_0 IN F_6- AND OSCP-DEPLETED SUB-MITOCHONDRIAL PARTICLES

G. Sandri[*], L. Wojtczak[**] and L. Ernster
Department of Biochemistry, Arrhenius Laboratory, University of Stockholm, S-106 91 Stockholm, Sweden

The binding of F_1 to F_0 in F_1-depleted membrane preparations of chloroplasts (1) and mitochondria (2) requires the presence of cations. Both mono- and divalent cations are effective, and their role seems to be to neutralize the negative surface charge of the membrane, as indicated by measurements of the ζ-potential (2).

In mitochondria, the binding of F_1 to F_0 requires the presence of an "oligomycin sensitivity conferring protein" (OSCP) (3) and coupling factor 6 (F_6) (4). In submitochondrial particles depleted of F_1, OSCP and F_6, the addition of F_6 alone promotes the binding of F_1, but both F_6 and OSCP are needed to restore oligomycin sensitivity of the ATPase activity (5-7). These particles also require the presence of cations for reconstitution, but in this case NH_4^+, K^+, Rb^+ and Cs^+ are more effective than Na^+ and Li^+, and divalent cations (Ca^+ and Mg^{2+}) are ineffective (8). F_1 bound to the particles in the presence of monovalent cations alone is oligomycin insensitive and OSCP induces only a partial ($< 60\%$) oligomycin sensitivity. Both F_6 and OSCP, together with a monovalent cation, are required for full oligomycin sensitivity. Interestingly, the maximal amount of F_1 bound to the particles in the presence of monovalent cations alone is equal to that bound in the presence of cations, F_6 and OSCP and can be rendered oligomycin sensitive by the subsequent addition of the two proteins.

These findings suggest that F_0 has a binding site for F_1 in addition to the binding site(s) provided by F_6 and OSCP.

---

[*] Permanent address: Istituto di Chimica Biologica, Università di Trieste, Trieste, Italy

[**] Permanent address: Department of Cellular Biochemistry, Nencki Institute for Experimental Biology, Warsaw, Poland

This work has been supported by the Swedish Natural Science Research Council.

References

1. Telfer, A., Barber, J. and Jagendorf, A.T. (1980) Biochim. Biophys. Acta 591, 331-345.

2. Sandri, G., Suranyi, E., Eriksson, L.E.G., Westman, J. and Ernster, L. (1983) Biochim. Biophys. Acta 723, 1-6.

3. MacLennan, D.H. and Tzagoloff, A. (1968) Biochemistry 7, 1603-1610.

4. Fessenden-Raden, J.M. (1972) J. Biol. Chem. 247, 2351-2357.

5. Russell, L.K., Kirkley, S.A., Kleyman, T.R. and Chan, S.H.P. (1976) Biochem. Biophys. Res. Commun. 73, 434-443.

6. Vadineanu, A., Berden, J.A. and Slater, E.C. (1976) Biochim. Biophys. Acta 449, 468-479.

7. Norling, B., Glaser, E. and Ernster, L. (1978) in Frontiers of Biological Energetics, Vol. I (Dutton, P.L., Leigh, J.S. and Scarpa, A. eds.) Academic Press, New York, pp. 504-515.

8. Sandri, G., Wojtczak, L. and Ernster, L. (1981) in Vectorial Reactions in Electron and Ion Transport (Palmieri, F., Quagliariello, E., Siliprandi, N. and Slater, E.C., eds.) Elsevier/North Holland, Amsterdam, pp. 197-207.

PHOTOAFFINITY CROSS-LINKING OF F_1ATPase FROM THE THERMOPHILIC BACTERIUM PS3

H.-J. Schäfer[+], G. Rathgeber[+], K. Dose[+], and Y. Kagawa[§]
[+]Institut f. Biochemie, Universität Mainz, D-6500 Mainz, W. Germany
[§]Department of Biochemistry, Jichi Medical School, Tochigi-ken, Japan

F_1ATPase (EC 3.6.1.34) possesses three catalytic nucleotide binding sites on the ß subunits and three noncatalytic nucleotide binding sites on the α subunits (1).

Irradiation of F1ATPase from Micrococcus luteus (the enzyme contains 2.5-3.0 tightly bound nucleotides) in the presence of 8-azido ATP ($8-N_3$ATP) or 8-azido-1,N[6]-etheno ATP ($8-N_3 \epsilon$ATP) results in a specific labeling of the ß subunits (2,3). This labeling is dependent on the presence of Mg^{2+} ions.

In contrast to these results photoaffinity labeling of F_1ATPase (TF_1) from the thermophilic bacterium PS3 (the enzyme does not contain tightly bound nucleotides (4)) leads to a specific labeling of the α subunits (5). This labeling is almost independent of the presence of Mg^{2+} ions. Ohta et al. (4), however, have demonstrated that the catalytic site of TF_1 is located on the ß subunit, too.

There are two explanations for the diverging results of the different F_1ATPases:

1. The labeling takes place at different nucleotide binding sites: F1ATPase from Micrococcus luteus is labeled at the catalytic sites on ß, whereas TF_1 is labeled at the noncatalytic sites on α.

2. The nucleotide binding sites are situated at the interface between the α and the ß subunits. In this case $8-N_3$ATP can label either the α or the ß subunits.

These two possibilities are not mutually exclusive.

To prove the localisation of nucleotide binding sites between two subunits we have synthesized the bifunctional (cross-linking) 3'-aryl-azido-8-azido ATP (DiN_3ATP) (6). For the F_1ATPase from Micrococcus luteus the application of DiN_3ATP as photoaffinity label has demonstrated the vicinity of the catalytic nucleotide binding site on a ß subunit to an α subunit (7).

TF_1 hydrolyzes $Mg \cdot DiN_3$ATP. The hydrolysis of $Mg \cdot$ATP is competitively inhibited by $Mg \cdot DiN_3$ATP. These experiments demonstrate the specific interaction of DiN_3ATP with a catalytic site of TF_1 and thus the suitability of DiN_3ATP as photoaffinity label for this enzyme.

Irradiation of TF_1 in the presence of DiN_3ATP results in a reduction of ATPase activity and in a formation of cross-linked proteins of higher molecular masses (m > 100 kD). The formation of these cross-links is partially dependent on the presence of Mg^{2+} ions. Addition of ATP or ADP prior to the labeling procedure decreases the yield of formed cross-links. Addition of AMP does not influence the cross-link formation. Hydrolytic cleavage of the most intense cross-link proves its composition of α-β. This indicates the localisation of nucleotide binding sites at the interface between the α and the β subunits.

Besides the cross-links formed by two of the major subunits α and β a small amount of even higher molecular weight cross-links is observed. The formation of this cross-link is nucleotide specific and depends on the presence of Mg^{2+} ions. This suggests the presence of at least two nucleotide binding sites between three subunits.

The localisation of several nucleotide binding sites between α and β subunits favors these models for ATP synthesis/hydrolysis which require strong subunit-subunit interactions (8,9). Our results are the first direct proof for the model of site-site interactions in F_1 as proposed by Senior and Wise (9).

1 Cross, R.L., and Nalin, C.M. (1982) J. Biol. Chem. 257, 2874-2881.
2 Scheurich, P., Schäfer, H.-J., and Dose, K. (1978) Eur. J. Biochem. 88, 253-257.
3 Schäfer, H.-J., Scheurich, P., Rathgeber, G., and Dose, K. (1980) Anal. Biochem. 104, 106-111.
4 Ohta, S., Tsuboi, M., Oshima, T., Yoshida, M., and Kagawa, Y. (1980) J. Biochem. (Tokyo) 87, 1609-1617.
5 Schäfer, H.-J., Scheurich, P., Rathgeber, G., Dose, K., and Kagawa, Y. (1982) Second Bioenergetics Conference, pp. 57-58, L.B.T.M.-C.N.R.S. Edition, Villeurbanne.
6 Schäfer, H.-J., Scheurich, P., Rathgeber, G., Dose, K., Mayer, A., and Klingenberg, M. (1980) Biochem. Biophys. Res. Commun. 95, 562-568.
7 Schäfer, H.-J., Scheurich, P., Rathgeber, G., and Dose, K. (1980) First European Bioenergetics Conference, pp. 175-176, Pàtron Editore, Bologna.
8 Boyer, P.D., Kohlbrenner, W.E., McIntosh, D.B., Smith, L.T., and O'Neal, C.C. (1982) Ann. N. Y. Acad. Sci. 402, 65-83.
9 Senior, A.E., and Wise, J.G. (1983) J. Membrane Biol. 73, 105-124.

Work supported by the Deutsche Forschungsgemeinschaft.

ATP-SYNTHESIS BY RECONSTITUTED CF_oF_1-LIPOSOMES

Günter Schmidt and Peter Gräber

Max-Volmer-Institut für physikalische und biophysikalische Chemie, Technische Universität Berlin,
D-1000 Berlin 12, Germany

The chloroplast ATPase CF_oF_1 can be isolated and reconstituted into liposomes (1). However, the ATP-yield
and the rate of ATP-synthesis was only about 0.1 % of
the maximal values obtained under native conditions.
This might be due to principal reasons (e. g. the reconstituted enzyme is structurally or functionally different from the native one) or to the fact that the isolation, reconstitution and/or assay conditions were not
optimal. Therefore, we tried to optimize these conditions.

1) Assay conditions: We measured the ATP-synthesis
catalyzed by the reconstituted CF_oF_1 after energization
by an acid-base transition in the presence of a K^+/vali-
nomycin diffusion potential. The vesicles were found to
be highly sensitive to acidic environment, but rather
stable to basic pH. Under optimal pH_{in}, pH_{out} and $\Delta\psi$
the vesicles yielded 20 to 50 ATP/CF_1.

2) Reconstitution conditions: Reconstitution was performed with the dialysis method (2). The type of dialysis membrane, the dialysis time, the protein/lipid ra-

tio, and the temperature was optimized. Under the best reconstitution and assay conditions the rate of ATP-synthesis was measured using a rapid-mixing quenched flow apparatus. The maximal rate obtained was 200 ATP/$(CF_1 \cdot s)$. This is about half of the maximal rate observed under native conditions.

(1) Pick, U. and Racker, E. (1979) J. Biol. Chem. 254, 2793 - 2799

(2) Sone, N., Yoshida, M., Hirata, H. and Kagawa, Y. (1977) J. Biochem. (Tokyo) 81, 519 - 528

THE ROLE OF THE ESSENTIAL TYROSINE IN MITOCHONDRIAL ATPASE WHICH IS
MODIFIED BY 4-CHLORO-7-NITROBENZOFURAZAN

R. Sutton and S.J. Ferguson

Department of Biochemistry, University of Birmingham, P.O. Box 363,
Birmingham B15 2TT, U.K.

Mitochondrial F_1ATPase is inhibited by covalent modification with
4-chloro-7-nitrobenzofurazan (Nbf-Cl). F_1ATPase is completely inhibited by a single Nbf bound to ATPase; the Nbf being bound to a
tyrosyl residue [1]. All F_1ATPases isolated to date have shown a
similarly reactive tyrosine, suggesting conservation of the tyrosyl
residue [2,3]. The tyrosyl residue thus modified is unusually reactive, as Nbf-Cl reacts only very slowly with N-actyltyrosine ethyl
ester under comparable conditions [1,4]. If the tyrosyl modified
ATPase is incubated at pH 9.0 the Nbf undergoes an intramolecular
shift to a lysyl residue; the enzyme remaining inhibited [5].

The above data suggest that the Nbf-reactive tyrosyl residue may
play an active role in acid-base catalysis of ATP synthesis and hydrolysis. We have attempted to ascertain whether the Nbf-Cl modified F_1-
ATPase was capable of catalysing any part of the catalytic cycle by use
of quench-flow methods. The results suggested which step in the catalytic cycle was inhibited by Nbf-Cl modification of the reactive
tyrosyl residue, and gave some indication of whether Nbf-Cl reactive
tyrosine was situated at the active site.

The relative position of the Nbf-reactive tyrosyl residue and the
lysyl residue to which Nbf transfers was investigated. F_1ATPase which
had been modified on lysine by incubation of the tyrosyl-modified
enzyme at pH 9.0 was reacted once more with Nbf-Cl. It was found that
the lysyl modified ATPase was capable of reacting with an Nbf-Cl group
on a tyrosyl residue in a manner indistinguishable from the native
enzyme. This indicated that the lysyl ánd tyrosyl residues were sufficiently spatially separated and that no steric hindrance to the binding
of a second Nbf group could be found [6].

These results are correlated with recent results obtained on the
protective effect of phosphate on the inhibition of F_1ATPase by Nbf-
Cl. This arises from a discrepancy between the work of Ferguson et al.
(1975) [1] who obtained no protection by phosphate of the Nbf-Cl
reactive tyrosine and Ting and Wang [7] who did. This was resolved;
the protective effect of phosphate was shown to be pH dependent. If
the experiments were carried out at pH 7.5 no phosphate protection of
inhibition by NbfCl could be found, whilst at pH 8.0 phosphate provided marked protection. At pH 8.0 a K_d of 2 mM was calculated for
phosphate binding. The membrane bound ATPase of Paracoccus

<u>denitrificans</u> shows phosphate protection of reaction with Nbf-Cl at pH 8.0 in a similar manner to that described for the corresponding enzyme from <u>Rhodospirillum rubrum</u> [8]. Comparison of the same reaction with the ATPase of submitochondrial particles will be of value. All these data show that binding of phosphate can prevent reaction of the tyrosine with Nbf-Cl. That the tyrosine is at a phosphate binding site is one interpretation, but conformational changes caused by phosphate binding elsewhere cannot be ruled out as the possible basis for the protective effect of phosphate.

This work was supported by the SERC, U.K.

References

1.   Ferguson, S.J., Lloyd W.J., Lyons, M.H. and Radda, G.K. (1975) Eur. J. Biochem. 54, 117-123.
2.   Deters, D.W., Racker, E., Nelson, N. and Nelson, H. (1975) J. Biol. Chem. 250, 1041-1047.
3.   Lunardi, J., Satre, M., Bof, M. and Vignais, P.V. (1979) Biochemistry 18, 5310-5316.
4.   Aboderin, A.A., Boedefeld, E. and Luisi, P.C. (1973) Biochim. Biophys. Acta 238, 20-30
5.   Ferguson, S.J., Lloyd, W.J. and Radda, G.K. (1975) Eur. J. Biochem. 54, 127-133.
6.   Sutton, R. and Ferguson, S.J. (1984) Eur. J. Biochem. in the press
7.   Ting, L.P. and Wang, J.H. (1980) J. Bioenerg. Biomembr. 12, 79-93
8.   Cortez, N., Lucero, H.A. and Vallejos, R.H. (1983) Biochim. Biophys. Acta 724,-403.

INTERACTIONS OF NUCLEOTIDE ANALOGUES AT THE HIGH AFFINITY BINDING SITES
OF MITOCHONDRIAL F_1-ATPASE
H. Tiedge and G. Schäfer, Institut für Biochemie, Medizinische Hoch-
schule Lübeck, D-2400 Lübeck 1

Cooperative binding phenomena at the catalytic sites of F_1-type
ATPases have been demonstrated by use of a variety of nucleotide analogues. In particular, studies with aromatic 2'/3'-ethers and -esters
of adenine nucleotides have shown mutual interactions of the high affinity binding sites on the mitochondrial enzyme (1, 2). Further investigations on this type of analogues are described in this report:
1. binding of the 2'/3'-ether 2', 3'-0-(2,4,6-trinitrophenyl)-ADP
   (TNP-ADP) to isolated, nucleotide depleted F_1-ATPase from beef
   heart mitochondria,
2. exchange kinetics (e. g. bound TNP-ADP versus free 3'-0-naphthoyl
   -(1)-ADP (N-ADP)) as an indicator of site-site interactions.

   1. Binding of TNP-ADP to F_1-ATPase
The binding process was monitored both by the increase of the fluorescence yield of the ligand and by the decrease of the tyrosine-type
fluorescence of the enzyme. A stoichiometric titration based on the
first method is shown in fig. 1. Upon binding, the fluorescence of
TNP-ADP is increased drastically (by a factor of 8 for the first site,
by a factor of 2 for the next two sites). Simultaneously, the fluorescence of F_1 is decreased by 22 %. Binding is clearly biphasic: saturation is reached at a stoichiometry of 3 mol ligand/mol enzyme, while
one more abrupt change of slope occurs at a stoichiometry of 1 mol
ligand/mol enzyme. This kind of titration curve can be explained most
easily by a successive occupation of binding sites with largely differing affinities, binding to the first site resulting in a more drastic
change of fluorescence than binding to the next two sites of comparatively lower affinity. In contrast to TNP-ADP, only one (out of three)
binding site for ADP can be detected fluorometrically (2). Attempts to
determine exact K_d-values failed so far, however; even in a nanomolar
range of concentrations titration curves resulted to be stoichiometric
while in the subnanomolar range the signal/noise ratio becomes insufficient for quantitative evaluation. Therefore, K_d-values can only be
estimated to be in the nanomolar range or below.
   2. Exchange kinetics TNP-ADP/N-ADP
Whereas TNP-ADP bound to the enzyme is displaced very slowly (compared
with the catalytic halflife of F_1) even by an excess of ADP or N-ADP,
bound ligands such as N-ADP are readily exchanged by TNP-ADP (fig. 2).
Up to a saturation point of 3 mol TNP-ADP/mol enzyme, exchange kinetics are first order with respect to TNP-ADP (i. e. second order overall), the rate constant decreasing abruptly after one binding site
has been occupied by TNP-ADP. These exchange kinetics are believed to

reflect the fact that the first molecule of TNP-ADP not only binds much stronger to F_1 than for example N-ADP does, but by the binding interaction itself converts this site into a site of very high affinity ($K_d < 10^{-9}$ M), simultaneously decreasing the affinities of the other two sites (irrespective of whether these sites are occupied by other ligands or not). In short, induced fit at the first site would slow down binding to the second one and third one.

These results 1) suggest that 2'/3'-ether and 2'/3'-ester analogues of ADP occupy the same class of binding sites on mitochondrial F_1-ATPase and 2) add further evidence to the hypothesis that binding to these sites is accompanied by negatively cooperative interactions.

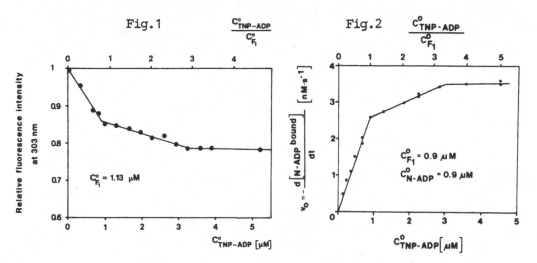

Fig. 1: Stoichiometric fluorescence titration of F_1 with TNP-ADP

Fig. 2: Exchange of bound N-ADP by TNP-ADP. Stoichiometric amounts of F_1 and N-ADP were incubated for 20 min. After addition of the indicated amount of TNP-ADP, release of N-ADP (initial rate) was monitored via fluorescence anisotropy.

References

(1) Grubmeyer, C. and Penefsky, H.S. (1981)
    J. Biol. Chem. 256, 3718-3727
(2) Tiedge, H., Lücken, U., Weber J. and G. Schäfer (1982)
    Eur. J. Biochem. 127, 291-299

THE EFFECT OF THE DETERGENT LAURYLDIMETHYLAMINE OXIDE ON THE ATPASE
COMPLEX OF E. COLI

M. Tommasino and R. A. Capaldi
Institute of Molecular Biology, University of Oregon, Eugene, Oregon
97403 USA

Recently it has been shown that the hydrolytic activity of the isolated
extrinsic sector of the ATPase complex from E. coli (F_1 ATPase) is increased 4-6 times in the presence of the amphipatic detergent lauryldimethylamine oxide (LDAO) (1).
This observation could be explained in at least two ways:

1) increase of the single site turnover of the catalytic high
    affinity site;

2) increase of the cooperativity among the catalytic sites.

To examine these possibilities we have measured the single-site turnover of ATP hydrolysis at low substrate concentrations in the presence
and absence of LDAO. An approximate two-fold increase in the hydrolytic activity was observed in the presence of this detergent. To obtain
further insight into the effect of LDAO, F_1 ATPase purified from a
mutant strain of E. coli (unc A 401, AN 718) that shows full single-
site activity but essentially no enhanced (cooperative) turnover has
been used. The single-site hydrolytic activity of this mutant was
identical to that of the wild type F_1 ATPase as reported by Wise et al.
(2). This single-site turnover was stimulated two-fold by LDAO as for
the wild type enzyme. The steady-state hydrolysis rate was only enhanced two-fold in marked contrast to wild type F_1 ATPase which in
parallel experiments was increased 4-6 fold in the presence of LDAO.
These results indicate that LDAO affects both the single-site turnover
and the intersubunit cooperativity.

References

(1) Lotscher, H. R., deJong, C. and Capaldi, R. A., Biochemistry,
     in press

(2) Wise, J. G., Latchney, L. R., Ferguson, A. M. and Senior, A. E.
     (1984), Biochemistry 23, 1426-1432.

PREPARATION OF SUBUNIT DEFICIENT F₁, RECONSTITUTION AND CHEMICAL MODIFICATION OF ISOLATED SUBUNITS

R. Tuttas and W.G. Hanstein

Institut für Physiologische Chemie I der Ruhr-Universität Bochum, Universitätsstr. 150, 4630 Bochum, BRD

Preparation of subunit deficient F_1. Ribipress vesicles of E.coli strains K 12 or ML 308/225 were extracted with chloroform by a modification of the method of Beechey et al. [1]. Inclusion of ATP and mercaptoethanol in the extraction buffer increased the yield and specific activity of F_1. ATPase was further purified by ion exchange chromatography (DEAE-Sepharose CL 6 B) and gel filtration (Biogel A 1.5 m). F_1 lacking δ or δ and ε were prepared by the same method from E.coli strains K 12 and ML 308/225, respectively.

Preparation of subunits δ and ε . F_1 from E.coli strain ML 308/ 225 was prepared essentially by the procedure of Vogel and Steinhart [2]. Isolation of subunits δ and ε was performed according the general procedure of Hager and Burgess [3]. After PAGE on 13% Laemmli gels protein bands were visualized by KCl and cut out. After elution of δ and ε by diffusion into SDS-containing buffer and dialysis overnight, the protein was concentrated by ultrafiltration. The subunits δ and ε were precipitated with acetone and treated with 6 M guanidine-HCl. Renaturation was achieved by diluting the sample with buffer and dialyzing overnight against 100 volumes of buffer.

Inhibition of ATPase activity by subunit ε. ε decreased ATPase activity of δ,ε-deficient F_1 by about 90%. 50% of maximal inhibition needed less than 1 mol ε per mol F_1. ATPase activity of complete F_1 could be reduced by 30-40% and δ deficient F_1 was inhibited by about 70%. A Hill plot of inhibition data gave a Hill coefficient of 1.3 to 2.0 for δ and δ,ε deficient F_1, which suggest more than one binding site for ε .

Reconstitution of subunit deficient F_1 with δ and ε . Energy dependent reactions such as the quenching of ACMA fluorescence or transhydrogenation in the presence of subunit deficient F_1 could be restored by the isolated subunits. δ and ε showed high reconstitutive activity since amounts of δ and ε subunits similar to those present in complete F_1 were necessary to get 50% of maximal transhydrogenase activity.

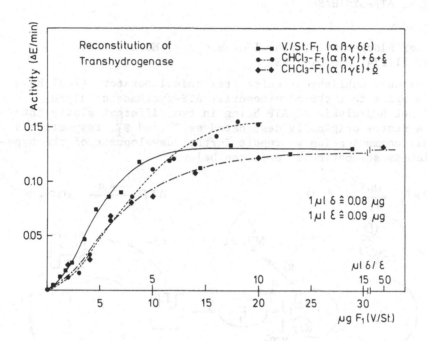

Reconstitution of Transhydrogenase

- ■—■ V./St.F₁ (α β γ δε)
- ●···● CHCl₃-F₁ (α β γ) + δ + ε
- ◆—◆ CHCl₃-F₁ (α β γ ε) + δ

1 µl δ ≅ 0.08 µg
1 µl ε ≅ 0.09 µg

Chemical modification of δ subunit. The effect of TNM, DTNB and NEM on the reconstitutive activity of δ was determined by means of ACMA fluorescence and transhydrogenation in the presence of δ deficient F₁, modified δ and depleted particles. Incubation of δ (0.13 mg/ml) with 0.47 mM TNM for 4 hours abolished both reactions to 0%. 1 mM DTNB reduced ATP driven quenching of ACMA fluorescence significantly but did not affect transhydrogenation. Transhydrogenase reconstitution of δ treated with 0.4 mM NEM was 100% inhibited, and ATP driven quenching of ACMA fluorescence reduced by more than 60%, a value which corresponds to 20% fully active δ molecules.

Further data of chemical modification of δ and ε will be presented.

[1] Beechey, R.B., Hubbard, S.A., Linnett, P.E., Mitchell, A.D., and Munn, E.A. (1975) Biochem. J. 148, 533-537

[2] Vogel, G., and Steinhart, R. (1976) Biochemistry 15, 208-216

[3] Hager, D.A., and Burgess, R.R. (1980) Anal. Biochem. 109, 76-86

This work was supported by DFG grant Ha 1124.

FURTHER DEVELOPMENT OF THE CONCEPT OF PSEUDOREVERSIBILITY OF
MITOCHONDRIAL ATP-SYNTHASE

A.D. Vinogradov

Department of Biochemistry, School of Biology, Moscow State
University, 117234, Moscow, USSR

The previously published results from this laboratory (1-8) have
led us to propose that the mitochondrial ATP-synthase catalyzes synthesis and net hydrolysis of ATP being in two different slowly inter-
convertible states originally designated as F_1^S and F_1^H, respectively
(3). In this communication we report further development of the hypothesis which is schematically depicted below:

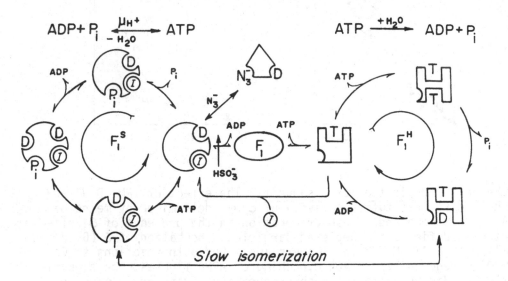

The following are some new findings relevant to the problem of the
reversibility of oxidative phosphorylation.

Both azide, which is known to stabilize the $F_1 \cdot ADP$ complex and inhibit the net ATP hydrolysis, and sulphite which destabilizes $F_1 \cdot ADP$
complex thus activating ATP hydrolysis (6) are shown to be the inhibitors of succinate-supported ATP-dependent NAD^+ reduction by AS-submi-
tochondrial particles artificially "coupled" by low concentrations of
oligomycin. This indicates that the presence of ADP at the specific
tight site of F_1 (1) is a prerequisite of an operation of the enzyme
as the reversible "coupled" ATPase (left part of the Scheme). In har-
mony with this finding we observed an inhibition of ATP-dependent reverse electron transfer by the ATP-regenerating system (phosphoenol

pyruvate plus pyruvate kinase) which maintains very low level of free ADP in the reaction mixture.

After incubation of AS-particles with ATP and protein inhibitor of Pullman and Monroy (I) (9) an ATPase activity becomes almost completely inhibited without any loss of capacity for ATP synthesis. An inhibited ATPase and phosphorylation capacity are not sensitive to trypsin independently of $\Delta \mu_{H^+}$. No correlation between the rate of net ATP hydrolysis and the rate of ATPase inhibition by an excess of (I) exists, although the presence of ATP is necessary for the inhibition. The apparent K_m values for ATP hydrolysis and ATP-dependent inhibition by (I) differ by at least one order of magnitude, suggesting the presence of 2 ATP-sites in ATP-hydrolase (right part of the Scheme).

In extension of our data on the effect of P_i on ADP-specific tight site and lack of activation or inhibition of the steady-state net ATP hydrolysis (7) we found saturation dependence of K_i^{ADP} on P_i with an apparent K_m for P_i about 100 μM. This suggests that the cooperative binding of ADP and P_i gives rise to the actively operating ATP-synthase (left part of the Scheme).

One important feature of our proposal is that the studies on the net ATP hydrolysis (left part of the Scheme) may lead to some false conclusions on the mechanism of energy coupling in the mitochondrial membrane.

References

1. Fitin, A.F., Vasilyeva, E.A. and Vinogradov, A.D. (1979) Biochem. Biophys. Res. Commun. 86, 434-439.
2. Minkov, I.B., Fitin, A.F., Vasilyeva, E.A. and Vinogradov, A.D. (1979) Biochem. Biophys. Res. Commun. 89, 1300-1306.
3. Minkov, I.B., Vasilyeva, E.A., Fitin, A.F. and Vinogradov, A.D. (1980) Biochem. Int., 1, 478-485.
4. Vasilyeva, E.A., Fitin, A.F., Minkov, I.B. and Vinogradov, A.D. (1980) Biochem. J. 198, 807-815.
5. Vasilyeva, E.A., Minkov, I.B., Fitin, A.F. and Vinogradov, A.D. (1982) Biochem. J. 202, 9-14.
6. Vasilyeva, E.A., Minkov, I.B., Fitin, A.F. and Vinogradov, A.D. (1982) Biochem. J. 202, 15-23.
7. Yalamova, M.V., Vasilyeva, E.A. and Vinogradov, A.D. (1982) Biochem. Int. 4, 334-344.
8. Vinogradov, A.D., Vasilyeva, E.A. and Yalamova, M.V. (1982) 2nd EBEC-Reports, vol. 2, Lyon, pp. 113-114.
9. Pullman, M.E. and Monroy, G.C. (1963) J. Biol. Chem., 238, 3762-3769.

NUCLEOTIDE BINDING OF F_1-ATPase AS STUDIED BY
PHOTOAFFINITY-SPIN-LABELED DERIVATIVES OF ATP AND ADP.
P. Vogel[a], I.G. Atchatchloui[a], R. Philipp[a], G. Rathgeber[b],
H.-J. Schäfer[b], and W.E. Trommer[a]
a) Fachbereich Chemie, Universität Kaiserslautern, 6750
Kaiserslautern, West Germany
b) Institut für Biochemie, Universität Mainz, 6500 Mainz, West
Germany

F_1-ATPases from various sources contain up to six nucleotide
binding sites: rapidly exchanging catalytic sites and non-catalytic
sites which are either slowly or rapidly exchanging. Binding of ATP
to the catalytic sites is rather weak, whereas ADP is bound very
tightly to the slowly exchanging sites which may play a structural
and/or regulatory role (1). Although there is general agreement that
these sites are located on the α and β subunits (2), little is known
about their relative spatial arrangement. Further information may be
obtained, e.g., by means of spin-spin interaction between bound
spin-labeled (SL) nucleotides, as has been successfully applied in
case of glyceraldehyde-3-phosphate dehydrogenase (3). Hence, a photoaffinity SL-ATP (II,III) could be covalently linked to the loosely
binding catalytic sites and subsequently, SL-ADP to the tight sites.
In order to differentiate between their ESR signals, the ^{14}N in one
of the labels should be replaced by ^{15}N (4,5).

3'-SL-ATP (I; R=H)
8-Azido-3'-SL-ATP (II; R=N_3)

3'-Azido-8-SL-ATP (III)

3'-SL-ATP (I) as prepared according to (6) with slight modifications is hydrolyzed to the corresponding ADP derivative by F_1-ATPase
from Micrococcus luteus. Mg^{2+} is required whereas Ca^{2+} inhibits the reaction. In the complex with the enzyme 3'-SL-ATP exhibits
an ESR spectrum typical for highly immobilized species. Mg^{2+}
significantly enhances the association constant of this complex.

8-Azido-ATP can be covalently bonded to the catalytic site of M.
luteus F_1-ATPase (7). We have now prepared the ATP derivative (II)
which contains both the spin-label and the photoaffinity group. It is
hardly hydrolyzed by F_1-ATPase but, in agreement with (7), it can be
covalently bonded to the enzyme by irradiation at 366 nm. The modified enzyme has lost its ATPase activity and again exhibits a strongly
immobilized ESR spectrum.

Due to ester exchange, nucleoside-3'-esters are contaminated with
the corresponding 2'-ester. Hence, in 3'-SL-derivatives additional
spectral components may arise from their structural isomers. Since
2'-desoxy-ATP is efficiently hydrolyzed by the M. luteus enzyme, we
are now synthesizing 2'-desoxy-SL-nucleotides and their azido derivatives, thus, avoiding this ambiguity.

(II) is highly active as substrate of hexokinase as shown by the
coupled assay with glucose-6-phosphate dehydrogenase. Besides (II) we
have recently prepared the corresponding ADP derivative. Moreover,
(III) was synthesized in order to allow for triangulation in case of
spin-spin interaction. Detailed enzymic studies are presently carried
out.

1. Di Pietro, A., Godinot, C., and Gautheron, D.C. (1981) Biochem.
   20, 6312-6318
2. Kozlov, I.A., and Novikova, I.Yu. (1982) FEBS Lett. 150,
   381-384
3. Glöggler, K.G., Balasubramanian, K., Beth, A.H., Park, J.H., and
   Trommer, W.E. (1982) Biochim. Biophys. Acta 706, 197-202
4. Davoust, J., Seigneuret, M, Hervé, P., and Devaux, P.F. (1983)
   Biochem. 22, 3137-3145
5. Philipp, R., McIntyre, J.O., Robinson, B.H., Huth, H., Trommer,
   W.E., and Fleischer, S. (1984) Biochim. Biophys. Acta, submitted
6. Streckenbach, B., Schwarz, D., and Repke, K.R.H. (1980) Biochim.
   Biophys. Acta 601, 34-46
7. Schäfer, H.-J., Scheurich, P., Rathgeber, G., and Dose, K. (1980)
   Anal. Biochem. 104, 106-111

ISOLATION AND PROPERTIES OF F_oF_1-ATPase FROM BEEF HEART MITOCHONDRIAL INNER MEMBRANE

Joachim Weber and Günter Schäfer
Institut für Biochemie, Medizinische Hochschule, D-2400 Lübeck, Germany

Whereas the functional and structural properties of F_1-ATPase(s) have been intensively studied, the actual knowledge about the F_oF_1-ATP-synthase system is far away from this state, especially in the case of the enzyme from beef heart mitochondria. In the study presented here we describe a modified method for the isolation of this complex in high yield and some properties of this preparation, particularly binding of nucleotides and subunit arrangement.

For the preparation of the F_oF_1-complex submitochondrial vesicles were solubilized in a medium containing 1.3% of the zwitterionic detergent CHAPSO, 0.25% cholate, 20% glycerol, 0.5 M Na_2SO_4; by this treatment more than 90% of the ATPase-activity could be recovered. Further purification was achieved by differential precipitation by PEG 6000 / Mg^{2+}; all ATPase-activity precipitated between 6.5 and 10% PEG (w/w), resulting in an enzyme with a purity of more than 85% at a yield of about 90% (10% of the total protein mass of the submitochondrial particles). At lower yield the purity could be increased by cutting the fraction between 7.5 and 9.5%; part of the preparation appeared to be in a vesicular or aggregated state (the fraction between 9 and 9.5%).

By SDS-gel electrophoresis of the resulting preparation (after staining by Coomassie Blue) the following composition was obtained: Besides the five subunits of the F_1-part (α-ϵ) subunit a and b of F_o were present, OSCP, and - at least - two polypeptides in the range of a molecular weight between 8000 and 11000 daltons; the latter turned out to be soluble in chloroform/methanol , revealing them as membrane proteins. In addition, a substoichiometric amount of the inhibitory peptide (IF_1) could be detected. As determined by the luciferin/luciferase method, this preparation of the F_oF_1-complex contained about 1 mol endogeneous nucleotides per mol F_oF_1 (based on an estimated molecular mass of 500000 daltons).

For maximal ATPase-activity addition of phospholipids was an absolute requirement: with lysolecithin (1 mg/mg protein) a specific activity of 35-45 U/mg was achieved; oligomycin diminished this value to a residual activity of 20-25%, which could not be decreased further by addition of OSCP. Like F_1-ATPase (1), F_oF_1 gave non-linear Lineweaver-Burk plots upon variation of the concentration of ATP, unless CO_3^{2-} was added. The K_m-value was in the same range as in the case of F_1-ATPase, 100-150 μM (in presence of CO_3^{2-}); ATPase-activity was inhibited competitively by 3'-O-naphthoyl(1)-ADP with an inhibition constant of 10μM

(5 μM for nucleotide depleted F_1-ATPase). When incorporated into liposomes by the method given in (2), the F_OF_1-complex showed an ATP/P_i-exchange activity of about 25 nmol/mg·min, which could be inhibited by oligomycin, DCCD, and naphthoyl-ADP.

For investigation of the nucleotide binding sites of F_OF_1 two specific ligands were chosen: naphthoyl-ADP, a typical ligand for equilibrium binding at the high affinity sites of F_1-ATPase (3), and azidonitrophenylpropionyl-ADP, which was found to tag the low affinity sites by photolabeling (4). In contrast to F_1 (depleted of endogeneous nucleotides), that revealed two binding sites for naphthoyl-ADP, F_OF_1 bound only 1 mol/mol enzyme; this result might be correlated with the finding of 1 mol endogeneous nucleotides per mol F_OF_1. As with F_1, the dissociation constant was found to be about 20 nM. Addition of lysolecithin did not change the thermodynamic binding data, however accelerated binding of naphthoyl-ADP as well as its exchange against ADP.

The behaviour of the F_OF_1-complex towards photolabeling by azidonitrophenylpropionyl-ADP did not differ significantly from the one of F_1-ATPase: Complete inhibition of ATPase-activity was obtained by insertion of 2-3 mol photolabel per mol F_OF_1.

In order to obtain some information about the subunit arrangement of the F_OF_1-complex, the "vesicular" (or aggregated) preparation was treated with increasing amounts of NaBr (up to 4.5 M) or urea (up to 6 M), causing partial dissociation of the complex. After centrifugation, SDS-gel electrophoresis of sediments and supernatants gave the following results: 1) Incubation with NaBr led to dissociation of subunits α, β, and δ from the F_1- part, while γ and ε stayed attached to the "membrane part". 2) Upon treatment by urea from the F_1-subunits only γ sedimented with the membrane in higher amounts, while subunit b of F_O and OSCP also dissociated.

References:

1. Krull, K.W., Schuster, S.M. (1981) Biochemistry 20, 1592-1598
2. Kagawa, Y. (1979) Methods Enzymol. 55, 711-715
3. Tiedge, H., Lücken, U., Weber, J., Schäfer, G. (1982) Eur.J.Biochem. 127, 291-299
4. Weber, J., Lücken, U., Schäfer, G. (1984) Eur.J.Biochem., submitted

DIFFERENT TYPES OF NUCLEOTIDE BINDING TO CHLOROPLAST COUPLING FACTOR

Indra Willms and Andreas Löhr
Botanisches Institut der Tierärztlichen Hochschule Hannover
Bünteweg 17d, D-3000 Hannover 71, W.-Germany (RFA)

Chloroplast coupling factor (CF_1) contains one tightly bound ADP per molecule. This ADP becomes exchangeable with the water phase after energization of the thylakoid membranes (1). This second conformational state of the enzyme is able to bind added ADP and to catalyze ATP formation. After de-energization of the thylakoids, the enzymatically active conformation of CF_1 becomes active in ATP hydrolysis. This ATPase activity is inhibited by tightly binding of one ADP per CF_1 (2).

A correlation between non-exchangeable binding of ADP, enzymatical activity, and energy supply to the thylakoids has been found (3). The total number of nucleotide binding sites is not affected by illumination intensity. Variations in affinity to nucleotides have been observed. During illumination with saturating light intensities three ADP binding sites per CF_1 molecule with a $K_s > 10\mu M$ have been found. When dissipating transmembrane proton gradient in the dark, one site per CF_1 is able to bind ADP with high affinity ($K_s < 10\mu M$). This bound ADP becomes inaccessible to the incubation medium and remains bound to isolated CF_1 protein during column chromatography. Even more, after binding of three ADP per CF_1 in the light, only one ADP per CF_1 can be found non-exchangeably bound in the dark. The remaining two ADP molecules appear to be bound with $K_s > 10\mu M$ in the dark. Though the total number of binding sites is not affected, K_s values are greatly varied with pH of incubation medium.

Nucleotide exchange rate on high affinity binding sites is low as
compared to phosphorylation rate. In contrast nucleotide exchange
rate on low affinity sites seems to come up with phosphorylation rate
or even exceed this rate.

Employing PS II conditions with poisoned PS I no light induced
exchange on high affinity binding sites was found. This incapacity in
nucleotide exchange was diminished with aging of thylakoid membranes.
We understand this last result as indicating a compartimentation in
intact membranes. This leads to a restriction of charges (ions / protons) to sequestered domains (3). Thus a local ion gradient may induce conformational changes in nearby proteins resulting the observed
changes in nucleotide affinity of the CF_0CF_1-complex.

Legends

1) H. Strotmann, S. Bickel, and B. Huchzermeyer (1976)
   FEBS Lett. 61, 194 – 198
2) J. Schumann and H. Strotmann (1980)
   Proc. V. Int. Congr. Photosynth. 881 – 892
3) H. Strotmann, S. Bickel-Sandkötter, U. Franek, and V. Gerke (1981)
   Energy Coupling in Photosynth. 20, 187 – 196
4) Y.Q. Hong and W. Junge (1983)
   Biochim. Biophys. Acta 722, 197 – 208

THIOLS IN OXIDATIVE PHOSPHORYLATION

T. Yagi and Y. Hatefi
Scripps Clinic & Research Foundation, La Jolla, CA 92037 U.S.A.

Multiple forms of thiols appear to be involved in ATP synthesis and hydrolysis. In this study we present results with regard to two types of dithiols, which appear to be involved in energy communication through the ATP synthase complex (complex V). These are factor B and a putative dithiol whose modification is energy-potentiated and results in uncoupling.

1. Factor B Factor B was discovered by Sanadi and coworkers (1) and first purified by You and Hatefi (2). The role of factor B is not clear. It appears to reconstitute energy transfer to and from F_1-ATPase when added to AE-SMP (SMP extracted with ammonia-EDTA). Treatment of factor B with mono- and vicinal dithiol modifiers, or oxidation of its vicinal dithiol to disulfide, results in inhibition (3), which under appropriate conditions can be reversed by further treatment of the modified or oxidized factor B with various mono- and dithiol compounds (1). The latter by themselves do not replace factor B in restoring to AE-SMP the capability for energy transfer to and from F_1-ATPase, but low levels of oligomycin added to AE-SMP can do so. The restorative effect of oligomycin may be related in part to the fact that ammonia-EDTA extraction renders SMP somewhat deficient with respect to F_1-ATPase and partially uncoupled. Addition to AE-SMP of factor B or low levels of oligomycin restores the ability of the particles to maintain a membrane potential, while addition of F_1-ATPase + OSCP has only a partial ($\sim$ 60%) effect (4). If we assume that factor B is a required component of the ATP synthase complex, then the fact that oligomycin alone can restore ATP-driven functions to AE-SMP suggests that these particles are only partially depleted with respect to factor B. Thus, once the proton leakiness of AE-SMP is repaired by low levels of oligomycin then the factor B-containing ATP synthase complexes are able to energize ATP-driven reactions at high rates. This interpretation agrees with the differential effect of ammonia-EDTA extraction on the various ATP-driven reactions of SMP.

2. A dithiol whose modification is energy-potentiated and results in uncoupling SMP, AE-SMP, F_1-depleted ASU-particles, and ATP synthase preparations appear to contain a dithiol which can be modified by incubation of the particles at $30^{\circ}C$ with various mono- and dithiol reagents, including phenylarsine oxide, Cd^{2+}, diamide, p-chloromercuriphenylsulfonate, monobromobimane, mono and bifunctional maleimides, and Cu-orthophenanthroline. The modification of this site is energy-potentiated, and results in uncoupling (4). Unlike the partial uncoupling of AE-SMP particles, the uncoupling by the above reagents

is not blocked by oligomycin (or DCCD). However, by addition of membrane-permeable mono- and dithiol compounds, the uncoupling is reversed nearly completely when it has been achieved by dithiol oxidizing reagents such as diamide and Cu-orthophenanthroline (4), or partially when the SH groups have been modified by reaction with mercurials or Cd^{2+}. The fact that the uncoupling by the above reagents is not blocked by oligomycin distinguishes this effect from the defect of AE-SMP. The demonstration of this effect in complex V, as well as in reconstitutable ASU-particles suggests that the target for the above mono- and dithiol reagents is located in the F_0 segment of the ATP synthase complex. Phenomenologically, the energy-potentiated uncoupling of SMP and complex V by the above reagents is analogous to the uncoupling of chloroplasts by o-phenylenedimaleimide (5). However, in the latter system, the dimaleimide is known to modify SH groups in the γ subunit of CF_1, and as expected the resultant uncoupling is blocked by DCCD (5).

3. In addition to the above, treatment of SMP or complex V with high concentrations of thiol compounds causes uncoupling as marked by the inhibition of energy-linked reactions, loss of membrane potential, and increase of oligomycin-sensitive ATPase activity (4). The order of potency of the thiol compounds tested was dihydrolipoamide > 2,3-dimercaptopropanol >> β-mercaptoethanol and dithiothreitol. The increased ATPase activity of SMP which accompanied uncoupling was itself also inhibited when the thiol concentration was increased further (6). The uncoupled particles could not be recoupled by multiple Sephadex filtration and further treatment with diamide or Cu-orthophenanthroline, in an attempt to reform the putative disulfide which the thiol compounds might have reduced. A likely possibility is that the thiol compounds simply impregnate the membrane and cause uncoupling by acting as protonophores. While this possibility is currently being examined, it should be pointed out that once again CF_1 contains a dithiothreitol-reducible disulfide, whose reduction leads to stimulation of ATPase activity (7).

(1) Sanadi, D.R. (1982) Biochim. Biophys. Acta 683, 39-56
(2) You, Y.S. and Hatefi, Y. (1976) Biochim. Biophys. Acta 423, 398-412
(3) Stiggall, D.L., Galante, Y.M. Kiehl, R., and Hatefi, Y. (1979)
    Arch. Biochem. Biophys. 196, 638-644
(4) Yagi, T. and Hatefi, Y. (1984) Biochemistry, in press.
(5) Weiss, M.A. and McCarty, R.E. (1977) J. Biol. Chem. 252, 8007-8012
(6) Yagi, T. and Hatefi, Y. (1984) International Workshop on Membranes
    and transport in Biosystems, in press.
(7) Arana, J.L. and Vallejos, R.H. (1982) J. Biol. Chem. 257, 1125-1127

KINETIC STUDIES ON THE REACTION MECHANISM OF THE PROTON TRANSLOCATING ATPase FROM NEUROSPORA

J.Ahlers[o], R.Addison[''] and G.A. Scarborough['']
[o]Institut für Biochemie und Molekularbiologie, Fachbereich Biologie, FU Berlin, Ehrenbergstr.26-28, 1 Berlin 33
['']Department of Pharmacology, School of Medicine, University of North Carolina, Chapel Hill, NC 27514, USA

Different degradation patterns of the Neurospora plasma membrane ATPase (EC 3.6.1.3) by trypsin in the presence of various reaction cycle participants or analogues thereof suggested the existence of at least 3 different conformational states of the ATPase during its catalytic cycle (1). In the presence of Mg^{2+} and vanadate, the native enzyme is degraded to a $M_r = 95,000$ form, which has approximately 60 % higher activity than the undegraded $M_r = 105,000$ form. The kinetic properties of these forms have been compared with especial interest in the effects of Mg^{2+}, MgATP, MgADP and vanadate. It was found that the enhanced activity of the $M_r = 95,000$ enzyme is the consequence of an increase in V_{max}. The K_m and K_i for vanadate were not altered but the sensitivity for the competitive inhibitor MgADP was found to be diminished 4 - 5 fold indicating that MgADP binds either to a different site than MgATP or to a different conformation.

Low concentrations of free Mg^{2+} ions activate both forms of the enzyme by increasing V_{max}, whereas the affinity of the enzyme for the substrate is decreased. Above 0.1 mM Mg^{2+}, V_{max} remains constant but K_m continues to increase leading to an inhibition of enzyme activity, which has previously been described as being pseudocompetitive in the case of the analogous enzyme from yeast (2). In agreement with the studies on the protection against trypsin inhibition (1) these results can best be explained by the assumption that the ATPase undergoes (a) conformational change(s) during the catalytic cycle.

Mg^{2+} ions play a key role in two separate steps of this cycle:
a) at low magnesium concentrations the ATPase is activated via a random mechanism, in which the Mg^{2+} ions are not

necessary for substrate binding but rather enhance the
rate in the limiting step of the reaction;
b) at higher magnesium concentrations a conformation of
the enzyme displaying decreased affinity for MgATP is
formed.

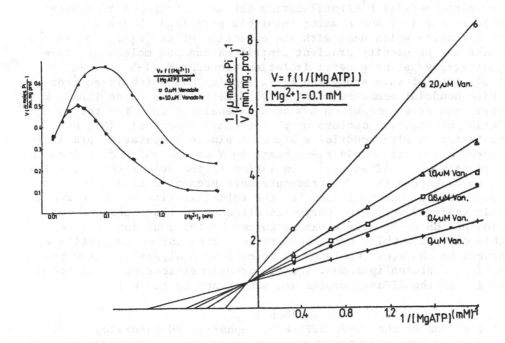

Figure 1: Influence of vanadate.

Vanadate acts as a mixed-type (largely competitive) inhibitor. At high magnesium concentrations the affinity of
the enzyme for vanadate is an order of magnitude larger
than at low concentrations, suggesting that vanadate binds
to the low affinity form of the ATPase.

A model describing the interactions of enzyme, MgATP,
MgADP, Mg^{2+} and vanadate is discussed.

(1) Addison, R. and Scarborough, G.A. (1982)
    J.Biol.Chem.257, 10421 - 10426
(2) Ahlers, J. (1981) Biochim.Biophys.Acta 649, 550 - 556.

SEPHAROSE-AH CHROMATOGRAPHY FOR THE PURIFICATION OF THE
MITOCHONDRIAL F_1-F_0 ATPase.

Georges Dreyfus and Jorge Ramírez
Departamento de Bioenergética, Centro de Investigaciones en Fisiología
Celular, UNAM. Apartado Postal 70-600, 04510, México, D.F. MEXICO.

The mitochondrial H^+-translocating ATPase is composed of a water
soluble sector (F_1) and a water insoluble part (F_0). There are
several reports which deal with the obtention of the F_1-F_0 complex
and make use of density gradient centrifugation and molecular sieve
chromatography as the general isolation procedures (1-5). At an
initial phase of this study the F_1-F_0 complex was solubilized from
the mitochondrial membrane with Lauryl dimethylamino oxide (LDAO) a
zwitterionic detergent which has been extensively used for the
isolation of reaction centers of photosynthetic bacteria (6), but
never tested on mitochondrial systems. A simple preparation procedure
has been devised to obtain F_1-F_0 based on a method which describes the
purification of F_1-ATPase (7), consisting of the selective adsorption
of F_1 to Sepharose-AH. The procedure here presented allows the
exchange of the detergent used for the solubilization of the enzyme.
This is achieved while the enzyme is attached to the Sepharose beads.
Elution of the F_1-F_0 ATPase was performed by the addition of 1M KCl.
In this case LDAO which has a low critical micellar concentration was
exchanged by cholate. Cholate was removed by dialysis to allow the
formation of proteoliposomes. The different purification steps for the
obtention of the ATPase complex are summarized in table 1.

TABLE 1
Purification of the F_1-F_0 ATPase by Sepharose-AH chromatography.

Fraction	Total Protein (mg)	Total Activity[a]	Specific Activity[b]	Yield (%)
Submitochondrial particles	22.3	30.14	1.35	100
LDAO extract	5.6	18.2	3.25	60.4
Sepharose-AH	0.58	5.51	9.25	18.3

[a]Total activity expressed as μmol of Pi liberated by the amount of
protein indicated.

[b]Specific activity expressed as μmol Pi liberated $min^{-1} mg^{-1}$.

The purification factor was 7.5 and the yield 18%. Homogeneity of the
preparation was assessed by SDS-PAGE 12%. It is composed of

approximately 18 bands, 5 correspond to the F_1-ATPase, 2 to OSCP and the DCCD binding proteolipid respectively, the presence of the latter was confirmed by the specific binding of [^{14}C]DCCD to a 14 000 dalton polypeptide. The reconstituted enzyme complex displayed ATP-Pi exchange sensitive to Oligomycin and DCCD, and hydrolytic activity 62% sensitive to oligomycin, Figure 1 shows the energy-dependent fluorescence quenching of 9-aminoacridine displayed by the liposomes containing the F_1-F_0 ATPase (Fig. 1a). It can be observed that the proton pumping activity of this preparation is sensitive to Oligomycin and Triphenyltin (fig. 1 b, c). Figure 1d shows the membrane energization of submitochondrial particles showing Oligomycin sensitivity. The procedure seems to be reliable and rapid, the use of other detergents to solubilize the ATPase complex will be discussed.

FIGURE 1.

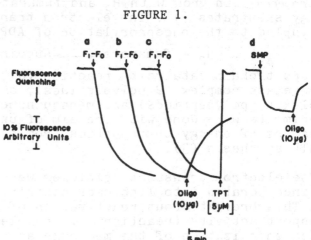

REFERENCES

1) Stigall, D.L., Galante, Y.M. and Hatefi, Y. (1976)
   J. Biol. Chem. 253, 956-964.
2) Serrano, R., Kanner, B.I. and Racker, E. (1976)
   J. Biol. Chem. 251, 2453-2461.
3) Galante, Y.M., Wong, S.Y. and Hatefi, Y. (1979)
   J. Biol. Chem. 254, 12372-12378.
4) Hughes, J., Joshi, S., Torok, K. and Sanadi, R.D. (1982)
   J. Bioenerget and Biomemb. 14, 287-295.
5) Penin, F., Godinot, C., Comte, J. and Gautheron, D.C. (1982)
   Biochim. Biophys. Acta 679, 198-209.
6) Feher, G. and Okamura, N.Y. (1978) in the Photosynthetic Bacteria
   (Clayton, R.K. and Sistrom, W.R. eds). p. 349 Plenum Press,
   New York.
7) Tuena de Gómez-Puyou, M. and Gómez-Puyou, A. (1971)
   Arch. Biochem. Biophys. 182, 82-86.

ELECTRON TRANSPORT COUPLED PHOSPHORYLATION IN A RECONSTITUTED LIPOSOMAL SYSTEM

M. Graf, M. Bokranz, E. Mörschel, H. Mell, R. Böcher, G. Unden, and A. Kröger

FB Biologie/Mikrobiologie, Philipps-Universität Marburg, Karl-von-Frisch-Straße, D-3550 Marburg, W.-Germany

The anaerobic bacterium *Wolinella* (formerly *Vibrio*) *succinogenes* can grow with H_2 and fumarate as the sole energy substrates (1). The electron transport reaction (a) is coupled to the phosphorylation of ADP (2). The electron

$$H_2 + \text{fumarate} \longrightarrow \text{succinate} \qquad (a)$$

transport chain catalyzing reaction (a) consists of a hydrogenase complex (2 polypeptides), a fumarate reductase complex (3 polypeptides) and menaquinone. The following experiments were done with the aim of understanding the mechanism of energy transduction from electron transport to ATP synthesis (3).

The electron transport complexes were isolated (4,5) and incorporated into liposomes containing menaquinone (6). This brought about reactivation of the electron transport activity (reaction a). The electron transport caused energization of the membrane as shown by quenching of the fluorescence of an acridin dye.

The ATP synthase was extracted from the membrane of *W. succinogenes* and purified 20-fold. The enzyme (MW 400 000) was found to consist of at least 6 different polypeptides including the DCCD-binding subunit. In electron micrographs after negative staining the enzyme exhibited the typical structure: sphere, stalk, and base piece. The enzyme catalyzed light-driven ATP synthesis, when incorporated into liposomes together with bacteriorhodopsin.

Liposomes containing the ATP synthase and the electron transport components in about equal amounts (20 molecules/liposome) catalyzed the phosphorylation of ADP as a function of reaction (a). The phosphorylation was abolished by uncoupling agents. The yield ($P/2e \leqslant 0.1$) appeared to be a function of the proportion of the hydrogenase molecules oriented towards the inside of the lipo-

somes and decreased with increasing activity of electron
transport. The rate of phosphorylation was approximately
20 µmol ATP/ min per g ATP synthase. These results are
consistent with the hypothesis (3) that energy transduction occurs via a $\Delta\mu_H$ which is formed by transmembrane
electron transport.

References:

1. Bronder, M., Mell, M., Stupperich, E., and Kröger, A.
   (1982) Arch. Microbiol. 131, 216-223

2. Kröger, A., and Winkler E. (1980) Arch. Microbiol. 129,
   100-104

3. Kröger, A., Dorrer, E., and Winkler, E. (1980) Biochim.
   Biophys. Acta 589, 118-138

4. Unden, G., Böcher, R., Knecht, J., and Kröger, A. (1982)
   FEBS Letters 145, 230-234

5. Unden, G., and Kröger, A. (1982) Eur. J. Biochem. 120,
   577- 584

6. Unden, G., Mörschel, E., Bokranz, M., and Kröger, A.
   (1983) Biochim. Biophys. Acta 725, 41-48

SOLUBILIZATION AND PURIFICATION OF THE H+ -TRANSLOCATING ATPase
FROM HEVEA LATEX (RUBBER TREE)

B. MARIN*, E. KOMOR**
* O.R.S.T.O.M., Unité de Recherches sur les Mécanismes Biochimiques
et Physiologiques de la Production Végétale, Department F, 24, rue
Bayard, F-75008 Paris, France.
** Universität Bayreuth, Lehrstuhl Pflanzenphysiologie,
Universitätsstrasse 30, D-8580 Bayreuth.

The latex of Hevea brasiliensis contains a membranous fraction
(lutoids) which in its properties resembles the vacuoles from higher
plants (1). A tonoplast-bound ATPase has been described as an
electrogenic proton pump which translocates protons from the cytoplasm
to the vacuoles (2-4).

The ATPase from tonoplast membranes was solubilized using different
detergents described as non-denaturating (Triton X-100 ; dodecyl
octaethylene glycol monoether ; polyoxyethylene-9-lauryl ether ;
different bile salts ; n-octyl-β-D-glucopyranoside ; various
sulfobetaines derivatives such as the different zwittergents and
CHAPS). Zwittergent-14 at low concentrations (3-5mM) produced maximum
solubilization of tonoplast membranes but it inactivated ATPase
activity at high concentrations. Octylglucoside only solubilized
tonoplast ATPase at very high concentrations (20mM). CHAPS seems to
be relatively ineffective.

Nevertheless, the most successfull procedure was the use of
dichloromethan, a very simple method which allows purification of the
tonoplast Mg^{2+}-ATPase with about 80 % recovery of the original
activity (5). ATPase was purified 200-fold by two differential
precipitation steps with ammonium sulfate and gel filtration on
Sephacryl S-200. Negative staining electron microscopy indicated
that the solvent-extracted material consisted of amorphous protein
particles, while the initial membrane preparation was composed of
homogeneous vesicles with an average diameter of 250 ± 50 nanometers.

The molecular weight of the solubilized enzyme, estimated by gel
filtration on a Sephacryl S-200 or S-300, was 200 ± 20 kD. In sodium
dodecyl sulfate-polyacrylamide gel electrophoresis, several bands
were seen. This enzyme is probably composed of (at least) four
different polypeptides with the following molecular weights : 11,
27, 63 and 88 kD. No trace of a 100 kD-protein was found, so that a
phylogenetic relationship to the plasmalemma-bound ATPase is unlikely
(6).

The properties of the solubilized ATPase were more or less the same as of the membrane-bound ATPase with respect to K_m value for MgATP, substrate specificity, inhibitor sensitivity and effect of anions, whereas the stimulation by ionophores was lost (7). Reduced activity of solubilized enzyme was partially restored with phospholipids.

Taken together, as described for the vacuolar membrane ATPase of Neurospora crassa (8), all these results support the idea that this enzyme belongs to a new class of proton-translocating ATPases, an intermediate type between the F_1F_0-type enzyme (such as the reversible, proton-translocating ATPase of mitochondria, chloroplasts and bacteria) and the other proton (or ion)-translocating enzymes, as rewieved by Maloney (9). The tonoplast-bound ATPase of fungi and higher plants closely resembles the proton-translocating of chromaffin granules in its characteristics (see differents papers in 10).

References :
1. D'AUZAC, J., CRETIN, H., MARIN, B. and LIORET, C. (1982) Physiol. Vég. 20, 311-331.
2. CRETIN, H. (1982) J. Membrane Biol. 65, 175-184.
3. MARIN, B., MARIN-LANZA, M. and KOMOR, E. (1981) Biochem. J. 198 365-372.
4. MARIN, B. (1983) Planta 157, 324-330.
5. MARIN, B. and KOMOR, E. (in press).
6. GOFFEAU, A., and SLAYMAN, C.W. (1981) Biochim. Biophys. Acta 639, 197-223.
7. MARIN, B. and KOMOR, E. (in press).
8. BOWMAN, B.J. (1983) J. Biol. Chem. 258, 15238-15244.
9. MALONEY, P.C. (1982) J. Membrane Biol. 67, 1-12.
10. MARIN, B. (1985) Biochemistry and function of vacuolar adenosine-triphosphatase in fungi and plants. Springer-Verlag, Heidelberg, Berlin, New-York.

MONOCLONAL ANTIBODY EFFECTS ON THE INTERACTIONS OF FUNCTIONAL SITES
OF THE Na$^+$,K$^+$-ATPASE

William J. Ball, Jr., Department of Pharmacology and Cell Biophysics,
University of Cincinnati College of Medicine, Cincinnati, OH 45267
USA

Monoclonal antibodies can serve as unique probes of specific
regions of membrane bound enzymes. Using the methods of Galfre et
al.[1], in this laboratory several monoclonal antibodies have been
isolated which are directed against the catalytic subunit of the
purified lamb kidney Na$^+$,K$^+$-ATPase. This enzyme is of interest
because it actively regulates the transport of Na$^+$ and K$^+$ across the
cell membrane, and it is the pharmacological receptor for cardiac
glycosides[2].

Studies of antibody binding employing an ELISA type-surface
adsorption assay have shown these monoclonal antibodies to be sensitive to some conformational changes and species differences in the
Na$^+$,K$^+$-ATPase (Ball et al. BBA 719:413, 1982). In addition, two of
these antibodies have been found to inhibit the Na$^+$,K$^+$-ATPase activity by altering ATP binding to the enzyme even though they appear to
bind to different antigenic regions of the catalytic subunit.
Antibody M7-PB-E9 acts like "a partial competitive inhibitor" with
respect to ATP and acts as an uncompetitive or mixed inhibitor with
respect to the Na$^+$ and K$^+$ dependency of ATP hydrolysis. This antibody also does not alter the cooperativity of the Na$^+$ and K$^+$ sites.
Antibody M10-P5-C11 acts as a noncompetitive inhibitor with respect
to ATP. Both antibodies partially inhibit the Na$^+$ and MgATP dependent phosphoenzyme intermediate formation but have no effect on
either the ADP + Enz ~ P $\rightleftharpoons$ ATP + Enz exchange or the K$^+$-stimulated
dephosphorylation. In addition, the K$^+$-dependent p-nitrophenylphos-
phatase activity was not affected. Thus, the phosphorylation but not
the dephosphorylation steps are altered. These data suggest that
both antibodies are binding at or near the ATP binding site.

Further studies have also shown that these antibodies affect
ouabain binding to the glycoside receptor site. Antibody M7-PB-E9
does not alter equilibrium levels of ouabain bound but in the presence of Mg^{2+} antibody M7-PB-E9 can stimulate the rate of ouabain
binding to the enzyme. The addition of ATP largely abolishes this
increase. In contrast, in the presence of Mg^{2+} ATP antibody
M10-P5-C11 can block about 75% of the expected ATP-induced stimulation in the rate of ouabain binding. One antibody can induce an
"ATP-like" conformational change in the enzyme while the other antibody either physically blocks ATP binding or it prevents the ATP
induced conformational change.

These data demonstrate the existence of an antigenically distinct portion of the catalytic subunit which is involved in or regulates nucleotide binding and phosphorylation of the enzyme but not dephosphorylation. This agrees with chemical modification studies that suggest that the nucleotide binding and the phosphorylation sites are separate but in physical proximity due to protein folding[3]. They also demonstrate that specific mechanisms for antibody effects on enzyme function can be identified and that monoclonal antibodies can alter the interactions between distinct regulatory sites on the protein.

[This work was supported by a grant from the American Heart Association and by NIH grant RO1-24941.]

EFFECT OF MONOCLONAL ANTIBODIES ON OUABAIN BINDING

	Relative Percent [^{3}H]Ouabain Bound Ligands Present			
	Mg^{2+}	$Mg^{2+}ATP$	$Na^+,K^+-Mg^{2+}ATP$	$Mg^{2+}P_i$
Equilibrium Binding				
control	100%	100%	100%	100%
plus antibody M7-PB-E9	127%	104%	101%	105%
plus antibody M10-P5-C11	96%	67%	75%	101%
Relative Rates of Binding				
control	100%	100%	100%	100%
plus antibody M7-PB-E9	243%	121%	95%	102%
plus antibody M10-P5-C11	100%	26%	61%	100%

Figure 1: The Na^+,K^+-ATPase was preincubated with antibody at 37°C and then [^{3}H]ouabain was added to give 1×10^{-7} M ouabain. Equilibrium binding times were for 2-6 hrs depending upon ligands present. The relative rates of binding are given for control (in absence) or sample (in presence of added antibody) for each particular binding condition. Binding times were 5-15 min with ouabain binding rates: MgP_i > Na^+,K^+-MgATP > MgATP > Mg.

References:
(1) Galfre, G., Howe, S., Milstein, C., Butcher, G., and Howards, J. (1977) Nature 266, 550-552.
(2) Robinson, J.D., and Flashner, M.S. (1979) Biochim. Biophys. Acta 549, 145-176.
(3) Jorgensen, P.L. (1982) Biochim. Biophys. Acta 694, 27-68.

RELATIONSHIP OF SPECTRAL AND FLUORESCENCE PROPERTIES OF TNP-ATP BOUND
TO THE Ca^{2+}-ATPase of SARCOPLASMIC RETICULUM TO ENERGY COUPLING
DURING ACTIVE TRANSPORT

M C Berman° and Sybella Meltzer"
° Department of Chemical Pathology, University of Cape Town Medical
School, Observatory 7925, Cape Town, South Africa
" Department of Biochemistry, University of Melbourne, Australia

A number of procedures[1], including mild acid conditions[2] and EGTA
treatment[3], irreversibly uncouple calcium transport from Ca^{2+}-
dependent ATPase activity by isolated sarcoplasmic reticulum (SR)
vesicles.   The defect appears to be 'intramolecular' and not to
involve simple increase in permeability of the lipid bilayer to
transported Ca^{2+}.   The ATP analogue, TNP-ATP, which binds to the
Ca^{2+}-ATPase to the extent of one mole per mole, shows several-fold
fluorescence increase during enzyme turnover that has been related to
catalytic enzyme phosphorylation (E-P formation)[4].   Dupont, on the
basis of this enhanced fluorescence has suggested that exclusion of
H$_2$O molecules from the catalytic site is linked to their transfer and
hydration of transported Ca^{2+} species and has proposed that the
Ca^{2+}-ATPase might in fact operate as a calcium pump[5].   In this study
we have sought to correlate calcium pump activity of the Ca^{2+}-ATPase
with turnover-dependent spectral and fluorescence changes of bound
TNP-ATP.

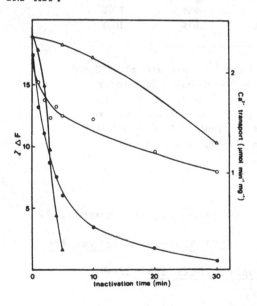

Fig. 1   Effects of EGTA
inactivation on Ca^{2+} transport
and turnover-induced
fluorescence of bound TNP-ATP.
SR vesicles, 1 mg/ml, were
inactivated in 1 mM EGTA,
pH 7.0 and 37°C in the absence
(closed symbols) and presence
(open symbols) of 1 mM CaCl$_2$.
After incubation, samples were
removed for assay of calcium
transport (▲,△) and of
(ATP + Ca^{2+})-dependent TNP-ATP
fluorescence (●,○).

The irreversible effects of preincubation of SR vesicles in EGTA on ATP plus Ca^{2+}-dependent TNP-ATP fluorescence are shown in Fig. 1. EGTA abolished transport within 5-10 min and enhanced ATPase activity (not shown). In the same time scale maximum TNP-ATP fluorescence decreased 60-70% and further preincubation resulted in a further slow decline. Fluorescence excitation and emission maxima were unaltered. Titration experiments with TNP-ATP showed that in the 95% uncoupled state, binding of TNP-ATP both to the static and turning over Ca^{2+}-ATPase were not decreased by more than 15%.

The spectral absorbance properties of bound TNP-ATP have also been studied by means of difference spectroscopy. ATP, in the presence of Ca^{2+}, induces a spectral shift in the absorbance maximum from 510 to 491 nm. Upon hydrolysis of added ATP, the spectrum reverts and the cycle may be repeated several times. Progressive uncoupling of the Ca^{2+}-ATPase modified the ATP-dependent shift such that at 90% inactivation of transport no spectral shift was apparent. Further preincubation in EGTA resulted in a decrease in TNP-ATP binding as gauged from the magnitude of ΔA_{510} in the absence of ATP.

It appears from spectral and fluorescent properties of bound TNP-ATP that its environment is rendered relatively hydrophobic during turnover. Uncoupling of transport is associated with increased access of H_2O to this nucleotide binding site. TNP-ATP may serve as a useful probe of conformational changes linking catalytic and vectorial events during Ca^{2+} pump cycles.

1. Berman, M.C. (1982) Biochim. Biophys. Acta 694, 95-121
2. Berman, M.C., McIntosh, D.B. and Kench, J.E. (1977) J. Biol.Chem. 252, 994-1001
3. McIntosh, D.B. and Berman, M.C. (1978) J. Biol. Chem. 253, 5140-5146
4. Watanabi, T. and Inesi, G. (1982) J. Biol. Chem. 257, 11510-11516
5. Dupont, Y. (1983) FEBS Lett. 161, 14-20

TISSUE-SPECIFIC MODULATION OF EXTRAMITOCHONDRIAL FREE CALCIUM

Paolo Bernardi and Giovanni Felice Azzone
CNR Unit for the Study of Physiology of Mitochondria and Institute of
General Pathology, Via Loredan 16, I-35131 Padova (Italy)

The pathway for mitochondrial Ca^{2+} efflux at high membrane potential
operates at a very low rate (see, e.g., Ref.1), whereas the Vmax of
the uniporter catalyzing Ca^{2+} uptake is higher than the Vmax of the
respiratory chain (2). Thus, if the extramitochondrial free Ca^{2+} is
suddenly raised by one order of magnitude (e.g. from 10^{-6} to 10^{-5} M),
it takes only a few seconds to regain the steady state, while if the
change is in the opposite direction (e.g. from 10^{-6} to 10^{-7} M) several minutes are required to regain the original level. It is therefore likely that cells subjected to high calcium traffic (such as kidney
cells) or to cyclic increase in cytosolic free Ca^{2+} (such as cardiac
cells) have developed a defense mechanism against excess mitochondrial
Ca^{2+} accumulation. Mg^{2+} is a well known inhibitor of the initial rate
of Ca^{2+} uptake in mitochondria from various sources (2,3). Below we
study whether Mg^{2+} affects the steady state Ca^{2+} distribution.

Fig. 1 shows that the steady state pCa_o ($- \log \left[Ca^{2+} \right]$ outside the
mitochondria) is greatly affected by Mg^{2+}. Furthermore, heart and
kidney mitochondria are far more sensitive than liver mitochondria.
At 1 mM Mg^{2+} the pCa_o maintained by liver mitochondria is almost unchanged, whereas the pCa_o maintained by heart and kidney mitochondria
is shifted to about 5.8, which is much higher than the estimated free
Ca^{2+} concentration in these cell types (4). Note that at low $\left[Mg^{2+} \right]$
the steady state pCa_o is maintained to the same level by mitochondria

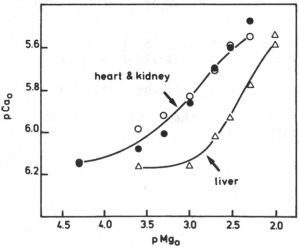

Fig.1 - Effect of Mg^{2+} on
steady state pCa_o.
The incubation medium contained 0.14 M sucrose, 40 mM cho
line, 10 mM Tris-Mops pH 7.0,
5 mM succinate, 0.5 mM Pi, 1
mg/ml BSA, 2 µM rotenone, 1
µg/ml oligomycin, 3 µM Ca^{2+},
and Mg^{2+} as indicated. Fin.
vol. 2 ml, 30°C. The experiment was started by the addition of 2 mg of mitochondria,
and values on the ordinate re
fer to pCa_o at steady state.
(o) heart; (●) kidney; (▲) li
ver mitochondria.

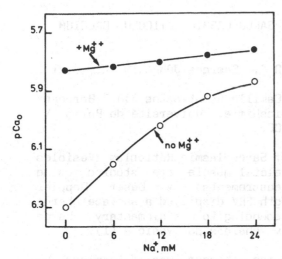

Fig.2 - Relationship between pCa_o and Na^+ concentration in heart mitochondria: effect of Mg^{2+}.
Experimental conditions as in Fig. 1, with Na^+ as indicated. (o) no Mg^{2+}; (●) 1 mM Mg^{2+}. The experiment was started by the addition of 4 mg of rat heart mitochondria, and values on the ordinate refer to the steady state pCa_o.

from all sources, indicating that the shift does not depend on intrinsic differences of the uniporter activity.

It has been shown that Na^+ addition stimulates the rate of Ca^{2+} efflux in heart mitochondria (5). Since this shifts the steady state extramitochondrial free Ca^{2+} to higher levels, we investigated the relative effect of Na^+ and Mg^{2+} on the pCa_o maintained by rat heart mitochondria. Fig.2 shows that, in the absence of Mg^{2+}, increasing Na^+ concentrations were able to shift the pCa_o to lower levels. When the pCa_o was already shifted by Mg^{2+}, on the other hand, the effect of Na^+ was greatly reduced. This may represent an indication that the Na^+-stimulated Ca^{2+} efflux catalyzed by heart mitochondria is inhibited by $Mg2+$. Studies directed to substantiate this hypothesis are in progress in our laboratory.

REFERENCES

1) Bernardi, P. and Azzone, G.F. (1983) Eur.J.Biochem. 102, 555-562
2) Bragadin, M. Pozzan, T. and Azzone, G.F. (1979) Biochemistry 18, 5972-5978
3) Vinogradov, A. and Scarpa, A. (1973) J.Biol.Chem. 248, 5527-5531
4) Tsien, R.Y. and Rink, T.J. (1983) in Current Methods in Cellular Neurobiology (Barber, J.L. Ed.) Vol.3, Chapter 9, John Wiley Inc.
5) Crompton, M., Capano, M. and Carafoli, E. (1976) Eur.J.Biochem. 69, 453-462

SURFACE CHARGE PROPERTIES OF SARCOPLASMIC RETICULUM CALCIUM PUMP : FUNCTIONAL ASPECT.

Daniel BRETHES [o], Bernard ARRIO ", Georges JOHANNIN " , Didier DULON [o] and Jean CHEVALLIER[o].
[o] I.B.C.N. du C.N.R.S., 1 rue Camille Saint-Saëns 33077 Bordeaux Cedex FRANCE - " Institut de Biochimie, Université de Paris XI Centre d'Orsay 91405 Orsay FRANCE.

The electrical properties of Sarcoplasmic Reticulum Vesicles (SRV) obtained from rabbit skeletal muscle were studied using electrophoretic mobility measurements by Laser Doppler Velocimetry. At low ionic strength SRV displayed a surface charge density of $5.4 \ 10^{-3}$ C/m^2 corresponding to one elementary charge per 30 nm^2 or about 2900 negative charges per vesicle (1).

The origin of these negative charges was determined by modifying the phospholipid content of reconstituted vesicles (rSRV) obtained by using a technique slightly modified from (2). This method allowed to obtain active vesicles, homogeneous in size, with a symmetrical repartition of the Ca^{2+}-ATPase molecules (as shown by cryofracture analysis). When phosphatidylcholine (PC) was the only phospholipid added during the reconstitution process the rSRV displayed a negative charge due to the ATPase : 3 e^- per subunit were found. This result was close to the value found by Haynes and Mandveno using the fluorescent probe ANS- (2 e^- per ATPase) (3). Mixture of phosphatidylserine (PS) and PC gave more negatively charged rSRV. Their surface charge density was linearly dependent of the PS/PC ratio but an important part of the negative charges added (as PS molecules) were neutralized by divalent cations (Ca^{2+}, Mg^{2+}) bound to rSRV.

In conditions of active calcium transport, upon addition of ATP (1mM) in the presence of 50 uM of both $CaCl_2$ and $MgCl_2$ and at low ionic strength, the electrophoretic mobility of SRV increased rapidly, reached a maximum value stable several minutes and then slowly decreased to its initial value. The pH dependence of the plateau value had an optimum at pH 7.2. No modification was observed when the calcium ionophore A23187 was added before ATP. These results suggested that the increase of the surface charge density observed was not due to the phosphorylation of the ATPase but, instead, should be related to the calcium uptake process (i.e.: calcium depletion phenomenon from its external binding sites).

BIBLIOGRAPHY :

1- ARRIO, B., JOHANNIN, G., CARRETTE, A., CHEVALLIER, J. and BRETHES, D. (1984) Arch. Biochem. Biophys. 228, 220-229.

2- BRETHES, D., AVERET,N., GULIK-KRZYWICKI, T. and CHEVALLIER, J. (1981) Arch. Biochem. Biophys. 210, 149-159.

3- HAYNES, D.M. and MANDVENO, A. (1983) J. Memb. Biol. 74, 25-40.

Ca^{2+}-DEPENDENT UNCOUPLING OF OXIDATIVE PHOSPHORYLATION IN COLD-ACCLIMATED RAT LIVER MITOCHONDRIA

N.N.Brustovetsky, E.I.Maevsky, and S.G.Kolaeva
Institute of Biological Physics, USSR Academy of Sciences, Pushchino, Moscow Region, 142292, USSR

Earlier it was shown that cold-acclimation of rats activates respiration and decreases ADP/O ratio in liver mitochondria (RLM) /1/. These events seem to be due to increase in concentration of free fatty acids (FFA) upon cold-acclimation of rats /2,3/. It was shown that albumin binding to FFA eliminates the changes in oxidative phosphorylation induced by cold-acclimation /2/. However, these changes are partially eliminated with EDTA which binds to divalent cations /4/. Basing on these facts we suppose that not only FFA but divalent cations (*e.g.*Ca^{2+}) play an important role in both activation of respiration and decrease of ADP/O ratio in cold-acclimated RLM. It is known that energy linked uptake and electroneutral efflux of Ca^{2+} form on inner RLM membrane a steadily functioning Ca^{2+}-cycle which leads to partial energy dissipation from respiratory chain and facilitates RLM respiration /5/. Enhancement of Ca^{2+}-cycle activity must increase the respiration rate and decrease ADP/O value, *i.e.* it induces alterations similar to those in RLM upon cold-acclimation. Intensity of Ca^{2+}-cycle functioning seems mainly to be determined by the rate of electroneutral Ca^{2+} efflux which provides continuous transport of Ca^{2+} through the inner RLM membrane. That is why we studied the effect of cold-acclimation on the rate of electroneutral Ca^{2+} efflux and its relationship with oxidative phosphorylation in RLM of control and cold-acclimated rats.

Wistar rats were kept for 2-3 weeks at $+2^{o}C$ (cold) and $+23^{o}C$ (control). RLM were prepared using differential centrifugation in sucrose medium in the absence of EGTA. The rate of O_2 consumption was determined using Clark electrode. The rate of Ca^{2+} efflux from RLM was measured using Ca^{2+}-electrode after RLM were preloaded with 200 nM Ca^{2+} per mg protein^{-1} in the presence of 1 µM ruthenium red (RR).

Table 1 demonstrates the experimental data on increase of respiration rate and decrease of ADP/O in RLM induced by cold-acclimation. The rate of Ca^{2+} efflux increases several times. This fact points to activation of Ca^{2+}-cycle. If this is the case, inhibition of Ca^{2+}-cycle functioning must result in greater decrease of respiration rate and increase of ADP/O in RLM of cold-acclimated animals. In fact, in the presence of RR the respiration rates of RLM decrease (especially in the cold-acclimated group) and level off, and ADP/O increases in cold-acclimated RLM. At the same time RR has no effect on RLM prepared in the presence of 1 mM EDTA (data not shown), this indicates that RR effect on RLM respiration is mediated by Ca^{2+}. Thus, cold-acclimation of rats leads to activation of Ca^{2+}-cycle due to increase in the rate of electroneutral Ca^{2+} efflux from RLM, and as a result, the rate of respiration increases and ADP/O ratio decreases. Increase of the rate of Ca^{2+} efflux seems to be due to elevation of concentration of FFA which substantially activate Ca^{2+}/H^+-exchange upon cold-acclimation /6/.

Table 1. Some parameters of oxidative phosphorylation and the rate of electroneutral Ca^{2+} efflux from RLM in control and cold-acclimated rats.

	Respiration rate, ng-atoms $O_2 \cdot min^{-1} \cdot$ mg protein^{-1}		ADP/O	Rate of Ca^{2+} efflux ng-ion$\cdot min^{-1} \cdot$ mg protein^{-1}
	+ADP	−ADP		
Control	233	48	1.68	5.73
Cold	298	67	1.47	32.22
Control + RR	143	38	1.71	−
Cold + RR	137	38	1.57	−

Succinate was used as substrate. No reliable difference between Control + RR and Cold + RR was observed, P > 0.05, in other cases P < 0.01, n = 6.

References

1. Hannon J.P. (1960) Fed.Proc., 19, Suppl. 5, 139-144.
2. Skulachev V.P. (1972) Energy transformation in biomembranes. Moscow, Nauka, 203 p.
3. Smith R.E., Hoijer D.J. (1962) Physiol.Rev., 42, 60-142.
4. McBurney L.J., Radomski M.W. (1973) Comp.Biochem.Physiol., 44B, 1219-1233.
5. Stucki J.W., Ineichen E.A. (1974) Eur.J.Biochem., 48, 365-375.
6. Roman J., Gmaj P., Nowicka C., Angielski S. (1979) Eur. J.Biochem., 102, 615-623.

THE 35 K DCCD REACTIVE PROTEIN FROM PIG HEART MITOCHONDRIA IS THE
MITOCHONDRIAL PORINE.

V. De Pinto, R. Benz[+] and F. Palmieri
Institute of Biochemistry, Faculty of Pharmacy, Bari, Italy
[+]Fakultät für Biologie, Universität Konstanz, F.R.G.

Dicycloexylcarbodiimide (DCCD) has been shown to bind to a variety
of different ion transport systems in mitochondria: to the CF_o
component of the H^+-ATPase, to the cytochrome c oxidase, to the
cytochrome b-c_1 complex, to the uncoupling protein from brown adipose tissue mitochondria and more recently to the K^+/H^+ antiporter.
All these systems were labelled at high DCCD concentrations.
At low DCCD concentrations (i.e. 2 nmol/mg protein) only three bands
of Mr 9 K, Mr 16 K and Mr 35 K react with DCCD. The two bands of
lower molecular weights correspond to the proteolipid of the CF_o
component and to an aggregate of it, whereas it has been proposed
that the Mr 35 K DCCD protein could be identical to the phosphate
carrier (1).
Here we will present evidence that the 35 K DCCD reactive protein
is not identical to the phosphate carrier. This finding is based on
the following:
(i) Both proteins behave different on a HTP column in the presence or
    absence of cardiolipin (according to 2)
(ii) Both protein can be separated by passing the HTP eluate through
    an organomercurial-agarose column, which retains the phosphate
    carrier (3)
(iii) A further purification step using a dry HTP/celite column, which
    gives in the eluate a electrophoretically pure protein of about
    0.2% of the total mitochondrial protein.
Furthermore, the 35 K DCCD reactive protein is not identical to the

Fig.1 Stepwise increase
of the membrane conduc-
tance in the presence
of the 35 K DCCD reacti-
ve protein. Membrane
from asolectin/choleste-
rol. 1 M KCl; Vm: 5mV;
25 C.

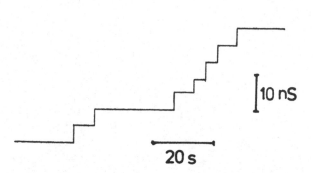

subunit II of the cytochrome c oxidase or to cytochrome b because it
does not comigrate with these proteins on different SDS polyacrila-
mide gel systems.

The incorporation of the 35 K DCCD reactive protein into lipid bilayer
membranes from asolectin/cholesterol resulted in a stepwise increase
of the membrane conductance at a given low transmembrane potential
(Fig.1). The single channel conductance of these steps was about 4.5
nS in 1 M KCl. Furthermore, the pore was found to be voltage dependent
and switched to lower conductance levels if the voltage across the
membrane was increased above 30 mV. The whole characteristics of the
pore induced by the 35 K DCCD reactive protein was very similar to
those which have been found previously with rat liver mitochondrial
porine (4). From this close analogy we concluded that the 35 K DCCD
reactive protein in pig heart and the mitochondrial porine are identical.

1. Houstek, J., Pavelka, S., Kopecky, J., Drahota, Z. and Palmieri, F.
   (1981) FEBS Lett. 130, 137-140
2. Bisaccia, F. and Palmieri, F. Biochim. Biophys. Acta, in press
3. De Pinto, V., Tommasino, M., Palmieri, F. and Kadenbach, B. (1982)
   FEBS Lett. 148, 103-106
4. Roos, N., Benz, R. and Brdiczka, D. (1982) Biochim. Biophys. Acta
   688, 204-214

CALCIUM TRANSPORT IN MEMBRANE VESICLES OF STREPTOCOCCUS CREMORIS
FUSED WITH BACTERIORHODOPSIN PROTEOLIPOSOMES

A.J.M. Driessen, K.J. Hellingwerf and W.N. Konings
Department of Microbiology, University of Groningen, Kerklaan 30,
9751 NN HAREN, The Netherlands

Lactic acid bacteria lack functional electron transfer chains
since these organisms are impaired in the synthesis of porphyrins
and therefore cannot synthesize cytochromes. Consequently these
organisms are not able to generate a proton motive force across
their cytoplasmic membrane by electron transfer. In membrane
vesicles of lactic acid streptococci a membrane potential can be
generated by artificial methods, such as valinomycin mediated
potassium efflux. This potential lasts only a few minutes and does
not allow detailed quantitative studies of the effects of the proton
motive force or solute uptake (1). For the generation of a proton
motive force for a longer period of time a primary proton pump has
been incorporated into Streptococcus cremoris membrane vesicles.
This has been achieved by fusion of S. cremoris membrane vesicles
with Bacteriorhodopsin proteoliposomes containing negatively charged
phospholipids. This fusion occurs when both preparations are mixed
at low pH (< 6.0). Analysis of the fused membranes on a sucrose
gradient revealed a peak with a density intermediate of the membrane
vesicles and the proteoliposomes. The density of this peak decreased
with increasing cardiolipin content of the proteoliposomes.

The orientation of Bacteriorhodopsin in the fused membranes is
inside-out with respect to the in vivo orientation in Halobacterium
halobium. In the light a $\Delta\psi$, interior positive and a ΔpH interior
acid, is generated.

The proton motive force generated by Bacteriorhodopsin upon illumination can drive calcium uptake in the fused membranes. Uptake of
calcium increased with increasing light intensity and substrates at
high light intensities. Addition of the ionophore valinomycin stimulated calcium uptake and led to an increase of the ΔpH. A low level
of calcium uptake was observed in the dark and in the light in the
presence of nigericin or CCCP, S-13 or both valinomycin and nigericin. Bleaching of Bacteriorhodopsin in the presence of hydroxylamine
also abolished calcium uptake.

REFERENCES

1. Otto, R., Lageveen, R.G., Veldkamp, H. and Konings, W.N. (1982)
   J.Bacteriol. 149, 733-738

MECHANISM OF INTERACTION BETWEEN CATION AND LIGAND BINDING SITES IN NA$^+$,K$^+$-ATPASE.

E. Grell°, E. Lewitzki°, G. Krause°, W.J. Richter" and
F. Raschdorf"

°Max Planck Institute for Biophysics, D 6000 Frankfurt 71
"Ciba Geigy AG, CH 4000 Basel.

Cardiac glycosides, such as ouabain, act as specific and powerful inhibitors of Na$^+$,K$^+$-ATPase. The binding affinity of these inhibitors depends on the ionic composition of the medium. In order to study the mechanism of binding of cardiac glycosides to this enzyme (isolated from pig kidney (1)), kinetic studies have been carried out.

To allow detection of the binding of such an inhibitor to the enzyme by spectroscopic techniques, a fluorescent derivative of ouabain has been synthesized. The labeled compound (DEDO) contains the DANSYL (5-dimethylamino-naphthalene-1-sulfonic acid) residue and is chemically stable in aqueous solution. The structure of DEDO has been elucidated on the basis of FAB mass spectrometry (cf. Fig.1).

Prior to kinetic studies, the binding affinity between DEDO and the enzyme has been determined by means of enzymatic activity in media of different ionic composition. The log K value obtained for DEDO at 37°C in the presence of 5 mM MgCl$_2$ (extrapolated to zero K$^+$conc.) is about 6.4 at pH 7.5.

Kinetic studies have been performed by employing fast chemical reaction techniques, eg. a three component stopped-flow instrument. Binding of DEDO to Na$^+$,K$^+$-ATPase leads to an increase of fluorescence intensity if Mg^{++} is present as a minimum requirement for binding. The formation rate constant has been determined based on a model of reversible complex formation. With Mg^{++}, a k$_{on}$ value of 4.3 x 10^3 M^{-1}sec^{-1} is found at 37°C and pH 7.5. The dissociation rate constant has been obtained upon addition of excess unmodified ouabain which leads to a dissociation of the enzyme-DEDO complex. Only evidence for a single exponential decay of the enzyme-DEDO complex in the presence of Mg^{++} is found. The corresponding k$_{off}$ value is in the range of (1.8 ± 0.3) x 10^{-3}sec^{-1} and is higher than those reported in the literature, determined mainly by fast filtration techniques with [^{3}H]-ouabain

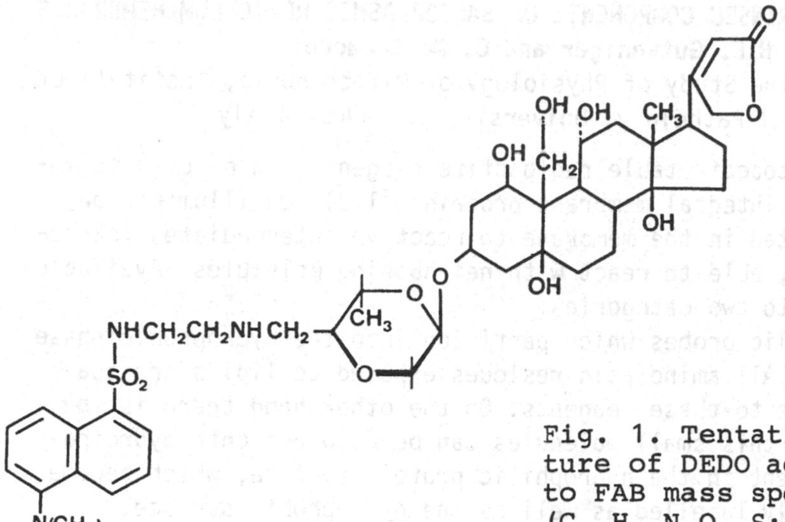

Fig. 1. Tentative structure of DEDO according to FAB mass spectrometry ($C_{43}H_{61}N_3O_{12}S$; MW = 843)

(2,3,4) or one of its derivatives (5) and enzyme preparations from other sources. The ratio of the experimentally observed formation over that of the dissociation rate constant (log K = 6.5) agrees well with the value of the binding affinity obtained from enzymatic activity studies. These kinetic parameters have been determined for various ionic compositions as well as in the presence of phosphate and vanadate. The results reveal that phosphate and vanadate lead to a substantial increase of the formation rate constant.

References

(1) Jørgensen, P.L. (1974) Biochim. Biophys. Acta 356:36
(2) Schoner, W., Kirch, N., Halbwachs, C. (1980) In: Skou, J.C., Nørby, J.G. (eds.) Na,K-ATPase. Academic Press, London, pp 421-430
(3) Wallick, E.T., Pitts, B.J.R., Lane, L.K., Schwartz, A. (1980) Arch. Biochem. Biophys. 202:449
(4) Hansen, O. (1971) Biochim. Biophys. Acta 233:122; Hansen, O. (1979) Biochim. Biophys. Acta 568:265
(5) Moczydlowski, E.G., Fortes, P.A.G. (1980) Biochem. 19:969

LABELLING OF INTRINSEC COMPONENTS OF SARCOPLASMIC RETICULUM MEMBRANES

H.E. Gutweniger and C. Montecucco

C.N.R. Unit for the Study of Physiology of Mitochondria, Institute of
General Pathology, University of Padua, Italy

Several photoactivatable radioactive reagents can be used to label and identify integral membrane proteins (1,2). On illumination, they are converted in the membrane to reactive intermediates (carbenes or nitrenes), able to react with neighboring molecules. Available reagents fall into two categories:

a) small lipophilic probes which partition into the hydrophobic phase of the membrane. All amino acid residues exposed to lipids are spacially accessible to these reagents. On the other hand there is the possibility that this small molecules can be absorbed onto hydrophobic pockets present in the hydrophilic protein surface, which become then radioactively labelled as well as the hydrophobic surface.

b) the second group includes lipid analogs carrying the photoreactive group at a selected position of a fatty acid chain. This restricts the freedom of motion of the group, whose labelling can be restricted to a limited portion of the membrane, giving rise to a more reliable result.

In the present study we have labelled rabbit and rat sarcoplasmic reticulum membranes (3), with both a carbene-generating type a) reagent, 3-(trifluoromethyl)-3-(m-iodophenyl)diazirine ([^{125}I] TID) (4) and two nitrene generating lipid probes: 1-myristoyl,2-[12-amino(4N-3-nitro-1-azidophenyl)]-dodecanoyl-sn-glycero-3-[^{14}C]phosphocholine(azidophospholipid) and 12-amino-N(2-nitro-4-azidophenyl)dodecanoyl-[6-^{3}H]glucosamine(azidoglycolipid) (5,6). Their specific radioactivities were respectively 10 Ci/mmol, 174 Ci/mol and 38 Ci/mol. All these probes are able to transfer efficiently into the SR membrane upon incubation. Both kinds of reagents label the 160000 glycoprotein, the Ca^{2+} ATPase, the 53-55000, 30000 and 20000 proteins in both rabbit and rat sarcoplasmic reticulum. These results provide strong evidence that these proteins are integral components of the sarcoplasmic reticulum membranes, as well as a rabbit SR polypeptide of M.W. 6000, which is labelled both by TID and by the lipid analogs. [^{125}I]TID labels additional components: calsequestrin in rabbit but not in rat SR, a rabbit 35500 component, and two rat proteins of M.W. 32000 and 22500 dalton, which are not modified by the photoreactive lipid analogs. These dif-

ferences in the patterns of labelling will be discussed in the light
of the different properties of the probes.

References

1) Bayley, H. (1983) Photogenerated reagents in Biochemistry and Molecular Biology, Vol. 12, Elsevier Science Publ., Amsterdam
2) Bisson, R. and Montecucco, C. (1984) in "Techniques for the Analysis of Membrane Proteins" (R. Cherry and C.I. Ragan, Eds.), Chap-
   mann and Hall, London, in press
3) Tada, M., Yamamoto, T. and Tonomura, Y. (1978) Physiol. Rev. 58,
   1-79
4) Brunner, J. and Semenza, G. (1981) Biochemistry 20, 7174-7182
5) Bisson, R. and Montecucco, C. (1981) Biochem. J. 193, 757-763
6) Gutweniger, H.E. and Montecucco, C. (1984) Biochem. J., in press

ION TRANSPORT AND ENERGY CONSERVATION IN ECTOTHIORHODOSPIRA SPECIES

J. F. Imhoff[1], B. Meyer[1], K. J. Hellingwerf[2], and W. N. Konings[2]

[1] Institut für Mikrobiologie, Universität Bonn, Meckenheimer Allee 168, D-5300 Bonn, Federal Republic of Germany

[2] Department of Microbiology, University of Groningen, Biological Centre, 9751 NN Haren, The Netherlands

Ectothiorhodospira species are phototrophic procaryotes and obligate anaerobes that use reduced sulfur compounds as electron donors for photosynthesis; sulfide and thiosulfate are oxidized to elemental sulfur and sulfate (1). In nature these bacteria are found in marine and hypersaline to extremely saline environments, preferably with alkaline pH (2). Under conditions of cultivation in the laboratory they grow best at pH 7.5-9.2 and at salt concentrations of 3% to 30%, depending on the strain. In particular the extremely halophilic species have to cope with several molar salts and alkaline conditions in their surroundings. Unlike Halobacteria, which accumulate potassium as the major osmotic active solute in their cells (3), Ectothiorhodospira species have organic solutes (predominantly betaine) as osmotic compatible solutes (4). Also potassium, which is in the media in concentrations below 5 mM, is accumulated in the cells.

In Ectothiorhodospira halochloris the energy generating system is apparently well adapted to its halophilic properties: The rate of light mediated ATP synthesis is the highest close to the salt concentration of optimal growth at 2.5 M NaCl. In dependence on the external pH light stimulated proton uptake could be observed, which contrasts findings with Chromatium vinosum under similar conditions. Also artificially created proton gradients in the dark showed interesting deviation from the effects observed with Chromatium vinosum. Uptake experiments with benzoic acid and methylamine by flow dialysis gave evidence for a more acidic pH inside the cells of about 2 units, if suspended in medium at pH 10.5.

The membrane-bound ATPase has an easily solubilized factor with a molecular weight comparable to the F_1-factor of E. coli (5). In Ectothiorhodospira halochloris the enzyme was inhibited by low concentrations of sodium and potassium chloride. The membrane-bound enzyme is activated by trypsin and shows neglegible inhibition by oligomycin and DCCD.

The membrane potential of <u>Ectothiorhodospira mobilis</u> was measured by the distribution of tetraphenylphosphonium ions (TPP^+) with a TPP^+-sensitive electrode. Corrections for unspecific binding of TPP^+ to the cells was made according to Lolkema et al. (6). The rate of TPP^+-uptake and the value of the membrane potential were dependent on the mineral composition and its concentrations in the medium used for suspending the cells. Similar to the findings with the rate of ATP synthesis, the highest membrane potential was found at conditions close to those for optimal growth of this strain.

References
(1) Imhoff, J.F. (1984) Intern. J. System. Bacteriol. 34, 338-339
(2) Imhoff, J.F., Sahl, H.G., Soliman, G.S.H., and Trüper, H.G. (1979) Geomicrobiol. J. 1, 219-234
(3) Christian, J.H.B., and Waltho, J.A. (1962) Biochim. Biophys. Acta 65, 506-508
(4) Galinski, E.A., and Trüper, H.G. (1982) FEMS Lett. 13, 357-360
(5) Futai, M., and Kanazawa, H. (1983) Microbiol. Rev. 47, 285-312
(6) Lolkema, J.S., Hellinwerf, K.L., and Konings, W.N. (1982) Biochim. Biophys. Acta 681, 85-94

SYNTHESIS OF ATP CATALYZED BY THE (Ca^{2+}, Mg^{2+})-ATPase FROM ERYTHROCYTE GHOSTS AND CARDIAC SARCOLEMMA. EFFECT OF CALMODULIN.

J. Mas-Oliva. Departamento de Bioenergética, Centro de Investigaciones en Fisiología Celular, Universidad Nacional Autónoma de México, AP 70-600. 04510 México, D.F. México.

As shown by the number of reports published in the last few years, the plasma membrane (Ca^{2+}, Mg^{2+})-ATPase has been demonstrated to be an important molecule whereby excitable (1) and non-excitable cells (2) couple the hydrolysis of ATP to the withdrawal of calcium from the cytoplasm to the extracellular space. Although the hydrolytic component of the reaction has been extensively studied in different cell types, the reversal of the ATP driven calcium pump has not been investigated in plasma membranes (3) as deeply as in cytoplasmic membranes (4). Even though the effect of calmodulin has been traditionally studied only upon the hydrolytic sequence of the reaction, it has been recently suggested that calmodulin might also enhance the reverse reaction or ATP $\longleftrightarrow$ Pi exchange apparently stimulating the overall reaction sequence of the plasma membrane (Ca^{2+}, Mg^{2+})-ATPase (5). During the present investigation an effort was made in order to find out if the calmodulin sensitive ATP $\longleftrightarrow$ Pi exchange mechanism carried out by the ATPase may be a general feature of other plasma membrane (Ca^{2+}, Mg^{2+})-ATPases.

Rabbit erythrocyte ghosts were prepared employing several modifications to the method described in (6). In order to separate the inside-out from the right side out vesicles, a wheat germ agglutinin Sepharose-6MB column was employed as previously described by us (7). The basal (Ca^{2+}, Mg^{2+})-ATPase activity found in the calmodulin-depleted inside-out vesicles was dependent upon the external calcium concentration of the reaction medium (Fig. 1). Although no ATP $\longleftrightarrow$ Pi exchange was detectable at low calcium concentrations, when the concentration of this cation in the medium was raised in the presence of constant Pi and ADP concentrations, an activation of the ATP $\longleftrightarrow$ Pi exchange reaction was observed (Fig. 1). The addition of bovine brain calmodulin to the calmodulin depleted vesicles, increased the affinity of the ATPase for calcium and the Vmax of the reaction. Moreover, when these vesicles were restored of the activator protein calmodulin, the calcium dependent ATP $\longleftrightarrow$ Pi exchange reaction was also enhanced (Fig. 1). It was also observed that the optimal hydrolytic activity was achieved when the ATP concentration of the medium was progressively increased (Fig. 2). In contrast, if optimal conditions for the measurement of the ATP $\longleftrightarrow$ Pi exchange were used employing depleted and calmodulin added vesicles, an increase in the ATP concentration of the medium slightly decreased the capacity of the enzyme to reverse the reaction. Under these conditions, calmodulin

seemed to interact with the enzyme in a way optimizing at low ATP
concentrations the ATP ⬌ Pi exchange reaction. The study of calcium
transport under optimal conditions for the expression of the ATP ⬌ Pi
exchange reaction in sarcolemmal vesicles and erythrocyte ghosts will
be discussed. Supported by CONACYT Grant ICSAXNA-001872.

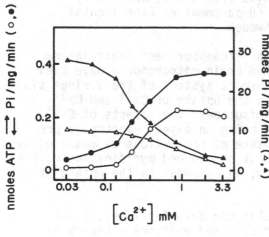

Fig. 1. ATP hydrolysis (△,▲).
ATP ⬌ Pi exchange (o,●).
Calmodulin free vesicles (△,o).
Vesicles incubated with 1 µg
calmodulin / ml (▲,●).
Vesicles were preincubated at
4°C for 3 h + 0.01 % TX-100.
The assay was performed for
5' at 37°C and the assay
medium contained 50mM Tris/
maleate buffer (pH 7.4);
10mM $MgCl_2$; 4mM Pi; 0.2mM ADP;
6mM ATP; 10uM ouab; 0.01%
TX-100 and final $CaCl_2$ as
indicated.

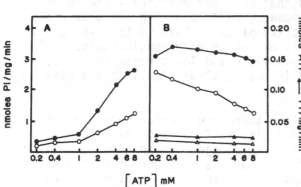

Fig.2 A)ATP hydrolysis
B) ATP ⬌ Pi exchange.
Calmodulin free vesicles
(o). Vesicles added of
1 µg of calmodulin / ml
(●). Vesicles incubated
in the absence of TX-100
minus calmodulin (△)
and plus calmodulin (▲).
assay medium as in Fig.1.

References: 1) Dipolo, R. (1978) Nature (London) 274, 390-392.
2)Nellans, H.N. and Popovitch, J.E. (1981) JBC 256, 9932-9936. 3) Rossi
J.P.F.C., Garrahan, P.J. and Rega, A.F. (1978) J. Membrane Biol. 44,
37-46. 4) de Meis, L. (1981) The Sarcoplasmic Reticulum. Transport
and Energy Transduction. In: Transport in Life Sciences (Vol. 2) E.
Bittar ed. John Willey and Sons. New York. 5) Mas-Oliva, J. de Meis,
L. and Inesi, G. (1983) Biochemistry 22, 5822-5825. 6) Waisman, D.M.,
Gimble, J.M., Goodman, D.B.P. and Rasmussen, H. (1981) JBC 256, 409-
414. 7) Mas-Oliva, J., Williams, A. and Nayler, W.G. (1980) Anal.
Biochem. 103, 222-226.

TRANSPORT OF Ca^{2+} AND Mn^{2+} BY MITOCHONDRIA FROM RAT LIVER, HEART AND BRAIN

[1]G. Şandri*, [1,2]A. Montag, [1]V. Konji**, [1]K. Nordenbrand, [2]L. Hillered and [1]L. Ernster
[1]Department of Biochemistry, Arrhenius Laboratory, University of Stockholm, S-106 91 Stockholm and [2]Department of Experimental Medicine, Pharmacia AB, Uppsala, Sweden

In these experiments Ca^{2+} and Mn^{2+} transport and their interaction in isolated rat liver, heart and brain mitochondria have been studied. We have compared these transport systems of the various tissues using the following parameters: the uptake of Ca^{2+} and Mn^{2+}, alone or in combination, using radioisotopes; the effects of Ca^{2+}, Mn^{2+} or both on State 4 respiration using an oxygen electrode; proton ejection associated with the uptake of the cations as measured by a glass electrode; and the enhancement of reduced pyridine nucleotide oxidation, measured fluorimetrically, in response to the uptake of the two cations.

In agreement with previously published data for liver, heart (1-7), and brain (8), we found that Ca^{2+} and Mn^{2+} are taken up by mitochondria in a cooperative manner, Ca^{2+} enhancing the uptake of Mn^{2+}, and Mn^{2+} retarding the uptake of Ca^{2+}. This is true for all three tissues although the rates of uptake vary considerably. Ca^{2+} uptake is most rapid in brain and liver and least rapid in heart. Mn^{2+} uptake is also most rapid in brain, slower in liver and even slower in heart.

When measuring proton ejection there are similar responses to Ca^{2+} uptake in liver, heart and brain. Mn^{2+} uptake, however, is accompanied by a slower and smaller proton ejection in heart and liver than in brain. The addition of both cations together gives a greater response than the sum of the component reactions. Similar effects were observed when measuring the stimulation of State 4 respiration by the two cations.

Reduced pyridine nucleotide oxidation accompanying Ca^{2+} uptake is qualitatively similar in all three sources of mitochondria although

---

* Permanent address: Istituto di Chimica Biologica, Università di Triste, Trieste, Italy

**Fellow of International Seminar in Chemistry, University of Uppsala, on leave of absence from the Department of Biochemistry, University of Nairobi, Nairobi, Kenya.

it is less pronounced in brain than in heart or liver mitochondria. There is little response to Mn^{2+}. When added together, Mn^{2+} increases the time span of the Ca^{2+} response, without significantly affecting the amplitude.

The results are consistent with the existence of a cooperative mechanism for Ca^{2+} and Mn^{2+} uptake by the three sources of mitochondria. We are currently investigating the efflux of Mn^{2+} from the various types of mitochondria and its relationship to Ca^{2+} efflux.

This work is supported by the Swedish Natural-Science Research Council.

References

1. Ernster, L. and Nordenbrand, K. (1967) Abstr. 4th FEBS Meeting, Oslo, p. 108.

2. Mela, L. and Chance, B. (1968) Biochemistry 7, 4059-4063.

3. Ernster, L., Hollander, P., Nakazawa, T. and Nordenbrand, K. (1969) in The Energy Level and Metabolic Control in Mitochondria (Papa, S., Tager, J.M., Quagliariello, E. and Slater, E.C., eds.) Adriatica Editrice, Bari, pp. 97-113.

4. Vinogradov, A. and Scarpa, A. (1973) J. Biol. Chem. 248, 5527-5531.

5. Ernster, L., Nakazawa, T. and Nordenbrand, K. (1978) in The Proton and Calcium Pump (Azzone, G.F., Avron, M., Metcalfe, J.C., Quagliariello, E. and Siliprandi, N., eds.) Elsevier North Holland, Biomed. Press, Amsterdam, pp. 163-176.

6. Hunter, D.R., Komai, H., Haworth, R.A., Jackson, M.D. and Berkoff, H.A. (1980) Circ. Res. 47, 721-727.

7. Hughes, B.P. and Exton, J.H. (1983) Biochem. J. 212, 773-782.

8. Hillered, L., Muchiri, P.M., Nordenbrand, K. and Ernster, L. (1983) FEBS Lett. 154, 247-250.

ION SELECTIVITY IN DIVALENT CATION ACCUMULATION IN YEAST IS ABOLISHED
BY INHIBITORS OF THE PLASMA MEMBRANE ATPase

A.P.R. Theuvenet, B.J.W.M. Nieuwenhuis, J. van de Mortel and G.W.F.H.
Borst-Pauwels
Department of Chemical Cytology, University of Nijmegen,
6525 ED Nijmegen, The Netherlands

Accumulation of divalent cations in yeast is a complicated process.
Ultimate accumulation levels attained for Mg^{2+} and Mn^{2+} are far more
greater than for Ca^{2+} and Sr^{2+} (1, 2). They are determined by the operation of at least three transport systems: an influx system with no
or hardly any ion selectivity (2), an efflux system in the plasmalemma (2, 3) and a transport system in the vacuolar membrane (4), all
involved in the regulation of the cytoplasmic concentration of the
divalent cations. After an 1 h incubation period e.g. Mn^{2+} is accumulated ten times more than Sr^{2+} (2). Apparently the differential accumulation of these cations is caused by the efflux pump in the plasmalemma, operating more effectively for Sr^{2+} than for Mn^{2+}. We will now
show that disruption of the plasmalemma under isotonic conditions
with DEAE-dextran as well as the presence of the inhibitors of the
plasmamembrane ATPase ethidium bromide (EB) (5) or 1-ethyl-3-(3)-
dimethylaminopropyl)-carbodiimide (EDAC) (6) abolishes the differential accumulation of Mn^{2+} and Sr^{2+}, but in different ways.

Under isotonic conditions the polybase DEAE-dextran disrupts the
plasmalemma of yeast but leaves the vacuoles intact (7). The initial
rate of uptake of Mn^{2+} and Sr^{2+} is more than an order of magnitude
increased in these polybase-treated cells whereas for both cations
almost the same ultimate accumulation level is reached as is found
for Mn^{2+} in intact cells (compare Figs. 1 and 2).

EB also increases the initial rates of Mn^{2+} and Sr^{2+} uptake in intact cells. The accumulation of Mn^{2+} is reduced slightly while that
of Sr^{2+} is increased considerably. The ultimate accumulation level
that is reached is almost the same for both cations (Fig. 2). The
lower Mn^{2+} accumulation may reflect a loss in ability of the cells
to concentrate Mn^{2+} in their vacuoles. This is supported by the
finding that in DEAE-dextran-treated cells accumulation of Mn^{2+}, as
well as of Sr^{2+}, is almost completely prevented by EB (Fig. 1). The
increased Sr^{2+} accumulation in the intact cells found in the presence
of EB is compatible with an inhibition of the Sr^{2+} efflux pump in the
plasmalemma.

The carbodiimide EDAC reduces the accumulation of Mn^{2+} to the level
of that of Sr^{2+} in control cells. In addition, contrary to EB, EDAC
is without effect on the ultimate Sr^{2+} accumulation level. Moreover,
it does not significantly affect the initial rate of uptake of both
Sr^{2+} and Mn^{2+} (Fig. 2). As carbodiimides are well known carboxylgroup

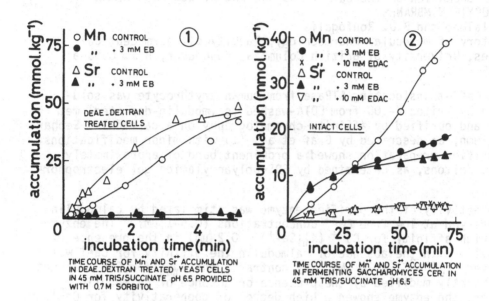

TIME COURSE OF Mn⁺⁺ AND Sr⁺⁺ ACCUMULATION
IN DEAE-DEXTRAN TREATED YEAST CELLS
IN 45 mM TRIS/SUCCINATE pH 6.5 PROVIDED
WITH 0.7M SORBITOL

TIME COURSE OF Mn⁺⁺ AND Sr⁺⁺ ACCUMULATION
IN FERMENTING SACCHAROMYCES CER. IN
45 mM TRIS/SUCCINATE pH 6.5

modifying reagents and their attack may alter the conformation of membrane proteins (8), it may be hypothesized that EDAC modifies the efflux pump in the plasmalemma and thereby abolishes the ion selectivity of this transport system.

The further unraveling of the complicated process of divalent cation uptake in yeast needs experiments of ion uptake in isolated membrane vesicles of both the tonoplast and the plasmalemma. The application of EB and EDAC should be helpful in the characterization of the transport systems of both membranes.

1. Rothstein, A., Hayes, A., Jennings, D. and Hooper, D. (1958) J. Gen. Physiol. 41, 585-594.
2. Nieuwenhuis, B.J.W.M., Weijers, C.A.G.M. and Borst-Pauwels, G.W.F.H. (1981) Biochim. Biophys. Acta 649, 83-88.
3. Eilam, Y. (1982) Biochim. Biophys. Acta 687, 8-16.
4. Lichko, L.P., Okorokov, L.A. and Kulaev, I.S. (1982) Arch. Microbiol. 132, 289-293.
5. Peters, P.H.J., unpublished results.
6. Fuhrmann, G.F., Wehrli, E. and Boehm, C. (1974) Biochim. Biophys. Acta 363, 295-310.
7. Huber-Wälchli, V. and Wiemken, A. (1972) Arch. Microbiol. 120, 141-149.
8. Godin, D.V. and Schrier, S.L. (1970) Biochemistry 9, 4068-4077.

MODE OF OPERATION OF THE Ca^{2+}-TRANSLOCATING ATPase FROM HUMAN ERYTHROCYTE MEMBRANES.

A. Villalobo and B.D. Roufogalis
Laboratory of Molecular Pharmacology, Faculty of Pharmaceutical
Sciences, University of British Columbia, Vancouver, B.C. Canada
V6T 1W5

The Ca^{2+}-translocating ATPase from human erythrocyte was solubilized by Triton X-100 from EDTA-washed (calmodulin-depleted) membranes and purified by affinity-chromatography on a calmodulin-Sepharose column, as described by Graf et al. (1) with minor modifications. The purified preparation showed a prominent band of approximately 135,000 daltons, as determined by SDS-polyacrylamide gel electrophoresis.

The activity of the purified enzyme was stimulated by calmodulin when assayed at low free Ca^{2+} concentrations (0.5-1 µM). The enzyme had a high affinity for Ca^{2+} ($K_m'(Ca^{2+}) \simeq 0.2$ µM) in the presence of calmodulin. In the absence of calmodulin the affinity for Ca^{2+} was decreased ($K_m'(Ca^{2+}) \simeq 10$ µM). By contrast, the V_{max} for Ca^{2+} was not significantly modified by the presence or absence of calmodulin. Moreover, the enzyme showed a high degree of cooperativity for Ca^{2+}, with a Hill coefficient of nearly 4 in the presence of calmodulin and close to 2 in the absence of calmodulin. Magnesium increased the V_{max} of the enzyme. However, the $K_m'(Ca^{2+})$ was significantly lower at low Mg^{2+} concentrations, both in the presence and in the absence of calmodulin.

The purified enzyme also exhibited a high affinity for calmodulin ($K_m'(calm) \simeq 0.1$ µg.ml^{-1}) at high concentrations of $MgCl_2$ (5 mM). However, the apparent affinity for calmodulin was decreased dramatically ($K_m'(calm) \simeq 2-3$ µg.ml^{-1}) when the enzyme was assayed at low concentrations of $MgCl_2$ (0.5 mM).

The purified enzyme also exhibited complex kinetics towards ATP, which was consistent with two binding sites for ATP. One of the sites was of high affinity ($K_m'(ATP) \simeq 1$ µM), while a low affinity site exhibited a K_m' for ATP several orders of magnitude higher.

The purified enzyme was reconstituted into artificial phospholipid vesicles of asolectin or phosphatidylcholine, by a cholate dialysis method (2). In both cases the reconstituted enzyme actively transported Ca^{2+} and formed a chemical Ca^{2+} gradient across the membrane with very high efficiency, as demonstrated by an ATP hydrolysis control ratio (ratio of the rate of ATP hydrolysis in the presence and in the absence of ionophore) of 9-10 induced by the $Ca^{2+}/2H^+$ exchanger, A23187.

The activity of the reconstituted enzyme was also stimulated by the K^+/H^+ exchanger, nigericin, or the $Na^+(K^+)/H^+$ exchanger, monensin (2-3 fold), probably because they collapsed a H^+ concentration gradient (ΔpH) formed across the membrane during ATP hydrolysis, as suggested by Niggli et al. (3). In order to further substantiate this mechanism, a new $Ca^{2+}(Na^+)$ ionophore, CYCLEX-2E (which, in contrast to A23187, does not contain ionizable protons (4)) was used. CYCLEX-2E alone maximally stimulated the activity of the reconstituted ATPase only about 3 or 4 fold. When CYCLEX-2E was used in the presence of H^+ conducting agents such as FCCP, CCCP or 2,4-DNP, the stimulation induced by the combination of ionophores was larger than that by each ionophore alone, as expected for a Ca^{2+}/H^+ exchange mechanism. A typical experiment is shown in Table 1.

Table 1

Addition	ATPase activity (nmoles.min^{-1}per mg protein)
None	168
FCCP	204
CYCLEX-2E	494
CYCLEX-2E + FCCP	646
A23187	1442

However, the formation of a membrane potential could not be excluded, since K^+ (in the presence of valinomycin), used as a charge compensating cation, consistently stimulated the rate of ATP hydrolysis of the reconstituted Ca^{2+} transport ATPase by 30-80%.

The present work favours a model of operation of the Ca^{2+} pump in which Ca^{2+} is exchanged for nH^+ during ATP hydrolysis.

References:
1. Graf, E., Verma, A.K., Gorski, J.P. Lopaschuk, G., Niggli, V., Zurini, M. Carafoli, E. and Penniston, J.T. (1982) Biochemistry 21, 4511-4516.
2. Kagawa, Y. and Racker, E. (1971) J. Biol. Chem. 246, 5477-5487.
3. Niggli, V., Sigel, E. and Carafoli, E. (1982) J. Biol. Chem. 257, 2350-2356.
4. Drobnies, A.E. and Deber, C.M. (1982) Biochim. Biophys. Acta 691, 30-36.

(Supported by the Canadian Heart Foundation and the Medical Research Council of Canada. The skillful assistance of Ms. Laura Brown is greatly appreciated.)

THE EFFECTS OF THE NATURE OF EXTRAMITOCHONDRIAL ANIONS ON RUTHENIUM RED INSENSITIVE Ca^{2+} EFFLUX FROM RAT LIVER MITOCHONDRIA

I. Warhurst[o], A.P. Dawson" and M.J. Selwyn".
[o]Department of Biological Sciences, University of Keele, Keele, Staffs U.K.
"School of Biological Sciences, University of East Anglia, Norwich NR4 7TJ, U.K.

Previous work from this laboratory has shown that rat liver mitochondria possess a high electrogenic permeability to anions such as Cl$^-$ and NO$_3^-$ at alkaline pH [1].   In addition, electrogenic anion uniport has been shown to be dependent on the level of intra-mitochondrial Ca^{2+}[2], and is inhibited by low concentrations of DCCD [3].   Other studies [4, 5] have shown that ruthenium red insensitive Ca^{2+} efflux and electrogenic anion uniport are inhibited by local anaesthetics, suggesting a possible link between the two systems.

Further experiments using Arsenazo III as an indicator of extra-mitochondrial Ca^{2+} indicate that ruthenium red insensitive Ca^{2+} efflux is increased in a Hepes (impermeant species) based medium compared to that seen in Cl$^-$ or NO$_3^-$ media.   The dependence of Ca^{2+} efflux on the external anion might be taken as evidence for the view that ruthenium red insensitive Ca^{2+} efflux and anion uniport are associated in some way.   However, it was found that $\Delta\Psi$ was larger in KHepes (212 mV) than in KCl media (160 mV) and that other studies [6] have shown the rate of Ca^{2+} efflux to increase with $\Delta\Psi$ in the 130-180 mV range.   The coupling between the two systems may therefore be of an electrical rather than a molecular nature.

Ruthenium red insensitive Ca^{2+} efflux in the presence of uncoupler was also dependent on the nature of the external anion, but was additionally dependent on the intramitochondrial Ca^{2+} content. However, at a set internal Ca^{2+} load, Ca^{2+} efflux was greater in K Hepes than in KCl or KNO$_3$.   The results strongly suggest that ruthenium red insensitive Ca^{2+} efflux in the presence of uncoupler is also correlated with the permeability of the extramitochondrial anion (NO$_3^-$> Cl$^-$ > Hepes, as indicated by swelling experiments).

Since DCCD inhibits electrogenic anion uniport at low concentra-tions [3], the effect of the inhibitor on ruthenium red insensitive Ca^{2+} efflux was studied.   At the low concentrations used to inhibit anion permeability [5 n moles/mg], DCCD also inhibited Ca^{2+} efflux to a similar extent in K hepes and KCl media, but had little effect in KNO$_3$.   Other experiments have shown that DCCD decreases $\Delta\Psi$. This indicates that the effects of DCCD on Ca^{2+} efflux could be accounted for by the ability of DCCD to reduce $\Delta\Psi$.

In the presence of uncoupler, DCCD stimilated Ca^{2+} efflux at a set intramitochondrial Ca^{2+} loading, the extent increasing as the permeability of the external anion decreased. This effect of DCCD could be due to the inhibition of anion movements which in some way oppose Ca^{2+} efflux under these conditions. The finding that DCCD markedly increased Ca^{2+} efflux in a Hepes (impermeant species) medium could be due to the blockage of the movement of endogenous anions such as Pi or HCO_3^-.

Thus, these relationships between Ca^{2+} efflux and the permeability of the external anion provide further evidence for a link between the two systems, although the coupling may be electrical rather than molecular in nature.

[1] Selwyn, M.J., Dawson, A.P. and Fulton, D.V. (1979) Biochem.Soc. Trans. 7, 216-219.

[2] Warhurst, I.W., Selwyn, M.J., Dawson, A.P. and Fulton, D.V. (1982) EBEC Rep. 2, 549-550.

[3] Warhurst, I.W., Dawson, A.P. and Selwyn, M.J. (1982) FEBS Lett. 149, 249-252.

[4] Dawson, A.P. and Fulton, D.V. (1980) Biochem. J. 188, 749-755.

[5] Selwyn, M.J., Fulton, D.V. and Dawson, A.P. (1978) FEBS Lett. 96, 149-151.

[6] Bernardi, P. and Azzone, G.F. (1983) Eur.J.Biochem. 134, 377-383.

Lipid Mechanism of Membrane Transport and its Enzymatic Control

Konrad Kaufmann
Dept. Neurobiology, The Weizmann Institute of Science, Rehovot, Israel

It is evident that specific proteins induce specific transport across biological protein-lipid membranes. Autonomous protein ionophores have been proposed, although such a mechanism creates a causality paradoxon: *how could ligand binding cause free energy coupling while enzymatic hydrolysis has to produce the free energy first?*

The lipid mechanism resolves this paradoxon. It is evident from observable monolayer phase diagrams. It has been experimentally established in vitro for ACh: the ligand is first hydrolysed by the enzyme. The protons thus produced induce ion channel fluctuations in the phospholipid bilayer.

Chemiosmosis is a direct consequence for surface localized hydrolysis. The chemomechanical coupling by phospholipid monolayer protonation reversibly stores free catalytic energy. Proton dissipation into the bulk is thus avoided for lipid relaxation times. Asymmetric surface hydrolysis actively pumps ions through proton-induced ion channels by local protomotive force.

1. Fluctuation mechanism of ion channels

The lipid monolayers inavoidably fluctuate. In thermal equilibrium, the strength of area per lipid fluctuations obeys the isothermal compressibility $\partial a/\partial \pi$ measurable according to Langmuir:

$$\langle (\delta a)^2 \rangle = - kT (\partial a / \partial \pi)_{T, \mu \ldots}$$

These thermodynamic fluctuations are macroscopically controlled by temperature T, surface pressure π, electrochemical potentials of protons or calcium, and likewise by pH and the electrostatic membrane potential $\Delta \psi$. The associated fluctuations in local bilayer thickness exponentially increase the Boltzmann probability for dielectric break-down and induction of local defects such as ion channels in the else impermeable lipid lattice. The fluctuations diverge at a critical point, become extremal at onset of first order phase transition or separation, in particular near the phospholipid pK even for heterogeneous lipid bilayer membranes.

The observation of discrete ion channels in pure phospholipid bilayers near the pK (Fig.1) demonstrates the lipid mechanism in the absence of ionophores. In the presence of enzymatic proton sources, ligand-induced ion channels of similar conductivities and relaxation times appear even at neutrally buffered bulk pH, in the presence of purified AChase, and likewise of AChR containing residual AChase activity. Any one fully active AChase protein is capable to lower the local pH by up to 3 units, sufficient to control the lipid mechanism even at neutral and buffered bulk pH.

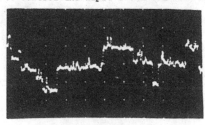

Fig.1. Trace of membrane currents across a synthetic methyl phosphatidic acid bilayer in 1 M KCl, at +77 mV and pH 2 cis side, pH 6.5 trans. Scale units 12.5 pA / 100 msec.

2. Local protomotive forces cross bilayer channels

Enzymatic phospholipid protonation chemomechanically couples free catalytic energy into the lipid monolayer associated. Efficiency is 1 for reversible adiabatic coupling. Local pH changes by 3 units store above 1 kcal per mole protonated lipids, but only near the pK, according to methyl phosphatidic acid phase diagrams. Dissipation of protons into the bulk is thus avoided for lipid relaxation times.

Asymmetrically protonated bilayers exert a local protomotive force for surface diffusion, which is fast against lipid protonation relaxation. Passive local diffusion thus transports protonic free energy along or across the bilayer in presence of ion channels. I expect from hydrodynamic theory and lipid relaxation times adiabatic pulses in local density, temperature, surface pressure, and electrostatic potential resembling those observed in vivo during axonal and photoreceptor action potentials, with propagation velocities below ca. 20 m/sec calculated from Langmuir diagrams, and slowing-down to 0 m/sec at any critical point.

Any asymmetrically localized hydrolase sufficient to reach the boundary lipid's pK, thus inducing ion channels, must actively transport protons toward the non-catalytic side. Specific cotransport of Na follows from the electrostatic term in the protomotive force, but only for Na-activated hydrolysis. Uncoupling residual channels in the bilayer establish the K equilibrium potential equal to the resting potential. Recently, according proton transport has been found at Na-K-ATPase. Uncoupling "slips" violating integer ion stoichiometries have also been observed. The theory is compatible with both the global and the local view-point on chemiosmosis. Initial global symmetry in equilibrium with the surface,

$$\mu = \mu_0 = \mu_0' = \mu'$$

local, asymmetric nonequilibrium by hydrolysis $\mu \neq \mu_0$, local equilibration during critical conductive state, $\mu_0 = \mu_0'$ establish independent global thermodynamic equilibria on both membrane sides after termination of the enzymatically induced conductivity :

$$\mu = \mu_0 \neq \mu_0' = \mu'$$

Asymmetric msec membrane current pulses can be induced by protons, together with passive ion channels, and the associated order transitions in the phospholipids are observed as discrete steps in the membrane capacitance, in the following reconstituted bilayers:
- pure synthetic methylated phosphatidic acid in bidestilled water, acidified asymmetrically by HCl to pH 2; proton transport is thus obvious;
- presence of 1 M KCl, variation of lipid chain length;
- pure synthetic diphytanoyl lecithin in 1 M KCl (proton-induction pH 2.5);
- soybean lecithin lipid mixture in 1 M KCl (below pH 3.8);
- Torpedo membrane fragments active in ATPase and AChase, in 0.1 M NaCl.
In the latter system, the same phenomenology was likewise induced by ATP (Fig.2), producing pH 5, and reversibly quenched by alkali neutralization.

Fig.2. ATP-induced ion channels of 1000 pS and 1300 pS of 50 msec lifetime, across Torpedo electric organ membrane fragment/ soybean lecithin bilayer membrane. ATP lowered the pH to 5 at the cis side, trans pH 6.5, 0.1 M NaCl, +38 mV.

ANTIGENIC RELATIONSHIPS BETWEEN PROTEINS INVOLVED IN
PHOSPHATE TRANSPORT IN YEASTS.

S. Bedu[*], R. Jeanjean[*], F. Blasco[*] and M. Hirn[o].
[*]Laboratoire de Physiologie Cellulaire, Faculté des Sciences
de Luminy, 13288 Marseille Cédex 9, France.
[o]Centre d'Immunologie, Faculté des Sciences de Luminy, 13288
Marseille Cédex 9, France.

The phosphate translocation from the external medium to
the cytoplasm is mediated by several proteins in the yeast
Candida tropicalis. This uptake is an energy-dependent
process using the protonmotive force generated by the plasma
membrane ATPase (1). Phosphate transport is sensitive to
osmotic shock which reduces the transport velocities on the
whole cells. Two Pi-binding proteins (PiBP$_1$ and PiBP$_2$)
released by osmotic shock from the cell wall have been
identified and characterized (2). PiBP$_1$ has a low for Pi (Km
= 17 to 22 M) whereas PiBP$_2$ has a high affinity (Km = 5 to
6). The inhibitory effect of antibodies raised against the
PiBP$_2$ suggests that this protein is involved in Pi uptake at
physiological pH values (3). The high affinity Pi transport
system is thought to be composed of Pi-binding proteins,
located near the cell surface within the cell wall and a Pi
specific plasma membrane carrier (4). The role of the
binding proteins could be to keep the Pi pool constant in
the cell wall or to load the plasma membrane carrier.

By immunoblotting technique, we have shown that the
antibodies raised against PiBP$_2$ recognize antigenic
determinants on two plasma membrane proteins: a protein with
a Mr = 30 kDa (MP 30) and, to a lesser extent, the other one
with a Mr = 52 kDa. The protein with Mr = 52 kDa is likely a
membranous form of PiBP$_2$. For this reason,the MP30 protein
has been more specifically studied and purified by affinity
chromatography on a Sepharose 4B CN column coupled with
anti-PiBP$_2$ antibodies and by SDS-PAGE. Antibodies have been
raised against the purified and denatured form of the
protein MP30. They inhibit the phosphate transport on
protoplasts but not on whole cells and their corresponding
Fab fragments inhibit the transport on both whole cells and
protoplasts. From these results, we conclude that the
antibodies can reach the plasma membrane protein MP30 when
the cell wall is absent-i.e. in protoplasts - or if they are

on lower size – under Fab fragments form –. In conditions where the antibodies can bind to the MP30 protein, the phosphate uptake is inhibited. It is suggested that the MP30 protein may be the specific membranous Pi carrier and that the polyclonal antibodies bind near the active site (5).

Homologies between proteins from different yeasts genera, involved in phosphate transport, have been investigated. Antibodies raised against Candida tropicalis MP30 protein have been tested, by immunoblotting, on membranes proteins from Saccharomyces cerevisiae and Schizosaccharomyces pombe. No proteins from S. pombe membrane are recognized by polyclonal anti-MP30 antibodies but one protein from S. cerevisiae, with Mr = 30 kDa in SDS-PAGE, shares common antigenic determinants with MP30 from C. tropicalis. Moreover, on the same fashion than for C.tropicalis, antibodies inhibit the phosphate uptake on S. cerevisiae protoplasts and the Fab fragments inhibit the transport on whole cells. These findings hint at the possibility that the phosphate transport across the plasma membrane is mediated by the same type of carrier (a protein with Mr = 30 kDa) on both yeasts genera. These antigenic homologies could be expected since Nieuwenhuis and Borst-Pauwels have shown that the high-affinity Pi transport system in S. cerevisiae (called H^+-phosphate cotransport) presents similar characteristics with that of C. tropicalis. However, some attempts in order to isolate Pi-binding proteins after osmotic shock or after protoplast formation have not been successful in S. cerevisiae (6).

References.
1 Goffeau A. and Slayman C.W. (1981) Biochim. Biophys. Acta 639, 197-223.
2 Jeanjean R. and Fournier N. (1979) Febs Lett. 105, 163-166.
3 Jeanjean R., Attia A., Jarry T. and Colle A. (1981) Febs Lett. 125, 69-71.
4 Jeanjean R., Bedu S., Rocca-Serra J. and Foucault C. (1984) Arch. Microbiol. 137, 215-219.
5 Jeanjean R., Blasco F. and Hirn M. (1984) Febs Lett. 165, 83-87.
6 Nieuwenhuis B., Borst-Pauwels G.W. (1984) Biochim. Biophys. Acta 770, 40-46.

THE ACTIVE Pi CARRIER FROM HEART MITOCHONDRIA CORRESPONDS TO A SINGLE BAND OF THE HYDROXYLAPATITE ELUATE.

F. Bisaccia, A. Rizzo and F. Palmieri

Institute of Biochemistry, Faculty of Pharmacy and CNR Unit for the Study of Mitochondria and Bioenergetics, University of Bari, via Amendola 165/A, 70126 Bari, Italy

When high resolution SDS gel electrophoresis is applied, the hydroxylapatite eluate of Triton X-114 solubilized pig heart mitochondria is shown to contain 5 protein bands in the Mr-region of 30000-35000, which are differently distributed in the various fractions. In this work the role of cardiolipin in the purification of the mitochondrial phosphate carrier by hydroxylapatite has been investigated. Without added cardiolipin, the reconstituted phosphate transport activity in the hydroxylapatite eluate is small and only confined to the first two fractions. With cardiolipin added to the extract, the eluted activity is much higher and present until fraction 6. Addition of cardiolipin directly to the hydroxylapatite eluate, i.e. after the purification step, causes a much smaller increase of the reconstituted phosphate transport activity, which is also limited to the first two fractions. The activity retained by hydroxylapatite in the absence of cardiolipin is eluted after addition of this phospholipid to the column. The requirement of added cardiolipin diminishes on increasing the concentration of solubilized mitochondria. In all these conditions, among the 5 protein bands present in the whole hydroxylapatite eluate, only the presence and the relative amount of band 3 of Mr 33000 corresponds to the phosphate transport activity. Cardiolipin is the only phospholipid tested which causes elution of band 3 from hydroxylapatite; on the other hand, it prevents the elution of band 2 and retards that of band 5 (the ADP/ATP carrier). Band 1 starts to appear in the second fraction even without cardiolipin. On increasing the concentration of cardiolipin, in the first fraction of the hydroxylapatite eluate band 3 increases and the contamination of band 4 decreases. Under optimal conditions a preparation of band 3 about 90% pure and with high reconstituted phosphate transport activity is obtained (F. Bisaccia and F. Palmieri, BBA in the press). It is concluded that the elution of the phosphate carrier from hydroxylapatite requires cardiolipin and that the phosphate carrier is identical with (or with part of) band 3 of the hydroxylapatite eluate.

PURIFICATION OF THE MITOCHONDRIAL TRICARBOXYLATE CARRIER BY HYDROXYAPATITE.

I. Stipani, V. Iacobazzi and F. Palmieri

Institute of Biochemistry, Faculty of Pharmacy, and CNR Unit for the Study of Mitochondria and Bioenergetics, University of Bari, via Amendola 165/A, 70126 Bari, Italy

The mitochondrial tricarboxylate carrier has been extracted from rat liver mitochondria or submitochondrial particles with Triton X-100, in the presence of 1,2,3-benzenetricarboxylate and cardiolipin, and partially purified by chromatography on hydroxyapatite. The purified fraction, which also contains the ADP/ATP carrier and the phosphate carrier, after incorporation into liposomes catalyzes a 1,2,3-benzenetricarboxylate-sensitive ^{14}C-citrate/citrate exchange. The reconstituted ^{14}C-citrate/citrate exchange activity of the hydroxyapatite eluates is very low when crude extracts of mitochondria or submitochondrial particles from heart and brain (instead of liver) are applied to hydroxyapatite. ^{14}C-citrate exchanges not only with internal citrate but also with cis-aconitate, phosphoenolpyruvate and malate. On the other hand, virtually no ^{14}C-citrate is taken up by the proteoliposomes when they contain no anion, or anions which are not substrates of the tricarboxylate carrier like phosphate. The reconstituted citrate exchange is strongly inhibited by 1,2,3- -benzenetricarboxylate, p-iodobenzylmalonate, 0,2 mM mersalyl and p-hydroxymercuribenzoate, but only slightly by NEM. The tissue and substrate specificity and the inhibitor sensitivity of citrate transport in liposomes are therefore similar to those described for the citrate transport in mitochondria. Further similarities are shown by the kinetic parameters, e.g. the Km for citrate and the Ki for 1,2,3-benzenetricarboxylate in liposomes are close to those found in mitochondria. The maximal rate of citrate exchange in the reconstituted system is 338 μmol $\cdot$ min^{-1} $\cdot$ g protein^{-1}, at 30°C and pH 7.0 (I. Stipani and F. Palmieri, FEBS Lett. 161, 1983, 269). Inclusion of cardiolipin in the solubilization buffer increases the reconstituted citrate transport activity in the hydroxyapatite eluate and prevents the inhibition by high concentrations of Triton. Since it cannot be excluded that more tricarboxylate carrier is extracted in the presence of added cardiolipin, the effect of cardiolipin was directly tested on the hydroxyapatite eluate. The reconstituted citrate transport activity of the hydroxyapatite eluate is increased by cardiolipin and decreased by Triton. Both effects are concentration dependent and reversible. No other phospholipid tested including acidic and unsaturated ones can substitute for cardiolipin.

ON THE TWO CURRENT HYPOTHESES EXPLAINING THE ANOMALIES IN
DELOCALISED CHEMIOSMOSIS

M.G.L. Elferink, J.M. van Dijl, K.J. Hellingwerf, W.N. Konings.
Dept. of Microbiology, University of Groningen, Kerklaan 30,
9751 NN HAREN, The Netherlands

In studies on solute transport in Rhodopseudomonas sphaeroides we
have demonstrated that in addition to a proton motive force cyclic
or linear electron transfer is required for solute transport to take
place. At constant proton motive force values the uptake of solutes
is a linear function of the rate of cyclic or linear electron
transfer (1, 2). This is also valid for the lactose transport
protein from Escherichia coli introduced in the cytoplasmic membrane
of the phototrophic bacterium via genetic manipulation (3).

Regulation of solute uptake by electron transfer through a linear
electron transfer chain is also observed in E. coli, both in intact
cells and in isolated cytoplasmic membrane vesicles. However,
differences in the regulation of various secondary transport systems
are observed. For glutamate uptake in addition to a proton motive
force, respiration is obligatory, like for solute transport in Rps.
sphaeroides. On the other hand lactose and proline uptake takes
place at anaerobic conditions, but initiation of linear electron
transfer results in higher initial uptake rates and steady state
accumulation levels.

In cytoplasmic membrane vesicles of E. coli the rate of lactose
uptake correlated with the rate of linear electron transfer when the
oxidation rate of lactate was progressively inhibited with
increasing concentrations of the lactate dehydrogenase inhibitor
oxamate or with the oxidase inhibitor KCN. The proton motive force
under these conditions remained constant or decreased slightly. In
parallel experiments the proton motive force was titrated with a
combination of nigericin and valinomycin and the rate of lactose
uptake was measured as a function of the proton motive force. It was
clear that the slight decrease in proton motive force observed with
the electron transfer chain inhibitors could not account for the
observed decrease in uptake rate. Moreover, when the proton motive
force was very low ($\leq$ 10 mV) there is still a significant uptake
rate and accumulation level of lactose at high respiration rates. It
should be noted that our experimental conditions were such that all
parameters under investigation were recorded simultaneously.

The interaction between the electron transfer chains and
transport cariers is not restricted to H^+/symport systems but also
occurs in Na^+/methyl ß-D-thiogalactopyranoside (= TMG) symport via

the melibiose transport carrier.

To explain our observations two hypotheses are considered. (i) The interaction between electron transfer chains and solute carriers may be caused by localized chemiosmotic phenomena (4). The observation that not only H^+/solute but also Na^+/solute symport systems are regulated by the rate of the electron transfer chain, however, argues against this explanation. (ii) The second hypothesis is that regulation of electron transfer is exerted by an influence on the redox state of the transport protein via an interaction with one or more electron transfer intermediates (5). With this interpretation it is difficult to explain the significant rate of lactose uptake and accumulation level when the bulk phase proton motive force is low unless the redox interaction can provide free energy for solute transport.

We favour the redox interaction model and experiments are currently in progress to investigate this working hypothesis.

METHODS

All measurements were done under conditions that the proton motive force was composed of the membrane potential only. The rate of uptake, the electrical potential across the membrane ($\Delta\psi$) (with an ion selective TPP^+ electrode) and the rate of oxygen consumption (with a Clark-type oxygen electrode) were measured simultaneously in a thermostated 5 ml vessel. A correction for TPP^+ binding was aplied as described (6).

REFERENCES
1. Elferink, M.G.L., Friedberg, I., Hellingwerf, K.J. (1983) Eur.J.Biochem. 129, 583-587
2. Elferink, M.G.L., Hellingwerf, K.J., van Belkum, M.J., Poolman, B., Konings, W.N. (1984) FEMS Lett. 21, 293-284
3. Elferink, M.G.L., Hellingwerf, K.J., Nano, F.E., Kaplan, S., Konings, W.N. (1983) FEBS Lett. 164, 185-190
4. Westerhoff, H.V., Melandri, B.A., Venturoli, G., Azzone, G.F., Kell, D.B. (1984) FEBS Lett. 165, 1-5
5. Konings, W.N., Robillard, G.T. (1982) Proc.Natl.Acad.Sci. U.S.A. 79, 5480-5484
6. Lolkema, J.S., Hellingwerf, K.J., Konings, W.N. (1982) Biochim.Biophys.Acta 681, 85-94

ATP DRIVEN MALATE TRANSPORT INTO VACUOLES ISOLATED FROM BARLEY MESOPHYLL PROTOPLASTS

U. I. FLÜGGE[+], E. MARTINOIA[°], G. KAISER[°], U. HEBER[°] and
H. W. HELDT[+]
[+]Lehrstuhl für Biochemie der Pflanze, Untere Karspüle 2,
 D-3400 Göttingen, FRG
[°]Lehrstuhl Botanik I, Mittlerer Dallenbergweg 64, D-8700 Würzburg, FRG

In photosynthesis of C3 plants such as wheat, spinach and barley the main
part of the fixed carbon is converted into sucrose (1, 2). Significant
amounts of the assimilated carbon are also found in malate, which exhibits
pronounced diurnal concentration changes (3). In assimilating barley
mesophyll protoplasts it could be shown that malate is transported rapidly
into the vacuoles (4). In spinach leaves, the vacuolar malate concentration
at the end of the day was found to be increased to about 40 mM, with the
cytosolic malate concentration remaining at a very low level (3). This
suggested that the malate transport into the vacuoles is an energy driven
process. An active transport of malate into the vacuole has been postulated
also for CAM plants, where malate plays a central role as an assimilatory
metabolite. There, recently a vacuolar ATPase has been identified, with
properties similar to that of H^+ transporting ATPases, although its
participation in malate transport is speculative so far (5). In the present
paper, evidence is presented for an ATP driven active transport of malate
into vacuoles of C3 plants.

Fig 1 shows the kinetics for the uptake of malate into isolated vacuoles.
In the absence of ATP, the uptake of malate slows down after about 5 - 7
min, leading to an internal malate concentration which is only slightly
higher than the external one. In the presence of MgATP, the uptake of
malate is largely accelerated, proceeding in a linear mode for a period of
20 min. After that time, the concentration of malate inside exceeds the
external concentration by a factor of 15. A double reciprocal plot of the
initial rates of malate transport at various concentrations (Fig 2)
reveals, that ATP increased the V(max) from 2.6 µmol/mg Chl.h in the
absence of ATP by a factor of 3 - 4, whereas the half saturation of the
transport (2.5 mM) was not markedly affected.

For further characterization of the transport, we tested the effects of
various inhibitors known to inhibit transport ATPases and of ionophores on
the MgATP stimulated malate transport. Both diethylstilbestrol and
N',N' dicyclohexylcarbodiimide proved to be effective inhibitors of the
energized malate uptake, whereas p-hydroxymercuribenzene sulfonic acid
inhibited the MgATP dependent and the non-energized transport as well.
Other inhibitors such as sodium orthovanadate, an inhibitor of plasma
membrane ATPases and oligomycin and azide, inhibitors of the

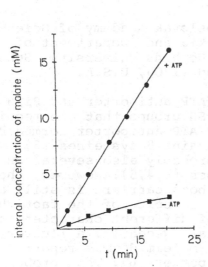
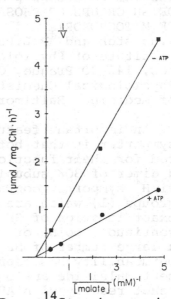

Fig 1,2   Effect of ATP (8 mM) added as MgATP on [^{14}C] malate uptake
into isolated vacuoles (20°C). Fig 1 (left) time course with 1.1 mM malate.
Fig 2 (right) concentration dependence.

mitochondrial ATPases, had no or only little effects on the energized
malate uptake. In addition, the MgATP dependent malate uptake was
completely prevented in the presence of the ionophores nigericin and
valinomycin/CCCP in the presence of external potassium.

From this it can be concluded that the active component of the malate
uptake is coupled to the action of a tonoplast ATPase establishing an H^{+}
gradient and/or a membrane potential which leads to the observed
accumulation of malate inside the vacuoles.

References
(1) Giersch, C., Heber, U., Kaiser, G., Walker, D. A., Robinson, S. P.
      (1980) Arch. Biochem. Biophys. 205, 246 - 259
(2) Stitt, M., Wirtz, W., Heldt, H. W. (1980)
      Biochim. Biophys. Acta 593, 85 - 102
(3) Gerhardt, R. and Heldt, H. W. (1984) Plant Physiol., in press
(4) Kaiser, G., Martinoia, E., Wiemken, A. (1982)
      Z. Pflanzenphysiol. 107, 103 - 113
(5) Smith, I. A. C., Uribe, E. G., Ball, E., Heuer, S., Lüttge, U.
      (1984) Eur. J. Biochem., in press

Supported by the Deutsche Forschungsgemeinschaft.

THE USE OF EOSINE-5-MALEIMIDE TO ESTABLISH THE LOCATIONS
OF SH GROUPS OF PHOSPHATE AND ADENINE NUCLEOTIDE CARRIERS
OF MITOCHONDRIA

J.Houštěk and [+]P.L.Pedersen
Institute of Physiology, Czechoslovak Academy of Sciences, 142 20 Prague, Czechoslovakia and [+]Department of
Physiological Chemistry, Johns Hopkins University, School
of Medicine, Baltimore, Maryland 21205, U.S.A.

An important feature of ADP/ATP antiporter and Pi/H^+
symporter is that both contain SH groups that are required for their function (1). ADP/ATP antiporter formed by
a dimer of 30K subunits (2) contains 8 cysteines (3).
Pi/H^+ symporter contains most probably also several SH
groups (1) which are of two types (1,4,5). However, the
exact location of SH groups in both carriers is still a
continuous point of contention, in spite of the fact that
a large number of SH reagents of different character and
permeability has been used so far. Recently, a fluorescent maleimide analog, eosine-5-maleimide was reported to
react readily with ADP/ATP antiporter (6). This probe is
relatively large in size and bears several charged groups.
Therefore, it was of interest to investigate to what extent eosine-5-maleimide permeates the mitochondrial membrane in order to use it to assess the side locations and
reactivity of SH groups in these transport systems.

At concentrations up to 300 uM eosine-5-maleimide was
found to be impermeable to the inner mitochondrial membrane. In contrast to N-ethylmaleimide, eosine-5-maleimi-
de failed to reduce matrix glutathione or to inhibit
ß-hydroxybutyrate dehydrogenase located on matrix side of
the inner mitochondrial membrane.
Eosine-5-maleimide readily labeled ADP/ATP antiporter
in submitochondrial particles (SMP) where the inner membrane orientation is inversed but it did not label this
carrier in intact mitochondria. Labeling was reduced when
SMP were prepared from carboxyatractyloside (CAT) loaded
mitochondria but unaffected when CAT was added to SMP.
Neither ADP, nor N-ethylmaleimide influenced the labeling
in SMP. In mitochondria where eosine-5-maleimide labeling
was absent even in the presence of ADP, ADP induced the
reactivity of the transporter towards [^{3}H]-N-ethylmalei-
mide. This labeling was abolished by CAT.
Unlike ADP/ATP transporter Pi/H^+ symporter could be

modified by eosine-5-maleimide in intact mitochondria as shown both by a complete inhibition of phosphate induced swelling and by prevention of $[^3H]$-N-ethylmaleimide labeling of Pi/H^+ symporter. Unlike inhibition of Pi/H^+ symporter by N-ethylmaleimide, inhibition by eosine-5-maleimide was strongly temperature dependent. Thus, a change in incubation conditions from $0^\circ C$ to $25^\circ C$ increased the effectivity of eosine-5-maleimide more that four times (20% and 90% inhibition after 5 min at 24 nmoles inhibitor / mg protein). Significantly, at $25^\circ C$ eosine-5-maleimide remains impermeable to the inner mitochondrial membrane.

Results of experiments referred to here, in addition to the demonstration that eosine-5-maleimide is an impermeant membrane probe, show that eosine-5-maleimide-reactive SH groups associated with Pi/H^+ symport and ADP/ATP antiport systems are located on opposite sides of the inner mitochondrial membrane. The SH groups of Pi/H^+ symporter labeled with eosine-5-maleimide are essential for transport activity and are located on the cytosolic side of the membrane. They are identical with those attacked by N-ethylmaleimide and organic mercurials. In contrast, the eosine-5-maleimide-reactive SH group(s) of ADP/ATP carrier is located on the matrix surface of the inner membrane and is most probably different from SH groups which react with N-ethylmaleimide.

REFERENCES
1. Fonyo, A. (1978) J.Bioenerg.Biomembr. 10, 171-194.
2. Hackenberg, H. and Klingenberg, M. (1980) Biochemistry 19, 548-555.
3. Aquilla, H., Misra, H., Eultz, M. and Klingenberg, M. (1982) Hoppe-Seyler's Z.Physiol.Chem. 363, 345-349.
4. Griffith, D.G., Partis. M.D., Sharp, R.W. and Beechey, R.B. (1981) FEBS Lett. 134, 261-263.
5. Houštěk, J., Bertoli, E., Stipani, I., Pavelka, S., Megli, F.M. and Palmieri, F. (1983) FEBS Lett. 154, 185-190.
6. Müller, M., Krebs, J.J.R., Cherry, R.J. and Kawato, S. (1982) J.Biol.Chem. 257, 1117-1120.

THE NONIDENTITY OF PROTON AND CHLORIDE CONDUCTING PATHWAYS IN BROWN ADIPOSE TISSUE (BAT) MITOCHONDRIA

J.Kopecký, F.Guerrieri[+], P.Ježek, Z.Drahota and J.Houštěk
Institute of Physiology, Czechoslovak Academy of Sciences,
142 20 Prague, Czechoslovakia and [+]Institute of Biological
Chemistry, University of Bari, 70124 Bari, Italy.

Thermogenesis-facilitating, regulatable H^+-conductance
of BAT mitochondria, independent of H^+-ATPase function,
is related to the specific 32K uncoupling protein which
is also involved in the high conductance for Cl^- and
some other anions (1). While both the H^+- and Cl^--conduc-
tance are inhibited by purine nucleotides (2), only the
former process is suppressed by the removal of FFA (2).
FFA probably serve as an acute regulator of energy dissipation (3). According to (2), Cl^--transport competes
with H^+-transport, thus indicating a common channel. Recently, a decoupling between the two processes by FFA was
suggested (4). However, the relationship between the two
transport activities remains unclear.

In this report, attempts were made to follow simultaneously the transport of H^+ (pH measurements) and of anion
(swelling) in BAT mitochondria using valinomycin-induced
potassium diffusion potential as a driving force. As
shown in Fig.1, valinomycin induced in mitochondria suspended in KCl medium not only Cl^--uptake, as proposed in
(2), but also H^+-extrusion. The H^+-extrusion was signifi-

Fig.1

A - 150mM KCl
B - 110mM K_2SO_4

A,B - pH 7.0,
8 uM rotenone,
1 uM antimycin,
2.5 uM oligomy-
cin, 1mg prot./
ml, 25°C

H^+- pH of the
medium, L.S. -
light scatte-
ring, Val - 1 ug/
ml.

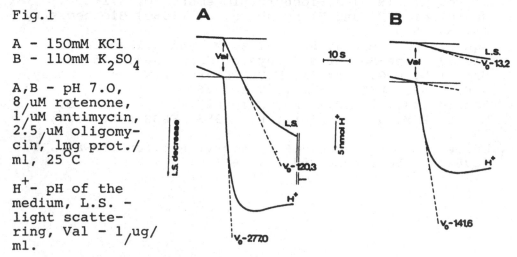

cantly more intensive (Vo - 277nmol/min/mg prot.) than Cl^- -uptake under these conditions (70nmol/min/mg prot.- ref. 2). The time course of the two processes was different (Fig.1). Based on the T1/2 values (H^+- 2s; Cl^- -20s), the H^+-extrusion was completed during a ten times shorter time period than the parallel Cl^--uptake. When Cl^- was replaced with SO_4, valinomycin induced an analogous H^+- -extrusion which was not accompanied by anion uptake (Fig.1). The rate of this H^+-extrusion (Vo - 141nmol/min/ /mg prot.) was lower than that during a parallel anion uptake. The H^+-extrusion was equally inhibited by GDP (maximal 90% inhibition at 200 uM) or by BSA (maximal 75% inhibition at 2mg/ml) in both types of media indicating that the sensitivity of H^+-transport is not modulated by a parallel anion transport. The titration of GDP-induced inhibition of H^+ extrusion and of Cl^- -uptake revealed that anion transport is more sensitive than H^+-transport (apparent Ki 2.2 and 6.0 uM, respectively).

The results presented here show that transport of H^+ and anion related to BAT uncoupling protein are independent processes. Based on the rates of H^+-transport and its sensitivity to GDP and BSA in the two types of media it is concluded that H^+-transport and anion transport are not competitive. Moreover, as a difference in the rate of H^+-extrusion in the two media exists even in the presence of BSA (equal inhibition), it is not necessary to assume any decoupling by FFA (4). Therefore, it is suggested that H^+- and Cl^--conducting pathways of BAT mitochondria are formed by independent transport entities which could be represented either by two transition states of a single channel or by two distinct channels.

REFERENCES
1. Nicholls, D.G. and Locke, R. (1984) Physiol.Rev. 64. 1-64.
2. Nicholls, D.G. and Lindberg, O. (1973) Eur.J.Biochem. 37, 523-530.
3. Locke, R.M., Rial, E., Scott, I.D. and Nicholls, D.G. (1982) Eur.J.Biochem. 129, 373-380.
4. Rial, E., Poustie, A. and Nicholls, D.G. (1983) Eur. J.Biochem. 137, 197-203.

PARTIAL PURIFICATION AND RECONSTITUTION OF THE ASPARTATE/GLUTAMATE CARRIER FROM MITOCHONDRIA

R. Krämer

Institut für Physikalische Biochemie, Universität München, Goethestrasse 33, 8000 München 2, FRG

The aspartate/glutamate exchange is an integral part of the malate-aspartate cycle, which facilitates the transport of reducing equivalents from the cytosolic to the mitochondrial compartment. Numerous reports on kinetics and regulation of the aspartate/glutamate carrier have been published (for review see ref. 1).

Isolation and characterization of this transport protein has been hampered by the lack of a specific inhibitor. Therefore, in the present study, functional reconstitution was used as a tool to identify the isolated aspartate/glutamate carrier. Reconstitution can be achieved either by detergent removal from cholate-solubilized proteins or by direct incorporation into preformed liposomes with an additional freeze-thaw-sonication procedure.

The partial purification consists of a preextraction of beef-heart mitochondria with tridecyl-decaoxyethylene-ether and solubilization with dodecyl-octaoxyethylene-ether in the presence of high salt concentrations. The high speed supernatant is then fractionated on hydroxylapatite.

By this procedure the specific transport activity is increased about 30-fold as compared to the original mitochondrial extract. However, the preparation still contains a high amount of co-purified ADP/ATP carrier protein.

In the liposomal system the reconstituted aspartate/glutamate transport can be measured giving reliable kinetics; it shows specificity for L-aspartate and L-glutamate and is not inhibited by a variety of inhibitors known for other mitochondrial transport systems. Furthermore, the aspartate/glutamate transport in the reconstituted system can be shown to be regulated by the same factors ($\Delta\Psi$ and ΔpH) as determined in mitochondria (2,3). Apparently the aspartate/glutamate carrier can be isolated in functionally active form and can be reconstituted by this method, which therefore represents a suitable tool for further purification and characterization of this important transport system.

(1) La Noue, K.F. and Schoolwerth, A.C. (1979) Ann.Rev.Biochem.48, 871-922
(2) La Noue, K.F., Duszynski, J., Watts,J.A. and McKee,E. (1979) Arch. Biochem.Biophys. 195, 578-590
(3) Murphy,E., Coll,K.E., Viale,R.O., Tischler,M.E. and Williamson,J.R. (1979) J.Biol.Chem. 254, 8369-8376

MEASUREMENT OF THE ELASTICITY COEFFICIENT OF THE ADENINE NUCLEOTIDE
TRANSLOCATOR TOWARDS THE EXTRAMITOCHONDRIAL ATP/ADP RATIO
P.J.A.M. Plomp, C.W.T. van Roermund, A.K. Groen, R.J.A. Wanders,
A.J. Verhoeven and J.M. Tager Lab. of Biochemistry, Un. of Amsterdam

The distribution of control among the enzymes in a metabolic pathway is determined by the relative properties of the enzymes participating in the pathway. In quantitative control analysis (1,2) the relationship between the properties of the individual enzymes in the pathway and the flux control coefficients of those enzymes is given by the
connectivity theorem (1). In this theorem, the enzymic properties are
represented by the elasticity coefficients. An enzyme has an elasticity coefficient for each factor (substrate, product or effector) that
influences its rate. Whereas flux control coefficients are simply a
measure of the distribution of control, elasticity coefficients provide a means of understanding the mechanisms underlying the distribution of control. Unfortunately measurement of the elasticity coefficient of an enzyme towards a factor is not easy, since all other factors influencing the activity of the enzyme must be kept constant. For
this reason the elasticity coefficient of the adenine nucleotide translocator towards the extramitochondrial ATP/ADP ratio ($\varepsilon^{tr}_{(T/D)_{ex}}$) was
calculated from flux control coefficients (3).

We have now devised a method for measuring ($\varepsilon^{tr}_{(T/D)_{ex}}$) directly.
This method is based on the assumption that at one rate of
oxygen uptake there is only one value for the intramitochondrial ATP/
ADP ratio and membrane potential (4). Two conditions with different
fluxes through the adenine nucleotide translocator and different extramitochondrial ATP/ADP ratios, but with the same rate of oxygen consumption are compared. As indicated above, the other two factors influencing the rate of the translocator, i.e. (ATP/ADP)$_{in}$ and $\Delta\psi$, will be
equal. In isolated rat-liver mitochondria the two conditions can be
obtained by stimulating oxygen uptake, either by an extramitochondrial
ATP consuming process (glucose + hexokinase) or by simultaneous ATP
utilisation inside the mitochondria (citrulline synthesis) and outside
the mitochondria (glucose + hexokinase), such that the rate of oxygen
consumption is the same under both conditions. The values for $\varepsilon^{tr}_{(T/D)_{ex}}$
obtained with this method at three different rates of respira-
tion are shown in Table 1; they are similar to the calculated values
under comparable conditions (3).

Also given in Table 1 is the effect of a change in (ATP/ADP)$_{ex}$ on
O$_2$ consumption, i.e. the "overall" elasticity coefficient of O$_2$ uptake
towards (ATP/ADP)$_{ex}$ ($\varepsilon^{ox}_{(T/D)_{ex}}$). The "overall" elasticity coefficient
is the lower limit for the elasticity coefficient of the transloca-
tor towards (ATP/ADP)$_{ex}$ (3). Especially at high rates of respiration

the difference between $\varepsilon^{tr}_{(T/D)_{ex}}$ and $\varepsilon^{ox}_{(T/D)_{ex}}$ is small, indicating that the response of respiration towards (ATP/ADP)$_{ex}$ is mainly determined by the adenine nucleotide translocator. When the "overall" elasticity coefficient is determined in intact hepatocytes, values >2 are obtained. Thus the elasticity coefficient of the adenine nucleotide translocator towards (ATP/ADP)$_{ex}$ in intact rat hepatocytes must be much higher than in isolated mitochondria, suggesting that the adenine nucleotide translocator operates much closer to equilibrium in isolated hepatocytes than in isolated mitochondria.

Table 1. The elasticity coefficient of the adenine nucleotide translocator towards the extramitochondrial ATP/ADP ratio.

Rate of respiration (% of State 3)	$\varepsilon^{tr}_{(T/D)_{ex}}$	$\varepsilon^{ox}_{(T/D)_{ex}}$
50	1.70	0.82
60	1.27	0.69
70	0.77	0.59

Isolated rat-liver mitochondria (approx. 1 mg protein/ml) were incubated at 25 ^{o}C in glass stoppered vessels in a medium containing 100 mM KCl, 25 mM Mops, 10 mM potassium phosphate, 10 mM $MgCl_2$, 1 mM EGTA, 10 mM succinate, 1 mM malate, 0.5 mM ATP, 17 mM HCO_3^- and 0.5 mg/ml rotenone. The pH was 7.0 and the gas phase 95% O_2 and 5% CO_2. After 2 min, samples were taken at 1 min intervals to measure the rate of glucose-6-phosphate synthesis. After 4 min, a sample was taken to measure the rate of O_2 uptake in an oxygraph vessel equipped with a Clark-type electrode.

References
1. Kacser, H. and Burns, J.A. (1973) in: Rate Control of Biological Processes (D.D. Davies, ed.), pp. 65-104, Cambridge University Press London.
2. Heinrich, R. and Rapoport, T.A. (1974) Eur. J. Biochem. 42, 89-95.
3. Wanders, R.J.A., Groen, A.K., Van Roermund, C.W.T. and Tager, J.M. (1984) Eur. J. Biochem., in press.
4. Duszynski, J., Bogucka, K., Letko, G., Küster, U., Kunz, W. and Wojtczak, L. (1981) Biochim. Biophys. Acta 637, 217-223.

Laboratory of Biochemistry, B.C.P. Jansen Institute, University of Amsterdam, P.O. Box 20151, 1000 HD Amsterdam, The Netherlands

IMMUNOLOGICAL IDENTIFICATION OF THE PHOSPHATE TRANSPORT PROTEIN AND
THE ADP/ATP CARRIER IN TOTAL MITOCHONDRIAL PROTEINS FROM DIFFERENT
TISSUES

U.B.Rasmussen and H.Wohlrab
Boston Biomedical Research Institute, Boston, MA 02114, USA, and
Department of Biological Chemistry, Harvard Medical School, Boston,
MA 02115.

Antisera were raised in rabbits against the purified beef heart
mitochondrial phosphate transport protein (PTP) and against the
purified ADP/ATP carrier (AC). Both proteins were purified as
described in (1).

Mitochondria were prepared from beef, pig and rat hearts and
from rat liver and blowfly flight muscle. The mitochondrial
proteins were separated by SDS-polyacrylamide gel electrophoresis
(SDS-PAGE) and then immediately electro-blotted onto
nitrocellulose. The purified PTP and AC were also blotted. The
nitrocellulose sheets were stained with two different antisera
against PTP and one antiserum against AC (2,3).

The results show that one PTP-antiserum stains PTP specifically
from all tissues but the electrophoretic mobilities (all in the
30-35kDa region) are different. The second PTP-antiserum does only
stain PTP from the heart tissues. The AC-antiserum stains AC
specifically from all tissues except flight muscle, but differences
in mobilities are seen. Species as well as organ differences are
suggested. No cross-reactivity is seen between PTP-antiserum and AC
and vice versa (Fig.1), although these two membrane proteins are
similar in molecular weight (4), amino acid composition (1,5),
behavior in various purification steps (1,4,5), and both possess
phosphate (organic vs. inorganic) binding sites. Both PTP and AC
react with the transport inhibitor N-ethylmaleimide, which causes a
change in mobility, but does not affect antibody binding. Other
weakly stained bands are seen, which indicates that there are
antigenic sites in other mitochondrial proteins similar to either
PTP or AC.

The advantage of the method used here is, that the antigens can
be identified from among a mixture of all mitochondrial proteins on
the basis of their 1) relative mobilities in SDS-PAGE and 2) cross
reactivity with various antisera. The method also permits SDS-PAGE
of total mitochondria immediately after SDS solubilization,
minimizing or eliminating modification of proteins (proteolysis)
other than unfolding due to SDS.

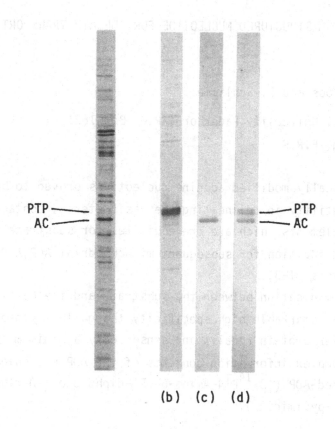

Fig.1. Total beef heart mitochondrial proteins (0.1nmol cytochrome b/ml), electro-blotted onto nitrocellulose, were stained with (a) Amido Black, (b) PTP-antiserum, (c) AC-antiserum, or (d) PTP-antiserum and AC-antiserum. The concentration of antiserum was 15ul/ml.

1. Kolbe,H.V.J., Costello,D., Wong,A., Lu,R.C. and Wohlrab,H. (1984) J.Biol.Chem. (in press).
2. Towbin,H., Staehelin,T. and Gordon,J. (1979) Proc.Natl.Acad.Sci. USA. 76, 4350-4354.
3. Gershoni,J. and Palade,G.E. (1983) Anal.Biochem. 131, 1-15.
4. Wohlrab,H. (1980) J.Biol.Chem. 255, 8170-8173.
5. Aquila,H., Misra,D., Eulitz,M. and Klingenberg,M. (1982) Hoppe-Seyler's Z.Physiol.Chem. 363,345-349.

MS-ADP : A MINIMUM STRUCTURED NUCLEOTIDE FOR ADP,ATP TRANSPORT IN
MITOCHONDRIA

H. Renz, K.-S. Boos and E. Schlimme

Lab. Biol. Chem., University Paderborn, P.O. Box 1621,
D-4790 Paderborn, F.R.G.

The use of chemically modified adenine nucleotides proved to be
of exceptional utility in delineating the basic steric, contact
and structural elements which are prerequisites for substrate recognition and in addition for subsequent mitochondrial ADP,ATP
transport catalysis (1-3).
The chemical communication between the substrate and the carrier
system ensures a remarkable high specificity though the cytosolic
orientated carrier protein recalls and senses only a minimum structure from the complex information contents of the ADP molecule, i.e.
Minimum Structured-ADP ($[2-^{14}C]$ 4-Amino-6-(5'-diphospho-β-D-ribo-
furanosylmethyl)-pyrimidin).

The postulated minimum structure (3) has now been synthesized chemically as the ^{14}C-labeled compound (4) and turns out to substitute ADP in mitochondrial ADP, ATP transport, although the translocation rate of MS-ADP is reduced to about one half compared to ADP under saturation conditions. The binding -and translocation properties of MS-ADP prove our postulate of a three point substrate attachment at the cytosolic active center (3,5).

The methylene-bridge in MS-ADP was introduced to gain the required distances between MS-ADP and the complementary carrier protein binding sites and to create a certain flexibility in MS-ADP allowing the adoption of the substrate over-all conformation.

Moreover, the results with MS-ADP prove the reliability of our strategy of chemical substrate modification (1-3), which - in combination with findings on metallochromic pseudosubstrates (6,7) - led to the proposal of a new model for metalloprotein-catalyzed ADP, ATP transport (5).

1) Schlimme,E., Boos,K.-S., Bojanovski,D. & Lüstorff,J. (1977) Angew.Chem. 89, 717

2) Boos,K.-S. & Schlimme,E. (1979) Biochemistry 18, 5304

3) Schlimme,E., Boos,K.-S. & de Groot,E.J. (1980) Biochemistry 19, 5569

4) Renz,H. & Schlimme,E. (1984) Liebigs Ann.Chem., submitted

5) Boos,K.-S. & Schlimme,E. (1983) FEBS-Lett. 160, 11

6) Boos,K.-S. (1982) Biochim.Biophys.Acta 693, 68

7) Boos,K.-S. (1984) Habilitationsschrift Universität Paderborn

SYNCHRONOUS INCREASE IN GDP-BINDING CAPACITY AND NUCLEOTIDE- SENSITIVE
PROTON CONDUCTANCE DURING COLD-ADAPTATION IN B.A.T. MITOCHONDRIA

Eduardo Rial and David G. Nicholls
Neurochemistry Laboratory, Department of Psychiatry, Ninewells Medical
School, Dundee University, Dundee DD1 9SY, Scotland, UK

The inner membrane of mitochondria from thermogenically active
brown fat possesses a 32 kDa integral protein (1) which functions as a
regulatable proton short-circuit capable of uncoupling respiration
from ATP synthesis (reviewed in 2). This uncoupling protein can be
inhibited in vitro by purine nucleotides (3). Low concentrations of
free fatty acids override this low conductance state (4) consistent
with being the acute physiological regulators of the conductance.

The protein is induced during cold adaptation, but the only animal
in which the time-course of induction of the protein has been followed
in any detail is the rat. The laboratory of Himms-Hagen (5) reported a
rapid increase in GDP-binding in the first hours of cold- adaptation
unaccompanied by a detectable increase in the 30-35kDa range of
proteins on SDS-gels and have proposed that pre-existing but latent
uncoupling protein in the inner mitochondrial membranes becomes
unmasked during the first few hours of cold-exposure, revealing
additional nucleotide binding sites without the necessity for protein
synthesis. In this paper we examine the time-course for the induction
of the uncoupling pathway in cold-adapting guinea-pigs. The protein
has been quantified from the capacity for high affinity binding of GDP
to the intact mitochondria, and has been compared with the functional
parameters diagnostic of the protein, namely the nucleotide-sensitive
proton conductance, and the sensitivity to uncoupling by low concentrations of fatty acids. A monophasic increase in nucleotide titre was
observed, with no evidence of an early "unmasking" of pre-existing
nucleotide binding sites (Fig. 1). The nucleotide-sensitive
conductance increased in precise synchrony with the nucleotide binding
capacity (Fig. 1A), but mitochondria from new-born animals, and those
acutely cold-adapted, showed anomalously low sensitivities to fatty
acid uncoupling (Fig. 1B).

This work was supported in part by a project grant from the MRC. E.R.
was supported by a grant from the Spanish Direccion General de
Politica Cientifica.

1. Heaton, G.M., Wagenvoord, R.J., Kemp, A. & Nicholls, D.G. (1978)
Eur. J. Biochem. 82, 515-521.
2. Nicholls, D.G. & Locke, R.M. (1984) Physiol. Rev.64, 1-69.
3. Nicholls, D.G. (1974) Eur. J. Biochem. 49, 573-583.

4. Locke, R.M., Rial, E., Scott, I.D. & Nicholls, D.G. (1982a) Eur. J. Biochem. 129, 373-380.
5. Desautels, M., Zaror-Behrens, G. & Himms-Hagen, J. (1978) Can. J. Biochem. 56, 378-383.

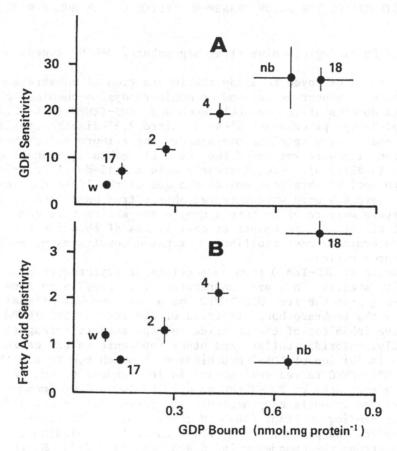

Correlation between GDP binding capacity, GDP-sensitive proton conductance and fatty-acid induced proton conductance
Symbols: nb, new-born animals; wa, warm-adapted animals; 17hr, 2d, 4d, 18d; animals cold-adapted for 17hr, 2 days, 4 days and 18 days respectively. GDP-sensitive conductance is defined as the decrease in conductance when GDP is added to depleted of fatty acids, and fatty-acid induced conductance as the increase when palmitate is added to achieve a 2:1 molar ratio to albumin.

SPINLABEL STUDIES ON THE ANION TRANSPORT SYSTEM OF THE HUMAN RED CELL
MEMBRANE
K. Schnell
Institut für Physiologie, Universität Regensburg, D-8400 Regensburg

Spinlabels can be employed to study the interaction of substrate anions
with the anion transport system and to monitor physico-chemical changes
of the lipid domains of the red cell membrane. NDS-TEMPO (N-4-(2,2,6,6
-tetramethyl-1-oxyl)piperidinyl-N'-4-(4'-nitro-2,2'-disulfonatostilbene)
thiourea, K-salt) is a specific non-penetrating spinprobe for the inorganic anion transport system of the red cell membrane (Schnell et
al. BBA 732 (1983) 266). 5-doxyl-stearic acid and 15-doxyl-stearic acid
were used to monitor physico-chemical changes of the lipid domains. All
experiments were executed with resealed ghosts from human red cells.
The ESR-spectra were resolved into a mobile and an immobile component
by spectral titration. The unidirectional fluxes of chloride and sulfate were determined under equilibrium exchange conditions by measuring the tracer efflux.
The ESR-spectra of NDS-TEMPO from suspensions of erythrocyte ghosts
are composite spectra. They are constituted of a three sharp line
signal arising from the free NDS-TEMPO and a weak immobile signal
arising from the membrane bound fraction of NDS-TEMPO. NDS-TEMPO is
a competitive inhibitor of the chloride and the sulfate transport.
Concomitantly, chloride, sulfate and other inorganic anions cause a
competitive inhibition of NDS-TEMPO binding. From pH 6.0 to 8.0 the
binding of NDS-TEMPO to red cell ghosts is independent of pH. The apparent binding constants of NDS-TEMPO amounts to 1.3 ± 0.2 µM (mean $\pm$ S.D.).
The shape of the immobile component of the ESR-spectra is not affected.
The chloride binding constant (16 ± 3 mM 20°C) from the NDS-TEMPO binding studies was independent of pH, while the sulfate binding constants
exhibited a strong pH-dependence (pH 6.4:30 mM, pH 7.2:75 mM, pH 8.0:
30 mM). The maximum number of NDS-TEMPO binding sites per cell was in
the range of $9.5 \cdot 10^5 - 1.1 \cdot 10^6$ and was not significantly affected by pH.
In contrast, the chloride and the sulfate fluxes exhibit a strong pH-
dependence of the maximal flux with pH maxima at pH 7.5 (0°C) and pH
6.4 (25°C). The apparent chloride half-saturation constants from the
flux concentration curves increase from 10 mM at pH 6.2 to 45 mM at
pH 8.5 (Dalmark, J.Physiol. 250 (1975) 39) and the apparent sulfate
half-saturation constants increase from 30 mM at pH 6.4 to 125 mM at
pH 8.5 (25°C). (Schnell et al., J.Membrane Biol. 30 (1977) 319).

Covalently binding, irreversible inhibitors of the anion transport such
as DIDS (4,4'-diisothiocyanato-2,2'disulfonatostilbene) or DNFB (1-
fluoro-2,4-dinitrobenzene) completely block the binding of NDS-TEMPO
to red cell ghosts. Competitive inhibitors of the anion transport (salicylate, benzoate, 2,4-dinitrophenol) cause a competitive inhibition of

NDS-TEMPO binding. Noncompetitive inhibitors (phlorhizin, dipyridamole) show either a noncompetitive (phlorhizin) or a competitive (dipyrida-mole) inhibition of NDS-TEMPO binding. In addition, phlorhizin causes a shifting of the outer extrema of the immobile component of the ESR-spectra which is not observed in the presence of dipyridamole.

The ESR-spectra of 5-doxyl-stearic acid and 15-doxyl stearic acid la-belled red cell ghosts show an almost complete immobilization of the lipid spinlabels. The addition of DIDS, DNDS, phlorhizin, dipyridamole or salicylate did not produce any shape change of the ESR-spectra. The chloride and the sulfate transport across the red cell membrane were not inhibited by the lipid-spinlabels.

The following conclusions can be drawn from our experiments: NDS-TEMPO binds to the anion transport site. The strong pH-dependence of the chloride and the sulfate flux is due to an activation or an inactiva-tion of the anion transport system rather than to a modification of the anion binding to the anion transport sites. Competitive inhibitors of the anion transport bind to the anion transport site. Noncompetitive anions bind to the anion transport protein and do not cause a physico-chemical change of the lipid domains of the cell membrane.

Supported by the DEUTSCHE FORSCHUNGSGEMEINSCHAFT

THE BIOCHEMICAL CHARACTERISATION OF AUTO-ANTIBODIES AGAINST THE ADENINE-NUCLEOTIDE-TRANSLOCATOR IN PRIMARY BILIARY CIRRHOSIS (PBC)

H.-P. Schultheiss", P. L. Schwimmbeck", P. A. Berg'
"Dept. of Medicine, Klinikum Großhadern, University of Munich, FRG
'Dept. of Medicine, University of Tübingen, FRG

Primary biliary cirrhosis is a progressive destructive liver disease of unknown cause (1). Besides the clinical and histological findings it is characterised by the presence of anti-mitochondrial antibodies (AB). In earlier examinations we identifed the adenine-nucleotide-translocator (ANT) as an auto-antigen in PBC (2). Further studies showed that the ANT is characterized immunochemically by an organ- and conformation-specificity (3,4). The ANT, an intrinsic, hydrophobic protein of the inner mitochondrial membrane, facilitates the transport of ATP to the cytosol respectively of ADP back to the inner mitochondrial space. According to its conformation state to the cytosolic side (c-conf.) or matrix side (m-conf.) the ANT can be fixed by the specific ligands carboxyatractylate (CAT $\hat{=}$ c-conf.) respectively bongkrekate (BKA $\hat{=}$ m-conf.).

We investigated the serum of 13 patients with proven PBC, whether AB have an influence on the function of the ANT and/or the binding of the specific ligands CAT and BKA. An indirect micro solid phase radioimmunoassay was established. It was performed on microtiter plates with iodinated protein A as a second antibody. Moreover, binding-studies with radioactiv labelled CAT and BKA and measurements of the nucleotide transport (NT) with the inhibitor-stop-method combined with the back exchange were performed.

Results: In PBC different AB-populations could be distinguished. They are characterised as follows:

1. The AB show a high organ- and conformation-specificity.
2. The binding of CAT was decreased by the AB.
3. The conformational change from "c"- to "m"-conf. was blocked by the AB as indicated by a decreased binding of BKA.
4. 6 of 13 PBC-sera inhibited the NT in liver mitochondria, while no inhibition of the NT was seen in heart or kidney mitochondria.
5. Several sera caused an increase of the nucleotide leakage. This could be prevented completely by the addition of CAT or BKA (fig 1).

The AB 4. and 5. were observed simultaneously in the same serum. However, the different functional activities of the AB were only seen at different serum dilutions (fig. 1). This concentration dependent inhibition of one AB by the other implies that both AB bind to overlapping antigenic determinants of the "translocation apparatus".

In conclusion, these results show the existence of different, functional active and high specific AB-populations against the ANT in PBC. The different concentrations and/or affinities of the AB give a good explanation for the supposed missing correlation between the AB-activity and the clinical degree of the disease.

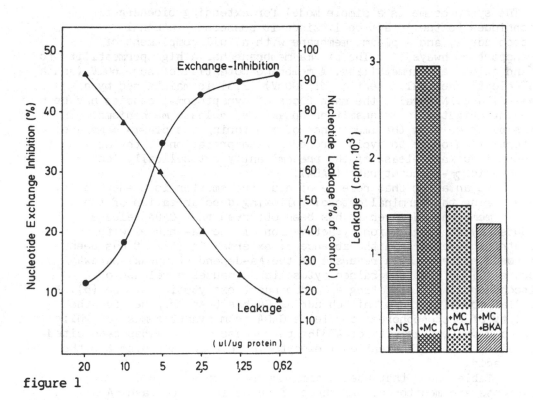

figure 1

References:
1. Thomas HC. and Epstein O. 1980. Springer Semin. Immunopathol. 3:375
2. Schultheiss H.-P., Berg P.A. and Klingenberg M. 1983. Clin. Exp. Immunol. 54,648
3. Schultheiss H.-P. and Klingenberg M. 1982. Second European Bioenergetic Conference, short reports, 471
4. Schultheiss H.-P. and Klingenberg M. 1984. Eur. J. Biochem., in press

SYNAPTOSOMAL BIOENERGETICS: PRIMARY ROLE OF THE PLASMA MEMBRANE
POTENTIAL IN REGULATING AMINO ACID TRANSMITTER TRANSPORT.

Talvinder S. Sihra and David G. Nicholls
Neurochemistry Laboratory, Department of Psychiatry, Ninewells Medical
School, Dundee University, Dundee DD1 9SY, Scotland, UK

The synaptosome is a simple model for extending bioenergetic
techniques to the intact cell, since it possesses internal
mitochondria, and a plasma membrane with a full complement of
transport pathways (1). The plasma membrane has a high permeability to
K^+ and a low Na^+-permeability. A membrane potential of some 60mV (and a
Na^+-electrochemical potential of 100mV) (2,3) is maintained by a
Na+K-ATPase. Naturally the main focus of synaptosomal studies has been
the investigation of transmitter uptake and release mechanisms. In
this paper we show the importance of monitoring the plasma membrane
potential in order to avoid erroneous interpretation of the action of
agents inducing release of the predominantly cytosolically located
amino acid γ-aminobutyrate (GABA).

It is an axiom that release of neurotransmitter is Ca-dependent,
Ca entering the terminal cytosol following depolarization of the
plasma membrane. However it has been observed that GABA release
induced by depolarization by activation of the Na-channel with
veratridine, occurs in the absence of external Ca (4). It has been
proposed either that a reversal of the Na-dependent uptake pathway
occurs (5) or that the raised cytosolic Na causes a release of
mitochondrial Ca initiating a Ca-dependent exocytotic release (6). A
similar mobilization of mitochondrial Ca has been proposed for the
ability of protonophores to release GABA from synaptosomes (6) while
the Ca-dependent ability of A23187 to release GABA (7) has been cited
as evidence for a conventional exocytotic release mechanism for the
amino acid.

The Table shows that when plasma membrane potentials and Ca
movements are monitored, all three of these agents release GABA as a
simple consequence of a lowered plasma membrane potential. Thus the
extent of GABA efflux induced by veratridine shows the same
correlation with the extent of plasma membrane depolarization in
synaptosomes with Ca-loaded mitochondria as with synaptosomes which
have been extensively depleted of Ca by incubation with EGTA for
35min. The release of GABA induced by FCCP again shows a dependency on
plasma membrane potential, but is independent of the Ca content of the
synaptosome. The release of GABA induced by ionophore A23187 is
dependent on the presence of external Ca, but since the ionophore
depolarizes the plasma membrane only in the presence of the cation,
once more the GABA release can be ascribed to plasma membrane
depolarization and a reversal of the Na-dependent uptake pathway,

rather than a manifestation of an exocytotic, Ca-dependent mechanism.

It is concluded that it is essential to monitor plasma membrane potentials in experiments in which amino acid efflux from synaptosomes is induced.

1. Nicholls, D.G. & Akerman, K.E.O. (1981) Phil. Trans. Soc. Lond. B. 296, 115-122.
2. Scott, I.D. & Nicholls, D.G. (1979) Biochem. J. 186, 21-33.
3. Åkerman, K.E.O. & Nicholls, D.G. (1981) Eur. J. Biochem. 117, 491-497.
4. Benjamin, A.M. & Quastel, J.H. (1972) Biochem. J. 128, 631-646.
5. Haycock, J.W., Levy, W.B., Denner, L.A. & Cotman, C.W. (1978) J. Neurochem. 30, 1113-1125.
6. Sandoval, M.E. (1980) Brain Res. 181, 357-367.
7. Redburn, D.A., Shelton, D. & Cotman, C.W. (1976) J. Neurochem. 26, 297-303.

TSS is supported by the SERC.

Synaptosomes were preincubated with ^{45}Ca, ^{86}Rb and ^{3}H-GABA in order to allow the simultaneous determination of plasma membrane potential, Ca-content and ^{3}H-GABA distribution. Plasma membrane potential was estimated from the ^{86}Rb distribution inserted into the Nernst equation (2,3). +Ca, incubation performed in the presence of 1.3mM Ca; +EGTA incubation in the presence of EGTA for at least 35min.

Additions		$\Delta\psi$(mV)	Ca nmol/mg	GABA in/out
Control	+Ca	56	9	251
	+EGTA	51	<0.2	211
100µM-veratridine	+Ca	6	18	53
	+EGTA	0	<0.2	45
100µM-FCCP	+Ca	40	4	120
	+EGTA	40	<0.2	120
7.5µM-A23187	+Ca	36	-	150
	+EGTA	65	-	210

THE MOVEMENT OF SULFOBROMOPHTHALEIN FOLLOWED DIRECTLY IN ISOLATED
PLASMA MEMBRANE VESICLES : IMPLICATION OF BILITRANSLOCASE

G.L. Sottocasa,G.Baldini, S.Passamonti, G.C. Lunazzi, and C.
Tiribelli.* Istituto di Chimica Biologica and * Patologia Medica,
University of Trieste, Trieste, Italy.

Liver can efficiently perform the extraction from plasma and the
excretion into the bile of a number of organic anions. The process
involves obligatorily a specific translocator isolated by us from
liver plasma membrane, named bilitranslocase.(1,2). The
thermodynamics ad the intimate mechanism by which bilirubin and other
organic anions are transfered from plasma to the hepatocyte cytoplasm
and therefrom to bile is still obscure. In order to clarify these
important metabolic steps, sulfobromophthalein, (BSP) instead of
bilirubin was used. BSP belongs to a class of organic anions able to
inhibit bilirubin uptake by the liver. Sulfobromophthalein as well as
bilirubin, is efficiently bound by albumin in plasma and by ligandin
in hepatic cytoplasm. Moreover, BSP has another metabolic step in
common with bilirubin at the plasma membrane level, namely the ability
to bind to and to be transported by bilitranslocase. Although there
are many similarities between bilirubin and BSP in the uptake
mechanism, the exogenous dye has practical advantages over the
physiological one. BSP is soluble in water and behaves as a pH
indicator. Above pH 7.0 a clear absorption peak at 580 nm appears
whose extintion coefficient is sharply dependent on pH in the alkaline
region. Taking advantage of this property it is possible to follow
directly by means of dual wavelength recording spectrophotometer, BSP
movements across sealed vesicles whose internal aqueous medium has a
pH value different from outside. This has been done successfully with
liposomes (3) where BSP translocation could be reconstituted in vitro
by addition of purified bilitranslocase. Under these conditions BSP
tranport was found to be electrogenic, namely it was greatly
accelerated by a membrane potential . In order to get insight into a
more physiological model for organic anion translocation, we decided
to attempt the kinetic study of BSP movements in rat liver plasma
membrane vesicles. The results of these study carried out on two
different types of plasma membrane preparations allowed to conclude

that a specific electrogenic dye movement could be found only with plasma membrane vesicles. Other subcellular organelles were unable of such a function. The specificity was confirmed also by subfractionation studies of the microsomal fraction (known to contain a large portion of plasma membranes) by density gradient centrifugation. In this way a subfraction could be isolated at the interface between 20% (w/v) and 39 % sucrose solutions in which BSP transporting activity was co-purified with Na-K-ATPase and, to a certain extent, also with 5' nucleotidase. As expected for a bilitranslocase-mediated process, the velocity measured was reduced by competitors such as Rifamycin SV and Nicotinate. The Ki values found were 20 μM and 50 nM for the two anions respectively, in excellent agreement with previous studies carried out on the isolated protein. Even more convincingly, a strong inhibition of the BSP-transporting capacity of the vesicles could be obtained by pretreatment with affinity-chromatography purified monovalent (Fab) anti-bilitranslocase antibodies. It is concluded that also in the plasma membrane vesicles BSP translocation occurs by an electrogenic process which involves bilitranslocase. This conclusion implies that an appropriate concentration gradient of the organic anion should be maintained across the sinusoidal plasma membrane to compensate for the negative inside potential present in vivo. On the other hand if the excretion of the modified dye (glucorono- or glutathione- conjugated) occurs also electrogenically, it is well possible that the membrane potential at the canalicular pole of the cell may be used as the driving force for the overall transhepatic process. These possibilities are currently under investigation in our laboratory.

1) Tiribelli,C., Lunazzi, G.C., Luciani, M., Panfili, E., Gazzin, B., Liut,G.F., Sandri, G., and Sottocasa, G.L., (1978), Biochim. Biophys. Acta, 532, 105-112.
2) Lunazzi,G.G., Tiribelli,C., Gazzin, B. and Sottocasa G.L., (1982) Biochim. Biophys. Acta, 685, 117-122.
3) Sottocasa, G.L., Baldini, G., Sandri, G., Lunazzi, G.C. and Tiribelli,C., (1982) Biochim. Biophys. Acta, 685, 123-128.

THE TRANSPORT OF SPERMINE ACROSS THE INNER MEMBRANE OF LIVER MITO-
CHONDRIA AND ITS INFLUENCE ON THE TRANSPORT OF PHOSPHATE AND OTHER
ANIONS.

A.Toninello, F.Di Lisa, D.Siliprandi and N.Siliprandi
Istituto di Chimica Biologica dell'Università e Centro Studio Fisiologia Mitocondriale CNR - Via F. Marzolo, 3 - 35131 PADOVA (ITALY)

At physiological concentrations spermine and analogous polyamines prevent the loss of respiratory control in heat aged mitochondria and have a restorative effect on their phosphorylative capacity (1). Moreover spermine prevents the fall of transmembrane potential ($\Delta\psi$) induced by Ca^{2+} and Pi cycling and also fully restores
collapsed $\Delta\psi$ provided that ATP is added together (2). This action
is compatible with the assumption that spermine might act as a physiological factor preserving and controlling the native permeability properties of mitochondrial membrane.

In the present communication we report that spermine upon binding to the mitochondrial membrane is transported into the inner
space provided that mitochondria are energized and inorganic phosphate (Pi) is present in the incubation medium. The successive
addition of FCCP, or Antimycin A, induces a release of the spermine
taken up (Fig. 1). On its turn spermine facilitates Pi transport
and accumulation in energized liver mitochondria. Furthermore the
release of spermine induced by FCCP is associated to a parallel
release of the over accumulated Pi.

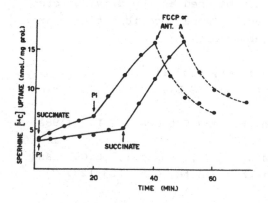

Fig. 1 - Spermine uptake by
rat liver mitochondria
(RLM). 1 mg RLM/ml suspended
at 20°C in the following medium: 200 mM sucrose, 10 mM
Hepes pH 7.4, 1.25 µM rotenone, 0.1 mM ^{14}C spermine. Additions: 5 mM succinate, 0.5
mM Pi, 0.1 µg/mg protein
FCCP or 1 µg/mg protein Antimycin A.

Addition of mersalyl prevents the uptake of both spermine and Pi. In the presence of Pi spermine also affects the transport of succinate into the mitochondria and peculiarly this polyamine is capable to remove the block of the transport of both succinate and adenine nucleotides induced by palmitoyl CoA (3) (Fig. 2). Also the oxidation of pyruvate and citrate, very likely owing to a facilitated transport into mitochondria, is enhanced by spermine.

Conclusions: 1. Spermine uptake by rat liver mitochondria is energy and Pi dependent; 2. Spermine transport and accumulation in mitochondria significantly enhances the uptake of Pi and other anions (succinate, citrate, pyruvate); 3. It is assumed that the positive action of spermine on Pi transport and accumulation is due either to a facilitated binding of the anion to the mitochondrial membrane or to an increased electropositivity created by spermine accumulation in the inner mitochondrial space.

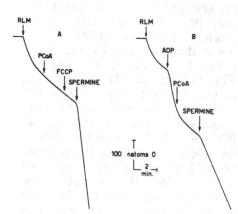

Fig. 2 – Effect of spermine on succinate oxidation. 2.5 mg RLM/ml were incubated at 20°C in the following medium: 200 mM sucrose, 10 mM Hepes pH 7.4, 5 mM succinate, 1.25 µM rotenone, 0.5 mM Pi. Additions: 50 µM palmitoylCoA in A and 10 µM in B; 0.1 mM spermine, 1 mM ADP.

References.

1) Phillips,J.E. and Chaffee,R.R.J. (1982) Biochem.Biophys.Res. Comm. 108, 174–181.
2) Toninello,A., Di Lisa,F., Siliprandi,D. and Siliprandi,N. (1984) in "Advances in Polyamines in Biomedical Science" (Caldarera C.M. and Bachrach V. eds.) pp. 31–36 CLUEB, Bologna.
3) Morel,F., Lauquin,G., Lunardi,J., Duszynski,J. and Vignais,P.V. (1974) FEBS Letters, 39, 133–138.

TOWARDS STRUCTURE AND FUNCTION OF THE MITOCHONDRIAL PHOSPHATE
TRANSPORT PROTEIN (PTP)

H. Wohlrab, U.B. Rasmussen, H.V.J. Kolbe, A. Collins, and
D. Costello.
Boston Biomedical Research Institute, Boston, MA 02114, and
Department of Biological Chemistry, Harvard Medical School, Boston,
MA 02115 USA

The mitochondrial phosphate transport protein (PTP) catalyzes
the net transport of inorganic phosphate across the inner
mitochondrial membrane. This is a function essential for steady
state oxidative phosphorylation as carried out by mitochondria in
the cell. We have purified the protein as a reconstitutively active
preparation from beef heart, rat heart, pig heart, blowfly flight
muscle, and rat liver (1). The protein from these different sources
has basically the same CNBr digestion pattern as analyzed by sodium
dodecylsulfate polyacrylamide gel electrophoresis (SDS-PAGE). While
small differences in their CNBr digests are difficult to identify,
the PTP's from these various sources do show definite small
differences in mobility in SDS-PAGE (about 34 kDa) (1,2). PTP's
from beef heart and pig heart migrate as a double band (α and β) in
our SDS gels (2), while PTP's from rat heart and rat liver migrate
as a single band (1). PTP from blowfly flight muscle appears to
migrate as two bands of practically identical mobilities. Since PTP
from rat liver has a CNBr digestion pattern very similar to that
from beef heart and assuming similar Coomassie Blue staining, we
have calculated the concentration of PTP in rat liver mitochondria
to be 0.28 nmol/mg mitochondrial protein. Using the Vmax determined
for phosphate transport in rat liver mitochondria by Coty and
Pedersen (3), we calculate a turnover number for PTP in rat liver
mitochondria of 750 min^{-1} (0°C)(1). The turnover number for PTP
from beef heart in our reconstituted system (net transport, ΔpH
driven) is 7×10^3 min^{-1} (22°C)(4). These results suggest that
the reconstituted activity must be close to that in the
mitochondrial membrane.

In determining the turnover number of the reconstituted beef
heart PTP, we have identified lipids that generate proteoliposomes,
at low lipid to protein ratios, with high phosphate uptake activity
(4). The plant phosphatidylethanolamine : phosphatidylcholine
(highly purified lipids) ratio is important and egg calcium
phosphatidate dramatically stimulates SH reagent-sensitive net
phosphate uptake (4). We find no dramatic effect by beef heart
cardiolipin (CL) on PTP's specific transport activity (possibly due
to sufficient copurified CL). The addition of CL to some

solubilization media however, contrary to a published report (5), increases dramatically the amount of PTP solubilized from mitochondria without changing its specific transport activity as assayed under our conditions (4). We have reconstituted phosphate transport activity with PTP that was exposed to SDS and urea before incorporation into proteoliposomes. Experiments under these types of conditions are expected to help us identify those lipids that may be associated directly with PTP and thus stimulate its activity or even be an essential part of its catalytic entity.

The purity of the beef heart PTP has been established by quantitative N-terminal analysis (H_2N-Ala-Val-Glu-Glu-Glx-Tyr-) of the highly purified (2,6) and reconstitutively active protein. A comparison of this partial sequence with the beef heart mitochondrial DNA (7) suggests that PTP is coded for by a nuclear gene.

Supported by grants from the USPHS NIH.

1. Wohlrab, H., Kolbe, H.V.J., Rasmussen, U.B., and Collins, A. (1984) In: Epithelial Calcium : Phosphate Transport (Bronner, F., and Peterlik, M., eds.) Alan R. Liss, Inc., New York, N.Y. (in press).
2. Kolbe, H.V.J., Costello, D., Wong, A., Lu, R.C., and Wohlrab, H. (1984) J. Biol. Chem. (in press).
3. Coty, W.A., and Pedersen, P.L. (1974) J. Biol. Chem. 249 2593.
4. Wohlrab. H., Collins, A., and Costello, D. (1984) Biochemistry 23 1057.
5. Kadenbach, B., Mende, P., Kolbe, H.V.J., Stipani, F., and Palmieri, F. (1982) FEBS Lett. 139 109.
6. Kolbe, H.V.J., and Wohlrab, H. (1984) Fed. Proc. 43 2055.
7. Anderson, S., DeBruijn, M.H.L., Coulson, A.R., Eperon, I.C., Sanger, F., and Young, I.G. (1982) J. Mol. Biol. 156 683.

PROLINE DEHYDROGENASE OF ESCHERICHIA COLI K12: REQUIREMENTS FOR
MEMBRANE ASSOCIATION.

Janet M. Wood.
Department of Chemistry and Biochemistry,
University of Guelph, Guelph, Ontario, Canada   N1G 2W1

   To utilize L-proline as carbon or nitrogen source, Escherichia
coli and Salmonella typhimurium must express genes putP and putA
(1,2). Proline dehydrogenase, the putA gene product, is a
flavoprotein of molecular weight 130,000 which associates with
the inner surface of the cytoplasmic membrane. Mutations at putP
inactivate proline porter I, which catalyzes active proline uptake
powered by the proton-motive force. Genetic data suggest that,
in addition to its enzymatic role, the putA gene product serves as a
repressor controlling the divergent transcription of putP and putA
(2). Roth and his colleagues suggest that saturation of specific
membrane sites by proline dehydrogenase leads to its accumulation in
soluble form, interaction with an operator sequence, and repression
of the put genes (2,3).

   Proline dehydrogenase can be readily extracted from the bacterial
membrane fraction. The solubilized enzyme transfers electrons from
L-proline to exogenous electron acceptors (4,5). It has a high K_m
for L-proline (105 mM) and it is insensitive to the respiratory
chain inhibitors Amytal and cyanide (4-6). The membrane associated
enzyme can be examined using inverted membrane vesicles prepared
with the French Pressure Cell (7). It transfers electrons from
proline to O_2 via the respiratory chain with coupled transmembrane
proton translocation (7). Unlike the soluble enzyme it has a low K_m
for L-proline (3 mM) and it is inhibited by Amytal and cyanide (7).
Proline: O_2 oxidoredactase activity identical to that of native
membranes can be reconstituted by mixing purified enzyme with enzyme
deficient membranes from a putA mutant strain (6). The reconstituted
activity is a saturable function of enzyme concentration at constant
membrane concentration, but the activity approached is 20-fold
higher than that of membranes prepared from putA[+] bacteria induced
for proline utilization (6). Furthermore, bacteria harbouring
multiple copies of gene putA yield membranes with elevated proline
oxidative activity. Thus although saturation of the membrane with
proline dehydrogenase is possible, it probably does not occur during
induction of the put genes in vivo.

   The O_2-dependent oxidation of L-proline to Δ^1-pyrroline-5-
carboxylate (P5C) by membrane associated proline dehydrogenase can be

detected without interference from the soluble enzyme. The association of proline dehydrogenase with inverted membrane vesicles can
therefore be monitored by observing the chromogenic reaction of P5C
with o-aminobenzaldehyde (6). Enzyme/membrane association requires
the presence of $MgCl_2$ and the reduction of a membrane constituent.
The latter requirement can be met by electron flow from L-proline via
proline dehydrogenase, D-lactate via D-lactate dehydrogenase or NADH
via NADH dehydrogenase. Proline is effective under aerobic conditions
whereas D-lactate and NADH promote enzyme/membrane association only
anaerobically. The rate and extent of association are unaltered in
the presence of the proton ionophore, carbonyl cyanide m-chloro-
phenylhydrazone. These observations suggest that proline
dehydrogenase associates with a specific cytoplasmic membrane
receptor. Reduction of the receptor is prerequisite to that
association. Either it is more readily reduced by L-proline than by
D-lactate or NADH, or reduction of the receptor by proline dehydrogenase is accompanied by additional molecular interactions that
promote formation of a stable aggregate. We are currently attempting
to identify the receptor and to more precisely define the chemistry
of the enzyme-receptor interaction.

1. J.M. Wood, J. Bacteriol., 146 (1981) 895-901.
2. R. Menzel and J.R. Roth, J. Mol. Biol., 148 (1981) 21-44.
3. S.R. Maloy and J.R. Roth, J. Bacteriol., 154 (1983) 561-568.
4. R.C. Scarpulla and R.L. Soffer, J. Biol. Chem., 253 (1978) 5997-
    6001.
5. R. Menzel and J.R. Roth, J. Biol. Chem., 256 (1981) 9762-9766.
6. S.B. Graham, J.T. Stephenson and J.M. Wood, J. Biol. Chem., 259
    (1984) 2656-2661.
7. J.L.A. Abrahamson, L.G. Baker, J.T. Stephenson and J.M. Wood,
    Europ. J. Biochem., 134 (1983) 77-82.

HEPATIC AND CARDIAC TRANSPORT OF GLUTATHIONE DISULFIDE

T.P.M. Akerboom, T. Ishikawa, and H. Sies
Institut für Physiologische Chemie I, Universität Düsseldorf,
Moorenstraße 5, D-4000 Düsseldorf, FRG

In the hepatic turnover and the interorgan metabolism of glutathione the efflux from the liver cell plays a significant role. Increased efflux rates of glutathione disulfide are observed during increased hydroperoxide metabolism both in liver and in heart.

Perfused organs. In liver, GSSG is excreted mainly into the biliary compartment at a rate of 0.4 nmol x min^{-1} x g liver^{-1}. Transport occurs against a concentration gradient of about 20 to 50, implying an active transport process (1). No transport maximum could be found at values up to 80 nmol x min^{-1} x g liver^{-1} and intracellular contents as high as 300 nmol x g liver^{-1}. In contrast to liver, efflux of GSSG from perfused heart showed saturation kinetics, with a V_{max} of 7.5 nmol x min^{-1} x g heart^{-1} and an apparent K_m of 30 nmol x g heart^{-1} (2). In both organs GSSG transport can be inhibited by glutathione conjugates. In the perfused liver, transport of other bile constituents like taurocholate or bilirubin is inhibited under conditions of increased GSSG release (3).

Isolated canalicular membrane vesicles. Experiments with plasma membrane vesicles of canalicular origin revealed carrier-mediated transport of GSSG showing saturation kinetics and an apparent K_m of 0.4 mM (4). GSSG transport has also been characterized in inside-out vesicles from human erythrocytes (5) and two independent transport processes were proposed, one with a K_m of 0.1 mM and one with a K_m of 7.1 mM. However in canalicular vesicles the high K_m transport system probably represents simple diffusion.

1. Akerboom, T.P.M., Bilzer, M. and Sies, H. (1982) J. Biol. Chem. 257, 4248-4252
2. Ishikawa, T. and Sies, H. (1984) J. Biol. Chem. 259, 3838-3843
3. Akerboom, T.P.M., Bilzer, M. and Sies, H. (1984) J. Biol. Chem. 259,
4. Akerboom, T.P.M., Inoue, M., Sies, H., Kinne, R. and Arias, I.M. (1984) Eur. J. Biochem. 141, 211-215
5. Kondo, T., Murao, M. and Taniguchi, N. (1982) Eur. J. Biochem. 125, 551-554

TIGHT COUPLING BETWEEN SECRETION AND ENERGY CONSUMPTION IN HUMAN
PLATELETS

J.W.N. Akkerman, and A.J.M. Verhoeven
Dept. of Haematology, University Hospital Utrecht, The Netherlands

Secretion responses by a variety of secretory cells depend on a
sufficient support of metabolic energy in the form of ATP (1). In
blood platelets extra metabolic energy is consumed in parallel with
secretion of three types of granules. This energy consumption is
quantitated by the analysis of the initial hydrolysis of radiolabeled
ATP and ADP following abrupt arrest of ATP resynthesis induced by
2-deoxy-D-glucose (30 mM), D-glucono-1,5-lactone (10 mM) and cyanide
(1 mM) (2,3). At 37°C unstimulated platelets consume 3.5 µmol ATPeq/
min/10[11] cells, which is immediately accelerated to 16 (same units)
upon stimulation with 5 U/ml thrombin. At 10 s after stimulation,
dense, α- and lysosomal granule secretion is 85,50 and 30% of maximal
secretable amounts, respectively. Energy consumption is strongly
temperature-dependent, both in stimulated and unstimulated cells
(Fig.1). Between 10 and 42°C energy consumption in stimulated cells
is always 3.5-4.5 fold higher than in unstimulated cells, suggesting
that the secretion responses require the extra metabolic energy.
Secretion is also temperature-dependent, and varies in parallel with
energy consumption (Fig.2). Extrapolation to 100% secretion reveals
that complete secretion coincides with the consumption of an extra
amount of energy of 2.5, 4.2 and 6.7 µmol ATPeq/10[11] cells, respectively. These findings indicate that the three secretion responses
have different energy requirements.

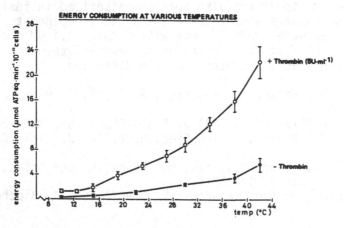

Fig.1. Energy consumption at various temperatures in platelets incubated without (●) and with (o) 5 U/ml thrombin (means ± SD; n = 7).

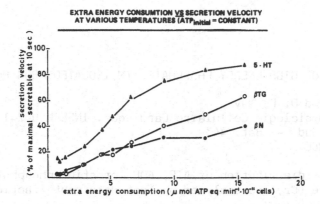

EXTRA ENERGY CONSUMPTION VS SECRETION VELOCITY
AT VARIOUS TEMPERATURES (ATP_initial = CONSTANT)

Fig.2. Secretion velocities and the extra energy consumption in platelets stimulated with 5 U/ml thrombin at various temperatures. Serotonin (5 HT), β-thromboglobulin (βTG) and N-acetyl-β-D-glucosaminidase (βN) were the markers for dense, α- and lysosomal granule secretion, respectively. Means from 7 experiments.

Similar secretion versus energy consumption plots are obtained at a constant temperature (37°C) when (I) platelets are stimulated with 5 U/ml thrombin at different initial levels of ATP (2); (II) platelets are stimulated with different doses of thrombin at a constant, high initial ATP level (4), and (III) different intervals during the responses are investigated after stimulation with a constant dose of agonist (4). Hence, secretion is tightly coupled to the simultaneous consumption of energy. The agreement between these approaches also indicate that potentially disturbing factors in the free energy change for ATP hydrolysis $\Delta G = \Delta G° + RT \ln[(ADP)(Pi)/(ATP)]$, such as variations in temperature and initial ATP level, hardly affects these measurements.

1. Verhoeven, A.J.M, Mommersteeg, M.E. and Akkerman, J,W.N. (1984). Biochim. Biophys. Acta, in press.
2. Akkerman, J.W.N., Gorter, G., Schrama, L., and Holsen, H. (1983). Biochem. J. 210, 145-155.
3. Verhoeven, A.J.M., and Akkerman, J.W.N. (1984). This abstract book.
4. Verhoeven, A.J.M, Mommersteeg, M.E., and Akkerman, J.W.N. (1984). Biochem. J., in press.

COMPARTMENTATION OF HIGH ENERGY PHOSPHATES IN ISOLATED FROG HEART
CELLS

M. Arrio-Dupont and D. De Nay
Laboratoire de Physiologie Cellulaire Cardiaque, INSERM U-241
Université Paris-Sud - Bât. 443
91405 Orsay FRANCE.

The subcellular distribution of ATP, ADP, creatine phosphate
(CrP) and creatine (Cr) was analyzed by fast detergent fractionation.

Ventricular myocytes were prepared by collagenase and trypsin perfusion of frog heart. In Ringer medium 1 mM Ca^{2+}, 80 - 90 % of the
of the cells are spindle shaped (200 μm X 10 μm) and do not beat
spontaneously. Beating (15 - 20 beats min^{-1}) is induced by 0.5 mM
BA Cl_2 and contracture by 80 mM KCl. Concentrations of ATP, ADP or
CrP were determined by bioluminescence assays, using purified firefly
luciferase ; Creatine was assayed by fluorimetry.

Digitonine fractionation (1) (1 mg/ml, 10 sec at 4° C., in
20 mM MOPS, 3 mM EDTA, 230 mM mannitol medium) was used to separate
mitochondria and myofilaments from cytosol. To separate myofilaments
from the other cellular compartments, Triton X 100 was used
(2 %, 15 sec at 4° C. in the same medium as digitonine).
The digitonine or Triton concentrations were those giving 95 % lactate dehydrogenase leakage.

The compartmentation of high energy metabolites in cells resting
or beating, incubated with 5 mM pyruvate as substrate is indicated in
the table :

!Subcellular distribution of ATP, ADP, CrP and Cr in resting (r)!
!or beating (b) heart cells. Values are expressed in nmoles/mg !
!cellular protein.

		cytosol	mitochondria	myofilaments
ATP	r	13 ± 2	0.9 ± 0.1	$0 - 0.1$
	b	13 ± 2	0.8 ± 0.1	$0 - 0.06$
ADP	r	≤ 0.25	0.7 ± 0.1	1.1 ± 0.1
	b		$0.8 \quad 0.1$	$1.1 \quad 0.1$
CrP	r	28 ± 3	0	0
	b	28 ± 3	0	0
Cr	r	16 ± 3		
	b	16 ± 3		

The differences between mitochondrial contents of beating and resting cells are not significant. Most of the extramitochondrial ADP is
bound to myofilaments. The cytosolic concentrations of ATP, ADP, CrP
and Cr show that the creatine kinase reaction CrP + ADP $\rightleftharpoons$ Cr + ATP
is in near equilibrium in the cytosol of heart cells.

1 - Zuurendonk, P.F., Tischler, M.E., Akerboom, T.P.M.,
    Van der Meer, R., Williamson, J.R. and Tager, J.M. (1979)
    Methods in Enzymology 56, 207-223.

A COMPARATIVE STUDY BETWEEN PHOTOFERMENTATIVE AND FERMENTATIVE H_2 PRODUCING SYSTEMS

Cs. Bagyinka, N. Kaddouri, A. Dér, K.L. Kovács
Inst. Biophysics, Biological Research Center, Hung. Acad. Sci. Szeged, Hungary H-6701, POB. 521.

H_2 production by biological systems represents a great challenge in the search for alternative, renewable energy sources. There are two alternative ways to achieve this goal:
- H_2 photoevolution from water, i.e. biophotolysis
- fermentative H_2 production.

Biophotolysis is an attractive possibility, but due to substantial lack in our understanding of the underlying basic molecular processes it holds the promise of large scale utilization only in long term.

Fermentative H_2 production involves conversion of biomass into H_2, an ideal fuel. As energy conversion yields are usually low only simple systems with low energy input can be taken into account. Photosynthetic bacteria can be cultivated inexpensive in large quantities using sunlight as energy source. Photosynthetic bacteria thus can serve as biomass material and/or fermentative systems.

In order to produce H_2 the fermentative system should contain the enzymes pyruvate-formate lyase and formate-hydrogenlyase in addition to the usual fermentative enzymes. It has been found previously [1] that *Chlorobium limicola forma thiosulfatofilum L.* possesses the enzymes required for H_2 evolution from carbohydrates. Indeed, H_2 evolution from glucose and intermediates of glycolysis has been ob-

served in growing *C. limicola* cultures as well as in washed cell suspensions. The energy conversion yield, however,
is rather small (0.4 mol H_2/mol glucose) at least when
compared to other fermentative systems.

Dark fermentation of biomass and carbohydrates has been
performed by an *Enterobacter cloaceae* strain isolated in
this laboratory and showing high formate hydrogenlyase activity. The cell-free extract from photosynthetic bacteria
(e.g. *Thiocapsa roseopersicina* strain BBS) has been used
as biomass as well as growth medium for *E. cloaceae*. A
continuous H_2 evolution can be followed for at least several days with an energy conversion yield of 2.6 moles of
H_2/mol glucose. The anaerobic fermentative system is un-
sensitive to traces of O_2, H_2 partial pressure and can u-
tilize a wide range of mono- and disacharides. The yield
is not competitive with biogas production from the same
row materials but it seems superior than fermentation to
alcohol.

1. Kovács, K.L., Bagyinka, Cs., Serebriakova, L.T. /1983/
   Curr. Microbiol. 9 215-218.

THE SIGNIFICANCE OF MITOCHONDRIAL β-OXIDATION OF FATTY ACIDS NOT
LINKED TO ATP SYNTHESIS

M.N. Berry, R.B. Gregory, A.R. Grivell and P.G. Wallace
Department of Clinical Biochemistry, Flinders University School of
Medicine, Bedford Park, South Australia, 5042, Australia.

Most of the O_2-uptake of animal cells occurs within their
mitochondria and it is generally accepted that under normal
conditions this O_2-consumption is associated with ATP formation.
Consequently, theories concerning the regulation of respiration
invariably address the question of what controls the rate of
cellular ATP turnover (1-3).

In intact mammalian cells, however, it has been observed that the
rate of O_2-consumption can be stimulated under circumstances where
no increased demand for ATP is apparent. In brown adipose tissue
the proton gradient generated across the inner mitochondrial
membrane during fatty acid oxidation can be physiologically
dissipated with production of heat (4). In liver also, some of the
O_2-uptake induced by substrate addition appears to be associated
with heat production rather than ATP synthesis (5). We have
demonstrated that this is particularly the case for fatty acid
β-oxidation. Whereas the oxidation of acetyl CoA in the Krebs cycle
is obligatorily linked to ATP synthesis, the formation of this
acetyl CoA through the β-oxidation of short chain or long chain
fatty acids need not be coupled to phosphorylation of ADP, but
rather can be linked to some alternative mitochondrial energy-
transforming process, possibly reversed electron transfer (6).

Because there is some dispute concerning the primary locus of
fatty acid oxidation within the hepatocyte, we have examined this
question further using isolated rat liver mitochondria. In the
absence of a glucose-hexokinase trap, the bulk of the O_2-uptake was
related to β-oxidation and little Krebs cycle activity was
observed. The stimulation of respiration induced by addition of a
trap for ATP was associated with a marked increase in Krebs cycle
activity and a much smaller rise in the rate of β-oxidation.

In an alternative approach the respiration of isolated
mitochondria was inhibited with oligomycin. Whereas the oxidation
of malate was strongly impaired, β-oxidation of palmitate was much
less affected, although again Krebs cycle activity was depressed.
Addition of the uncoupling agent, FCCP relieved malate and Krebs
cycle oxidations to some extent with little effect on the rate of
β-oxidation.

By limiting the availability of 4-carbon Krebs cycle intermediates during hepatic mitochondrial palmitate metabolism, it can be readily shown that β-oxidation can be coupled to ATP synthesis. We argue, however, that when the phosphorylation potential is high, β-oxidation of fatty acids can be dissociated from ATP formation. This may have physiological significance in relation to mammalian thermogenesis. Although it is generally assumed that the rate of cellular ATP turnover is always sufficient to generate the heat necessary to maintain body temperature at 37°C there is no self-evident reason why this should be so. Indeed, a comparison of mammalian and poikilothermic metabolism suggests that as much as 80% of mammalian basal metabolism may be directly concerned with maintenance of body temperature, rather than with metabolic needs for ATP synthesis (7). It now seems clear that fatty acid can be the fuel for this heat production, not only in brown fat (4) but in other organs such as liver. A tendency to cooling could trigger receptors within the hypothalamus or elsewhere, leading to an enhanced sympathetic nervous system activity and release of catecholamines (8). This in turn would promote lipolysis and an increased flow of fatty acid substrate to the liver, leading to the formation of ketone bodies which are transported to the periphery for further metabolism.

Under normal conditions the acetyl CoA derived from these ketone bodies will be utilized by the peripheral tissues in reactions linked to ATP turnover. However, in cases of severe insulin lack associated with diabetes there will be a marked decline in lipid and protein synthesis and hence in cellular ATP turnover. Under these circumstances, the massive ketoacidosis that can occur may reflect a marked stimulation of β-oxidation as a compensatory mechanism for generating sufficient heat to maintain normothermia.

1. Chance, B. and Williams, G.R. (1956) Adv. Enzymol. 17, 65-134.
2. Tager, J.M., Wanders, R.J.A., Groen, A.K., et al. (1983) FEBS Lett. 151, 1-9.
3. Erecinska, M. and Wilson, D.F. (1982) J. Membrane Biol. 70, 1-14.
4. Nicholls, D.G. (1979) Biochim. Biophys. Acta 549, 1-22.
5. Hems, R., Ross, B.D., Berry, M.N. and Krebs, H.A. (1966) Biochem. J. 101, 284-292.
6. Berry, M.N., Clark, D.G., Grivell, A.R. and Wallace, P.G. (1983) Eur. J. Biochem. 131, 205-214.
7. Bennett, A.F., Dawson, W.R. (1976) in: Biology of the Reptilia, (Gans, C. ed.), vol. 5, 127-223, Academic Press, London.
8. Jansky, J. (1973) Biol. Rev. 48, 85-132.

AXIAL RESPONSE THROUGH CYTOKININS IN CONTROLLING PHOTO-
ELECTRON TRANSPORT DURING GREENING OF WHEAT LEAVES.

B. Biswal and U.C. Biswal
School of Life Sciences, Sambalpur University,
Jyoti Vihar-768019, Orissa, INDIA.

Role of axis in controlling greening process is known
in principle [1] . Excision of leaves from the axis
results in a lagphase of chlorophyll accumulation and reduced rates of accumulation of photosynthetic pigments [1].
However, treatment of the excised leaves with minerals,
organic substrates and plant hormones [2-4] stimulate
pigment accumulation and therefore mimic the action of
axis. The present work is an attempt to compare the kinetics of electron transport development during greening of
excised and intact leaves. The excised leaves are treated
with kinetin and the data are extrapolated to examine the
axial role in controlling electron transport through the
hormone.

Materials and Methods: Etiolated wheat (Triticum aestivum
Linn. emend. Thell CV. Sonalika) leaves (6d old) intact or
excised were taken as the experimental material. Whenever
necessary, the intact leaves were sprayed and the excised
ones were floated in the kinetin solution (50 μM). Chlorophyll was extracted and estimated as per Arnon[5] . The
DCPIP Photoelectron transport by isolated chloroplasts of
intact or excised wheat leaves was measured spectrophotometrically[6]

Results and Discussion: A comparative kinetics of chlorophyll accumulation during greening of intact and excised
leaves with and without kinetin is summarised in Table 1.
Excision causes a lag phase and a reduced rate of chlorophyll accumulation. Treatment of excised leaves with
kinetin removes the lag phase and stimulates chlorophyll
accumulation.

The development of electron transport as measured by
DCPIP photoreduction is conducted under these experimental
conditions. Excision results in a retardation of photoelectron transport development (Fig.1). However kinetin
treatment stimulates the rate of developmentsignificantly.

TABLE I

Effect of Kinetin on the time course of chlorophyll accumulation (μg/gm fresh wt) during greening of excised and intact leaves.

Time in hr	Control		Kinetin(50 μM)	
	Intact	Excised	Intact	Excised
2	50.00	45.09	105.00	110.30
8	297.00	60.41	455.00	415.60
12	400.00	400.00	742.30	505.80
48	325.50	256.00	1797.70	485.60

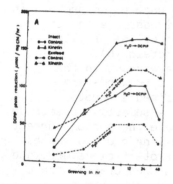

Fig.1. Comparative kinetics of development of electron transport during greening of excised and intact wheat leaves treated with kinetin.

These results, therefore, would suggest that in excised leaves, absence of axis results in a limitation in the supply of the hormone leading consequently to a reduced rate of photochemical reaction during plastid formation.

References:
1.Hardy,S.I.,Castelfranco,P.A. and Rebeiz,C.A. (1970) Plant Physiol.46, 705-707.
2.Knypl,J.S. and Rennert,A.(1970) Z.Pflanzenphysiol. 62, 97-107.
3.Hole,C.C. and Dodge,A.D.(1975)Physiol.Plant.34,22-25.
4.Biswal,B. and Biswal,U.C.(1981)Experientia 37,138-139.
5.Arnon,D.I.(1949) Plant Physiol. 24, 1-15.
6.Biswal,B.,Choudhury,N.K.,Sahu,P. and Biswal,U.C.(1983) Plant & Cell Physiol.24, 1203-1208.

BIOENERGETIC CHARACTERIZATION OF HUMAN BROWN FAT MITOCHONDRIA.

S. Cunningham, E. Rial, and D.G. Nicholls
Neurochemistry Laboratory, Department of Psychiatry, Ninewells Medical
School, University of Dundee, Dundee, DD1 9SY, U.K.

Brown adipose tissue is rich in mitochondria possessing a 32 kDa uncoupling protein (1). It has been suggested (2) that the tissue may play a role in weight regulation in animals and man. Bioenergetic evidence for the function (as opposed merely to the presence) of the uncoupling protein in human tissue has only been obtained for phaeochromocytoma patients with chronically elevated noradrenaline (3). We have now analyzed adolescent perinephric fat tissue and shown the consistent presence of areas of functional brown fat by morphological, structural and bioenergetic criteria.

Mitochondria were prepared from brown areas of each tissue sample, and also from brown adipose tissue of warm- and cold-adapted guinea-pigs. Binding of GDP is diagnostic of the uncoupling protein in the membrane of brown fat mitochondria (4), and is associated with an inhibition of proton conductance (5). Low concentrations of fatty acids reverse the low conductance state of the protein (6).

Results obtained from a 14 year old subject are shown in the Table; binding of GDP is intermediate between that to warm- and cold-adapted guinea-pigs. Proton conductance (defined as proton current per mV of $\Delta\psi$) was monitored using a TPP - selective electrode to quantify $\Delta\psi$ and an oxygen electrode to estimate the proton current (6). The proton conducting capacity of the uncoupling protein can be defined as the decrease in conductance when the protein binds GDP, and the sensitivity to fatty acids as the increase in conductance when palmitate is added to give a 2:1 ratio of palmitate to albumin (equivalent to 1.4µM unbound palmitate).

GDP caused an increase in $\Delta\psi$ of mitochondria from the cold- and warm-adapted animals, and from each human mitochondrial preparation so far examined. Results from a 14 year old subject and the guinea-pigs are shown in the Table. The increase in $\Delta\psi$ is accompanied by an inhibition of respiration, indicating an inhibition of proton conductance. The conductance decrease and fatty acid sensitivity of the human mitochondria were each intermediate between that of the warm and cold-adapted guinea-pig mitochondria.

This work was supported by a grant from the MRC to DGN and RTJ.

1. Heaton, G.M., Wagenvoord, R.J., Kemp, A. & Nicholls, D.G. Eur. J. Biochem. 82, 515-521 (1978).
2. Rothwell, N.J. & Stock, M.J. Nature (London) 281, 31-35 (1979).
3. Ricquier, D., Nechad, M. & Mory, G. J. Clin. Endocrin. Metab. 54,

803-807 (1982)
4. Nicholls, D.G. Eur. J. Biochem. 62, 223-228 (1976).
5. Nicholls, D.G. Eur. J. Biochem. 49, 573-583 (1974).
6. Locke, R.M., Rial, E. & Nicholls, D.G. Eur. J. Biochem. 129, 381-387 (1982)

Bioenergetic parameters of mitochondria prepared from human perinephric fat and from the dorsal brown fat of cold-adapted and warm-adapted guinea-pigs.

Mitochondria were prepared (5) from the human tissue and from the dorsal brown fat of guinea-pigs maintained at 28-31°C (WA-GP) or 4-7°C (CA-GP) for at least 2 weeks. For the measurement of membrane potential, respiration and proton conductance, mitochondria were incubated in a combined oxygen electrode / TPP - selective electrode assembly (6) in the presence of a-glycerophosphate, pyruvate and malate. Additions were made of 3mM-GDP and 32µM-palmitate (sufficient to produce 1.4µM unbound palmitate in the presence of the albumin) as detailed below. Proton conductance was calculated as described previously (6).

	Mitochondrial source		
	Human	WA-GP	CA-GP
GDP-bound			
(nmol/mg protein)	0.205	0.079	0.710
Respiration			
(nmol O_2. min -1. mg protein -1)			
control	258	149	537
+GDP	93	50	35
+GDP + palmitate	159	73	132
Membrane potential			
(mV)			
control	188	209	125
+GDP	214	222	228
+GDP + palmitate	188	211	203
Proton conductance			
(nmol H^+. min -1. mg protein -1. mV -1)			
control	12.3	6.4	38.6
+GDP	3.9	2.1	1.4
+GDP + palmitate	7.6	3.1	5.9
Decrease with GDP	8.4	4.3	37.2
Increase with palmitate	3.7	1.0	4.5

PERIODIC CHANGES IN THE MITOCHONDRIAL Ca^{2+}-TRANSPORTING SYSTEM OF THE CILIATE PROTOZOAN *TETRAHYMENA PYRIFORMIS*

Yu.V.Evtodienko, Yu.V.Kim, I.S.Yurkov and V.P.Zinchenko
Institute of Biological Physics, Academy of Sciences of the USSR,
142292 Pushchino, Moscow Region, USSR

During the cell cycle of *Tetrahymena pyriformis* the oscillations of Ca^{2+} and Mg^{2+} contents (1,2) as well as the cyclic changes of the respiration and adenine nucleotide levels (3) have been observed. Recently we have shown that *Tetrahymena* mitochondria have an energy-dependent Ca^{2+}-transporting system, whose properties are close to mammalian one (4). In order to elucidate the role of mitochondria in the above-mentioned changes during the cell cycle the simultaneous measurements of the various mitochondrial parameters have been performed in the heat-shocked synchronously-dividing cultures of *T.pyriformis*, both intact and digitonin-treated. It has been registered that during the cell cycle mitochondrial Ca^{2+} uptake rate synchronously oscillates with the membrane potential and respiration rate.

One of the causes of the observed changes can be the existence within the cells of autonomous systems functioning in the oscillatory regime, and, thereby, setting a consequence of the intracellular events. One of the candidates on such a role can be the mitochondria in which the ion transport oscillations are the well-known fact (5-7). In this connection of great interest is the observation of continuous, virtually undamped oscillations of the mitochondrial Ca^{2+} fluxes in digitonin-treated asynchronous cultures of *T.pyriformis* (Fig. 1). The cyclic changes of $[Ca^{2+}]_{out}$ and of the membrane potential ($\Delta\varphi m$) were induced by an addition of Ca^{2+} in the presence of trace amounts of a fluorescent Ca^{2+} chelator chlortetracyclin. The oscillations have a period from 2 to 10 minutes depending on the incubation conditions and able to continue one hour and more. In a model explaining the generation of oscillations across the mitochondrial membrane the key roles are assigned to the cyclic activation of the potential-independent Ca^{2+} efflux, whose activity is out of phase with the Ca^{2+} transport through Ca^{2+} uniport, and to the cyclic partial deenergization of mitochindria due to the increase of Ca^{2+} recycling.

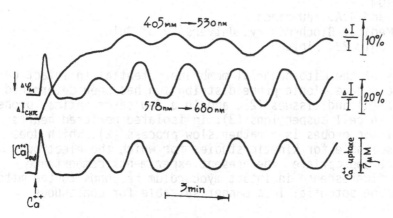

Fig. 1. Ca^{2+}-induced oscillations of $[Ca^{2+}]_{out}$ and of the fluorescences of the potential-sensitive dye, 3,3'-dipropylthiacarbocyanine iodine (dis-C_3-(5)) and Ca^{2+}-chlortetracyclin (CHTC) complexes in mitochondria of digitonin-treated *T.pyriformis* GL. The incubation medium contained 300 mM mannitol, 5 mM succinate, 4 mM glutamate, 2 mM potassium phosphate, 10 mM Tris-MOPS buffer, pH 7.0 at 25°C. *T.pyriformis* cells (2 mg cellular protein/ml) were treated by digitonin (0.4 mg/ml) in the presence of 5×10^{-6} M CHTC and preincubated 2 min before 0.05 mM $CaCl_2$ was added. The dis-C_3-(5) concentration was 10^{-6} M. The pairs of numbers show the excitation and fluorescence wavelengths of dis-C_3-(5) (578-680) and CHTC (405-530).

References

(1) London, J.F., Charp, P. and Whitson, G.L. (1979) J. Cell Biol. 83, 9a

(2) Walker, G.M. and Zeuthen, E. (1980) Exp. Cell Res. 127, 487-490

(3) Lloyd, D., Phillips, C.H. and Statham, M. (1978) Gen. Microbiol. 106, 19-26

(4) Kim, Yu.V., Kudzina, L.Yu., Zinchenko, V.P. and Evtodienko, Yu.V. (1984) Cell Calcium 5, 29-41

(5) Boiteux, A. and Hess, B. (1974) Faraday Symp. Chem. Soc. 9, 202-214

(6) Gooch, V.D. and Packer, L. (1974) Biochim. Biophys. Acta 346, 245-260

(7) Gylkhandanyan, A.V., Evtodienko, Yu.V., Zhabotinsky, A.M. and Kondrashova, M.N. (1976) FEBS Lett. 66, 44-47

CONTINUOUS OPTICAL MONITORING OF MITOCHONDRIAL MEMBRANE POTENTIAL IN
INTACT MYOCARDIUM

I.E. Hassinen, and R.A. Kauppinen
Department of Medical Biochemistry, University of Oulu,
Oulu, SF-90220 Oulu 22, Finland

Measurements of the mitochondrial membrane potential in intact cells
and tissues are scarce. Ionic probe distribution has been determined
in intact cells (1) and tissues (2), and in a few cases optical probes
have been used in cell suspensions (3). In isolated perfused hearts
the uptake of ionic probes is a rather slow process (2), which does
not lend them suitable for kinetic studies for which the electrochromic
dyes may be more appropriate. The present experiments demonstrate
that safranine fluorescence in intact myocardium responds to the mitochondrial membrane potential in a manner suitable for continuous
monitoring.

Hearts isolated from Sprague-Dawley rats were perfused by the
Langendorff procedure with Krebs-Ringer bicarbonate solution containing 10 mM glucose. The sinus node was excised to lower the endogenous
beating frequency to 1.5 Hz and to allow its control by electrical
pacing. Reflectance spectrum and safranine fluorescence changes were
simultaneously monitored with a dual-wavelength spectrophotometer-
surface fluorometer (4). Fluorescence was excited at 522.5 nm, and the
emission was measured above 575 nm.

At a 20 µM concentration, safranine was rapidly taken up by the
perfused heart, and the concentration in the tissue reached a saturating value within 8 min. There was no wash-out of safranine from the
heart after discontinuation of its infusion. 20 µM safranine had a
reversible positive chronotropic effect on the heart, and it disappeared upon discontinuing the dye infusion.

To compensate for fluorescence quenching due to endogenous chromophores, reflectance signal at 590 or 435.5 nm, isosbestic wavelengths for 522.5 nm in an unstained heart, was subtracted from the
fluorescence signal. Although an intense absorption band appeared in
the reflectance (absorbance) spectrum of the heart upon safranine
loading, the absorbance changes were not large enough to allow selective monitoring of safranine-specific spectral changes.

An increase of the beating frequency from 1.5 to 5 Hz produced on
increase in the safranine fluorescence. Increasing the K^+ concentration to 18 mM or Ca^{2+} deprivation caused a decrease in the fluorescence. Infusion of 20 µM carbonylcyanide m-chlorophenylhydrazine (CCCP)
caused an extensive increase in the safranine fluorescence. Perturba-

tion of the plasma membrane potential by changing the extracellular K^+ concentration in a heart, kept quiescent by omission of Ca^{2+}, did not affect safranine fluorescence. These findings indicate that the safranine signal mainly reflects changes in the mitochondrial energy state or membrane potential.

A large fluorescence increase, similar in magnitude to that caused by CCCP was observed when Ca^{2+} was readmitted after a 15-min Ca^{2+}- free perfusion. The phenomenon of "calcium paradox" may therefore be related to a disappearance of the mitochondrial membrane potential.

In contrast to the absorbance spectrum changes, the fluorescence of safranine has not obtained previous use in measurement of mito- chondrial membrane potential in vitro. Therefore, the reliability of the fluorescence measurements was tested in isolated mitochondria during imposition of graded K^+ diffusion potentials in the presence valinomycin. The fluorescence quenching was linearly proportional to the membrane potential ($r=0.984$, $P<0.001$).

The quantitative relation between safranine fluorescence and mito- chondrial membrane potential in intact tissue was tested by comparing the fluorescence changes with the membrane potential differences be- tween metabolic states of which data obtained by distribution of endogenous or exogenous ionic probes is available. The correlation between the changes of safranine fluorescence and membrane potential is linear ($r=0.9999$, $P<0.01$).

In conclusion, an isolated perfused rat heart can be irreversibly stained by safranine without functional impairment, and safranine fluorescence, as measured from the epicardial surface can be employed for continuous monitoring of the mitochondrial membrane potential changes in the tissue. However, the high concentration of endogenous chromophores necessitates compensation for colour quenching by means of simultaneous monitoring of the reflectance spectrum and sub- tracting of appropriate reflectance signal from the fluorescence.

References:

(1) Hoek, J.B., Nicholls, D.G., and Williamson, J.R. (1980) J. Biol. Chem. 255, 1458-1464
(2) Kauppinen, R.A. (1983) Biochim. Biophys. Acta 725, 131-137
(3) Åkerman, K.E.O. (1979) Biochim. Biophys. Acta 546, 341-347
(4) Hassinen, I.E., and Jämsä, T. (1982) Anal. Biochem. 120, 365-372

LACTATE DEHYDROGENASE FROM CONTROL AND TRANSFORMED HEPATOCYTES:
REACTION WITH ALKALINE PHOSPHATASE

Ann E. Kaplan
Laboratory of Experimental Pathology
National Cancer Institute
Frederick Cancer Researach Factility
Frederick, Maryland, 21701, U.S.A.

Kinetic and molecular modifications have been identified in a neoplastic, chemically-transformed cell line from rat hepatocytes compared with its control cell line (1). Studies with alkaline phosphatase were carried out to clarify molecular differences observed in separations by gel electrophoresis and isoelectric focussing.

Epithelial cells are used in these studies because they give rise to carcinomas. These tumors are identified in over 90% of the cancer patient population. The control hepatocyte, TRL, was established from ten day old rats (2), and its neoplastic cell line, NMU-3, was developed from TRL by exposure in vitro to nitrosomethylurea (3). The neoplastic cell line produces carcinomas in vivo, and is also characterized by very rapid formation of lactic acid.

Comparison of lactate dehydrogenase (LDH) from the TRL and NMU-3 cell lines shows that the kinetic behavior of the enzyme from the neoplastic cell is altered, producing lactic acid at a much higher rate than TRL cells. This property characterizes many neoplastic cells.

Examining molecular properties, like hepatocytes in situ, both cell lines are characterized by LDH-4 and -5. However, the molecular distribution is such that the TRL cells have most of their activity in LDH-4 whereas the NMU-3 cells have most of their activity in LDH-5 (1). Separation of LDH by isoelectric focussing showed unexpected differences between LDH-4 and -5. The LDH-4 species precipitate at isoelectric points (pI) between pH 4-6, whereas LDH-5 precipitates at pH 8.8 (1).

This wide difference leads to experiments with alkaline phosphatase. LDH preparations from both TRL and NMU-3 cells were treated with alkaline phosphatase for 30-120 minutes at 37°C with no affect whatsoever on the LDH activity. The isozymes were then separated by gel electrophoresis and both LDH-4 and -5 were identified. However, the properties of LDH-4 in the TRL preparation was markedly reduced, whereas the activity in LDH-5 was markedly increased. The

distribution of LDH activity in the alkaline phosphatase-treated
enzyme from TRL cells now looks like that of NMU-3 cells. In contrast, the LDH preparation from NMU-3 is unaltered by incubation with
alkaline phosphatase.

These results suggest that most of the LDH-4 from TRL cells contains serine-O-phosphate groups which are sensative to alkaline phosphatase treatment, and that neoplastic transformation of TRL cells to
NMU-3 cells resulted in a marked reduction of these molecular species.
Thus in TRL cells, LDH is largely acidic in pI, whereas LDH in NMU-3
cells is largely alkaline. However, in both the alkaline phosphatase
treated LDH from TRL cells and the LDH from NMU-3 cells, a small, but
constant amount of LDH activity continues to migrate as LDH-4 by gel
electrophoresis.

These results are the first to indicate that serine-O-phosphate
groups may serve to regulate the behavior of an enzyme in the glycolytic pathway. Until now no post-translational modifications have been
reported to regulate enzyme activity in glycolysis.

1. Kaplan, A.E., Hanna, P., Hochstadt, B. and Amos, H. (1982) Biophys.
   J 37(2), 3a.

2. Idoine, J.B., Elliott, J.M, Wilson, M.J. and Weisburger, E.K.
   (1976) In vitro 12, 541.

3. Williams, G.M., Elliott, J.M. and Weisburger, J.H. (1973) Cancer
   Res. 33, 606.

MITOCHONDRIAL TRANSMEMBRANE PROTON ELECTROCHEMICAL POTENTIAL AND METABOLITE COMPARTMENTATION IN INTACT MYOCARDIUM

R.A. Kauppinen, J.K. Hiltunen, and I.E. Hassinen
Department of Medical Biochemistry, University of Oulu,
Oulu, SF-90220, Finland

Ultimate testing of models of metabolic regulation can be made only by use of knowledge of the conditions prevailing in vivo. The methodological progress in the field of subcellular compartmentation (1) has made possible the measurements of the transmembrane gradients even in intact tissues. Information about metabolite gradients is particularly important in determining whether control is exerted at the level of membrane transport. The present experiments were undertaken to estimate transmembrane gradients in intact tissue in the framework of the chemiosmotic principles and metabolic control.

Isolated rat hearts were perfused with Krebs-Ringer bicarbonate solution. The energy consumption of the myocardium was varied by lowering the beating frequency to 1.5 Hz by excision of the sinus node or arrest by 18 mM K^+. The heart was freeze-clamped and freeze-dried and fractionated in a discontinuous non-aqueous density gradient (1,2). Triphenylmethylphosphonium (TPMP) cation was used as a membrane potential probe. The solvent mixture used did not extract $TPMP^+$ from lyophilized powdered heart which had been perfused with $TPMP^+$ (3). Mitochondrial transmembrane ΔpH was estimated from the distribution of tracer amounts of [^{14}C]propionate.

The mitochondrial membrane potential ($\Delta\Psi$) calculated from the distribution of $TPMP^+$ was 125 mV (negative inside) in the heart beating at 5 Hz and 150 mV in hearts beating at 1.5 Hz. The transmembrane ΔpH was 0.63 units and 0.53 units (alkaline inside) in hearts beating at 5 and 1.5 Hz, respectively. Thus the proton electrochemical potential ($\Delta\mu_{H^+}$) increased from 15.5 to 17.1 kJ/mol in this transition which also decreased the oxygen consumption by 56% (3).

Calculated from the creatine kinase equilibrium and the tissue average P_i concentration the ΔG_{ATP} in the cytosol was -55.8 and -57.5 kJ/mol in the hearts beating at 5 or 1.5 Hz, respectively (3). The H^+/ATP ratio in the ATP synthesis and transport would then be $\Delta G_{ATPc}/\Delta\mu_{H^+} = 3.6$, which is within the range reported for mitochondria in vitro. The [ATP]/[ADP] ratio was 2.7 in the mitochondria and the $[ATP]_f/[ADP]_f$ in the cytosol 127, calculated from the creatine kinase equilibrium. An electrogenic adenylate translocator (4) would impose a diffusion potential of 103 mV (negative inside) in the beating heart. This value is comparable with the mitochondrial $\Delta\Psi$.

The $\Delta G_{o/r}$ of the reactions across the two first phosphorylation sites of the respiratory chain of an isolated perfused heart is -121 kJ/2e$^-$, when the E_h of cytochrome c is measured by reflectance spectrophotometry (5) and that of NADH$_f$/NAD$_f$ from the glutamate dehydrogenase reaction using 20 µM as the mitochondrial NH$_4^+$. The $\Delta G_{o/r}$ across the two phosphorylation sites is twice the G_{ATPc} indicating near-equilibrium in these reactions of oxidative phosphorylation plus adenylate transport.

The distribution of glutamate obeyed the mitochondrial transmembrane ΔpH, the concentration ratio being 4.0 (higher inside) and that of aspartate 11.4 (higher outside) in the beating heart (6). The glutamate-aspartate exchange translocator is electrogenic (4). Therefore, the asymmetry of glutamate and aspartate distribution would impose a diffusion potential of 102 mV (negative inside). This value is close to the mitochondrial $\Delta\Psi$.

Using the glutamate and lactate dehydrogenase equilibria (7) as indicators, E_h of NADH$_f$/NAD$_f$ was -354 and -226 mV in the mitochondria and cytosol, respectively. The difference between these values is comparable to the mitochondrial $\Delta\Psi$.

In conclusion, a network of thermodynamic near-equilibria of some reactions of energy metabolism in intact myocardium is emerging: A near-equilibrium exists between certain redox reactions of the respiratory chain and the cytosolic adenylate system. This implies near-equilibrium of the adenylate translocator, which was verified also by the degree of asymmetry of adenylate distribution. The glutamate-aspartate translocase is near-equilibrium with the mitochondrial $\Delta\Psi$. The transmembrane redox potential difference between the free NADH/NAD pools is equal to the mitochondrial $\Delta\Psi$, which suggests that also the malate aspartate is in near-equilibrium.

References:
(1) Elbers. R., Heldt. H.-E., Schmucker, P., Sobol, S., and Wiese, H. (1974) Hoppe-Seyler's Z. Physiol. Chem. 355, 378-393
(2) Kauppinen. R.A., Hiltunen, J.K., and Hassinen, I.E. (1980) FEBS Lett. 112, 273-276
(3) Kauppinen. R. (1983) Biochim. Biophys. Acta 725, 131-137
(4) LaNoue. K.F. and Schoolwerth. A.C. (1979) Ann. Rev. Biochem. 48. 871-922
(5) Hassinen. I.E.. and Hiltunen. K. (1975) Biochim. Biophys. Acta 408. 319-330
(6) Kauppinen. R.A.. Hiltunen. J.K., and Hassinen. I.E. (1983) Biochim. Biophys. Acta 725, 425-433
(7) Nuutinen. E.M. (1984) Basic Res. Cardiol. 79, 49-58

DEENERGETIZATION OF ESCHERICHIA COLI BY BACTERIOPHAGE T1

H. Keweloh and E.P. Bakker
Fachbereich Biologie, Fachgebiet Mikrobiologie, Universität
Osnabrück, Postfach 4469, D-4500 Osnabrück, F.R.G.

The infection of E. coli cells with many different phages leads to strong permeability changes of the cytoplasmic membrane: the gradients for cations like K^+ and the membrane potential collapse (1,2). We are interested in the nature of these membrane changes caused by coliphage T1. Therefore we examined the fluxes of different ions and the energetic parameters of infected cells. The second aspect of our work is the physiological significance of these effects for viral amplification. It was proposed that the permeability changes bring about the shut off of host protein synthesis by changing the ATP concentration in the cell infected by T1 (3,4).

Immediately after addition of T1, cells of E. coli lost their accumulated K^+ and took up a roughly equivalent amount of Na^+ (Fig. 1). The rapid collapse of the gradients took also place when the major cation in the medium was represented by Li^+ or organic cations. Also cholin, for which these bacteria do not possess an intrinsic way of entrance, became permeable after infection. The membrane potential, as measured by TPP^+ distribution, decreased with the same rapid kinetics, whereas Δ pH remained constant or even increased somewhat. These results of our experiments indicate that the infecting virus builds an aspscific cation channel in the cytoplasmic membrane. The purpose of channel formation may be to abolish the membrane potential, which is normally high and internally negative. Channel formation enables then the entrance of the viral DNA polyanion. This contradicts the hypothesis that the transport of viral DNA across the membrane is energized by the proton gradient (5).

The cation gradients were restored in a process which is only slightly inhibited by chloramphenicol, when an elevated Mg^{2+} concentration in the medium prevented Mg^{2+} losses of the infected cells (Fig. 1). The changes of membrane gradients after T1 infection affected the cellular ATP concentration in an appropriate manner (Fig. 2). Only after a reenergetization of the membrane and a replenishment of the ATP pool a successfull infection occurred as indicated by lysis of the infected cells. At a permanently reduced cellular ATP concentration the viral DNA replication was blocked, whereas viral proteins were still synthesized. These results indicate that the reduced ATP concentration caused by membrane deenergetization cannot be the mechanism of shifting

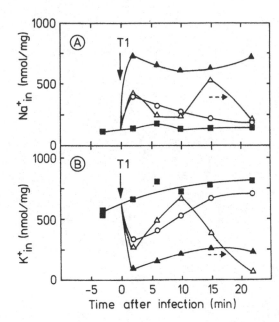

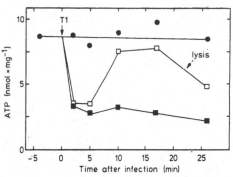

Fig. 2: ATP concentration in infected cells with 0.4 (■) or 5.4 (□) mM Mg^{2+} in the medium and in non infected cells (●); m.o.i. of added T1 was 20.

Fig. 1 :

Effect of T1 (m.o.i. = 20) on K^+_{in} and Na^+_{in} in cells of E. coli K-12 at 37°C. The Mg^{2+} concentration in the medium was 0.4 (▲) or 5.4 (△,O) mM; (O), cells preincubated with 100 μg/ml chloramphenicol; (■) , non infected cells.

macromolecular synthesis towards viral amplification. This was also observed in cells of the ATP-synthase negative mutant BH 273 (6). T1 did not lower the ATP concentration even immediately after phage addition to these cells, but could efficiently shut off host translation. Therefore we assume that the change in membrane permeability is only involved in the process of DNA invasion and not in the reorientation of macromolecular synthesis.

References:

1) Silver, S., Levine, E. and Spielman, P.M. (1968) J. Virol.2, 763-771

2) Labedan, B. and Letellier, L. (1980) Proc. Natl. Acad. Sci. USA 78, 215-219

3) Wagner, E.F., Ponta, H. and Schweiger, M. (1980) J. Biol. Chem. 255, 534-539

4) Wagner, E.F. and Schweiger, M. (1980) J. Biol. Chem. 255,540-542

5) Grinius, L. (1980) FEBS Lett. 113, 1-10

6) Schairer, H.U., Friedl, P., Schmid, B.J. and Vogel, G. (1976) Eur. J. Biochem. 66, 257-268

GLYOXYSOMAL FATTY ACID DEGRADATION IN NEUROSPORA CRASSA

Ch. Kionka, R. Thieringer, and W.-H. Kunau

Institut für Physiologische Chemie der Ruhr-Universität Bochum,
4630 Bochum, FRG

An inducible ß-oxidation system has been demonstrated in a particulate fraction from Neurospora crassa. The activities of all individual ß-oxidation enzymes were enhanced in cells after a shift from a sucrose to an acetate medium. The induction was even more pronounced by a transfer to a medium containing oleate as sole carbon and energy source. Since no acyl-CoA oxidase but an acyl-CoA dehydrogenase was detected, the latter enzyme seems to catalyze the first step of the ß-oxidation sequence in N.crassa. The intracellular particles housing the fatty acid degradation pathway have been identified as glyoxysomes by their equilibrium density of 1.21 g/cm^3 after isopycnic centrifugation in a linear sucrose gradient and by their content of isocitrate lyase and malate synthase. The lack of catalase, urate oxidase and acyl-CoA oxidase indicated that these glyoxysome-like microbodies [1] in N.crassa are clearly different from the various microbodies reported so far to contain a ß-oxidation pathway [2-4].

Enoyl-CoA hydratase, 3-hydroxyacyl-CoA dehydrogenase and 3-hydroxyacyl-CoA epimerase activities of this glyoxysomal ß-oxidation system have been copurified with one polypeptide by ion-exchange chromatography on phosphocellulose (P-11), dye-ligand chromatography on blue Sepharose and affinity chromatography on NAD-agarose. This protein gave a single band in polyacrylamide gel electrophoresis under denaturing conditions. Its subunit molecular weight was estimated to be 90,000. Its properties are compared with those of the trifunctional ß-oxidation enzyme of C. tropicalis.

The investigation of the biosynthesis and the intracellular transport of these ß-oxidation proteins in N.crassa should provide an attractive model by which the biogenesis of this unique type of microbodies can be studied.

[1] Wanner, G., and Theimer, R. (1982) Ann.N.Y.Acad.Sci. 386, 269-284
[2] Cooper, T.G., and Beevers, H. (1969) J.Biol.Chem. 244, 3514-3520
[3] Lazarow, P.B., and DeDuve, C. (1976) Proc.Natl.Acad.Sci. 73, 2043-2046
[4] Tanaka, A., Osumi, M., and Fukui, S. (1982) Ann.N.Y.Acad.Si. 386, 183-189

488

INTERACTION BETWEEN THE PLASMA MEMBRANE AND GLYCOLYSIS IN YEAST CELLS

Dietrich Kuschmitz and Benno Hess[o]
Max-Planck-Institut für Ernährungsphysiologie
Rheinlanddamm 201, 4600 Dortmund 1, FRG

In order to gain insight into the modes of dynamic coupling we analyzed the interaction of glycolysis and the cellular plasma membrane potential in yeast using glycolysis as the only ATP generating system (1), which drives the proton translocating proton ATPase of the plasma membrane (2). During this process the proton gradient is built up by active glycolysis and sets up a plasma membrane potential with or without companion movements of other ions. The simultaneous time analysis of the change of the activity of glycolysis and the plasma membrane potential is based on a record of NADH fluorescence as indicator of glycolysis and rhodamine 6G fluorescence as indicator of plasma membrane potential. Evidence that the plasma membrane proton translocating ATPase is directly involved in the generation of the rhodamine 6G indicated plasma membrane potential comes from a study in which the two processes have been uncoupled by appropriate inhibitors. Uncoupling agents such as sodium azide and the specific inhibitors of the ATPase diethylstilbestrol and vanadate ions (2) inhibit the formation of the plasma membrane potential and the rhodamine 6G response.

Upon activation of a suspension of yeast cells by addition of glucose glycolysis can be brought to a normal, non-oscillating steady state and to an oscillating state (3). The interaction between the plasma membrane potential and glycolysis was studied in the normal steady state by perturbation of either the membrane potential by electrogenically transported ions or of glycolysis by the addition of further amounts of glucose which decreases transiently the cellular ATP content. In both cases a quantitative coupling is found between the two systems as well as a threshold like on and off switch of the plasma membrane ATPase above and below respectively of a certain cellular NAD level which reflects the cellular ATP content.

Oscillating glycolysis drives the plasma membrane potential into an oscillating state. In a typical experiment the membrane potential oscillates between 50 and 100% of total rhodamine 6G fluorescence and the NADH fluorescence as indicator of glycolysis oscillates between 80 and 100% with a period of approximately 60 secs for both components. Comparing the maxima of the two oscillations it is obvious that both components are running with the same frequency. The relationship between rhodamine 6G and NADH fluorescence over one period can be studied by the phase plane plot technique. Here we distinguish three different phases during a limit cycle. In phase I, towards NAD and membrane potential maximum, both processes are synchronized. In the reverse reaction, phase II is characterized by a relatively fast rate of membrane potential decay but a slower rate of NAD reduction, whereas in phase III the two processes behave in the opposite way. At present time, a complete designation of the rate controlling steps in both coupled processes is not at hand.

On the other hand, the influence of the proton translocating system on glycolysis can be documented by affecting the plasma membrane potential from outside the cells using appropriate cations. Under non-oscillating conditions potassium, calcium and lanthanum ions induce oscillations of the plasma membrane potential and glycolysis.

It could well be that we are dealing here with a phenomenon of "chemical resonance" between two highly non-linear processes. The experiments illustrate furthermore a novel insight into the dynamic coupling as a time dependent phenomenon being of interest for understanding a number of other glycolyzing systems for which periodic states are known such as the heart muscle, smooth muscle and neural systems. Future experiments have to be designed to unveil the mechanisms of time pattern control under such in vivo conditions, when a large cellular network is coupled by an intercellular communication system.

1. Hess, B., Boiteux, A. and Kuschmitz, D. (1983) in Biological Oxidations. Springer Verlag Berlin, Heidelberg, New York, pp. 249-266.
2. Goffeau, A. and Slayman, C.W. (1981) Biochim. Biophys. Acta 639, 187-223.
3. Hess, B. and Boiteux, A. (1968) Hoppe-Seyler's Z. Physiol. Chem. 349, 1567-1574.

THE ROLE OF MITOCHONDRIA IN THE RELEASE OF TRANSMITTER SUBSTANCES FROM ISOLATED SYNAPTOSOMES

E. Ligeti[+] and V. Adam-Vizi[o]
[+]Department of Physiology and [o]2nd Institute of Chemistry and Biochemistry; Semmelweis Medical University, Budapest Hungary

Nerve terminals isolated from mammalian nervous tissue maintain high potential difference across their plasma membrane and they contain significant amount of different transmitter substances. The "calcium mechanism" (1) of neurotransmitter release seems to be valid also in the case of synaptosomes: depolarization of the plasma membrane opens specific voltage-sensitive calcium channels and the following increase of the cytosolic calcium concentration represents the main factor triggering the release of the neurotransmitter. However, various agents (ouabain, veratridine) were found to induce transmitter release even in the absence of extracellular calcium. These observations raised the possibility of participation of intracellular calcium stores in this process.

In the present study we manipulated the intracellular calcium content by different agents and followed the changes of the plasma membrane potential (2), of the liberation of ^{14}C-acetylcholine (ACh) (3) and the activity of the plasma membrane Na,K-ATPase enzyme. All the reported experiments were carried out in the absence of external calcium and the presence of 3 mM EGTA.

1. Effect of the calcium-ionophore A 23187

In the absence of extracellular calcium this ionophore was shown to mobilize intracellular calcium stores (4). As summarized in part A of Table I, under the above conditions A 23187 induced the release of ACh without any effect either on the membrane potential or on the activity of the Na,K-ATPase. The liberation of the transmitter was proportional to the concentration of the ionophore in the region of 10-50 μM. Above this value ACh release was not increased further.

2. Effect of mitochondrial uncoupling

Uncouplers increase the H^+-permeability of biological membranes; by this virtue they collapse mitochondrial membrane potential with the consequent inhibition of ATP synthesis, stimulation of mitochondrial ATPase activity and release of previously accumulated calcium. The effects of the uncoupler CCCP on synaptosomes are shown in part B of

Table I. In the concentration range of 50-500 nM, CCCP induced the release of transmitter substance but had no significant influence on the plasma membrane potential or the Na,K-ATPase activity. However, this enzyme was clearly inhibited by higher concentrations of the uncoupler, probably due to a drastic fall of the cellular ATP content. In this view the concomitant decrease of the plasma membrane potential is evident.

3. Relationship between ACh release and the extent of plasma membrane depolarization

It was found that ouabain (blocks Na,K-ATPase) and veratridine (opens Na-channels) induce already considerable ACh release when the plasma membrane potential is decreased by only a few mV. In contrast, when depolarization is carried out by high K^+ concentrations, no transmitter release can be detected even when the membrane potential is decreased by 20 mV. Mobilization of intracellular calcium by entering Na ions provides a possible explanation for the observed differences.

Table I. Effect of A 23187 (part A) and CCCP (part B) on isolated synaptosomes

	conc.	potential (mV)	ACh release (cpm/mg)	Na,K-ATPase (cpm)
A	--	71	580	2825
	10 μM	-	790	-
	25 μM	73	1010	2970
	50 μM	72	1150	2915
B	--	54	460	2880
	50 nM	51	700	-
	100 nM	50	950	-
	500 nM	51	1160	2810
	1 μM	45	1370	1950

References
(1) Katz,B. and Miledi,R. (1967) J.Physiol. 189 533-544
(2) Scott,I.D. and Nicholls,D.G. (1980) Biochem.J; 186 21-33
(3) Wonnacott,S. and Marchbanks,R.M. (1976) Biochem.J. 156 701-712
(4) Chen,J.L.J., Babcock,D;F. and Lardy,H.A. (1978) Proc.Nat.Acad.Sci.USA 75 2234-2238

INTERRELATIONSHIP OF Na$^+$ FLUXES AND INTERNAL pH ACIDIFICATION IN AN ALKALOPHILE.

D.McLaggan, M.J.Selwyn and A.P.Dawson
School of Biological Sciences, University of East Anglia,
Norwich, NR4 7TJ, England.

Our previous work, (1) has shown that when the external Na$^+$ concentration was very low, cells of the facultative alkalophile, Exiguobacterium aurantiacum, (2) were unable to regulate their cytoplasmic pH. However, on addition of Na$^+$ ions to a low-Na$^+$ high-pH incubation medium, the internal pH of the bacteria was rapidly adjusted to a lower level. Increasing the magnitude of the jump in Na$^+$ concentration increased the rate of pH$_i$ acidification and, in the presence of glucose, produced a transient overshoot of pH$_i$ acidification. These data were interpreted as being in accord with proposals, (3) that bacterial cytoplasmic pH is controlled by a Na$^+$ cycle consisting of a Na$^+$/H$^+$ antiport, (4,5) with a comparably rapid entry mechanism for Na$^+$ and, in addition, suggested that the Na$^+$ entry pathway is controlled by the internal pH.

Measurements have now been made of the Na$^+$ fluxes following jumps in external Na$^+$ concentration, which are accompanied by acidification of the cytoplasm. Addition of unlabelled Na$^+$ to cells incubated in buffer (pH 9.5) to raise the Na$^+$ concentration from 0.1mM to 0.7mM resulted in a transient 45% increase in ^{22}Na$^+$ in the cell fraction although the external specific activity dropped by a factor of 7. Assuming isotopic equilibrium before and after addition of 0.6mM unlabelled Na$^+$, the internal Na$^+$ concentration increased from 0.22mM to 2.5mM in approximately 2 minutes then subsequently declined to 1.9mM. Throughout this period, the cytoplasmic pH fell monotonically from pH 8.94, before the jump in external Na$^+$ concentration, to about pH 8.48 at the peak of internal Na$^+$ concentration and to pH 8.34 during the phase of net Na$^+$ efflux. Addition of glucose at this point caused a rise in the internal pH of nearly 0.2 pH units, accompanied by a rapid drop in internal Na$^+$.

These data indicate that Na$^+$ efflux does not occur solely via the Na$^+$/H$^+$ antiporter and suggest a minimum of three different routes for Na$^+$ transport. One coupled to proton translocation, one independent of proton translocation, and the third at least partially dependent on the presence of glucose. The first two modes of transport constitute the pH regulatory Na$^+$ cycle but these data do not distinguish between the alternatives of an Na$^+$ influx pathway with an Na$^+$/H$^+$ antiport and an Na$^+$/H$^+$ symport with an Na$^+$ efflux pathway. The latter mechanism is, however, difficult to reconcile with the previously reported overshoot in internal acidification following a jump in external Na^+ concentration (1). When the jump in external Na^+ was increased approximately 10-fold, to produce a final Na^+ concentration 6.1mM, the resulting Na^+ influx increased by a similar factor. Under these conditions (pH 9.5, minus glucose), the internal Na^+ concentration reached a peak of about 18mM 15 seconds after the addition of 6mM Na^+, then fell to about 9mM after a further 1.5 minutes. The internal pH fell monotonically by about 0.7 pH units to reach a steady state within 2 minutes of the addition of Na^+. Thus a 10-fold increase in the jump in external Na^+ concentration produced a similar increase in the maximal rise in internal Na^+ but only a comparatively small (<30%) increase in the change in internal pH. This lack of proportionality between Na^+ influx and H^+ influx is evidence in favour of a cycle involving a H^+-independent Na^+-influx pathway with an Na^+/H^+ antiporter rather than the Na^+/H^+ symport plus a H^+-independent Na^+-efflux cycle since the latter predicts proportionality between Na^+-influx and H^+-influx.

The rise in internal Na^+ to a concentration considerably greater than the external Na^+ concentration means that Na^+ entry must be an active transport process such as a Na^+ translocating ATPase or an electrophoretic movement in response to the membrane potential or an exchange of Na^+ for an internal solute. In the second case outward movement of a cation is required to maintain the membrane potential and in both this and the last case K^+ ions are the most likely candidate. This rapid reversal of the transmembrane Na^+ concentration gradient on addition of Na^+ to cells in a low-Na^+ medium thus permits rapid acidification of pH_i since the Na^+/H^+ antiporter could, depending on the stoichiometry, be driven by the membrane potential as well as by the Na^+ concentration gradient. These measurements of Na^+ fluxes thus provide further evidence for the control of cytoplasmic pH in this organism by the operation of a Na^+ cycle across the cell membrane in which one step couples transport of Na^+ and H^+. The energy requirement of this process is also shown to involve Na^+ transport and there appears to be at least one other mode of Na^+ transport.

1. McLaggan, D., Selwyn, M.J. and Dawson, A.P. (1984) FEBS Lett. 165, 254-258.
2. Collins, M.D., Lund, B.M., Farrow, J.A.E. and Schleifer, K.H. (1983) J. Gen. Microbiol. 129, 2037-2042.
3. Booth, I.R. and Kroll, R.G. (1983) Biochem. Soc. Trans. 11, 70-72.
4. Guffanti, A.A., Susman, P., Blanco, R. and Krulwich, T.A. (1978). J. Biol. Chem. 253, 708-715.
5. Padan, E., Zilberstein, D. and Rottenberg, H. (1976). Eur. J. Biochem. 63, 533-541.

INVOLVEMENT OF AN UBIQUINONE- AND CYTOCHROME b-RICH GRANULE IN SUPER-
OXIDE PRODUCTION BY HUMAN NEUTROPHILS.

F. Mollinedo and D. L. Schneider.

Department of Biochemistry, Dartmouth Medical School, Hanover, New
Hampshire 03756 (U.S.A.)

Human neutrophils display a sharp increase in oxygen consumption
and superoxide and hydrogen peroxide production ("respiratory burst")
when exposed to surface stimulation or ingestible particles. Although
the molecular mechanism for the reduction of molecular oxygen to superoxide, the first step in the generation of oxygen metabolites, remains to be determined, there is increasing evidence for an electron
transport chain. Cytochrome b and ubiquinone have been implicated as
two components of that putative chain (1,2). Subcellular fractionation studies in resting human neutrophils indicated that ubiquinone
and cytochrome b cofractionate in a novel tertiary granule (3). It
was also found that gelatinase, DCCD-sensitive, Mg^{2+}-dependent ATPase
and small amounts of acid hydrolases cosedimented with this tertiary
granule (3).

When cells were activated by phorbol myristate acetate (PMA), sucrose gradient fractionation indicated that the oxygen consumption
activity was found in the plasma membrane region as well as in dense
parts of the gradient (Fig. 1). Concomitantly, shifts of ubiquinone
and cytochrome b to these positions in the gradient were observed
(Fig. 1). Furthermore, 5'-nucleotidase (AMPase), a plasma membrane
marker, was also shifted to dense regions of the sucrose gradient.

Preincubation of neutrophils with cytochalasin E resulted in a
decrease of the lag time for oxygen uptake upon treatment with PMA.
Shifts in ubiquinone, cytochrome b and 5'-nucleotidase to deeper positions in the gradient were observed after cytochalasin E treatment
(Fig. 1). In addition, a rather specific release of gelatinase activity occurred.

These results indicate that both ubiquinone and cytochrome b are
involved in the respiratory burst activity, since they comigrate with
the oxygen uptake activity on gradients of extracts from PMA-activat-
ed cells. Also, the findings indicate that fusion of an ubiquinone-
and cytochrome b-rich tertiary granule with plasma membrane occurs
when cells are activated with PMA or treated with cytochalasin E.
However, in the cytochalasin E treated cells only a decrease in the
lag time was observed, not induction of the respiratory burst. A different degree of membrane fusion and the requirement of a further
signal arising from incubation with PMA are possible explanations.

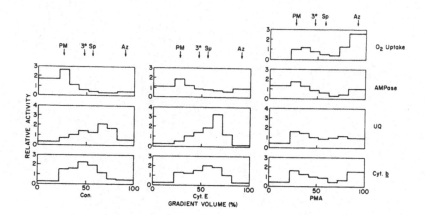

Figure 1. Sucrose gradient centrifugation of resting, cytochalasin E- and PMA-treated neutrophils.

Extracts from resting, cytochalasin E- and PMA-treated cells were fractionated by rate zonal centrifugation at 70,000 x g for 15 min. Plots of relative activity (RA) versus percent volume are given, where RA is the percent of activity in a fraction divided by the percent volume collected in that fraction. The specific and azurophilic markers were not significantly shifted under the experimental conditions used. PM, plasma membrane; 3°, tertiary granule; Sp, specific granule; Az, azurophilic granule.

On these grounds, it is reasonable to suggest that membrane fusion between tertiary granules and plasma membrane is a necessary but not sufficient condition for cell activation.

References

1. Segal, A.W. and Jones, O.T.G. (1978) Nature 276, 515-517.
2. Crawford, D.R. and Schneider, D.L. (1983) J. Biol. Chem. 258, 5363-5367.
3. Mollinedo, F. and Schneider, D.L. (1984) J. Biol. Chem. In Press.

MITOCHONDRIAL NADH——UBIQUINONE OXIDOREDUCTASE (COMPLEX I) DEFICIENCY IN AN INFANT WITH CONGENITAL LACTIC ACIDOSIS

R. Moreadith°, M. Batshaw", T. Ohnishi', B. Reynafarje°, and A. L. Lehninger°
°Department of Biological Chemistry, Johns Hopkins School of Medicine, Baltimore, Maryland, 21205, U.S.A.
"Department of Pediatrics, John F. Kennedy Institute, Baltimore, Maryland, 21205, U.S.A.
'Department of Biochemistry and Biophysics, University of Pennsylvania Philadelphia, Pennsylvania, 19104, U.S.A.

A white male, the product of a Gravida 2, Para 1 non-consanguineous 34 wk pregnancy presented in the first month of life with hypoglycemia, progressive lactic acidosis (5 → 31 mM, normal < 2.5) with increased lactate/pyruvate ratio (64:1) and hyperalaninemia (2.3 mM, nl < .44). Total plasma carnitine was 25 μM (normal > 37). GLC of urine showed no accumulation of organic acids. There was progressive hypotonia, hepatomegaly, brain CT abnormalities and respiratory insufficiency leading to death at 16 wks. Enzymatic determination of fibroblasts, liver and skeletal muscle biopsy samples ruled out pyruvate carboxylase, pyruvate dehydrogenase, PEPCK, glucose 6-phosphatase and fructose 1,6 bisphosphatase deficiencies. There was increased glycogen and lipid and giant mitochondria with concentric whorls in liver and muscle. Freshly isolated mitochondria from 4 tissues revealed a marked deficiency of NAD linked respiration but normal succinate linked respiration. Reduction of cyt c by NADH (rotenone sensitive) in permeabilized mitochondria revealed a deficiency of complex I: 5 nmol cyt c reduced/min/mg in skeletal muscle (control 206), 15 nmol cyt c reduced/min/mg in liver (control 131), 35 nmol cyt c reduced/min/mg in kidney (control 107), and 56 nmol cyt c reduced/min/mg in heart (control 349). However, complex I activity reactive with ferricyanide was identical to the normal control, suggesting the deficiency was in the Fe:S centers of complex I[1]. Electron paramagnetic resonance spectroscopy of liver submitochondrial particles revealed a selective absence of the Fe:S centers characteristic of complex I; the other mitochondrial Fe:S centers appeared normal in behavior. This represents the first case of a deficiency of the Fe:S centers of mitochondrial complex I[2].

(1) Ohnishi,T., Salerno,T.C.,(1982) in Iron-Sulfur Proteins, John Wiley and Sons, Inc., New York4, 283-327

(2) Moreadith,R.W. et al. (1984) T.Clin.Invest.74, in press.

RESPIRATORY CHAIN AND CALCIUM PUMP IN MUSCULAR CELLS FROM
MED AND CONTROL MICE.

Christian NAPIAS, Claude SARGER and Jean CHEVALLIER.
I.B.C.N. du C.N.R.S., 1 rue Camille Saint-Saëns 33077 Bordeaux
Cedex FRANCE.

Motor end-plate disease (MED) is an hereditary autosomal and
recessive neurological disorder (single gene mutation) in the
mouse (1). Two main allelic forms of this disease exist of which
the less severe Med^J/Med^J allows a biochemical study.
Pathological changes exhibited at the nerve endings and at the
skeletal muscle levels have been already described (2).

The degenerative changes of the skeletal muscles are obvious
eventhough the primary target of the mutation remains unknown. Up
to now, no biochemical study of the muscle has been undertaken.
Calcium ions, is thought to be involved in most of the
degenerative processes affecting the muscle cells (3). The aim of
this study is an attempt to detect alterations due to the
mutation at the levels of two internal membrane systems involved
in calcium pumping processes : mitochondria (MIT) and
sarcoplasmic reticulum (SR).

By using a procedure adapted to minute amounts of muscle
tissue (4) we isolated MIT and SR from both control (+/+ or
+/Med^J) and affected littermates. Possible cross contamination
and plasma membrane contamination of MIT and SR were
investigated. The fractions obtained by this way were pure enough
for such a biochemical study.

<u>Mitochondria</u> : Possible alteration of the respiratory chain
was investigated. Respiratory rates at states III and IV,
respiratory control ratios (RCR) and ADP/O ratios were measured.
They never showed significant modification of their respective
values for the affected animals when compared to their controls
whatever being the respiratory substrate tested. Morever, when
adding calcium (30 uM) to MIT maintained at state IV, it was
possible to show that the Ca^{2+} and the ATP synthesis were
competitive processes in a very similar way for both control and
affected littermates. The observed respiratory rate in presence
of ADP added after calcium was decreased to 60 % of the normal
value measured in the absence of calcium (state III).

Sarcoplasmic reticulum : The coupling between calcium uptake and ATP hydrolysis was investigated. Calcium transport was significantly lowered for the mutants. This phenomenon was related to partial uncoupling between transport and ATP hydrolysis as shown by an increased basal ATPase activity. This was occuring in spite of the large amount of EGTA (2 mM) added in the homogenization buffer. SR vesicles obtained from mutants behaved as normal SR in which one would have added detergent. In addition to this, freeze-thawing of such SR fractions always led to total loss of calcium transport activity and a related totally uncoupled Ca^{2+}-ATPase. This observation has been correlated to increased amounts of low molecular weight proteins.

From these studies, it is obvious that the mitochondria are not the main target of this mutation. For sarcoplasmic reticulum vesicles, further experiments are necessary for a better understanding of the alteration of calcium fluxes through this membrane.

BIBLIOGRAPHY :

1- DUCHEN, L.W., SEARLE, A.G. and STRICH, S.J., (1967) J. Physiol. (London) 189, 4P-6P.
2- PINCON-RAYMOND, M. and RIEGER, F., (1981) Biol. Cell. 40, 189-194.
3- WROGEMANN, K., HAYWARD, W.A.K. and BLANCHAER, M.C., (1979) Ann. N. Y. Acad. Sci. 317, 30-43.
4- ROCK, E., NAPIAS, C., SARGER, C. and CHEVALLIER, J., (1984) Biochimie (submitted).

CELLULAR DIFFERENTIATION AND PLASMA MEMBRANE H$^+$ ATPase IN DICTYOSTELIUM DISCOIDEUM

R. Pogge von Strandmann[+], J-P. Dufour[*]
+ Imperial Cancer Research Fund, London NW7, England
* Laboratory of Brewing Science and Technology, Place Croix du Sud, 3
  Catholic University of Louvain, B 1348 Louvain-la-Neuve, Belgium.

During development of Dictyostelium discoideum, individual amoeba are initially capable of differentiation into either stalk or spore cells and the choice between these appears to be regulated by an endogenous factor DIF, which diverts amoeba to stalk cell differentiation. From indirect in vitro experiments we proposed that DIF lowers intracellular pH in Dictyostelium which then acts as the switch between stalk and spore cell differentiation with low pHi favouring stalk cells (1).

In eukaryotes such as fungi and yeast, pHi appears to be regulated by an outwardly-directed plasma membrane proton pump which is sensitive to vanadate, the synthetic estrogen diethylstilbestrol (DES) and miconazole (2,3).

In Dictyostelium both DES and miconazole, as well as several natural estrogens, mimick the natural inducer DIF and induce stalk cell differentiation. It seemed likely that Dictyostelium possesses a fungal type proton pump which maintains the cytoplasmic pH. After inhibition of this pump by DES or miconazole, the drop in pHi could induce stalk cell differentiation.

Here we describe the discovery of such a proton pump in Dictyostelium, which can be assayed as vanadate sensitive ATPase or by ATP dependent pumping of protons into liposomes. Assayed as an ATPase, the particulate proton pump activity comigrates with a plasma membrane marker (iodine surface label) and is well separated from a mitochondrial marker (succinate dehydrogenase) and the lyzosomal marker acid phosphatase. The ATPase activity is similar to the yeast plasma membrane proton pump in its sensitivity to vanadate, DES and miconazole and the high Km for ATP. In addition it is sensitive to natural estrogens. Comparing biologically equipotent amounts of the synthetic DES and the natural DIF, then DIF is a much weaker inhibitor of the ATPase than DES, suggesting that DIF might act through a different route.

The enzyme can be solubilized from the plasma membrane with lysolethicin and reconstituted into asolectin liposomes. Using an acridine dye as a ΔpH probe, we can show ATP dependent proton pumping (Fig.1a) which is electrogenic, as the rate of pumping increases after

abolition of the membrane potential with valinomycin an K^+ (Fig.1b).
Pumping is sensitive to vanadate (Fig.1c), and less efficient with
CTP and UTP as substrates (Fig.1d,e).
However, it is not possible to demonstrate clearly the effects of
lipid soluble inhibitors like DES, estrogens and DIF.

1. Gross, J.D., Bradbury, J., Kay, R.R. and Peacey, M.J. Nature
   303, 244
2. Bowman, B.J., Mainzer, S.E., Allen, K.E. and Slayman, C.W. (1978)
   Biochim. Biophys. Acta 512, 13-28
3. Dufour, J-P., Boutry, M. and Goffeau, A. (1980) J. Biol. Chem.
   255, 5735-5741

ELECTROGENIC PROTON PUMPING

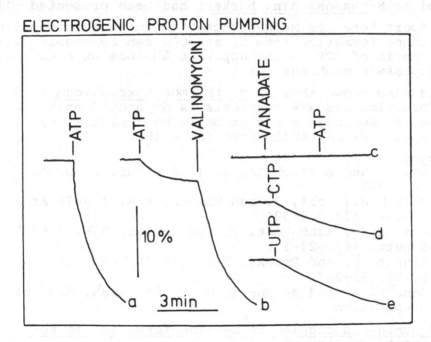

Figure 1. Quenching of a fluorescent acridine dye as a measure of the
ATP-dependent proton translocation (a); stimulation of the quenching
rate by abolition of the membrane potential (valinomycin + K^+) (b);
vanadate (500 µM) inhibition of the proton translocation (c); CTP and
UTP-dependent proton tranlocations (d,e).

THE FUNCTION OF $\Delta\mu H^+$ AND OF Na^+ IN THE ENERGY METABOLISM OF <u>METHANOBACTERIUM</u> <u>THERMOAUTOTROPHICUM</u>

P. Schönheit, and D.B. Beimborn
Mikrobiologie,FB Biologie, Philipps-Universität Marburg
3550 Marburg/Lahn

Methanogenic bacteria are strictly anaerobic archaebactheria that couple the exergonic process of methane formation with the synthesis of ATP. The mechanism of coupling is not yet fully understood. However, recently strong experimental evidence for a chemiosmotic coupling mechanism of ATP synthesis and methane formation from H_2 and methanol in <u>Methanosarcina</u> <u>barkeri</u> has been presented (1).

We report here for <u>Methanobacterium</u> <u>thermoautotrophicum</u> that methane formation from H_2 and CO_2 can be coupled with the synthesis of ATP in the apparent absence of a $\Delta\mu H^+$ over the cytoplasmic membrane (A).

It is also shown that in <u>M. thermoautotrophicum</u> both methane formation and ATP synthesis is dependend on Na^+ and that the Na^+ dependency can probably be explained by the presence of a Na^+/H^+ antiporter (2-5) (B).

References:
(1) Blaut, M. and Gottschalk, G. (1984) Eur.J.Biochem. 141,217-222.
(2) Perski, H.J., Moll, J. and Thauer, R.K. (1981) Arch. Mikrobiol. 130,319-321
(3) Perski, H.J., Schönheit, P. and Thauer, R.K. (1982) FEBS Lett. 143,323-326.
(4) Schönheit, P. and Perski, H.J. (1983) FEMS Microbiol. Lett. 20,263-267.
(5) Schönheit, P. and Beimborn, D. (1984) Arch. Microbiol. in preparation

(A) <u>ATP synthesis coupled to methane formation in the apparent absence of a $\Delta\mu H^+$</u>
1. Cell suspensions of M. thermoautotrophicum which synthesize CH_4 from H_2/CO_2 at a rate of 2.7 µmol/minx mg cells show a $\Delta\mu H^+$ across the cytoplasmic membrane of about 230 mV (inside negative) ($\Delta\psi$=-220 mV, $\Delta pH > -20$ mV) and an ATP content of 5-8 nmol/mg cells. The addition of low concentrations of the protonophore TCS (Tetrachlorosalicylanilide) resulted in an increase of the CH_4 formation rate and in a complete dissipation of $\Delta\mu H^+$, but had no effect on the ATP content of the cells.

2.Net ATP synthesis during methane formation from H_2/CO_2 was measured in cell suspension of M. thermoautotrophicum in the apparent absence of a $\Delta\mu H^+$. The dissipation of $\Delta\mu H^+$ was obtained by treatment of the cells with either TCS (Fig.1), with potassium in the presence of valinomycin (50 mM K^+,12 nmol valinomycin/mg cells), or with TPP (Tetraphenylphosphonium chloride) (3 µmol/mg).

Fig.1: ATP synthesis coupled to methane formation from H_2/CO_2 in M. thermoautotrophicum in the apparent absence of $\Delta\mu H^+$. (TCS concentration: 5nmol /mg)

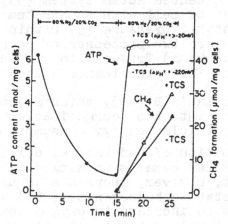

(B) Na^+ dependence of CH_4 formation and ATP synthesis

1.CH_4 formation in cell suspensions of methanogenic bacteria is dependend on low concentrations of Na^+. Li^+ can replace Na^+, other cations as K^+ and NH_4^+ had antagonistic effects on the Na^+ (Li^+)stimulated CH_4 formation.

2.The requirement for Na^+ (Li^+) is highest at acidic external pH (pH<6). Under these conditions the formation of a ΔpH is dependent on Na^+ in the medium(shown for M. thermoautotrophicum).

3.ATP synthesis driven by a potassium diffusion potential (in the absence of methane formation) is greatly stimulated by Na^+ or Li^+.

4.Cells of M. thermoautotrophicum contain a Na^+/H^+ antiporter which also works with Li^+. The antiporter is most active at acidic external pH (<pH6) and seemed to be inhibited by K^+. It is assumed that the Na^+(Li^+)/H^+ antiporter is functionally involved in both (Na^+ (Li^+) stimulated methane formation and ATP synthesis - driven by a potassium diffusion potential.

504

IS THE ENERGY FOR THE ACTIN NETWORK SUPPLIED FROM
GLYCOLYSIS ?

U. Tillmann and J. Bereiter-Hahn
Arbeitskreis Kinematische Zellforschung,
FB. Biologie der Johann Wolfgang Goethe-Universität,
Senckenberganlage 27, D-6ooo Frankfurt/Main, F.R.G.

The cellular network of actin filaments participates
in generation and maintainance of cell shape and locomo-
tion (1). ATP-dependent interactions between actin and
myosin (2) develop a tensed actin system (3), which determines the attachment of cell with substratum (4) and
maintains of hydrostatic pressure (5) within the cells.
Consequently, there are two processes of energy requirement of actin meshwork: a) Treadmilling of actin, b) Continuous contractions for maintainance a tensed actin
meshwork (6,7).
Therefore cytochalasin D (CD), which decomposes the
actin system and prevents its reorganisation (8), should
induce a change of cellular energy consumption (9).

1) Oxygen consumption of endothelial cells from tadpo-
le hearts (XHT-2 cells) remains unchanged on CD treatment
(Tab.). ATP-content, however, decreases about 25 % (Tab.).
2) Inhibition of lactate dehydrogenase (LDH) with 6 mM
oxamic acid induces an increase in oxygen consumption up
to twice the control value, but a reduction of ATP-content to approximatly 2o-25 % (Tab.). 3) In cells treated
with oxamic acid, CD leads to a reduction of oxygen consumption as well as ATP-level (Tab.).

Conditions	Respirations rate in %	ATP-content in %
Medium (control)	53	1oo
CD	52.5	75
Oxamic acid	1oo	78
Oxamic acid + CD	7o	56

We conclude that XTH-2 cells have a high glycolytic
ATP-production, which supplies the cytoskeletal energy
part.

CD-induced and ATP-dependent destruction of cytoskeleton, without influence on mitochondrial activity, is possible only when other energy producing processes exist.
Such phenomenon is described as Warburg-effect (1o). Thus

pyruvate is not only catalyized to acetyl-CoA, but is also used for lactate produktion, giving an oxygen-independent ATP-production sytem. Therefore, inhibition of LDH leads to an energy deficiency, which is compensated by an enhancement of respiration (Tab.). But the ATP-level reached only to 78 %, which is not clearly understood. Possible this could be due to a lack of mitochondrial capacity or a change in ATP-turn-over-time. However, under this condition whole ATP-quantity (i.e. 78 % of control value) is produced by mitochondria. Consequently, it may be presumed that in untreated cells, showing 5o % respiration, approximatly 4o % of ATP-production is supplied by mitochondria and remaining 6o % by glycolysis. This is supported by the observation that ATP-level goes down to 56 %, in cells treated with 1o uM antimycin A, which blocks mitochondrial respiration (unpublished).

CD-induced cytoskeletal destruction of cells with and without oxamic acid treatment, present equal ATP-content reduction. However in oxamic acid treated cells in addition oxygen consumption is also decreased to 3o % in relation to ATP-content.

In view of above observations we suggest, that 25 % of the whole cellular energy production is needed for maintainance of cytoskeleton in addtion to all other involved processes. Futher, the supply of the cytoskeletal energy part is reached by glycolysis, because under oxamic acid treatment the mitochondrial activity decreases.

1. Bereiter-Hahn, J. et. al. 1981. J.Cell Sci. 52:289-311
2. Kreis, T.E., Birchmeier, W. 198o. Cell 22:555-561
3. Strohmeier, R., Bereiter-Hahn, J. 1984 in press
4. Rees, D.A. et.al. 1977. Nature 267:124-128
5. Di Pasquale, A. 1975. Expl. Cell Res. 94:191-215
6. Simpson, P.A., Spudich, J.A. 198o. PNAS 77:461o-4613
7. Bereiter-Hahn, J., Tillmann, U. 1984 in press
8. Miranda, A. et.al. 1974. J. Cell Biol. 61:481-5oo
9. Schliwa, M. 1982.J. Cell Biol. 92:79-91
1o. Racker, E. 1976. Academic Press New York, San Francisco a. London

THERMODYNAMIC DESCRIPTION OF BACTERIAL GROWTH

K.van Dam and M.M.Mulder
Laboratory of Biochemistry, University of Amsterdam, Plantage Muidergracht 12, 1018 TV Amsterdam, The Netherlands

Bacterial growth can be adequately described, starting from a simple
model in which catabolism and anabolism are coupled by the intracellular ATP pool (1). The kinetics of the partial reactions in this model
are written in the form of equations, in which velocities are dependent on thermodynamic forces. Thus, an overall description in terms of
so-called Mosaic Non-Equilibrium Thermodynamics is obtained. Some experimentally easily obtainable situations lead to simplifications. For
instance, a steady state is rapidly reached in which the intracellular
concentration of ATP is constant. In this steady state it can be derived that a linear relation exists between the rate of catabolism and
the rate of bacterial growth. This has been experimentally confirmed.
The slope of the line and the intersection point with the ordinate
depend on both the growth conditions and the properties of the micro-
organism, i.e. which substrate is limiting, which enzymes are present,
etc.

Bacteria need special devices to balance the different metabolic
fluxes: in the steady state both the net flux of energy and the net
flux of reducing equivalents must be balanced. Thus, sometimes mechanisms must be found to dissipate energy, sometimes excess reducing e-
quivalents must be eliminated. To this end, the microorganisms must be
able to switch between different metabolic routes. For instance, the
induction of two uptake systems with different stoichiometries of energy utilisation may lead to the generation of a cycle of energy dissipation. The effect of different variables on the growth pattern is
discussed.

1.H.V.Westerhoff, J.S.Lolkema, R.Otto & K.J.Hellingwerf (1982) Biochim.
Biophys.Acta 683, 181-220

QUANTIFICATION OF ENERGY CONSUMPTION IN HUMAN PLATELETS
A.J.M. Verhoeven, and J.W.N. Akkerman
Dept. of Haematology, University Hospital Utrecht, The Netherlands

Human blood platelets are anucleated cells that when stimulated with specific agonists stick to one another (aggregation) and secrete the contents of three types of granules. Concurrently, anaerobic and aerobic energy generation accelerate, suggesting that extra metabolic energy is required for these functions. Platelets are ideally suited for the study of the energy cost of receptor-mediated responses, since metabolic ATP and ADP are the only known rapidly accessible intermediates for energy transduction (no phosphagens; very slow exchange with other nucleotides) and these metabolites can be precisely measured with isotopic tracer techniques (1). On this basis, we have developed a novel technique for the quantification of energy consumption in gel-filtered platelets (2). ATP regeneration is abruptly blocked and the decrease of radio-labeled ATP and ADP is determined in the subsequent 15 s interval (Fig.1). Metabolic arrest is induced by two techniques: A) by adding a mixture of 2-deoxy-D-glucose (30 mM) and glucono-1,5-lactone (GLAC; 10 mM) to platelets preincubated with 1 mM KCN, and B) by adding antimycin A (15 µM) plus GLAC (10 mM) to platelets suspended in a glucose-depleted medium. Both techniques give energy consumption rates that are close to the values calculated from lactate production plus oxygen consumption of uninhibited suspensions (Table I). However, energy consumption is decreased after preincubation with KCN (technique A).

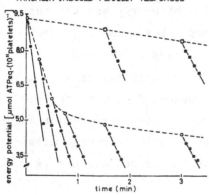

ASSESSMENT OF THE ENERGY CONSUMPTION DURING THROMBIN-INDUCED PLATELET RESPONSES

Fig.1. Platelets are stimulated with 5 U/ml thrombin at t = o (o;●). Antimycin A and GLAC (technique B) were added either simultaneously with thrombin or at 15, 30, 45, 90 and 180 s thereafter, and the subsequent ATP hydrolysis was analyzed (●). Similarly, energy consumption was measured in unstimulated cells (□ ; ■) by adding the inhibitors in the absence of thrombin.

Table I: Energy consumption in unstimulated platelets

Conditions	Method	Energy consumption (μmol ATPeq/min/10^{11} cells)
Glucose 1 mM Albumin 0.2%	lactate + O_2	6.5 ± 1.4 (9)
+ KCN	lactate	3.4 ± 0.8 (15)
	metabolic arrest (A)	3.5 ± 0.5 (7)
Glucose-free Albumin-free	lactate + O_2	7.3 ± 0.9 (5)
	metabolic arrest (B)	6.2 ± 0.9 (14)

mean ± S.D. (n)

Upon stimulation with thrombin (5 U/ml) energy consumption immediately accelerates to 17.7 ± 0.4 (4) μmol ATPeq/min/10^{11} cells (technique B); thereafter it decreases again and normalizes at 45 s after stimulation, which is when secretion is completed. The increase in energy consumption correlates with the velocities of the three secretion responses, both when different doses of thrombin are used and when other agonists (collagen, A23187) initiate these responses. No such correlation is found when the total energy consumption is compared with secretion. This indicates that secretion is tightly coupled to the increment in concurrent energy consumption.

More detailed analysis of the first 15 s after addition of the agonists reveals an increase in energy consumption that precedes the extracellular appearance of secretion markers. This phase can be studied by first stimulating the platelets with thrombin and at 2 s thereafter removing thrombin from its receptor by excess of hirudin (3). The initial receptor binding induces an increase in energy consumption that is paralleled by phosphorylation of 20 K and 40 K proteins, and alterations in the metabolism of (poly)phosphoinositides in the absence of liberation of granule markers. Hence, the increase in energy consumption in stimulated platelets reflects the energy cost of both induction and execution of secretory responses.

1. Daniel, J.L., Molish, I.R., and Holmsen, H. (1980). Biochim. Biophys. Acta 632, 444-453.
2. Akkerman. J.W.N., Gorter, G., Schrama, L. and Holmsen, H. (1983). Biochem. J. 210, 145-155.
3. Holmsen, H., Dangelmaier, C.A., and Holmsen, H.-K. (1981). J. Biol. Chem. 256, 9393-9396.

EFFECT OF SUCCINYLACETONE ON GROWTH AND RESPIRATION OF MALIGNANT, MURINE L1210 LEUKEMIA CELLS

E.C. Weinbach[*] and P.S. Ebert[+]

[*]Laboratory of Parasitic Diseases, National Institute of Allergy and Infectious Diseases, Bethesda, Maryland 20205, U.S.A.
[+]Laboratory of Molecular Oncology, National Cancer Institute, Bethesda, Maryland 20205 U.S.A.

Succinylacetone (4,6-dioxyheptanoic acid, SA) is a specific, irreversible inhibitor of δ-aminolevulinic acid dehydrase (1,2), the second enzyme of the heme biosynthetic pathway. The effects of SA on growth and respiration of L1210 cells were examined to determine the basis of its action on these vital physiological processes. Murine erythroleukemia (MEL) cells treated with SA exhibited a decreased heme content, and at a certain critical low heme concentration, these cells ceased dividing (1). In Walker 256 carcinoma and L1210 cells, however, SA did not decrease cellular heme content (3), yet these cells also stopped dividing. SA was not immediately toxic to L1210 cells. Cell division was restricted only after incubation with 2-3 mM SA for 2 days, without any detectable drop in heme levels.

As expected, L1210 cells in the stationary phase consumed less oxygen than cells in the log phase. Cells grown in the presence of 3 mM SA exhibited a modest decrease in respiration after 3 days, when cell division stopped. Both untreated and SA-treated cells responded similarly to mitochondrial inhibitors and uncoupling agents, showing that SA had no specific effects on electron transport, coupled phosphorylation, and respiration. SA may affect the production of a highly sensitive pool of heme-containing enzymes or cytochromes at levels not presently detectable. Our knowledge of SA's action at this time, accordingly, is limited to its inhibition of δ-aminolevulinic acid dehydrase. The mechanism by which SA inhibits growth of L1210 cells remains to be elucidated.

1. Ebert, P.S., Hess, R.A., Frykholm, B.C., Tschudy, D.P. (1979) Biochem. Biophys. Res. Commun. 88, 1382.

2. Tschudy, D.P., Hess, R.A., Frykholm, B.C. (1981) J. Biol. Chem. 256, 9915.

3. Tschudy, D.P., Ebert, P.S., Hess, R.A., Frykholm, B.C., Atsmon, A. (1983) Oncology 40, 148.

ON THE VARIATION OF THE HEAT DISSIPATION BY CHEMOORGANOTROPHIC
BACTERIA UNDER THE INFLUENCE OF PENTACHLOROPHENOL (PCP), PHENOL,
2-NITROPHENOL AND MERCURIC-CHLORIDE

Peter Weppen and Dieter Schuller
Universität Oldenburg, Fachbereich Chemie, Postfach 2503
D-2900 Oldenburg, Federal Republic of Germany

Biological thermodynamics and modern biological calorimetry have been
applied to study the bioenergetics of microbial populations on a large
scale [1]. In the field of ecotoxicology thermodynamic aspects of
the mode of action of environmental pollutants have been proposed as
an important subject of investigation [2] .

In a chemostat-culture the physiological state of a bacterial
population depends heavily on the dilution-rate (D). The D-rate
determines the mass- and energy-balance of the chemostat [3], the
'Available Reaction Potential' [4] or the nucleic acid composition
of the biomass. the former has been studied calorimetrically [5]
[6] and confirms the theoretical concept of the steady state
behaviour of the chemostat. Investigations on the interaction of
environmental pollutants and chemostat-cultures on the other hand have
been published scarcely [7].

We analyzed the action of four chemicals being standard substances
of environmental toxicity research on chemostat cultures of
Acinetobacter calcoaceticus and Escherichia coli calorimetrically as
has been published recently [8]. The technique allowes to record the
heat dissipation of the culture continuously and to analyse transient
effects quantitatively. Dose-response relations were obtained at fixed
D-rates and various initial concentrations of the particular chemical
as well as vice versa. Figure 1 presents the response of Escherichia
coli growing on glucose limited synthetic medium following the
addition   of 1.27 mMol PCP/l at three D-rates.

With the aid of direct calorimetry on chemostat cultures we found
a systematic dependence of toxic effects of chemicals on the D-rate
which allows to differentiate between decoupling agents and groth-
inhibiting chemicals (eg PCP and Mercuric-chloride). The calculation
of toxicity data from dose-response-analyses has been demonstrated to
be valid and comparable to standard test procedures [8], but the
correlation of physiological conditions and toxicity related to the
energy balance can not be analyzed by the simple tests.
The energetic yield-coefficient as defined by Prochazka et al [9]
was found to be an unifieing parameter to describe the behaviour of

decoupling agents. It described decoupling by means of decreasing values at test concentrations, when standard tests do not indicate a change of the metabolic state of a bacterial population.

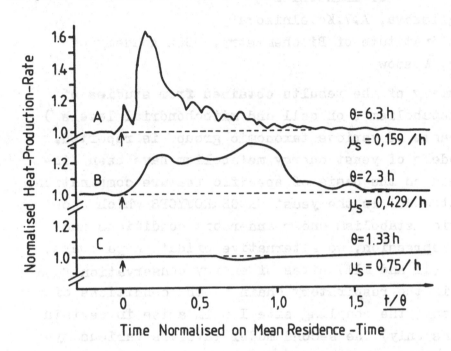

Figure 1: Action of 1.27 mMol PCP/l on E. coli.(Double dimensionless plot of relative Heat dissipation vers. t/residence time)

Literature cited:
1) A.E. Beezer (1980): Biological Microcalorimetry; Academic Press
2) J. Cairns Jr.: Water Research, 15 (1981) pp. 941-952
3) D. Herbert (1976): A.C.R. Dean, D.C. Ellwood, C.T.G. Evans & J. Melling (Eds.), Continuous culture 6: Application and new fields Ellis Horwood LTD., Pubs., Chichester, pp. 1-30
4) G.T. Daigger & C.P.L. Grady Jr.: Water Res., 16 (1982) pp. 365-382
5) J.M. Cardoso-Duarte, J.M. Marinho & N. van Uden (1976): For reference see (3), pp. 40-48
6) R. Brettel, I. Lamprecht & B. Schaarschmidt: Thermochimica Acta 49 (1981) pp. 53-61
7) C.I. Mayfield, W.E. Innis & P. Sain: Water, Air, and Soil Pollution 13 (1980) pp. 335-349
8) P. Weppen & D. Schuller: Thermochimica Acta, 72 (1984) 95-102
9) G.J. Prochazka, W.J. Payne & W.R. Mayberry: J. Bacteriology, 104 (1970) pp. 646-649

BIOENERGETICS OF THE YEAST CELL (MODELS OF YEAST ENERGY METABOLISM)

R. Zvjagilskaya, A.V.Kotelnikova
A.N.Bakh Institute of Biochemistry, USSR Academy of
Sciences, Moscow

A summary of the results obtained from studies of
energy metabolism (on cell and mitochondrial levels)
using yeasts of various taxonomic groups is reported.
Three models of yeast energy metabolism have been distinguished on the basis of specific feature combination.
The first model is the yeast SACCHAROMYCES which show
glycolytic metabolism under anaerobic conditions and
glucose repression, no alternative oxidation pathways,
with two (II and III) sites of energy conservation functioning in the respiratory chain in the conditions of de-
repression (the coupling site I will arise in certain
conditions only).The second model involves various types of CANDIDA and RHODOTORULA which show either a low
or no glucose repression and aerobic metabolism with all
three energy conservation sites in the respiratory chain
functioning at the stationary growth phase (in some species only two sites (II and III) may function at the exponential growth phase). This model is characterized by
switching on to an alternative oxidation pathway at the
stationary growth phase. The third model studied in detail in this laboratory is ENDOMYCES MAGNUSII cells
which display no glucose repression and will normally
show no alternative oxidation pathways (this alternative cyanide-resistant and SHAM-sensitive pathway can

be induced only in cells grown in the presence of antimycin A), their oxidative phosphorylation system is
stable and predominates in energy supply, with all three
sites of energy conservation in the respiratory chain
functioning at each growth phase. The studies of the model allowed specification of the structural organization
of the yeast respiratory chain and revealed regularities
of their energy coupling.

COUPLING OF ATP SYNTHESIS AND METHANE FORMATION IN METHANOSARCINA BARKERI

M. Blaut and G. Gottschalk
Institut für Mikrobiologie der Georg-August-Universität zu Göttingen
Grisebachstr. 8, D-3400 Göttingen, Federal Republic of Germany

Methanogenic bacteria are strictly anaerobic organisms that derive their energy from the conversion of CO_2 and H_2 to methane according to the following equation: $CO_2 + 4\ H_2 \rightarrow CH_4 + 2\ H_2O$. This process that is inhibited by protonophorous uncouplers [1] involves the transfer of 4 pairs of electrons. The terminal step in the reduction of CO_2 to methane involves the reduction of methyl coenzyme M ($CH_3-SCH_2CH_2SO_3^-$) by molecular hydrogen to methane and coenzyme M ($HSCH_2CH_2SO_3^-$) [2]. This step has the largest free energy change of all reactions involved in methane formation from CO_2 and H_2 and is the most suited site for energy conservation [3].

A few species including Methanosarcina barkeri are able to convert methyl group-containing substrates like methanol and methylamines besides H_2-CO_2 to methane. The conversion of these substrates occurs via the transfer of the methyl group to coenzyme M to form methyl coenzyme M [4] which is consequently reduced to methane. The methyl coenzyme M methyl reductase is the only reaction which is common to the conversion of all methanogenic substrates to methane. The electrons for this reaction may be generated either by oxidation of part of the methyl groups (equations 1-3) or by the oxidation of H_2 (equation 4):

$$CH_3OH + H_2O \rightarrow CO_2 + 6\ H \qquad (1)$$
$$3\ CH_3OH + 6\ H \rightarrow 3\ CH_4 + 3\ H_2O \qquad (2)$$

$$\overline{}$$

$$4\ CH_3OH \rightarrow 3\ CH_4 + CO_2 + 2\ H_2O \quad (3)$$

$$\text{or} \qquad CH_3OH + H_2 \rightarrow CH_4 + H_2O \qquad (4)$$

The use of methanol as methanogenic substrate allowed to investigate the role of the methyl coenzyme M methylreductase for the synthesis of ATP in methanogenic bacteria and to decide whether ATP is formed by substrate level or by electron transport phosphorylation in this process.

The addition of methanol to a cell suspension of M. barkeri resulted in the onset of methane formation and concomitantly in an increase of the intracellular ATP content from 1 to 10 nmol/mg of protein, under both N_2 and H_2. The protonmotive force (Δp) was estimated under identical conditions from the distribution of the lipophilic cation tetraphenylphosphonium ($\Delta\psi$) and of the weak acid benzoate (ΔpH) according to [5]. In the steady state of methane formation a Δp value of -130 mV (inside negative) was determined. The contribution of the pH gradient to Δp was negligible under these conditions.

The addition of the uncoupler tetrachlorosalicylanilide (TCS) to a cell suspension during methane formation from methanol caused a drastic decline of the intracellular ATP concentration and of the protonmotive force, under both N_2 and H_2. The effect of TCS on methane formation, however, was different: methane formation under H_2 continued whereas methanogenesis under N_2 was completely inhibited.

N,N' dicyclohexylcarbodiimide (DCCD), an inhibitor of the proton-translocating ATPase inhibited the synthesis of ATP in whole cells of M. barkeri driven by an artificially imposed pH-gradient. The addition of DCCD to a cell suspension of M. barkeri synthesizing methane from methanol and H_2 led to a decrease of the intracellular ATP concentation and to an inhibition of methanogenesis. The protonmotive force increased upon addition of DCCD and remained then at the same level as the control. The inhibiton of methanogenesis by DCCD was reversed by the addition of TCS. From these experiments [6] the following conclusions are drawn:
1. Methane formation from methanol and molecular hydrogen is coupled to the synthesis of ATP via the ATP synthase driven by the protonmotive force.
2. The methyl coenzyme M methylreductase reaction drives the translocation of protons across the cytoplasmic membrane thus establishing Δp.
3. There is a stringent coupling between ATP synthesis and electron flow from molecular hydrogen to methyl coenzyme M.
4. One of the steps of methanol oxidation (under N_2) depends on Δp or a product thereof.

References:

(1) Roberton, A.M. and Wolfe, R.S. (1970) J. Bacteriol. 102, 43-51
(2) Kell, D.B., Doddema, H.J., Morris, J.G. and Vogels, G.D. (1981) in Proceedings of the third international symposium on microbial growth on C-1 compounds (Dalton, H., Ed.) pp. 159-170, Heyden and Sons, London
(3) Taylor, C.D. and Wolfe, R.S. (1974) J. Biol. Chem. 249, 4879-4885
(4) Shapiro, S and Wolfe, R.S. (1980) J. Bacteriol. 141, 728-734
(5) Bakker, E.P., Rottenberg, H. and Caplan, R. (1976) Biochim. Biophys. Acta 440, 557-572
(6) Blaut, M. and Gottschalk, G. (1984) Eur. J. Biochem. 141, 217-222

518

SENSITIVE MITOCHONDRIA FROM SENSITIVE TISSUE
A.M.Babsky[o] and I.V.Shostakovskaya"
[o]Institute of Biological Physics, Academy of Sciences of the USSR,
142292 Pushchino, USSR
"I.Franko University, 290005 Lvov, USSR

Small intestinal epithelium is a tissue reactive to hormonal action
(1,2). The isolation of intact mitochondria from the tissue is made
difficult possibly by its reactivity (3). We have modified the isolation method of intact mitochondria from rat small intestinal epithelium
(IM) (4). IM obtained by this method phosphorylate ADP with respiratory control 3-3.5 (succinate as a substrate) and uptake Ca^{2+} effectively. Under succinate oxidation the maximal phosphorylation rate is
observed (fig.A, II), especially when isocitrate is added which eliminates oxalacetate inhibition of succinate oxidation (I). The phosphorylation rate in IM with succinate is three times greater as compared
with those in the presence of α-oxoglutarate (III). Phosphorylation is
absent with β-hydroxybutyrate as substrate (IV). In our conditions the
Ca^{2+} uptake in IM (fig.B) was observed only under succinate oxidation
(II). IM have considerably lower threshold of Ca^{2+} stimulation of respiration than liver mitochondria (LM). The addition of 45 µM $CaCl_2$
results in a 4-fold rise of respiration in IM with a good reverse to
initial level (metabolic state 4), whereas the same amount of $CaCl_2$
has no effect in rat LM. Similarly, IM are more sensitive to Mg^{2+}
than LM.

Adrenaline (AD) at small doses (5 µg per 100 g weight) administrat-
ed to rats was found to increase Ca^{2+} capacity in IM, whereas in LM
the effect was not observed (0.9% NaCl was administered to control animals). Administration of AD at a dose of 25 µg per 100 g weight stimulates phosphorylation of ADP (15 and 30 min) in mitochondria from both
tissues (Table). However, in IM the effect is stronger and more prolonged. It can be suggested that more sensitive and low-energized mitochondria such as IM compensate metabolic changes more slowly, than more resistant and high-energized LM. High reactivity of small intestinal epithelium to hormonal action *in vivo* is, probably, attributed for
high sensitivity of its mitochondria.

Abbreviations: IM, rat small intestinal epithelium mitochondria;
               LM, rat liver mitochondria; AD, adrenaline.

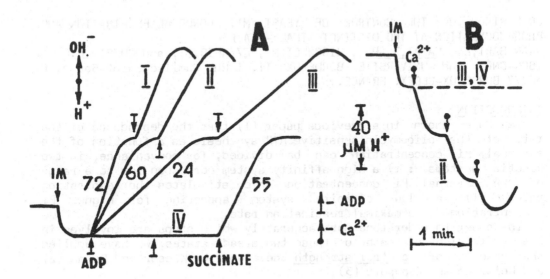

Figure. pH-registration of oxidative phosphorylation of ADP (A) and
Ca^{2+} uptake (B) of IM.
Substrates: I, succinate 6 mM + isocitrate 3 mM; II, succinate 6 mM;
III, α-oxoglutarate 6 mM; IV, β-hydroxybutyrate.
Media: 250 mM sucrose, 50 mM KCl, 1 mM KH_2PO_4, 3 mM tris-HCl pH 7.4.
ADP by portions of 150 nmoles/mg protein, $CaCl_2$ - by 30 nmoles/mg pr.
IM - 2.5 mg pr./ml. Phosphorylation rate is given on the curves in
nmoles H^+/min·mg pr.

Table. The effect of AD on the phosphorylation rate (in nmoles ATP/min·
·mg pr.) in IM and LM.

Min after AD administration	IM			LM		
	Control	AD	AD/control, %	Control	AD	AD/control, %
15	47.6	71.7	151*	86.2	112.0	130*
30	45.7	67.1	147*	86.9	100.4	115
180	47.5	57.0	120	87.6	85.8	98

* p≤0.05; n=5-7.

References
1. Selye, H. (1950) Stress. Acta Inc., Montreal.
2. Barbezat, G.O. and Reasbeck, P.G. (1983) in Endocrinology (Shizume,
   K. et al., eds.), pp. 161-165, Tokyo.
3. Iemhoff, W. and Hulsmah, W. (1971) Eur. J. Biochem. 23, 429-434.
4. Shostakovskaya, I.V. and Babsky, A.M. (1984) Ukr. Biochem. J.
   (USSR) 56, 137-141.

COMPARISON OF THE CONTROL OF YEAST MITOCHONDRIAL RESPIRATION AND
PHOSPHORYLATION AT TWO DISTINCT STEADY-STATES.
JEAN BART,E., MAZAT,J.-P. , RIGOULET,M., EZZAHID,Z., and GUERIN,B.
IBCN-CNRS and UNIVERSITE BORDEAUX II, 1 Rue Camille-Saint-Saens, F
33077 BORDEAUX-CEDEX. FRANCE.

INTRODUCTION :
    We have shown in a previous paper (1) that the dependance of the
rate of the oligomycin sensitive ATP synthesis as a function of the
external Pi concentration can be divided, for conveniance, in two
kinetic systems : i) a high affinity system corresponding to a range
of the external Pi concentration which stimulates the respiration
rate. ii) a low affinity system appearing for higher Pi
concentration, at a maximal respiration rate.
    In order to understand more accurately which steps are involved in
the definition of each of these two steady-states, we have applied
the concept of control strength developed by Kacser and Burns (2)
and Heinrich and Rapoport (3).
    The two main consequences of the theory, well exemplified in the
works of Groen et al. (4) and Tager et al. (5) , are the
following:
1) The control of a metabolic pathway can be shared among several
steps within the chain.
2) The distribution of the control strengths can vary according to
the different considered steady-states.
We have used this concept to compare :
1) the two steady-states encountered according to the Pi
concentration (Pi = 0.5mM which is representative of system I and
Pi=7.7mM where the variation in system II are preponderant).
2) the control of respiration and phosphorylation in previously
defined steady-states.

RESULTS and DISCUSSION
    When considering the control strength in ATP synthesis (Table
I), it is worth mentioning that:
    - The control strength in proton leak has large values in both
conditions.
    - The rest of the control strength is roughly shared between the
phosphate transport and the cytochrome c oxidase; but at low Pi
concentration (i.e. the respiratory chain is stimulable) the control
strength of the cytochrome oxidase is larger than the one of the
phosphate transport,while the reverse occurs at high Pi concentration
(System II where the respiration is no longer stimulable).

T A B L E I : Control strength in the ATP synthesis.
The respiratory substrate is EtOH 0.3% + Arsenite 10 mM
in order to block substrate-level phosphorylation

| | CONTROL STRENGTH | |
S T E P	System I Pi=0.5mM	System I Pi=7.7mM
AdN. translocator	0.12	0
Proton leak	0.35	0.41
Phosphate transport	0.17	0.45
ATP synthase	0	0
Cytochrome c oxidase	0.33	0.09
T O T A L	0.99	0.95

These results suggest that the Pi transport is mainly driven by
the $\Delta\mu_H+$ in system I, but by $\Delta\mu_{Pi}$ in system II.
    Considering the control strength of respiration in the same
conditions one can notice that :
    - the control strength in proton leak is much lower (0 in system
II).
    - the control strength of phosphate transport is not negligible
(0.2± 0.05).
    - the control strengths do not appreciably differ in the two
steady-states.
Finally, we would like to emphasize that the control strength of a
given step is not always the same if one consider either the
respiratory chain or the phosphorylation pathway.
    The result of this comparison raises some doubt on the common use
of respiration rate as reflecting the rate of phosphorylation.

                    B I B L I O G R A P H Y
(1) RIGOULET, M., EZZAHID, Z., and GUERIN, B. (1983) B.B.R.C. 113
751-756.
(2) KACSER, H. and BURNS, J.A. (1973) in Rate Control of Biological
Processes (Davies,D.D. ed) 65-104.Cambridge University Press.
London.
(3) HEINRICH, R. and RAPOPORT, T.A. (1974) Eur.J.Biochem. 42 89-95.
(4) GROEN, A.K., WANDERS, R.J.A., WESTERHOFF, H.V., VAN DER MEER, R.
and TAGER, J.M.(1982) J.Biol.Chem. 257 2754-2757.
(5) TAGER, J.M., WANDERS, R.J.A., GROEN, A.K., KUNZ, W., BOHNENSACK,
R.,KUSTER, U.,LETKO, G., BOHME, G.,DUSZYNSKI, J. and WOJTCZAK, L.
(1983) FEBS Letters 151 1-9.

ENERGETIC METABOLISMS (MITOCHONDRIAL AND CYTOPLASMIC) OF YEAST
STRAINS RESISTANT TO GLUCOSE REPRESSION

R. Caubet, N. Camougrand, A. Cheyrou, G. Velours, and M. Guérin
IBCN-CNRS, 1 rue Camille Saint-Saëns, 33077 Bordeaux cedex
(France)

Two sets of yeast strains resistant to glucose repression
were isolated. The first strain, SP_1, prototrophe, was selected
as resistant to thiophosphate on a lactate medium. The second
strain, dg, (ad⁻, lys⁻) was selected as resistant to allyl
alcohol and chloramphenicol.

Both strains were modified simultaneously in the fermentative
and in the oxidative pathways (loss of alcohol dehydrogenase I
and over production of cytochrome aa_3, the synthesis of which
being insensitive to glucose repression (1)).

In order to reoxidize cytoplasmic NADH, failing to be reoxidized via ADH_1, these strains developed a secondary mitochondrial hydrogen pathway, going from external NADH dehydrogenase
to cytochrome oxidase. This pathway was amytal sensitive,
antimicin insensitive and inhibited by high KCN concentration
(2).

In order to see whether the high NADH reoxidation capacity of
these mitochondria was related to a very effective glycolysis,
the glycolytic and gluconeogenic pathways were studied. Opposite
to this, glycolysis, when measured under optimal conditions,
showed a low level in all the enzymatic activities. Regulation
points of glycolysis, pyruvate kinase and phosphofructokinase,
were studied; in both cases, kinetics parameters and sensitivity to allosteric effectors were changed.

On the other hand, gluconeogenesis was very active and fructose 1-6,biphosphatase was insensitive to glucose repression,
leading to a glucose 6-P production.

Therefore, two enzymes of the hexose monophosphate pathway,
glucose 6-P-dehydrogenase and 6-P-glyconate dehydrogenase were
assayed. These activities were very high and by this way NADPH
was produced in excess.

It was shown that NADPH could be reoxidized by mitochondria in
an amytal sensitive pathway and work is underway to determine
whether phosphorylations are effective with this substrate.

1. Guérin, M., Camougrand, N., Velours, G. and Guérin, B., Eur.
   J. Biochem. (1982) 124, 457-463.
2. Camougrand, N., Caubet, R. and Guérin, M., Eur. J. Biochem.
   (1983) 135 367-371.

EVIDENCE FOR IN VIVO REGULATION OF THE ATP SYNTHASE IN INTACT CELLS
OF RHODOPSEUDOMONAS CAPSULATA

N.P.J. Cotton and J.B. Jackson

Department of Biochemistry, University of Birmingham, P.O. Box 363,
Birmingham B15 2TT, U.K.

The ATP synthase of energy-transducing membranes is known to be
highly regulated in vitro. The catalytic activity of the enzyme,
usually assayed in the direction of ATP hydrolysis, is modified by its
substrates, by the magnitude of Δ p and, in some cases, by reduced
thiols. The extent to which these effects are important in vivo is
not clear although the central position of this enzyme in energy
metabolism suggests that a degree of regulation is essential.

When intact cells of Rps. capsulata, suspended in fresh growth
medium under anaerobic conditions were illuminated, the membrane
potential, $\Delta\psi$, rose within 2 s to a high level and then, over the
next 20 s, declined slightly to a new steady-state. The presence of
the F_O inhibitor, venturicidin led to a slightly higher initial $\Delta\psi$
and prevented the subsequent decline. These results were obtained
either by following the re-distribution of butyltriphenylphosphonium
cation or by monitoring the electrochromic carotenoid shifts. The
latter respond sufficiently rapidly to measure also the ionic current
across the bacterial membrane [1]. After very short illumination
periods, the dependence of the dissipative ionic current, J_{DIS} upon
$\Delta\psi$ is non-ohmic [1]: J_{DIS} increases disproportionately with increasing $\Delta\psi$. Experiments in the presence of venturicidin show that
at high $\Delta\psi$ a large fraction of J_{DIS} passes through the ATP synthase
[2]. In the present experiments, the dependence of J_{DIS} on $\Delta\psi$ was
investigated after 20 s illumination. The non-ohmic nature was more
pronounced and an even greater fraction of J_{DIS} proceeded through the
ATP synthase. These results indicate that the rate of ATP synthesis
increased during the 20 s period.

The absolute value of J_{DIS} after 20 s illumination was only slightly
more than after 2 s. Because H^+/e^- ratio is independent of $\Delta\psi$, the
steady state value of J_{DIS} is proportional to the electron transport
rate [3]. It is concluded that the increased rate of ATP synthesis in
the cells is not accompanied by a large increase in the rate of
electron transport.

Adenine nucleotide levels were measured in cells quenched in
perchloric acid/EDTA after short periods of illumination. The ADP/ATP

ratio, taken as an index of the cytoplasmic phosphate potential, decreased markedly after 2 s illumination and continued to decrease over the next 20 s in the light. These results suggest that the increased rate of ATP synthesis is not due to the phosphate potential becoming less negative as a consequence of the induction of processes consuming ATP. It is more likely that the ATP synthase itself undergoes a change in activity during the transition from dark to illuminated conditions.

[1] Jackson, J.B. (1982) FEBS Lett. 139, 139–143.

[2] Clark, A.J., Cotton, N.P.J. and Jackson, J.B. (1983) Biochim. Biophys. Acta 723, 440–453.

[3] Cotton, N.P.J., Clark, A.J. and Jackson, J.B. (1984) Eur. J. Biochem., in press.

EFFECTS OF STARVATION AND STREPTOZOTOCIN DIABETES ON Ca^{2+} UPTAKE AND ATPase ACTIVITIES OF RAT LIVER MICROSOMES.

A.P. Dawson, School of Biological Sciences, University of East Anglia, Norwich, NR4 7TJ, U.K.

Ca^{2+} uptake into the microsomal fraction of rat liver has been shown to be enhanced by prior treatment of the animal(1) or hepatocytes(2) with glucagon. Injection of glucocorticoids into adrenalectomised rats has a similar effect(3), but pretreatment with insulin decreases Ca^{2+} accumulation(4). The role of glucagon, glucocorticoids and insulin in controlling Ca^{2+} accumulation by microsomes was studied using starvation and streptozotocin-diabetes to cause chronic changes in hormone levels.

Microsomes were isolated and Ca^{2+} uptake (in the presence of oxalate) measured as previously described(5). ATPase activities were assayed as in(6), at 10^{-9}M free Ca^{2+} to give Mg^{2+}-stimulated ATPase activity and at 10^{-5}M free Ca^{2+} to give Ca^{2+} plus Mg^{2+}-stimulated ATPase. The difference was taken as the Ca^{2+}-stimulated ATPase component. Streptozotocin (100 mg kg^{-1}) was injected sub-cutaneously in citrate-saline (pH 4.0).

In contrast to the situation where glucagon is injected into a physiologically normal animal, 36 hr starvation, which should cause

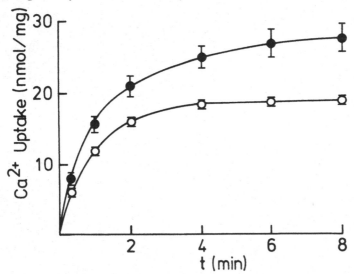

Fig 1 Ca^{2+} uptake by microsomes from normal (●) and 36 hr starved (o) rat livers. Each point is the mean ± SEM for 4 animals.

elevation of glucagon and glucocorticoid levels, results in a substantial depression of the initial rate and extent of Ca^{2+}
accumulation by microsomes (Fig 1). This appears to be unrelated to
changes in insulin level, since for streptozotocin-diabetic rats (5
days after injection) there was no difference in Ca^{2+} uptake between
the control and experimental groups. Table 1 shows that the
depression in Ca^{2+} uptake seen in starved animals can be explained by
a decrease in Ca^{2+} ATPase activity (measured in the presence of
A23187). In agreement with the results on uptake, there is no
difference between diabetic and control groups. A striking feature
of the data in Table 1 is the elevation of Mg^{2+} ATPase observed both
in starvation and diabetes, although the significance of this, and
the function of the enzyme, are unknown.

The finding that the effects of glucagon injection into a normal
animal and of starvation are opposite suggests that the former may
bring about a compensatory response to return the animal to a normal
state, or that in starvation some other change over-rides the effect
of glucagon. There is no evidence that insulin levels affect this
system in vivo.

Condition	n	Blood glucose (mM)	Mg ATPase (μmol Pi/ mg/10 min)	Ca stimulated ATPase (μmol Pi/mg/10 min)
Control	8	5.4 ± 0.25	0.29 ± 0.013	0.16 ± 0.01
36 hr starved	8	3.4 ± 0.2	0.61 ± 0.03**	0.10 ± 0.016*
Sham-injected	4	5.4 ± 0.3	0.31 ± 0.02	0.18 ± 0.006
5-day-Diabetic	4	29.2 ± 0.3	0.43 ± 0.015**	0.18 ± 0.006

Table 1   ** $p < 0.001$; * $P < 0.01$ Data are mean ± SEM

1 Bygrave, F.L. and Tranter, C.J. (1978) Biochem. J. 174, 1021-1030
2 Taylor, W.M., Bygrave, F.L., Blackmore, P.F. and Exton, J.H. (1979)
  FEBS Letters 104, 31-34
3 Friedman, N. and Johnson, F.D. (1980) Life Sciences 27, 837-842
4 Andia-Waltenbaugh, A.M., Lam, A., Hummel, L. and Friedmann, N.
  (1980) Biochim. Biophys. Acta 630, 165-175
5 Dawson, A.P. (1982) Biochem. J. 206, 73-79
6 Dawson, A.P. and Fulton, D.V. (1983) Biochem. J. 210, 405-410

PEROXISOMAL β-OXIDATION COMPLEX OF <u>CANDIDA</u> <u>TROPICALIS</u>

M. De la Garza, I. Kunigk, U. Schultz and W.-H. Kunau

Institut für Physiologische Chemie der Ruhr-Universität Bochum, Universitätsstraße, 4630 Bochum, FRG.

Candida tropicalis contains an inducible peroxisomal β-oxidation [1,2]. This complex is a suitable model system to study the biosynthesis and transport of proteins into peroxisomes. As a prerequisite the β-oxidation enzymes have to be isolated and characterized. Three individual proteins have been resolved and have been partially or completely purified.

1. Trifunctional protein

Enoyl-CoA hydratase, 3-hydroxyacyl-CoA dehydrogenase and 3-hydroxyacyl-CoA epimerase activities have been copurified with one protein from C.tropicalis grown on oleate as sole carbon and energy source. The ratio of the three β-oxidation enzyme activities remained constant during three purification steps: ionexchange chromatography on DEAE-cellulose (DE-52); dye-ligand chromatography on blue Sepharose CL-6B,and affinity chromatography on NAD-agarose. Characterization of the trifunctional protein by PAGE under denaturing and non denaturing conditions revealed a single band with a molecular weight of 102,000 and 185,000, respectively. These results suggest that the multifunctional protein is a dimer consisting of two identical subunits.

2. Thiolase

Thiolase activity has been resolved from the activities of the trifunctional protein by ion-exchange chromatography and has been further purified (hydroxylapatite, gel chromatography and affinity chromatography) to apparent homogeneity as judged by SDS-PAGE. Its properties are compared to those of known thiolases.

3. 2,4-Dienoyl-CoA Reductase

2,4-Dienoyl-CoA reductase have been reported to be obligatory for the degradation of unsaturated fatty acids [3]. Two 2,4-dienoyl-CoA reductase activities have been resolved from crude extracts of oleate-grown cells. These two activities have been partially purified and characterized.

[1] Tanaka, A., Osumi, M., and Fukui, S. (1982) Ann. N.Y. Acad.Sci. <u>386</u>, 183-189

[2] Dommes, P., Dommes, V., and Kunau, W.-H. (1983) J.Biol.Chem. <u>258</u>, 10846-10852

[3] Dommes, V. and Kunau, W.-H. (1984) J.Biol.Chem. <u>259</u>, 1781-1788

STEP INHIBITION (RESTRICTION) OF SUCCINATE DEHYDROGENASE UNDER
PROGRESSIVE PATHOLOGY
E.V.Grigorenko and M.N.Kondrashova
Institute of Biophysics, Academy of Sciences of the USSR, 142292
Pushchino, USSR

Using the method for physiological investigation of mitochondria
(M) (M.N.Kondrashova *et al.* in this book) the primary rise of succinate oxidation was shown to be replaced by its pronounced decrease
under immobilization stress and chronic pathological state of organism (1). This decrease is based on the inhibition of succinate dehydrogenase (I-SDH) as it is removed by SDH activators. The degree
of I-SDH depends on the strength and duration of the external influence on the organism. I-SDH in rat brain and liver M was determined on the base of different acceleration of phosphorylating
oxidation of succinate by various SDH activators conventional and
those found in our investigation. The stimulating effect of SDH
activators increases in the following row: glutamate (GLU) < isocitrate (ISO) < β-hydroxybutyrate (OB) < α-glycerophosphate (GP).
According to this we defined four steps of I-SDH which allowed us to
differentiate the degree of I-SDH under progressive pathology (Fig.).
Liver M of intact animals are characterized by small I-SDH (I). Under
24-hour immobilization stress the value of I-SDH rises depending on
the state of the organism. In liver M of stress-resistant Wistar rats
(W) the I-SDH is small and is easily removed by ISO and GLU (2). In
liver of stress-sensitive August rats (A) I-SDH is greater and is
better removed by ISO than by GLU (3). In brain M I-SDH is especially great and is removed under the addition of ISO rather than GLU
(6). Under chronic alcohol intoxication (4) and more deep stress of
August rats (A) (5) accompanied by rectal temperature fall per 5-10°C
stimulatory effect of GLU and ISO is not observed (both of them may
slightly inhibit respiration) whereas respiration is accelerated by
OB and GP.

The difference between the effects of various SDH activators may be
due to the different nature of inhibitors and to various redox state
of the respiratory chain. At the two first steps of I-SDH the main inhibitor is apparently oxalacetate which is generated by active succinate oxidation under stress. Stimulatory effect of both GLU and ISO is
related to the removal of oxalacetate by different mechanisms. More
deep I-SDH which is characterized by the absence of GLU and ISO stimulatory effect seems to be due to oxidized adrenaline and serotonin
accumulated under progressive pathology. Their inhibitory effect may
be abolished by reduction of the respiratory chain in flavin-Q region
caused by OB and GP oxidation.

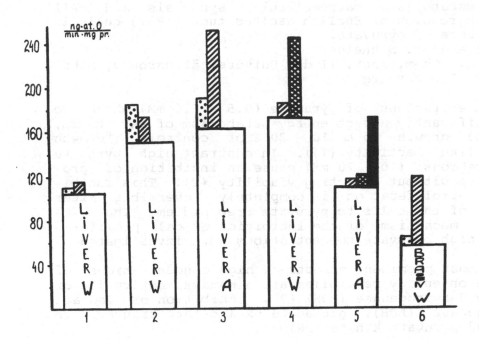

Figure. Step inhibition of succinate dehydrogenase.
Light part – succinate oxidation without activators;
shaded part – increment of succinate oxidation with activators: ▨ GLU; ▨ ISO; ▨ OB; ■ GP.

Physiological function of succinate oxidation acceleration is a
support of active resistance of organism whereas its step inhibition
prevents damage of MCH and provides passive resistance of organism.
Common inhibition of SDH initiated by inhibitors or by deenergization
in isolated M is manifested as decrease of the level of respiration in
intact animal M. The inhibition described here induced by physiological mechanisms in organism develops with concomitant stimulation of
succinate oxidation in M of excited animals as compared to intact animals. This inhibition restricts the magnitude of stimulation. In order
to differentiate physiological SDH inhibition it may be reasonably de-
nominated as restriction or step-restriction.

Reference

1. Kondrashova, M.N., Grigorenko, E.V., Guzar, I.B., Okon, E.B. (1982)
   in 2nd EBEC Conference (LBTM-CNRS ed.), pp. 589-590, Lyon, France.

Energy metabolism, macromolecular synthesis and cell cycle progression of Ehrlich ascites tumor (EAT) cells in the presence of pyruvate.

W. Kroll and Fr. Schneider
Physiolog. Chem. Inst. II der Universität Marburg, Lahn-berge, D-3550 Marburg.

Low concentrations of pyruvate (0.5 - 2.0 mM) which do not significantly affect energy metabolism of EAT cells, stimulate growth up to 10 - 20 % of controls ("growth factor like" activity (1)). In contrast high pyruvate concentrations (10 - 20 mM) cause an inhibition of pro-liferation without affecting viability (2). This inhibi-tion of proliferation is completely reversible after transfer of the cells to pyruvate free medium. The bio-chemical mechanisms of the inhibition of cell prolifera-tion by high pyruvate concentrations were investigated.

The most prominent effects of high concentrations of pyruvate on energy metabolism are a maximal activation of pyruvate dehydrogenase (PDH) (3), inhibition of lactate dehydrogenase (LDH), glutamate-oxalacetate transaminase (GOT) and pyruvate kinase (PK).

Inhibition of LDH and further of GOT, which is part of the malate-aspartate shuttle of transfer of cytosolic hydrogen into mitochondria, causes a shift in the redox state of the cells (NAD/NADH of controls 24, in the pre-sence of pyruvate 9). Inhibition of LDH and PK gives rise to an accumulation of intermediates of glycolysis above glyceraldehyde-3-phosphate dehydrogenase. ATP/ADP concen-tration ratios of whole cells and of cytosol were found in a normal rage (4 - 6).
DNA, RNA and protein synthesis were compatible with the growth rate of the cells.

Flow cytometric analysis of the cells grown in the presence of high pyruvate concentrations revealed a very slow cycle progression. As was shown by the BrdU-H33258 technique (4) of flow cytometry only G2M- and some late S-cells have divided within 24 h and are responsible for a slight increase in cell number.

Further experiments have shown that the proliferation inhibiting effect of high pyruvate concentrations depends on the presence of glucose: cell growth may be sustained in cultures grown in glucose free medium in the presence of 20 mM pyruvate and additional 10 mM uridine, which provides ribose-1-phosphate for the biosynthesis of nucleotides and nuleic acids (5).

From our experiments we conclude that the accumulation of glyceraldehyde-3-phosphate and the shift in the redox state of the cells are primarily responsible for the inhibition of cell proliferation by high pyruvate concentrations: Breakdown of glucose in glycolysis delivers NADH (which cannot be reoxidized fast enough) as well as glyceraldehyde-3-phosphate which accumulates because of the inhibition of PK and LDH; these effects explain the dependence of the pyruvate effect on the presence of glucose. Furthermore 500 uM exogenous glyceraldehyde-3-phosphate completely arrests proliferation of the cells, the mechanism of which is not quite clear.

1. Neumann, R. J. and McCoy, T. A. (1958) Proc. Soc. Exp. Biol. Med. 98, 303 - 306
2. Postius, S. and Schneider, Fr. (1979) Hoppe-Seyler's Z. Physiol. Chem. 360, 344 - 345
3. Postius, S. (1978) Anal. Biochem. 90, 534 - 542
4. Böhmer, R. M. (1979) Cell Tissue Kinet. 12, 101 - 10
5. Linker, W. and Schneider, Fr. (1983) Hoppe-Seyler's Z. Physiol. Chem. 364, 1172

REGULATION OF β-OXIDATION OF FATTY ACIDS: ROLE OF THE REDOX STATE OF MITOCHONDRIAL NAD(H).

P.M. Latipää, T.T. Kärki, J.K. Hiltunen, and I.E. Hassinen
Department of Medical Biochemistry, University of Oulu,
Kajaanintie 52 A, 90220 Oulu 22, Finland

Although the metabolic sequence of β-oxidation is well known, the regulation of its rate within mitochondria is still under dispute. In liver the ethanol oxidation results in an increased NADH/NAD$^+$ ratio both in the mitochondrial and cytosolic spaces, and it has been suggested that the inhibition of long chain fatty acid oxidation during ethanol oxidation is a consequence of the decreased mitochondrial redox potential (1). Also the β-oxidation of fatty acids can be inhibited by reconstructed malate-aspartate shuttle in isolated mitochondria, indicating a possible regulatory function of mitochondrial NADH/NAD$^+$ ratio in the β-oxidation (2). To study this further, we undertook experiments where the NAD(H) redox state was controlled during palmitoylcarnitine (PC) oxidation in rotenone-inhibited mitochondria in state 3.

Liver mitochondria isolated from fasted male Sprague-Dawley rats were incubated in a KCl based medium (3) containing 25 μM PC, 5 mM malonate and 2.5 mM ADP. Other additions to the incubation medium are given in Table 1. During palmitate oxidation it was possible to standardize the mitochondrial NADH/NAD$^+$ ratio using graded concentrations of acetoacetate (AcAc) or ketomalonate as hydrogen acceptors in the presence of rotenone. The intramitochondrial free NADH/NAD$^+$ ratio during the metabolic steady state was estimated from nicotinamide nucleotide fluorescence previously calibrated by β-hydroxybutyrate/AcAc ratios in equilibrium conditions. β-Oxidation rate was calculated from the rate of production of acid-soluble metabolites from [U-^{14}C]-PC and expressed as acetyl-units produced.

The results shown in Table 1 demonstrated that upon a decrease of the intramitochondrial NADH/NAD$^+$ the degradation of PC increased with a concomitant increase of the flux through the acyl-CoA dehydrogenase(s) as indicated by the increased rate of oxygen consumption. When the rate of β-oxidation decreased, long-chain acyl-CoA esters accumulated within mitochondria showing that the inhibitions of β-oxidation was located in the initial steps of this metabolic sequence. It has been demonstrated that the intramitochondrial acetyl-CoA/CoA ratio can regulate β-oxidation via thiolase (4) but this ratio did not change significantly under the present experimental conditions. The present results clearly demonstrate that the rate of mitochondrial β-oxidation is controlled by the redox state of mitochondrial NAD(H), but the mediators of this feed-back inhibition remains to be identified.

Table 1. Rate of oxygen consumption, free NADH/NAD$^+$ ratio and concentrations of some CoA metabolites during palmitoylcarnitine oxidation in isolated rat liver mitochondria.[a]

	Additions				
	Rotenone (4.5 μM)	Rotenone AcAc (0.5 mM)	Rotenone AcAc (8.0 mM)	Rotenone keto-malonate (5.0 mM)	None
Acetyl-units produced (nmol/min)	4.7 ±0.5(5)	7.3 ±0.4(5)	8.1 ±0.6(5)	10.5 ±0.8(5)	15.7 ±1.0(5)
O_2 consumption (natoms/min)	3.0 ±0.3(7)	4.1 ±0.2(7)	4.6 ±0.2(6)	6.4 ±0.4(7)	19.5 ±1.3(7)
$NADH_f/NAD_f^+$	0.103 ±0.005(3)	0.075 ±0.013(3)	0.054 ±0.017(3)	0.016 ±0.010(3)	N.D.
Acid-insoluble CoA derivatives (nmol)	1.4 ±0.1(7)	1.2 ±0.1(7)	1.1 ±0.1(7)	N.M.	0.6 ±0.1(7)
Acetyl-CoA (nmol)	0.29 ±0.09(3)	0.55 ±0.11(3)	0.39 ±0.02(3)	0.63 ±0.19(3)	0.79 ±0.09(3)
CoA (nmol)	0.17 ±0.02(3)	0.26 ±0.08(3)	0.46 ±0.06(3)	1.03 ±0.17(3)	0.86 ±0.31(3)

[a]Results expressed per mg mitochondrial protein, as means ± S.E. of the number of experiments in parentheses. N.D. not detectable, N.M. not measured.

Supported partly by grants from the Medical Research Council of the Academy of Finland and the Paolo Foundation and under a contract with the Finnish Life Insurance Companies.

References:

(1) Lindros. K. and Aro. H. (1969) Ann. Med. Exp. Biol. Fenn. 47,39-42
(2) Lumeng. L.. Bremer. J. and Davis. E.J. (1976) J. Biol. Chem. 251, 277-284
(3) Osmundsen. H. and Bremer, J. (1977) Biochem. J. 164,621-633
(4) Olowe. Y. and Schulz. H. (1980) Eur. J. Biochem. 109,425-429

ANALYSIS OF THE MEMBRANE POTENTIAL CHANGES INDUCED BY ADSORPTION OF
PHAGES TO E.COLI CELLS.

L. Letellier', P. Boulanger' and B. Labedan''

' Laboratoire des Biomembranes, Bat 433
''Institut de Microbiologie, Bat 409
Université Paris Sud
91405 Orsay Cedex , France.

Coliphages T4,T5 and BF23 differ in receptor specificity, stages of
adsorption and manner of DNA injection (1). Nevertheless, their
attachment to their respective receptors present in the host cell
outer membrane, triggers the same immediate and transient
depolarization of the cytoplasmic membrane, suggesting that a signal
must be transmitted between the two membranes (2,3). Moreover, this
process occurs independently of phage DNA injection as it occurs with
T4 ghosts as well (2,3).

Involvement of envelope-bound Ca++ in phage-induced depolarization (4)

The signal transmission was shown to be dependent on envelope-bound
calcium in the case of T4 but not in the case of T5 and BF23. Indeed,
addition of EGTA before T4 phages and ghosts specifically prevented
the membrane potential changes. This envelope-bound calcium becomes
accessible to the chelator only as a consequence of phage adsorption
and remains in this state during the depolarization and the
repolarization. Membrane potential changes will again occur if calcium
is added after the addition of EGTA and phage. The same concentration
(300 µM) of EGTA prevents the T4-induced depolarization between
multiplicities of infection of 6 and 30 although the amplitude of
depolarization is dependent on the multiplicity. As it is known that
T4 phages adsorb to the lipopolysaccharide (1) which bind Ca++ with
high affinity, we proposed (4) that upon attachment of T4 to the
lipopolysaccharide, Ca++ is released and become accessible to a
membrane component which would induce the depolarization. To reconcile
the opposite results obtained with T4, T5 and BF23 regarding Ca++, we
proposed (4) that, in the case of T5 and BF23, a direct interaction
between the inner and outer membrane is possible through adhesion
zones, since T5 and BF23 receptors (TonA and BtuB) are transmembrane
proteins (1).

Quantitation of the Ca++ released during the T4-induced depolarization using the fluorescent probe QUIN 2

Quin 2 is a fluorescent analog of EGTA which bind Ca++ with high affinity and selectivity (5). From the fluorescence changes of QUIN 2 it is thus possible to quantitate Ca++ in the nanomolar range. We have shown that addition of increasing concentrations of QUIN 2 up to 50 uM progressively decreases the T4-induced depolarization, in a manner similar to EGTA. Moreover, T4 ghost-induced depolarization was also prevented by QUIN 2, whereas QUIN 2 was without effect on T5 and BF23. From the fluorescence increase of QUIN 2 we have calculated that approximately 5 to 10% of the lipopolysaccharide-bound Ca++ was liberated after T4 phage and ghost adsorption.

Phage-induced release of the respiratory control

Release of the respiratory control in starved E.coli cells is a process related to the decrease of Δ pH+ and is, under certain conditions, associated with the entry of protons in the cytoplasm (6). We have shown that addition of phages T4, T5, BF23 and T4 ghosts induced an immediate release of the respiratory control of the starved cells and that the respiratory control ratio increased with multiplicity of infection. Moreover, addition of EGTA before T4 phages and ghosts prevented the increase of the rate of respiration. Thus, we propose that the release of respiratory control induced by phages and ghosts is a process related to the signal transmission, and not to DNA injection. Furthermore, these results suggest that phage attachment to the outer membrane induces an entry of protons in the cytoplasm.

References
(1) Goldberg, E. 1980. in "Virus receptors" L.Randall and L.Philipson (eds) ser B, vol 7. Chapman and Hall, London.
(2) Labedan, B. and Letellier, L. 1981. Proc.Natl.Acad.Sci. U.S.A 78: 215-219.
(3) Labedan, B. and Letellier, L. 1984. J.Bioenerg. and Biomembranes 16: 1-9.
(4) Letellier, L. and Labedan, B. 1984. J.Bacteriol. 157: 789-794.
(5) Tsien, R.Y., Pozzan, T. and Rink, T.J. 1982 J.Cell.Biol. 94: 325-334.
(6) Burnstein, C., Tlankova, L. and Kepes, A. (1979) Eur.J.Biochem. 94: 387-392.

MECHANISM OF THE HORMONAL STIMULATION OF GLUTAMINASE IN RAT LIVER
MITOCHONDRIA

J,D. McGivan
Dept. of Biochemistry, University of Bristol, University Walk,
Bristol, BS8 1TD, U.K.

The enzyme phosphate-dependent glutaminase (E.C.3.5.1.2) in liver
is loosely associated with the inner surface of the inner mitochondrial
membrane. It has been shown that on exposure of perfused liver or
isolated hepatocytes to the hormone glucagon glutamine metabolism is
markedly activated, and this activation is due primarily to an activation of glutaminase. The activity of glutaminase in liver mitochondria isolated from rats which have been injected with glucagon is
markedly increased over controls but this increase in activity is lost
on disruption of the mitochondria [1].

In order to gain some insight into the mechanisms involved in the
hormonal activation of glutaminase, this enzyme has been partially
purified from rat liver and its kinetic properties evaluated [2].
Glutaminase requires phosphate and ammonia as obligatory activators.
The dependence of the activity on glutamine concentration is highly
sigmoidal with half-maximum velocity at 22mM glutamine; the dependence on phosphate concentration is hyperbolic with half maximum activity at 5mM phosphate. On incubation of the enzyme with a mitochondrial membrane preparation, the kinetic properties are markedly altered. The dependence of the activity on glutamine concentration
becomes hyperbolic with half maximum activity at 6mM glutamine. Under
these conditions, 70% of the enzyme activity is independent of added
phosphate. In the presence of 10mM $MgCl_2$, the properties of glutaminase in this reconstituted system revert to those of the isolated
enzyme, suggesting that Mg^{2+} ions may inhibit the interaction of the
enzyme with the membrane.

The activity of glutaminase in isolated mitochondria at intermediate concentrations of phosphate and glutamine is increased on energisation of the mitochondria, and is further increased by incubation
of the mitochondria in hypotonic media [3] or by the addition of EDTA
[3], valinomycin, or Ca^{2+} ions. All these conditions lead to
swelling of the mitochondria. In slightly hypertonic media, the
kinetic properties of glutaminase as expressed in intact mitochondria
are similar to those of the isolated enzyme. As the osmolarity of
the medium is decreased the dependence of activity on glutamine concentration becomes progressively more hyperbolic and the enzyme becomes less phosphate-dependent. At osmolarities lower than 100mOsm,

the kinetics of the enzyme in intact mitochondria resemble those of
the enzyme in association with the mitochondrial membrane.

Glutaminase activity in intact mitochondria from glucagon-treated
rats is characterised by a less sigmoidal glutamine concentration
dependence and a lesser requirement for phosphate than is the
activity in control mitochondria at any given osmolarity. At very
low osmolarities the kinetics of glutaminase in mitochondria from
hormone-treated and control animals are identical.

On the basis of these and other data it is proposed that the
primary effect of glucagon is to cause mitochondrial swelling by an
unknown mechanism. Mitochondrial swelling leads to an increased
association of glutaminase with the inner mitochondrial membrane with
a consequential change in the kinetic parameters of this enzyme. The
result is a large activation of the enzyme at physiological concentrations of glutamine and phosphate.

[1] Lacey, J.H., Bradford, N.M., Joseph, S.K. and McGivan, J.D. (1981)
    Biochem. J. 194 29-33.
[2] Patel, M. and McGivan, J.D. (1984) Biochem. J. 220 583-590
[3] Joseph, S.K., McGivan, J.D. and Meijer, A.J. (1981) Biochem. J.
    194, 35-41.

RELATION OF THE ELECTRON FLOW AND $\Delta\psi$ IN MITOCHONDRIA OF AMOEBA IN SEVERAL CONSTRAINED CONDITIONS.

J.Michejda, K.Chojnicki, L.Hryniewiecka and P.Wolańska

Department of Bioenergetics, Poznań University, Fredry 10, Poznań, Poland.

Like in plant mitochondria /1/, the electron flow in mitochondria of Acanthamoeba castellanii, can utilize the cytochrome or the alternative pathway /2/ and the O_2 uptake is regulated by external NAD /3//bypassing site I/, AMP and GMP /in presence of CN/ while like in mammalian mitochondria efficient Ca^{2+}cycling is controled by $\Delta\psi$ /4/

The O_2 uptake was measured oxygraphically and $\Delta\psi$ monitored with the TPP^+ electrode /5/ in the same vessel in 2 ml of the incubation medium containing: 70 mM sucrose, 210 mM mannitol, 20 mM KCl, 10 mM TRIS buffer pH 7.4 and 10 μM TPP^+; additions: 5 mM $MgCl_2$, 2.5 mM phosphate.

With succinate or durohydrochinon/UQ/state 4 O_2 uptake $\sim$80 $nAtO_2$ x min^{-1} x mg^{-1} /RC $\sim$2.5, ADP:O $\sim$1.5/ developed $\Delta\psi$ 150 mV/P_i/ and 180 mV/+P_i/. Increase in $\Delta\psi$ due to P_i addition was not reflected in any increase in O_2 uptake. Similiar values were obtained with exogenous NADH, except 2 times higher O_2 uptake. With mitochondria from exponential phase of culture 0.3% BSA was necessary to keep at almost constant level and to obtain the highest RC and ADP:O ratio, while 0.1% BSA was sufficient with mitochondria from stationary phase. Nigericin in presence of P_i increased $\Delta\psi$ only by 10% in a fast transient manner, regardless the kind of substrate used, including ATP.

Ethanol and malate yielded lower O_2 uptake, 20 and 35 $nAtO_2$ x min^{-1} x mg^{-1}, respecitively, with three phosphorylating sites involved. P_i was always necessary to significantly increase $\Delta\psi$. With ethanol 10 μM rotenone excluded O_2 uptake by 90% but with malate by only 50%, whilst $\Delta\psi$ was kept at $\sim$ 150 mV.

In presence of UQ, O_2 uptake was blocked by CN to the level of endogenous respiration and not stimulated further by 1 mM AMP or 100 μM GMP. With succinate or NADH, addition of CN resulted in a strong decrease in O_2 uptake and in $\Delta\psi$, the former one /but not $\Delta\psi$/ beeing stimulated to the original state 4 level by further addition of GMP or AMP, regardless the presence or absence of rotenone. Malate /or ethanol/ supported respiration and $\Delta\psi$ generated by them were less decreased by CN, and both rose after addition of GMP or AMP, due to the stimulation of the alternative respiratory pathway, the site I still operating. Malate supported O_2 uptake in presence of CN was significantly stimulated by NAD with no increase in $\Delta \Psi$, however.

The stimulatory effect of GMP during malate oxidation in presence of CN was limited mainly to the increase in O_2 uptake but not in $\Delta \Psi$, when previously to GMP: a/ 0.5 mM EDTA was present, b/ one short state 3_{ADP} run was informed, c/ 100 μM ATP was added.

In presence of rotenone $\Delta \Psi$ of $\sim$160 mV developed in a sigmoidal course during few minutes after addition of 2 mM ATP /± P_i, -Mg/. At this stage exogenous P_i was not any more efficient in $\Delta \Psi$ increase, but the addition of Mg + oligomycin resulted in a significant increase in $\Delta \Psi$ /to 180 mV/. The lowering of this mounted $\Delta \Psi$ to 130 mV by CN suggests, that it was due to the increased input of electrons from endogenous substrates bypassing the rotenone-sensitive NADH dehydrogenase resulting in generation of $\Delta \Psi$ at sites II and III.

The results suggest that:
1/ external reduced UQ is oxidized almost exclusively via the cytochrome pathway /in contrast to mitochondria of Paramecium; 6/.
2/ GMP or AMP stimulated the O_2 uptake via the alternative pathway with all other substrates studied; this stimulation is paralelled by increase in $\Delta \Psi$ only when site I is operating /thus in the absence of rotenone/ and EDTA or ATP are absent: exogenous Mg is not necessary for this effect.

1. Moore A.L., Rich P.R. /1980/ TIBS 5, 284-288.
2. Hryniewiecka L., Jenek J., Michejda J.W. /1978/ in: Plant Mitochondria, Marseilles, Elsevier, 307-314.
3. Hryniewiecka L., Łojek U., Michejda J. /1982/ II Eur. Cong.Bioenergetics, Lyon, 571-572.
4. Michejda J., Domka-Popek A. /1982/ II Eur.Cong. Bioenergetics, Lyon, 243-244.
5. Kamo N., Muratsugu M., Hongoh R., Kobatake Y. /1979/ J.Membrane Biol. 49, 105-121.
6. Doussiere J. /1982/ These - Grenoble University.

HORMONAL REGULATION OF THE RESPIRATORY CHAIN IN INTACT HEPATOCYTES

P.T.Quinlan and A.P. Halestrap
Department of Biochemistry, University of Bristol,
University Walk, Bristol BS8 1TD.

Liver mitochondria isolated from rats treated with glucagon show enhanced rates of State 3 and uncoupled respiration from all substrates entering the respiratory chain before cytochrome c as a result of activation of electron flow in the b c_1 complex (1). This may be mimmicked by increasing the intramitochondrial volume (2). The stimulation of gluconeogenesis from L-lactate by glucagon is associated with increased mitochondrial ATP/ADP, O_2 consumption and matrix volume (see 3). These data suggest that hormonal stimulation of the respiratory chain may be occurring <u>in situ</u> and be important in the regulation of gluconeogenesis. The data presented here support this hypothesis.

Hepatocytes from 24 hour starved rats were suspended in Krebs Henseleit supplemented with 1% (w/v) defatted bovine serum albumin 10mM L-lactate and 1mM pyruvate and incubated at 37°C in a fluorimeter or split / double beam spectrophotometer with constant stirring and an atmosphere of O_2 : CO_2 (19:1). Data were collected and manipulated using a computer. For studying NAD(P)H fluorescence the excitation was at 350-395nm (using filters) and emission monitored at 420-480nm.

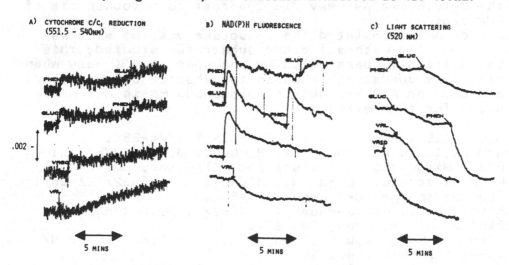

FIG. 1 THE EFFECTS OF HORMONES ON THE REDUCTION STATE OF CYTOCHROME C AND NADH AND THE MITOCHONDRIAL VOLUME IN INTACT HEPATOCYTES. CELLS WERE INCUBATED AT 12(A), 4(B) OR 7(C) MG PROTEIN/ML WITH CONTINUOUS STIRRING. HORMONES WERE ADDED WHERE INDICATED - GLUC = 10^{-7}M GLUCAGON, PHEN = 2×10^{-5}M PHENYLEPHRINE, VASO = 2×10^{-8}M VASOPRESSIN AND VAL = 2.5×10^{-9}M VALINOMYCIN.

Fig. 1 shows traces of the effects of glucagon (10^{-7}M) phenylephrine (2 x 10^{-5}M) vasopressin (2.5 x 10^{-8}M) and valinomycin (2.5 x 10^{-9}M) on the reduction state of cytochrome c/c_1 (a), NAD(P)H fluorescence (b) and light scattering (c). NAD(P)H and cytochrome c reduction increased in <1sec with both phenylephrine and vasopressin and after a delay of about 10sec with glucagon. The time course of these events follows that of the rise in intracellular [Ca^{2+}] measured using Quin2 (4) and may reflect activation of Ca^{2+} sensitive mitochondrial dehydrogenases (5). The rise in NAD(P)H is transient whilst cytochrome c reduction stays elevated, consistent with a subsequent activation of the respiratory chain induced by the increase in mitochondrial volume (Fig. 1c). In support of this valinomycin increased cytochrome c reduction and decreased NADH reduction with a time course similar to its effect on mitochondrial volume. Amytal (0.5mM) reversed the effects of glucagon or phenylephrine on cytochrome c reduction (Fig 2a) and concomitantly O_2 consumption and rates of gluconeogenesis also returned to control values (Fig. 2b). These data imply that hormonal activation of the respiratory chain is important in the regulation of gluconeogenesis in the hepatocyte.

Fig. 2 Effects of Amytal (0.5mM) on glucagon induced stimulation of gluconeogenesis and respiration

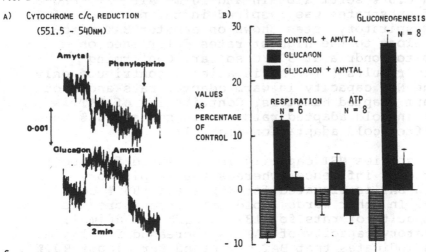

A) Cytochrome c/c_1 reduction (551.5 - 540nm)

References
1. Halestrap, A.P. (1982) Biochem. J. 204 37-47.
2. Armston, A.E. Halestrap, A.P. & Scott, R.D. (1982) Biochim Biophys. Acta 681 429-659
3. Quinlan, P.Q., Thomas, A.P., Halestrap, A.P. & Armston, A.E. (1983) Biochem. J. 214, 395-404.
4. Thomas, A.P., Alexander, J. & Williamson, J.R. (1984) J. Biol. Chem. In Press.
5. Denton, R.M. & McCormack, J.G. (1980) FEBS Lett. 119 1-8.

THERMOREGULATORY SIGNIFICANCE OF BROWN ADIPOSE TISSUE IN THE RAT AND THE DJUNGARIAN HAMSTER

J. Rafael, J. Patzelt and P. Vsiansky
Institut für Biochemie I, Universität Heidelberg, FRG

Brown adipose tissue (BAT) is estimated as the major site of non-shivering thermogenesis (NST) in the rat, contributing up to 60 % of the noradrenalin-induced heat production (1). Determination of the maximum respiratory capacity of the brown fat organ from its total content of mitochondria and the oxidative capacity of the isolated organelles leads to a contradictive view. In this study two non-hibernators with strongly different NST capacity, i.e. the rat and the Djungarian hamster (Phodopus sungorus), were investigated. The amount of mitochondria in the combined brown fat depots of the animals was determined from the specific activity of cytochrome oxidase in the tissue and in the isolated mitochondria respectively. Mitochondria were isolated in a medium containing 0.5 % serum albumin and 10 mM ATP (2), and a mixture of substrates was respired in the presence of 5 μM FCCP. Respiratory rates shown on hamster BAT mitochondria belong to the highest rates determined on isolated mitochondria at 37° C so far. Confirming preliminary results (3), BAT is able to contribute only 10 % of the NST capacity in warm adapted rats and about 20 % in warm adapted hamsters. Contribution of BAT is about 33 % in cold adapted rats and less than 45 % in hamsters after cold adaptation (Tab. 1).

In both species NST capacity is increased for about 43 % under cold influence, whereas the respiratory capacity in BAT is elevated for 365 % and 185 % concomitantly. In other words, cold influence increases the NST capacity of rats for 187 ml O_2/h per animal; the respiratory capacity of BAT is increased for 164 ml O_2/h. This indicates that BAT may stand for almost 90 % of the cold-induced NST increase. Similar results are obtained on Djungarian hamsters.

First comparative results on brown adipocytes and BAT mitochondria from guinea pigs indicate that the maximum respiration as measured on BAT mitochondria isolated in the described way is similar to that of the organelles in situ. It appears from our results that warm-adapted rats

Table 1: NST and maximum respiratory capacity of BAT in warm and cold adapted rats and Djungarian hamsters

	Warm adapted	Cold adapted	
RAT	n=11	n=12	
NST capacity (ml $O_2 \cdot h^{-1}$)	439	626	(+ 43 %)
BAT mitochondria			
mg total	54.9	250.0	(+ 353 %)
µmol $O \cdot min^{-1} \cdot mg^{-1}$	1.218	1.244	n.s.
ml $O_2 \cdot h^{-1}$ total	44.9	209.0	(+ 365 %)
Participation of BAT in NST	10.2 %	33.4 %	
DJUNGARIAN HAMSTER	n=11	n=18	
NST capacity (ml $O_2 \cdot h^{-1}$)	167	239	(+ 43 %)
BAT mitochondria			
mg total	30.1	88.8	(+ 195 %)
µmol $O \cdot min^{-1} \cdot mg^{-1}$	1.818	1.756	n.s.
ml $O_2 \cdot h^{-1}$ total	36.8	104.8	(+ 185 %)
Participation of BAT in NST	22.0%	43.8 %	

Warm adapted animals were kept under thermoneutral conditions (rats at 31° C and hamsters at 23° C), for cold adaptation animals were kept at 5° C for three weeks. Average body weights of warm and cold adapted rats were 300 g, of warm adapted hamsters 41.2 g and of cold adapted hamsters 35.0 g. NST capacity = maximum O_2-consumption as induced by injection of noradrenaline (0.4 mg/kg body weight) minus basal metabolic rate of the animal. Mitochondrial respiration was measured at 37° C as indicated in the text. Means $\pm$ SE are given.

and hamsters perform most of their NST independent from BAT. On the other hand, BAT seems of central importance during the adaptation of NST in response to cold-influence.

(1) Foster, D.O. and Frydman, M.L. (1978), Can. J. Physiol. Pharmacol. 56, 110-122.
(2) LaNoue, K.F., Koch C.D. and Meditz R.B. (1982), J. Biol. Chem. 357, 13740-13748.
(3) Rafael, J. (1983), J. Therm. Biol. 8, 410-412.

REDISTRIBUTION OF THE CONTROL STRENGTHS IN MITOCHONDRIAL OXIDATIVE PHOSPHORYLATIONS IN THE COURSE OF BRAIN EDEMA.

RIGOULET, M. °, AVERET, N." and COHADON, F."
°IBCN-CNRS, 1 Rue Camille-Saint-Saens, F 33077 Bordeaux-Cedex, France
"Université de Bordeaux II, ERA-CNRS 843, 146 Rue Leo-Saignat, BP 48, 33076 Bordeaux-Cedex, France

One of the best-known experimental model of brain insult leading to edema is the cryogenic lesion initially described by Klatzo (1). In this model, admittedly relevant to the understanding of a number of clinical situations in neurosurgical practice, the main cellular disorder seems to be disturbances in the lipid-protein organization, which affects several membrane-bound enzymes. The consequence of such events is a large modification in most, if not all, regulatory mechanisms and functionings of metabolic pathways. Therefore, in order to indentify the key enzymes involved in the pathophysiological development of edema, we propose to apply the concept introduced by Kacser and Burns (2) and Heinrich and Rapoport (3), leading to a quantitative evaluation of the various enzymatic steps control in a metabolic flux. We consider a description of the control distribution modifications during edema course as a useful approach to understand the initial functionnal alterations leading to irreversible damages.

As an illustration, we purpose quantifying in this report the participation of succinate dehydrogenase activity in the impairment of the oxidative phosphorylations during edema course.

In edematous tissue, several enzymatic activities (cytochromes, ATP-synthase complex and succinate-dehydrogenase) involved in mitochondrial oxidative phosphorylations decreased in different ways (4,5). As a consequence, the succinate dependant respiration (state 3) in mitochondria isolated from cerebral hemisphere decreases as edema increases. The fact that a drug (Naftidrofuryl) acting only in vivo at the succinate dehydrogenase level is able to re-establish a normal efficiency of oxidative phosphorylation seems to indicate that this enzyme catalyses an importent step for the respiratory activity control.

Succinate externally added as respiratory substrate is involved in two successive steps : transport by the dicarboxylate carrier and the succinate dehydrogenase activity. This situation is recognized as tricky (6); but by using two inhibitors, one (phenylsuccinate) acting only on the dicarboxylate carrier and the other (malonate) acting on both steps, it is possible to evaluate the different control strengths (Table I).

Table I : Evaluation of the control strengths of dicarboxylate carrier (C'_{i1}) and succinate deshydrogenase (C'_{i2}) as a function of edema development.

water/dry weight of cerebral tissue°	control strengths C'_{i1}	C_{i2}	Vo_2 (state 3)
4.0	0.35	0	740
4.5	0.32	0.05	550
4.8	0.32	0.22	450
5.0	0.30	0.45	400
5.5	0.28	0.50	350

°edema is defined as an increase in tissue water content
°° natoms O/min/nmol cyt.a+a

Concluding remarks

The variations of ATP-synthase, cytochromes and succinate--dehydrogenase activities according to edema evolution induce a redistribution of control strength in mitochondrial oxidative phosphorylations. The measure of these changes shows that the succinate-dehydrogenase activity dependant on edema level plays an important role in the control of mitochondrial energy producing mechanism as previously suggested (5).

BIBLIOGRAPHY

1) KLATZO, I. (1967) J. Neuropath. exp. Neurol., 26 , 1-14.
2) KACSER, H.and BURNS, J.A. (1973) in Rate Control of Biological Processes (Davies,D.D. ed) pp 65-104, Cambridge University Press London.
3) HEINRICH, R. and RAPOPORT, T.A. (1974), Eur. J. Biochem., 42 , 89-95.
4) RIGOULET, M., GUERIN, B., COHADON, F. and VANDENDREISSCHE, M., (1979), J. Neurochem., 32 , 535-541.
5) RIGOULET, M., AVERET,N., and COHADON,F. (1983), Neurochem. Pathol., 1 , 43-57.
6) GROEN A.K., VAN DER MEER, R., WESTERHOFF, H.V., WANDERS, R.J.A., AKERBOOM, T.P.M., and TAGER, J.M. (1982) in Metabolic Compartmentation (Sies ed.) pp 9-37 , Academic Press, London New York.

DETERMINATION OF THE PLASMA MEMBRANE POTENTIAL IN HUMAN FIBROBLASTS USING TPP$^+$ DISTRIBUTION

Michela Rugolo, Letizia Prosperi[*], Gianni Romeo[*] and Giorgio Lenaz
Istituto Botanico and *Laboratorio di Genetica-Clinica Neurologica,
University of Bologna, Italy.

Determination of the plasma membrane potential in human fibroblasts using intracellular microelectrode techniques appears to be complex. To overcome this technical difficulty we have adapted the method based on the distribution between cells and incubation medium of the lipophilic cation $[^{14}C]$-tetraphenylphosphonium (TPP$^+$) to cell cultures growing in monolayer. Lipophilic cations distribute according to a Nernst equilibrium across bilayer regions of the membrane. These cations are accumulated across both the plasma and mitochondrial membranes in hepatocytes, synaptosomes, fat cells, lymphocytes, (1-4), and their distribution is therefore a function of the magnitudo of both membrane potentials.

In the present communication we report that TPP$^+$ accumulation by quiescent human skin fibroblasts reaches a steady-state after 40 minutes and an accumulatio ratio of 60.4 ±15 is obtained. The large range of variability observed depends on the different cell lines employed.

Cultures derived from the same skin biopsy, utilized between 6-27th passages, give reproducible results. Optimal accumulation of the cation is observed at TPP$^+$ concentrations below 2 μM.

That TPP$^+$ is accumulated not only by the plasma membrane, but also by the internal mitochondria is confirmed by the data reported in Table 1.

In fact, the intracellular steady-state TPP$^+$ accumulation is greatly reduced by the addition of valinomycin. The K$^+$ ionophore can depolarize mitochondria by setting their potential to the K$^+$ diffusion potential between the cytosol and the mitochondrial matrix. Since K$^+$ in both compartments is ∼150 mM, the K$^+$ diffusion potential is zero. The uncoupler FCCP is the most effective in reducing TPP$^+$ accumulation ratio. The uncoupler CCCP or the combination of the respiratory chain inhibitor rotenone with the ATPase inhibitor oligomycin do not reduce TPP$^+$ accumulation ratio to the same extent as FCCP.

When the mitochondrial potential is completely abolished, TPP$^+$ accumulation can be utilized to quantify the plasma membrane potential, using

TABLE 1 - Effect of ionophores and inhibitors on TPP^+ accumulation by human skin fibroblasts.

Additions	$^{14}C-TPP^+$ accumulation ratio
none	53.2
valinomycin	38.8
CCCP	36.1
rotenone + oligomycin	31.6
FCCP	25.8

the Nernst equation, with correction for the potential-independent binding component. The value obtained for plasma membrane potential of human fibroblasts is 75 mV, in good agreement with the values obtained by microelectrode measurements. Recently Seemann et al.(5) have estimated $\Delta\Psi_p$ of 3T3 fibroblasts to be 78mV, using the lipophilic cation $TPMP^+$. They calculate $\Delta\Psi_p$ after subtraction of the $TPMP^+$ accumulation ratio measured at high K^+ medium from that measured in a low K^+ medium.

This correction does not seem to be appropiate and could lead to inaccuracy in the measured value of $\Delta\Psi_p$.

The release of total TPP^+ due to FCCP treatment could allow also an estimation for mitochondrial membrane potential, provided that the mitochondrial volume is determined. Unfortunately the contribution of the internal mitochondria to the total cellular volume is difficult to assess with any degree of precision, therefore TPP^+ accumulation can only be considered as a relative monitor of mitochondrial changes.

1) Hoek J.B.,Nicholls D.G.and Williamson J.R.(1980)J.B.C. 255,1458.
2) Scott I.D.and Nicholls D.G.(1980)Biochem.J.186,21.
3) Davis,J.D.,Brand,M.D.and Martin,B.R.(1981)Biochem.J.196,133.
4) Felber,S.M.and Brand,M.D.(1982)Biochem.J.204,577.
5) Seemann,D.,Furstenberger,G.and Marks,F.(1983)Eur.J.Biochem.137,485.
Acknowledgement: work supported by the Progetto Finalizzato Ingegneria Genetica e Basi Molecolari delle Malattie Ereditarie of the CNR., Rome.

DYNAMICS OF INSTABILITY OF MITOCHONDRIAL MEMBRANES DURING METABOLICALLY INDUCED HIGH AMPLITUDE SWELLING

D. Sambasivarao and V. Sitaramam
National Institute of Nutrition, I.C.M.R., Jamai Osmania (P.O.), Hyderabad-500 007 (AP), India.

The mechanism of swelling/contraction cycles in mitochondria is not clear, despite the early recognition of the importance of volume changes in mitochondria (1). We reported the 'beneficial' effect of hypotonicity on mitochondrial respiration and indices of oxidative phosphorylation (summarised in Fig.1) (2). The irreversible nature of swelling in nonelectrolyte media suggested that additional variables (indicated by '?' in Fig. 1) actively contribute to mitochondrial swelling.

Direct experimental evidence was obtained for the following train of events to occur during high amplitude swelling (HAS), i.e., on suspension of mitochondria in warm, buffered sucrose media: i. Respiration due to endogenous substrates leads to increased permeability of the inner membrane to polyols, as well as an increase in the net negative surface charge; ii. The swelling becomes auto-catalytic' due to the interactive, synergistic influence of swelling, respiration and surface charge density; iii. imbibition of water first leads to disruption of the outer membrane, as evidenced by leakage of intermembranous enzymes; iv. enhanced porosity of the inner membrane also leads to loss of NAD^+ and Mg^{2+} leading to an abrupt decrease in respiration corresponding to the onset of swelling. v. cristal unfolding sets in, contributing maximally to the observed turbidity changes; vi. the continued swelling and instability due to unscreened fixed charges on the surface of inner membrane leads to disruption of the inner membrane and loss of matrix enzymes; and vii. partial re-sealing sets in after disruption, such that the swollen mitochondria represent a mixture of open fragments and normally oriented inner membrane particles.

Further studies on osmotic pressure-volume relationships in mitochondria, in the presence of inert polyols (sucrose and mannitol) as external osmolytes, showed that respiration induces variable porosity of the inner membrane not only in HAS but also during oxidative phosphorylation per se. The experimental results reported here are consistent with i. HAS as an extreme case of respiration-driven osmotic swelling of mitochondria in nonelectrolyte media; ii. different agents that effect turbidity changes and swelling differ in their influence on swelling vs. the instability of the inner membrane, which in turn determines the reversibility of swelling and the functional status of the swollen mitochondria; iii. surface charge density on

the inner membrane contributes to the physical state and stability
of the inner membrane. The swelling/contraction cycles of mitochondria during oxidative phosphorylation appear to be directly related
to permeability changes in the inner membrane to polyols. This, in
turn, argues against the validity of $\Delta\psi$ and ΔpH measurements (3),
since the anomalous osmotic pressure-volume relationships in oxidative
phosphorylation indicate permeation of sucrose into the matrix space,
depending on the state of respiration.

1. Lehninger, A.L. (1965) The Mitochondrion. W.A. Benjamin Inc.
   New York.
2. Sambasivarao, D. and Sitaramam, V. (1983) Biochim. Biophys. Acta
   722, 256-270.
3. Sitaramam, V. and Sambasivarao, D. (1984) TIBS (In Press).

Fig. 1. Oxidative phosphorylation and volume regulation in mitochondria: Various coupling constants for heat production (K_1) and ATP/
ADP translocation (K_2) and ATP synthesis (K_3) are visualised to be
directly influenced by the osmotic stretch of the inner membrane, as
evidenced by variable stoichiometry of oxidative phosphorylation as
a function of the tonicity of the medium (2, 3). $\bar{K}_3$ and $\bar{K}_2$, additional volume (V)-dependent constants of ATP synthesis and translocase
activity. The query (?) symbolizes additional variables such as surface charge density and external osmotic pressure.

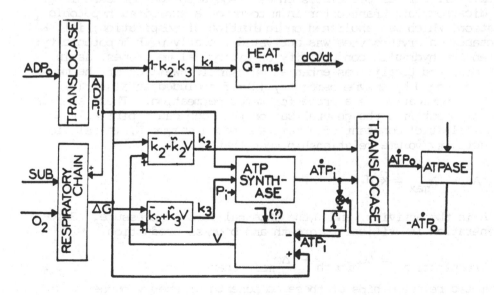

ENERGETIC BASIS OF OSMOTOLERANCE IN MICROBES AND DROUGHT/SALINITY RESISTANCE IN CULTIVARS.

V. Sitaramam and N.M. Rao, National Institute of Nutrition, I.C.M.R., Jamai Osmania (P.O.), Hyderabad-500 007 (AP), India.

Development of osmotolerant varieties of microbes and cultivars has immense industrial and agricultural applications. A common cellular mechanism would entail the identification of osmotic sensors as well as an elucidation of the link between osmotic pressure of the medium/soil and the metabolic rate and efficiency of the cell. We reported inhibition of respiration, activities of a number of membrane-bound enzyme systems, carrier proteins etc. as well as a deterioration of stoichiometry of oxidative phosphorylation (P/O ratio) with increasing tonicity of the medium (1). We reasoned that hypotonic activation of membrane-bound enzyme systems as well as efficiency of ATP synthesis could provide the energetic basis of osmotolerance in biomass production. The osmosensor would be the energy transduction mechanism itself.

Direct experimental evidence was obtained that: i. respiration due to endogenous substrates was inhibited with progressive hypertonicity of the medium in a variety of microbes, e.g., E.coli, Salmonella sp., nitrogen fixing bacteria, Saccharomyces sp., as well as plant tissues. ii. Carrier proteins such as -galactoside permease in E.coli, dicarboxylate transporter in mitochondria exhibited hypotonic activation, which was abolished on inhibition of respiration; iii. K^+ conductance in erythrocytes was activated linearly with hypotonicity; iv. Even the hydraulic conductivity of biological membranes, as in peroxisomes and E.coli, was enhanced on osmotic swelling of the membrane, as judged by enzyme osmometry (2) of occluded catalase (3), using H_2O_2 permeation as a probe for water permeation. These studies were consistent with the generalization that one can empirically define the 'elastic' constant of a membrane-bound enzyme $\tilde{K}$, in relation to activity/volume relationship such that,

$$A = A_{max} - \tilde{K}\,\pi \qquad\qquad \ldots 1$$

where A is the activity, and π, the external osmotic pressure. Since ATP generation is critical to growth and biomass production,

$$\Delta G_{respiration} = \Delta G_{growth} + \Delta G_{maintenance} \text{ etc.} \qquad \ldots 2$$

The osmotic relationships of these components of energy budget would be,

$$\Delta G_{resp.} = \Delta G_{resp.} \text{ (max)} - \overset{\smile}{K}_1 \pi$$

$$\Delta G_{growth} = \Delta G_{growth} \text{ (max)} - \overset{\smile}{K}_2 \pi$$

$$\Delta G_{maint.} = \Delta G_{maint.} \text{ (max)} + \overset{\smile}{K}_3 \pi \qquad \qquad \dots 3$$

where the energy requirements for maintenance and entropy may remain invariant or actually increase due to osmotic load. It can readily be shown that the critical external osmotic pressure at which growth ceases would be much less than that at which respiration ceases, insofar as G_{maint} and/or $\overset{\smile}{K}_3$ are finite. The phenotypic character for osmotolerance for any variety could thus be defined as a critical external osmotic pressure inhibitory for either respiration or growth.

We used the critical tonicity at which respiration ceases as a phenotypic marker in several microbes and cultivars such as dryland and wetland rice. Experimental data confirmed not only the energetic relationships of growth vs. respiration but also the predictive value of this phenotypic marker with regard to the relative drough resistence in traditional varieties of rice. Respiration was found to be a valid phenotypic marker in all species thus for tested, unlike the response to osmolytes such as proline and betaine (4), whose impact varied with varieties. A glaring example was found to be <u>Cereus</u> <u>hexaginosa</u> L, a cactus, which had very little osmolyte content in the sap (230 mOsm) but an inhibitory osmotic pressure of 2300 mOsm, whereas a carrot root has 550 mOsm in the cell sap but respiration was inhibited at 1100 mOsm. Methodology amenable for germ plasm screening and mutant identification is currently being developed.

1. Sambasivarao, D and Sitaramam, V. (1983). Biochim. Biophys. Acta. 722, 256-270.
2. Sitaramam, V. and Sarma, M.K.J. (1981). J. theor. Biol. 90, 317-336.
3. Sitaramam, V. and Sarma M.K.J. (1981). Proc. Natl. Acad. Sci. USA 78, 3441-3445.
4. Rudulier, D.L. and Valentine, R.C. (1982). TIBS 6, 431-433.

RAPID THYROID HORMONE STIMULATION OF ADENINE NUCLEOTIDE TRANSLOCA-
TION IN RAT LIVER

S.Soboll[+], M.J. Müller[o] and H.J. Seitz[o]

[+]Institut für Physiologische Chemie I, Universität Düsseldorf, Universtitätsstr. 1, D-4000 Düsseldorf 1, FRG

[o]Institut für Physiologische Chemie der Universität Hamburg, Martinistr. 52, D-2000 Hamburg 20, FRG

The effect of thyroid hormone on hepatic energy metabolism was investigated by measuring mitochondrial and cytosolic adenine nucleotide contents (1), redox state of NAD^+-systems (i.e. lactate/pyruvate; ß-hydroxybutyrate/acetoacetate ratios) and oxygen consumption. Measurements were performed in vivo in hypo- and hyperthyroid, 24h starved rats and in isolated perfused hypothyroid livers in the absence and presence of triiodothyronine (2).

Subcellular ATP/ADP ratios from hypothyroid rats were not significantly different from controls whereas hyperthyroidism resulted in a considerable decrease in mitochondrial ATP/ADP ratio and an increase in the cytosolic ratio.

The thyroid hormone stimulation of oxygen consumption was not due to uncoupling of oxidative phosphorylation which can be concluded from the observations (i) that it could be suppressed by oligomycin, (ii) the degree of coupling q according to (3) was unchanged and (iii) subcellular NAD^+-systems were not altered.

Therefore the decrease in mitochondrial ATP/ADP ratio- despite enhanced ATP-production- in hyperthyroidism and increase in cytosolic ratios is attributed to a stimulation of mitochondrial adenine nucleotide translocation, which should lead to a decrease in mitochondrial and increase in cytosolic ATP-content. This seems to be a direct and rapid effect of T_3 on the mitochondrial pathway, since it can also be produced in the isolated perfused liver within 30-120 minutes.

1 Soboll, S., Akerboom, T.P.M., Schwenke, W.D., Haase, R. & Sies, H. (1980) Biochem. J. 192, 951-954

2 Müller, M.J. & Seitz, H.J. (1980) Life Sci. 27, 827-835

3 Stucki, J. (1980) Eur. J. Biochem. 109, 269-283

INACTIVATION OF THE LACTOSE PERMEASE OF E. COLI BY THE RESPIRATORY ACTIVITY: POSSIBLE INVOLVEMENT OF THE SUPEROXIDE ION.
H. Thérisod, A. Ghazi and E. Shechter.

Laboratoire des Biomembranes, Université de Paris-XI, 91405 Orsay France.

We have recently shown that in E.coli, the lactose permease, which functions as a proton symport (1), becomes irreversibly inactivated during the transport of lactose when the cells are provided with an energy source (2). While in resting cells the uptake of lactose as a function of time is monotonous, in energized cells (i.e. cells in the presence of an exogenous energy source), the internal lactose concentration reaches a maximum and then declines to a steady state level 2 to 10 times lower than in resting cells.

This low steady state level of lactose accumulation is due to an irreversible decrease in the rate of influx and to an increase in the rate constant of efflux. It is accompanied by a large and irreversible decrease in the transmembrane electrical potential. In contrast, no such decrease in the transport parameters and in the transmembrane potential is observed in resting cells. The extent of these phenomena is dependent on the experimental conditions: it is maximal at alkaline pH, for low external potassium concentration and for relatively high external lactose concentrations (around or above the K(T) of transport); treatment of the cells with EDTA also enhances these phenomena.

The inactivation of the lactose permease is not related to the metabolism of the energy source: indeed, it is observed whatever the energy source, even with ascorbate-phenazine methosulphate which supposedly injects its electrons directly into the respiratory chain.

In energized cells, an artificial decrease in the rate of respiration brought on by the addition of inhibitors of the respiratory chain (cyanide, 2-heptyl-4-hydroxyquinoline-N-oxide) results in an uptake of lactose similar to the one observed in resting cells. Thus, it is the increase in the rate of respiration, induced by energization, which is responsible for the inactivation of the lactose permease.

While the presence of lactose enhances the susceptibility of the permease towards inactivation, the presence of thiodigalactoside (TDG, a competitive analogue of lactose) protects the carrier against inactivation: when cells are allowed to transport TDG, washed to eliminate the external TDG, and assayed for lactose transport in the presence of an energy source, no inactivation is observed.

When the cells are incubated 2 hours at room temperature and at dilute concentration, in the absence of lactose and energy source, and then assayed for lactose transport, one observes an impairment of lactose transport very similar to the one obtained upon immediate transport in energized cells. This inactivation induced by prolonged incubation is specifically related to the permease since it can be prevented by incubation in the presence of TDG.

The picture which emerges is thus the following: the activity of the respiratory chain in itself leads to an inactivation of the permease. The presence of lactose enhances the suceptibility of the permease towards inactivation, probably by immobilizing the protein in a conformation more accessible to the inactivating agent. In contrast, TDG protects the protein from the cytoplasmic side against inactivation. The irrerversible decrease of the membrane potential following the inactivation suggests that this inactivation consists in an uncoupling at the level of the permease, the protein being then able to catalyse a translocation of protons without its substrate.

Inactivation through the activity of the respiratory chain could be due to a direct interaction between the permease and an oxidized component of the respiratory chain. This hypothesis has not been tested. The alternative is the involvement of superoxide ions which are produced at the level of the respiratory chain and which are known to be at the origin of oxygen toxicity in cells (3). Addition of copper bis-salicylate, a complex with a high superoxide dismutase activity which is able to permeate the membrane, or of ferrous salts which increase the dismutase activity of the cytoplasm, partially suppress the inactivation of the lactose permease, thus suggesting that the superoxide ion could be the direct agent responsible for this phenomenon.

It can be expected that other permeation systems of E.coli are subject to the same kind of inactivation. In this respect, the data reported in (4) for the glucose PTS system could be interpreted in this way. If inactivation by superoxides of membrane energy transducing proteins, such as the lactose permease, was to be confirmed, it would provide a clue as to the mechanism of toxicity of this compound for the cell.

1 - West, I.C. and Mitchell, P. (1973) Biochem. J. 132, 587-592.
2 - Ghazi, A., Thérisod, H. and Shechter, E. (1983) J. Bacteriol. 154, 92-103.
3 - Fridovich, I. (1978) Science 201, 875-880.
4 - Hernandez-Asencio, M. and Del Campo, F.F. (1980) Arch. Biochem. Biophys. 200, 309-318.

GLYCOGEN PARTICLE BOUND PHOSPHORYLASE PHOSPHATASE: INFLUENCE OF
HORMONAL AND NUTRITIONAL CONDITIONS

E. Villa-Moruzzi

Istit. di Patologia Generale, Università di Pisa, 56100 Pisa, Italy

Skeletal muscle phosphatase activates glycogen synthase and inactivates phosphorylase, hence promoting glycogen synthesis. Phosphatase is present in the cytosol and bound to glycogen particles (1). While the former enzyme has been studied extensively (2,3), including the mechanism of its activation by the kinase F_A (3,4) and by trypsin-Mn^{2+} (3), much less is known of the regulation of the latter and whether it represents or not a different type of phosphatase.

In glycogen particles purified from rat skeletal muscle, up to 40% of the active phosphorylase phosphatase of the whole extract was present. Limited proteolysis with trypsin (3) almost doubled the activity (total activity), while Mn^{2+} (3) had no effect. Fasting, adrenaline and diabetes decreased these activities (see Table), but the ratio of active to total phosphatase did not change. These treatments induced also glycogen breakdown and shift of phosphorylase from glycogen particles to cytosol. These changes were partially reversible with glucose after fasting, but not with insulin in diabetes. Although the cytosolic enzyme was never affected by these treatments (values were always of 60-80 and 170-200 U/g w.w. for active and total activities respectively), the phosphatase activating kinase F_A (4) was less active in fasting and diabetes.

These results indicate that when glycogen is demolished the glycogen particle bound phosphatase, but not the cytosolic one, undergoes inhibition. This could be brought about by inhibitors, substrates or glycogen itself, and cannot be overcome by the means that are known to activate the cytosolic enzyme, namely trypsin or trypsin-Mn^{2+}, 3).

Whether this means that different enzymes are present is under investigation.

T A B L E

	Control	Fasting (24 h)	Fast.+glucose (120 min)	Adrenaline (4 min)	Diabetes	Diab.+insulin (100 min)
Phosphatase activity in glycogen particles (U/g w.w.):						
- active	51±9.5(6)	20±2.9(9)	29±1.6(6)	28±5.0(8)	27±4.0(6)	25±7.1(5)
- total	87±15 (6)	35±5.4(9)	46±5.0(8)	41±6.0(8)	45±6.8(6)	44±12 (5)
Phosphorylase a + b activity: b in glycogen particles	46±5.4(4)	8±0.7(4)	15±2.6(4)	22±5.3(5)	16± 10(2)	5±1.0(2)
Total muscle glycogen (mg/g w.w.)	6.8±0.8(5)	4.0±0.7(7)	3.4±0.6(6)	4.9±0.5(6)	4.7±1.4(4)	3.4±0.9(4)
Glycogen in glycogen particles (mg/g w.w.)	1.09±0.42(4)	0.04±0.01(4)	0.23±0.08(4)	0.25±0.11(4)	1.02±0.52(4)	0.32±0.16(4)
F_A activity in cytosol (arbitrary units)	34±3.0(4)	26±0.8(4)	25±2.5(4)	35±2.5(5)	27±3.1(7)	23±0.7(5)

Animals: rats of 190-210 g b.w.. Treatments: glucose:4 IP injections of 500 mg/kg; adrenaline:10 µg/kg IV and 400 µg/kg IP; diabetes: induced with streptozotocin 38 h before sacrifice; insulin: 5 IP injections of 10 U/kg, with glucose; Nembutal anaesthesia for 10 min. Extracts: as in (2) but using a Potter apparatus. Glycogen particles: by direct centrifugation of extract (1). Assay methods: (2,3) for phosphorylase phosphatase and F_A, (5) for glycogen and (6) for phosphorylase. Mean values ± S.E.; number of cases in parenthesis.

REFERENCES

1. Meyer,F.,Heilmeyer,Jr.,L.M.G.,Haschke,R.H. and Fischer,E.H. (1970) J.Biol.Chem. 245,6642-6648.

2. Ballou,L.M.,Brautigan,D.L. and Fischer,E.H. (1983) Biochemistry 22,3393-3399.

3. Villa-Moruzzi,E.,Ballou,L.M. and Fischer,E.H. (1984) J.Biol.Chem. in press.

4. Vandenheede,J.R.,Yang,S.-D.,Goris,J. and Merlevede,W. (1980) J. Biol.Chem. 255,11768-11774.

5. Hassid,W.Z. and Abraham,S. (1966) in: Methods in Enzymology, vol. III (Colowick,S.P. and Kaplan,N.O.,eds.), pp.34-37, Academic Press, New York.

6. Hedrik,J.L. and Fischer,E.H. (1965) Biochemistry 4,1337-1343.

THE pH DEPENDENCE OF CELLULAR ENERGY METABOLISM

D.F. Wilson, M. Erecińska, T. Kashiwagura, C. A. Deutsch and J. Taylor. Departments of Biochemistry and Biophysics, Pharmacology and Physiology, Medical School, University of Pennsylvania, Philadelphia, Pennsylvania, 19104, U.S.A.

Suspensions of isolated rat hepatocytes were incubated at extracellular pH values from 6.3 to 8.2. The intracellular pH, ATP, ADP, Pi, cytochrome c reduction and respiratory rate were measured as were 3-OH-butyrate and acetoacetate. The intracellular pH as measured by ^{19}F NMR (1) was equal to the extracellular pH at approximately 7.1 but changed only 0.45 pH unit for each pH unit of extracellular change. The [ATP] decreased and [ADP] increased with increasing pH while [Pi] remained essentially constant. The [ATP]/[ADP] decreased linearly from 8.5 to 1.5 as the extracellular pH increased from 6.3 to 8.3. Calculations based on the intracellular pH show that the free energy of hydrolysis of ATP (ΔG_{ATP}) was essentially independent of pH (2). Measurements of the 3-OH-butyrate/acetoacetate ratio and the state of reduction of cytochrome c showed that transfer of reducing equivalents from intramitochondrial NADH to cytochrome c was likewise independent of pH (2). This is consistent with the first two sites of mitochondrial oxidative phosphorylation being near equilibrium throughout the pH range (for review see 3). Cytochrome c was only 10% reduced at a pHe of 6.4 but was 35% reduced at a pHe of 8.3 although the respiratory rate and ΔG_{ATP} values were essentially the same. Suspensions of isolated rat liver mitochondria show similar behavior with increasing pH i. e. increasing cytochrome c reduction for equal ΔG_{ATP} and respiratory rate (4). This arises from a strongly pH dependent step in the mechanism of mitochondrial cytochrome c oxidase. Supported by GM-21524.

References
(1) Taylor, J.S. and Deutsch, C. (1984) Biophys. J. 43, 261-264.
(2) Kashiwagura, T., Deutsch, C.J., Taylor, J., Erecińska, M. and Wilson, D.F. (1984) J. Biol. Chem. 259, 237-243.
(3) Erecińska, M. and Wilson, D.F. (1982) J. Memb. Biol. 70, 1-14.
(4) Wilson, D.F., Owen, C.S. and Holian, A. (1977) Arch. Biochem. Biophys. 182, 749-762.

QUANTITATIVE EVALUATION OF MITOCHONDRIAL ATPASES DURING AND AFTER
ISCHEMIC STRESS IN WORKING RAT HEART

G. Zimmer[o], F. Beyersdorf["], J. Fuchs[o], H. Kraft[o], B.M. Heil[o] and
P. Veit[o]
[o]Gustav-Embden-Zentrum der Biologischen Chemie, and ["]Abt. für Thorax-
Herz- und Gefäss-Chirurgie, Universität Frankfurt am Main, Federal
Republik of Germany

The working rat heart (1) represents a significant model resembling conditions in vivo. In this model heart function (heart rate,
aortic and coronary flow) biochemical (CP, ATP, ADP, AMP) mitochondrial (ADP/O, OPR, RCR, q) function and morphological parameters
(ultrastructure and mitochondrial ATPase) were measured during
normoxic (I), hypoxic (II) and reoxygenation (reperfusion) (III)
phases. The influence of 2-mercaptopropionylglycine (oxidized) (ox-
MPG) on the above mentioned parameters was measured during reperfusion (IV).

Ultrastructure: During phase I the myocardial cells show a well
preserved ultrastructure; i.e. mitochondria exhibit a dark matrix,
tightly packed cristae and they contain numerous dark granules. The
nuclei have fine dispersed chromatin and show no swelling. Glycogen
granules are abundant throughout the cells. No signs of edema are
visible. In phase II subcellular changes of hypoxia are seen. Mitochondria have an electron-lucent matrix and many broken cristae. Normal granules are lacking. Nuclei exhibit chromatin margination and a
slight swelling. Glycogen granules are diminished. At the end of phase
III partial structural reconstitution is observed. Mitochondria
still show a clear matrix and damaged cristae, but intact cristae
have increased and a few dark granules have reappeared. By the end
of phase IV the mitochondrial dark granules have slightly increased,
the cristae are mostly normal and the matrix is less electron-lucent
than during phase III. The distribution of chromatin is almost normal, no edema is present. Glycogen granules have reappeared.

Heart and mitochondrial function: All cardiac function parameters
exhibit characteristic changes during the experimental phases I-IV.
Aortic flow, coronary flow and heart rate decrease precipitously in
phase II. After reoxygenation (phaseIII) an increase to about 50% the
original value is noted. Following addition of 1 mM ox-MPG a further
significant improvement of all 3 parameters is observed. The mito-

chondrial parameters OPR, RCR (see table 1) or q after a decrease in phase II significantly increase in phase IV, whereas ADP/O remains unchanged.

Mitochondrial ATPase: in Table 1 a quantitative determination of ATPase and of RCR values is presented. There is a drastic decrease of the amount of ATPases during hypoxia, which is clearly reversed during phase III and more intensely, during phase IV.

Table 1 Correlation of RCR in isolated mitochondria with quantitative determination of ATPase n= 4

Phase	length of i.m.m. counted	counted ATPases /100 μm i.m.m.	RCR
I	4250	17.6 ± 3.8	3.7 ± 0.2[+]
II	6200	2.2 ± 1.6	2.0 ± 0.1[+]
III	12000	8.5 ± 4.4	2.6 ± 0.2[+]
IV	11125	12.8 ± 4.5	3.9 ± 0.3[+]

[+]significant between phases ($p < 0.05$); means $\pm$ S.D.

It has been suspected previously (2) that MPG may improve anchoring of the ATPase molecules at their membrane sites. This idea is clearly corroborated here. Several ways of action of MPG (ox-MPG) are principally possible a) free radical scavenger mechanism b) restoration of thiol groups by SH-S-S- interchange reaction c) in add. to b) increase in polarity at the interface of the inner mitochondrial membrane (i.m. m.) thus improving reattachment (or reorientation of reattached) ATPase molecules.

It is concluded that the capacity for ATP production and myocardial cell recovery at the end of ischemia is dependent on high energy stores (functional ATPases (3)) at this time. ATP production is linked to a reversible attachment of the F_1F_o ATPases to the i.m.m.

References: 1) Neely, J.R., Liebermeister, H., Battersby, E.J., and Morgan, H.E. (1967) Amer.J.Physiol.212, 804-814; 2) Zimmer, G., Mainka, L. and Ohlenschläger, G. (1980) Arzneim.-Forsch/Drug Res. 30,632-635; 3) Beyersdorf, F., Gauhl, C., Elert, O. and Satter, P. (1981) Basic Res. Cardiol.76, 106-113
This work was supported by the Deutsche Forschungsgemeinschaft.

EXPRESSION OF MITOCHONDRIAL GENOME IN DROSOPHILA MELANOGASTER

S. Alziari, F. Berthier, M. Renaud and R. Durand
Laboratoire de Biochimie, Université de Clermont II, L.A. CNRS 0360,
B.P. 45, 63170 AUBIERE, France

The mt(mitochondrial) DNA of D. melanogaster has a length of
19 500 bp, 5 100 of which corresponding to a non coding zone rich in
Adenine and Thymine. The coding genes for 2 rRNAs, a few tRNAs, subunits 1,2,3 of cytochrome oxydase, URF A6L, subunit 6 of ATP synthase,
URF 1 and URF 2 have been localized on the genome. These genes are the
same as those present in the genome of mammals but their relative positions on mtDNA are different.
We have developped a method to purify mitochondrial RNAs free of
cytoplasmic ribosomal RNAs. In order to obtain an in vitro system able
to translate purified mRNAs, we have studied the protein synthesis in
whole or lysed isolated mitochondria.

MATERIALS AND METHODS

Isolation of mitochondria. Mitochondria were isolated from embryoes
(0-15 h) of D. melanogaster, strain B'$_2$.

Isolation and purification of mtRNAs. Mitochondria were treated by
a phenol-chloroform-isoamyl alcohol mixture. After an ethanol precipitation the pellet was treated by 2 M LiCl to eliminate mtDNA and mt
tRNAs. mt RNAs were fractionated in poly A$^-$ RNAs and poly A$^+$ RNAs by
chromatography on oligodT-cellulose column.mt RNAs were separated by
electrophoresis in denaturating agarose gel and revealed by ethidium
bromide.

Cartography of mt RNAs on the mt DNA. After electrophoresis mt RNAs
fractions were transfered onto nitrocellulose filters and hybridized
with clones of mt DNA restriction fragments (Eco RI, Hind III C) labeled with [32P] dCTP.

Uptake of aminoacids in whole and lysed mitochondria. Mitochondria
were lysed in an hypotonic medium containing Triton X 200 0.015 %.
Mitochondria were incubated in a medium optimized for maximal protein
synthesis (2) and containing cycloheximide, sulfanilamide, the 19
amino acids and ^{35}-Meth. After 50 min at 22°C the reaction was stopped by addition of SDS and β'mercaptoethanol. Mitochondrial proteins
were submitted to electrophoresis through polyacrylamide SDS gels.
Protein bands were identified by fluorography.

RESULTS AND DISCUSSION

Mitochondrial transcripts. By Northern hybridization 9 major poly A[+] RNAs (2 100 to 800 bases) including the larger rRNA and 15 minor poly A[+] RNAs (4 000 to 500 bases) were characterized and mapped on the mt DNA. The minor RNAs overlapping the major RNAs are distributed along the entire transcribed part of the mt DNA. If the sizes of the identified transcripts are added up, we obtained 40 kb, well in excess of the length of the mt DNA coding zone, 14.4 kb. There are transcripts which are either precursors covering several genes or transcripts of the non coding strand. Our results are consistant with those of Merten and Pardue (1).

Contamination of isolated mt RNAs by cytoplasmic rRNAs was detected by use of clones of nuclear DNA coding for cytoplasmic rRNAS. Our results show that purification of mitochondria on percoll gradient and treatment by digitonin strongly remove cytoplasmic rRNAs.

Mitochondrial translation. Attempts to translate mt poly A[+] RNAs in heterologous systems of cytoplasmic origin failed in success. We want to establish an homologous translation system of mitochondrial origin, but in a first time we have optimized neosynthesis of mito-chondrial proteins in whole and lysed mitochondria. Two types of fluorographic profiles could be obtained in each case. The first one shows 10 to 12 bands (65 to 10 kd) with less intermediate bands, the second one showing 25 to 30 bands (110 to 10 kd). Two facts will be discussed i) the number of protein bands exceeds in all experiments the potential capacity of the mitochondrial genome of D. melanogaster which is probably similar to that of the mammals (13 protein genes) ; a similar result was observed in HELA cell mitochondria (3). ii) protein bands of 70 to 110 kd exceed the coding potential of the larger gene (66 kd).

In lysed mitochondria, ^{35}S-Meth. incorporation in protein was greater than that observed in intact mitochondria and was aurin-tricarboxylic acid (50 μM) sensitive, a result which shows that lysed mitochondria are able to reinitiate protein synthesis. Those results enable us to consider that lysed mitochondria constitute an accessible mitochondrial system for exogenous mt mRNAS.

(1) MERTEN, S.H. and PARDUE, M.L. (1981) J. Mol. Biol., 153, 1-21.

(2) ALZIARI, S., STEPIEN, G. and DURAND, R. (1981) Biochem. Biophys. Res. Comm., 99, 1-8.

(3) CHING, E. and ATTARDI, G. (1982) Biochemistry, 21, 3188-3195.

NUCLEOTIDE SEQUENCE OF THE CLONED mRNA AND GENE OF THE ADP/ATP CARRIER
FROM NEUROSPORA CRASSA

Hermann Arends and Walter Sebald
Abteilung Cytogenetik, GBF-Gesellschaft für Biotechnologische Forschung
mbH, Mascheroder Weg 1, D-3300 Braunschweig, FRG

The ADP/ATP carrier catalyses the exchange of ADP and ATP across the
mitochondrial inner membrane. The functional carrier is a dimer of two
identical subunits of mol.wt. 30 000-33 000, depending on the species.
The amino acid sequence of the beef heart carrier has been determined[1].
The carrier protein is synthesized in the cytosol and imported post-
translationally into the mitochondria. The primary translation product
has the same apparent mol.wt. as the mature protein[2].

The present abstract describes the isolation of a cloned cDNA corresponding to the mRNA of the ADP/ATP carrier from Neurospora crassa. The
cDNA was identified by methods based entirely on the selection of mRNA
by hybridization and subsequent analysis of cell-free translation
products by antibodies[3,4]. These procedures were adapted so that 12
pools of 96 cDNA clones could be screened in one experiment. The mRNA
selected by three of these cDNA pools directed the synthesis of a polypeptide which is adsorbed by the antibodies against the ADP/ATP carrier.
The 96 cDNA plasmids of one of the positive pools were investigated
further, first in groups of 8 and then individually. Finally, the cDNA
of one clone was identified which hybridized specifically with only the
carrier mRNA. This cDNA consists of 109 bp corresponding to nucleotides
201-309 of the coding sequence. This fragment was used as a probe for
colony-filter hybridization. ADP/ATP carrier cDNA clones were found that
way with a frequency of 1 per 300-400. Two of them contained cDNA inserts
of 1136 bp and 1177 bp, respectively comprising the whole 3'end of the
mRNA including the poly(A) tail as well as the coding sequence up to the
triplet coding for amino acid 4. The gene was cloned and isolated as a
4.6 kb Eco RI fragment. The sequence of 1937 bp was determined wich contains the whole sequence of the mRNA. A comparison of cDNA and genomic
DNA revealed the presence of two short introns, one after the 33rd and
one after the 132nd base of the coding sequence.

The first 5' proximal AUG of the mRNA is assumed to determine the
start of the protein coding sequence for the following reasons. (i) It
represents the start of an open reading frame of 939 nucleotides, which
codes for a protein sequence which is identical in 148 positions with
the amino acid sequence of the beef heart carrier protein. (ii) The proposed Neurospora polypeptide is 16 amino acids longer than the beef heart
polypeptide. This is in accordance with the observation that the apparent
mol.wt. of 33 000 of the Neurospora carrier is slightly higher than that
of the beef heart carrier (mol.wt.30 000). (iii) The bases preceding this
ATG codon (ATATCACA) are similar to the corresponding segment of other

<u>Neurospora</u> mRNAs. This does not hold for all other in-phase ATGs of the sequence.

The homologies between the <u>Neurospora</u> and beef heart carrier are especially pronounced in certain segments of the polypeptide chain. The significance of these invariant features for the function of the ADP/ATP carrier remains to be established. Three domains of similar size and substructure can be discriminated, similar to but even more pronounced than those of the beef heart carrier. Each domain contains: (i) a lipo- philic segment of 18 - 24 residues; (ii) a large polar segment of 31 - 36 residues; (iii) a 26-residue segment of middle polarity containing two or three basic residues but no acidic residues; (iv) a short polar segment. There is marked homology between the amino acid sequences of the three domains, and it has been suggested that the whole polypeptide may have originated by triplication of an ancestral gene (Fig. 1).

REFERENCES

1 Aquila,H., Misra,D., Eulitz,M. and Klingenberg,M.(1982) Hoppe Seyler's
   Z.Physiol.Chem., 363 , 345-349
2 Zimmermann,R., Paluch,U., Sprinz,M. and Neupert,W.(1979) Eur. J.
   Biochem., 33 , 140-157
3 Viebrock,A., Perz,A. and Sebald,W.(1983) Meth.Enzymol., 97 , 254-260
4 Arends,H. and Sebald,W.(1984) EMBO J., 3 , 377-382

GENETIC ASPECTS OF pH REGULATION IN <u>ESCHERICHIA COLI</u>

I.R. Booth, G.C. Rowland, R.G. Kroll, and P.M. Giffard
Department of Microbiology, University of Aberdeen, Marischal College,
Aberdeen, AB9 1AS, Scotland, U.K.

Previous studies in this and other laboratories has shown that a
major component of the pH regulatory mechanism of <u>E. coli</u> is a
controlled proton pore (1,2). We have shown that this pore is feed-
back activated by the transmembrane pH gradient and causes a controlled acidification of the cytoplasm (1). As yet we have been
unable to pinpoint either a cation or an anion dependence which
would define the system as a classical antiporter or symporter,
respectively. We therefore initiated genetic studies of known cation
transport systems and energy-transducing complexes to ascertain
whether any of these were essential components of the pH-regulating
system.

The <u>phs</u>, <u>nha</u>, <u>trkB</u>, <u>trkC</u>, <u>unc</u> and <u>eup</u> genes have been investigated.
The <u>phs</u> mutation was originally proposed to affect the sodium-proton
antiport (2). However, we will present data to show that it is a
double mutation affecting a cysteine biosynthetic locus (<u>cysG</u>) and the
<u>crp</u> locus. The basis of the pH-sensitivity of the <u>phs</u> mutant is being
investigated. However, Na-linked functions (sodium-proton antiport,
proline uptake, and sodium-potassium exchange) are normal in this
mutant. Thus the primary lesion is not the antiport itself.

The <u>trkB</u> and <u>trkC</u> genes have been demonstrated not to be essential
to pH regulation by the isolation of both point and transposon
insertion mutants at these loci. These studies have also demonstrated
that <u>trkB</u> and <u>trkC</u> gene products do not interact to form a transport
system. Mutations at the <u>unc</u> locus have been used to study pH
regulation. While <u>uncA</u> mutants regulate their cytoplasmic pH poorly,
<u>uncAB</u> double mutants exhibit normal pH regulation. Deletion mutants
at the <u>eup</u> locus and point mutants at the <u>nha</u> locus are at present
under investigation.

1. Kroll, R.G. and Booth, I.R. (1983) Biochem. J. 216, 709-716
2. Zilberstein, D., Padan, E., and Schuldiner, S. (1980) FEBS letters
   116, 177-180.

MOLECULAR APPROACHES TO THE ASSEMBLY AND FUNCTION OF FUMARATE REDUCTASE OF ESCHERICHIA COLI

Gary Cecchini[1], Brian A. C. Ackrell[1], Robert P. Gunsalus[2], and Edna B. Kearney[1]

[1]Department of Biochemistry and Biophysics, University of California, San Francisco, California 94143 and VA Medical Center, San Francisco, California 94121, U.S.A.

[2]Department of Microbiology and Molecular Biology Institute, University of California, Los Angeles, California 90024, U.S.A.

The fumarate reductase complex (FRD) of E. coli is composed of four polypeptides (ABCD): the dissimilar subunits A and B coded for by the frdA and frdB genes, respectively, constitute the subunits of the enzyme (1-3), and the C and D polypeptides, encoded for by frdC and frdD, respectively, are required for attachment of the enzyme to the cytoplasmic membrane (4). We have used both in vitro and in vivo protein synthesis to study the assembly of the FRD complex in the cytoplasmic membrane. When plasmid pFRD23, which codes only for the A and B polypeptides of the complex, is used with an S-30 in vitro transcription/translation system, the two subunits are found in the soluble fraction, and are not assembled in the membranes; this confirms the findings of Weiner and co-workers with whole cells (4). When plasmid pFRD32 is used, which codes only for the C and D polypeptides, the two peptides are incorporated in the membranes, in the absence of A and B. Moreover, peptide D, synthesized when pFRD38 is used, is incorporated into the membrane without A,B or C. It is not yet known if A and B must be organized as fumarate reductase for assembly or if B can be incorporated followed by A. Assembly in the membrane can proceed post-translationally as shown by first using an S-100 transcription/translation system, which lacks membranes, and subsequently adding liposomes prepared from soybean phospholipids.

In vivo studies have been carried out with a minicell producing strain of E. coli (DS410). As evidence of membrane assembly, after fractionation of cells into cytoplasm and membranes (5), we have used in addition to SDS-PAGE an enzymatic assay with succinate as the substrate and the ubiquinone analog 2,3-dimethoxy-5-methyl-6-pentyl-1,4-benzoquinone (DPB) as electron acceptor. In recent work we have shown (6) that the FRD complex can function as a succinate-Q reductase and, through resolution and reconstitu-techniques, that the C and D polypeptides are needed for this function. Presence of the enzyme in the soluble fraction was detected by succinate-phenazine methosulfate (PMS) reductase activity (7). Succinate-DPB reductase activity was found only when cells were transformed with <u>both</u> pFRD23 and pFRD32, or pGC1002, which codes for the entire complex, ABCD, and all of the activity was found in the membrane fraction (Table 1). When pFRD23 was used alone, no DPB reductase activity was elicited although significant amounts of succinate-PMS reductase activity were found in the cytoplasm. Thus both <u>in</u> <u>vitro</u> and <u>in</u> <u>vivo</u> approaches underline the importance of peptides C and D for assembly and function of the FRD complex in the membrane.

Table 1

Distribution of PMS and DPB Reductase Activities in Transformed Minicells

	Membranes Reductase Activities[a]		Cytoplasm Reductase Activities[a]	
	PMS	DPB	PMS	DPB
DS410 + pBR322	7.9	4.8	*	*
DS410 + pGC1002	20.9	11.4	0.7	*
DS410 + pFRD23	8.0	4.8	7.2	0.5
DS410 + pFRD32	8.3	5.3	*	*
DS410 + pFRD23 and pFRD32	13.5	7.4	1.0	0.5

[a]Total units (μmol succinate oxidized/min at 30°). Values are normalized to 50 mg of cell protein before fractionation. *: values at limit of detection. DS410 + pBR322 is used as a control for the other plasmids.

This investigation was supported by the Veterans Administration, by NIH Grant HL-16251 and NSF Grant PCM 81-10585; and also by NSF Grant PCM 84-02974 (RPG).

References

1 Dickie, P., and Weiner, J.H. (1979) Can. J. Biochem. 57, 813-821
2 Cole, S.T. (1982) Eur. J. Biochem. 122, 479-484
3 Cole, S.T., Gründstrom, T., Jaurin, B., Robinson, J.J., and Weiner, J.H. (1982) Eur. J. Biochem. 126, 211-216
4 Lemire, B.D., Robinson, J.J., and Weiner, J.H. (1982) J. Bacteriol. 152, 1126-1131
5 Klionsky, D.J., Brusilow, W.S.A., and Simoni, R.D. (1983) J. Biol. Chem. 258, 10136-10143
6 Cecchini, G., Ackrell, B.A.C., Kearney, E.B., and Gunsalus, R.P. (1984) in (Flavins and Flavoproteins), Walter de Gruyter, New York, (in press)
7 Ackrell, B.A.C., Ball, M.B., and Kearney, E.B. (1980) J. Biol. Chem. 255, 2761-2769

LEVEL OF RAT MITOCHONDRIAL D-β-HYDROXYBUTYRATE DEHYDROGENASE (BDH)
IN HEPATIC AND EXTRAHEPATIC TISSUES UNDER VARIOUS METABOLIC CONDITIONS

C. Coquard, N. Latruffe, R. Duroc and Y. Gaudemer
Laboratoire de Biochimie ERA CNRS 1050 "Biologie Moléculaire des Membranes" Université de Franche-Comté 25030 Besançon France

Mammalian D-β-hydroxybutyrate dehydrogenase (BDH), an inner mitochondrial membrane marker protein (1), catalyses the interconversion of the two major ketone bodies (acetoacetate and D-β-hydroxybutyrate). Whereas for a long time one known that ketone bodies were mainly injurious for organism (in over-producing conditions i.e. diabete mellitus, acidosis...), now, we pay attention to the key role of ketone bodies, exclusively produced in liver, are energetic fuel ·for extrahepatic tissues, especially brain and cardiac muscle (3 ATP/BOH ✱ via NADH and respiratory chain) and are precursor for several biosynthesis pathways (production of acetyl CoA) (2).

According the importance of BDH in energetic metabolism we investigated the level of this enzyme in hepatic and extrahepatic tissue under various metabolic conditions using the rabbit monospecific anti BDH antibody raised against purified rat liver apoBDH (3). Quantification has been done using the Western electroblotting after PAGE-SDS slab gel of isolated mitochondria, then immunodecoration with 125I-protein A as it can be seen in figure 1.

Figure 1 Technique of immunoblotting used for quantification of BDH level in mitochondria isolated from different tissues

A mitochondrial protein separated on 12.5 % PAGE-SDS slab gel

B slab gel after electroblotting on 0.22 μ nitrocellulose sheet

C immunodecoration of nitrocellulose sheet with 125I-protein A after incubation with antiserum antiBDH

By this method we were able to estimate this amount of synthetized protein, which represents an instantaneous picture of the balance between the BDH gene expression and the protein turnover.

In normal physiological conditions the amount of BDH in liver mitochondria is about 20 μg/mg mito. protein. i.e. 0.65 nmole of BDH polypeptide chain. Whereas it is 5 times lower in kidney mitochondria, 7 times less in heart and 25 times less in brain.

After 68 h of rat fasting, condition when ketone bodies are over-produced, no significant variations in BDH level were observed.

However the BDH molecular activity from different tissues (figure 2) is the highest in heart and brain (two active ketone bodies metabolizing tissues) suggesting an adaptation mechanism by membrane phospholipid interactions (4).

The BDH level show variations during post natal development which it is high in new born animal and in suckling pups while it decreases regularely in adult and old rats.

Finally BDH level from the different tissues is not controlled by thyroid hormone which it is known to regulate positively gene expression of some primary dehydrogenases or some cytochromes of respiratory chain (5).

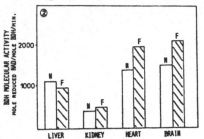

This preliminary study shows that the BDH gene (a nuclear gene) (6) is differently expressed according the tissue cell differenciation. Some variations can be induced by different metabolic conditions but the expression mechanism of this gene is still unknown.

Figure 2 BDH Molecular Activity in mitochondria isolated from Normal ☐ N or from 68 h Fasted rats ▨ F

References

1. Lehninger A.L., Sudduth M.C. and Wise J.B. (1960) J. Biol. Chem. 235, 2450-2455.
2. Page M.A., Krebs H.A. & Williamson D.K. (1971) Biochem. J. 121, 49-53.
3. Latruffe N., Brenner S. & Fleischer S. (1981) Biochem. Soc. Trans. 9, 249.
4. Gazzotti P., Bock H.G.O. et Fleischer S. (1975) J. Biol. Chem. 250, 5782-5790.
5. Nelson B.D., Mutwei A. et Joste V. (1984) Arch. Biochem. Biophys. 228, 49-53.
6. Coquard C., Latruffe N. et Gaudemer Y. (1982) Biol. Cell 44, 189-192.

ACKNOWLEDGEMENTS

Supported by a grant from the Fondation pour la Recherche Médicale

BIOSYNTHESIS OF BACTERIOCHLOROPHYLL-PROTEIN COMPLEXES OF THE BACTERIAL PHOTOSYNTHETIC APPARATUS

Roland Dierstein, and Gerhart Drews
Institute of Biology 2, Microbiology, Albert-Ludwigs-University , 7800 Freiburg, Fed. Rep. Germany

The synthesis of bacteriochlorophyll (Bchl) and of the Bchl-binding proteins is coordinately induced after lowering of oxygen partial pressure (in dark cultures) or lowering of light intensity (anaerobic culture) (1). These polypeptides are synthesized without signal sequences and immediately inserted into the membrane (2-5). The assembly of the polypeptides in the membrane is inhibited when Bchl synthesis is blocked by mutagenesis or inhibitors (1, 4). The level of mRNA specific for Bchl-binding polypeptides increased after induction (6). The half-life time of mRNA of B800-850 polypeptides is about 20 min, that of reaction center (RC) polypeptides about 8 min (3). Bchl-binding polypeptides which assemble without Bchl in the membrane are rapidly degraded (7). In presence of Bchl the pigment-protein complexes have no measurable turnover. The genes for Bchl-binding polypeptides of RC and light-harvesting complex B870 are clustered on the genome (8), those for the B800-850 complex are separately localized (9).

Supported by Deutsche Forschungsgemeinschaft.

References:
1) Ohad, I., Drews, G. (1982) In "Photosynthesis: Development, Carbon Metabolism and Plant Productivity" (Govindjee, ed.), pp. 90-140, Academic Press, New York
2) Dierstein, R., Schumacher, A., Drews, G. (1981) Arch. Microbiol. 128, 376-383
3) Dierstein, R.,(1984) Eur. J. Biochem. 138, 509-518
4) Dierstein, R., Drews, G. (1982) In "Cell Function and Differentiation", part B (Akoyunoglou, G. et al., eds.) pp. 247-256, Alan R. Liss, New York
5) Youvan, D.C., Bylina, E.J., Alberti, M., Begusch, H., Hearst, J.E. (1984) Cell, July issue
6) Clark, W.G., Davidson, E., Marrs, B.L. (1984) J. Bacteriol. 157, 945-948
7) Dierstein, R. (1983) FEBS Lett. 160, 281-286
8) Youvan, D.C., Alberti, M., Begusch, H., Bylina, E.J., Hearst, J. (1984) Proc. Natl. Acad. Sci. USA 81, 189-192
9) N. Kaufmann, G. Drews, unpublished.

KINETIC STUDIES ON THE BIOSYNTHESIS OF CYTOCHROME OXIDASE
FROM RHODOPSEUDOMONAS CAPSULATA UNDER LIMITED OXYGEN
CONCENTRATION.

Hendrik Hüdig and Gerhart Drews
Inst. für Biologie II, Mikrobiologie, Albert-Ludwigs-
Universität, Schänzlestr. 1, 7800 Freiburg, F.R.G.

Rhodopseudomonas capsulata, a facultative phototrophic
bacterium, produces ATP under chemotrophic conditions
by oxidative phosphorylation. Only b-type cytochromes
function in the two terminal oxidases, which differ in
their mid-point potential, sensitivity to KCN and to
CO (1).

The active preparation of the terminal high-potential
cytochrome c oxidase of Rps.capsulata, isolated from the
membrane fraction, contained a b-type cytochrome and one
polypeptide (M_r 65,000)(2). The active form of the cytochrome oxidase seemed to be a dimer of M_r 130,000 containing one molecule of protoheme (3). The oxidase was
incorporated into phospholipid vesicles to measure
proton extrusion with pulses of ferrocytochrome c for
one oxidase turnover. In accordance with the pH-shift
of its mid-point potential the purified oxidase showed
a proton extrusion of 0.24 H^+/e^- with uptake of 1 H^+/e^-
from the liposomes for the reduction of oxygen to water
(4). Nevertheless our results indicated that a DCCD-
sensitive proton-pumping subunit seemed to be separated
from the oxidase during enzyme purification.

Using crossed immunoelectrophoresis with antibodies
against the purified b-type cytochrome oxidase we
followed the biosynthesis of the cytochrome oxidase
when phototrophically grown cultures of Rps.capsulata
were shifted to chemotrophic growth conditions with a
final concentration of 10% air in the medium. The membrane-bound specific activity of the oxidase increased
more than 3 times, but the specific amount of oxidase
protein was nearly kept constant. These results were
the same for phototrophically grown cultures which
adapted to chemotrophic growth under high aeration.

With 10% air the fully adapted state was reached not
not before 4 hours after induction. So far the specific

oxidase activity and the amount of oxidase protein behaved competely different. The specific activity decreased about 90% whereas the amount of oxidase protein increased about 3 times. These changes in specific oxidase activity per oxidase protein corresponded to the growth rate and to the diminution of the bacteriochlorophyll content of the cells (a parameter for light adaptation).

Our results indicate that the high-potential b-type cytochrome oxidase is strictly regulated in Rps.capsulata during adaptation from phototrophic to chemotrophic growth. The biosynthesis of the active oxidase seems not only to be regulated by the incorporation of new oxidase protein into the cytoplasmic membrane but also by the separated incorporation of the protoheme.

References:

1) Zannoni,D., Melandri,B.A., Baccarini-Melandri,A.(1976) Biochim.Biophys.Acta 223, 413-430

2) Hüdig,H., Drews,G.(1982) Z.Naturforsch.37, 193-198

3) Hüdig,H., Drews,G.(1982) FEBS Lett.146, 389-392

4) Hüdig,H., Drews,G.(1984) Biochim.Biophys.Acta, in press

THE BIOGENESIS OF THE MITOCHONDRIAL OUTER MEMBRANE

E.C.HURT and T.HASE

Biocenter, University of Basel, CH-4056 BASEL,
SWITZERLAND

Import of proteins into mitochondria proceeds by several distinct routes (1). Polypeptides destined for internal mitochondrial compartments are made on cytoplasmatic ribosomes, generally as larger precursors. Translocation of these precursors into mitochondria is energy-dependent and accompanied by proteolytic removal of the presequence (1). In contrast, polypetides of the mitochondrial outer membrane lack a transient presequence and their insertion into the outer membrane does not require an energized membrane (2). Still, insertion into the outer membrane is a specific process (2, 3). In vitro-synthesized outer membrane proteins are only inserted into outer membranes, but not into microsomes (2) or the plasma membrane (3).

Here we report studies on the import of the 70 kDa protein of the yeast mitochondrial outer membrane. The nuclear gene of this protein had already been cloned (4) and sequenced (5). The protein contains a 60 kDa carboxy-terminal soluble domain which protudes into the cytosol. The amino-terminal 10 kDa part contains an uninterrupted stretch of 28 uncharged amino acid residues (from position 10 to 37) flanked on both sides by only basic amino acid residues (5). This domain probably constitutes the membrane anchor of the protein.

Are sequences anchoring the protein to the mitochondrial outer membrane identical to these that specifically direct it to that membrane during mitochondrial biogenesis?

By gene manipulation techniques, several deletions were introduced into the 70 kDa protein. Together, these deletions cover almost the entire polypeptide chain. Polypeptides lacking part of the C-terminal region still insert

into the mitochondrial outer membrane, even if the deletion involves 200 to 300 amino acids. Thus, targeting and
anchoring do not require the native conformation of the
entire protein. In contrast, deletions in the N-terminal
(10 kDa) region result in severe mistargeting. If amino
acid residues 14 to 106 are deleted, most of the mutant
protein molecules are found in the cytoplasm or associated with microsomes. A small fraction (fewer than 30 %
of the molecules) is still transported to the mitochondria, but now becomes localized on the matrix side of the
inner membrane. This suggests that the N-terminal deleted
70 kDa protein has lost its anchoring sequence, but still
retained part of its targeting sequence. This implies that
the sequences directing the protein to the mitochondria
are not the same as those through which the protein is
stably anchored to the outer membrane.

To study the import route of this protein in more detail,
it was synthesized in a combined transcription/translation
system: Its cloned gene was placed under control of a strong
bacteriophage promotor and transcribed with purified
E. coli RNA polymerase. The resulting mRNA was capped
and translated in a protein-synthesizing system derived
from wheat germ. This yielded 70 kDa protein in high
radiochemical yield and purity. The same method was also
used to obtain radiolabeled forms of the various deleted
polypeptides mentioned above. This powerful system has
opened the way for a detailed study of how this major
outer membrane protein is directed to the mitochondrial
surface and how it achieves its final localization.

1) Neupert, W. and Schatz, G. (1981) Trends Biochem.
   Sci. 6, 1-4
2) Gasser, S. M. and Schatz, G. (1983) J. Biol. Chem.
   258, 3427-3430
3) Freitag, H., Janes, M. and Neupert, W. (182) Eur.
   J. Biochem. 126, 197-2o2
4) Riezman, H., Hase, T., van Loon, A.P.G.M., Grivell,
   L.A., Suda, K. and Schatz, G. (1983) EMBO J. 2, 2161-
   2168
5) Hase, T., Riezman, H., Suda, K. and Schatz, G. (1983)
   EMBO J. 2, 2169-2172

THREE POLYPEPTIDES SPECIFICALLY ASSOCIATED WITH THE 23 AND 33 kDa
PROTEINS OF THE PHOTOSYNTHETIC OXYGEN EVOLVING COMPLEX
U.Ljungberg, H.-E. Åkerlund, C. Larsson, B. Andersson
Dept of Biochemistry, University of Lund, 220 07 Lund, Sweden

The polypeptide composition of PS II and the oxygen evolving complex
has until recently been largely unknown. The isolation of functional PS
II preparations has limited the number of involved polypeptides to
10-15 (1), of which two are the 23 and 33 kDa proteins, inferred to be
constituents of the oxygen evolving complex (2). To identify the polypeptides specifically associated with these two extrinsic proteins
immunoprecipitation of partly solubilized PS II particles, using monospecific IgG against the 23 and 33 kDa proteins respectively, was done.
The IgG did not only precipitate the antigenic proteins themselves, but
also three polypeptides of 24, 22, and 10 kDa. It is suggested that
these three polypeptides are subunits of the oxygen evolving complex.

PS II particles were prepared from spinach thylakoids according to
Berthold et al.(3). Part of the preparation was Tris-washed twice at
800 mM Tris-HCl, pH 8.4, which completely removed the 23 and 33 kDa
proteins. The untreated and the Tris-washed particles (200 ug/ml in 10
mM sodium phosphate buffer, pH 6.5) were further solubilized using a
detergent mixture (40% Triton X-100, 60% Zwittergent TM-314) at ratios
of 3, 4, 5, and 10 mg detergent/mg chlorophyll. After 30 min incubation
on ice, unsolubilized material was pelleted. Monospecific sera, as
revealed by "western" blotting, against each of the 33 and 23 kDa proteins were obtained from rabbits. IgG was purified by affinity chromatography on protein A-Sepharose CL-4B, and was further purified from
antibodies normally occurring in the pre-immune rabbit sera, and
cross-reacting with the galactolipids (4), by adsorbing the IgG fractions with solubilized Tris-washed PS II particles, devoid of the 23
and 33 kDa proteins. Immunoprecipitation was performed by mixing the
final IgG fraction with normal PS II particles, solubilized at the same
detergent to chlorophyll ratio. After 2h incubation on ice, the precipitate was pelleted and analysed by SDS-PAGE. The gels were stained in
Coomassie Brilliant Blue R-250 and quantified with an LKB gel scanner.

When the IgG against the 23 kDa protein was added to the PS II particles, solubilized at a detergent to chlorophyll ratio of 3, the polypeptide composition of the precipitates was markedly different from that
of the starting material. The percentage coprecipitation of each polypeptide is shown in Fig 1. Three polypeptides of 22, 24, and 10 kDa
showed a very high coprecipitation, while the 26-27 kDa apo-polypeptides of LHCP and the 43 and 47 kDa apo-polypeptides of PS II reaction

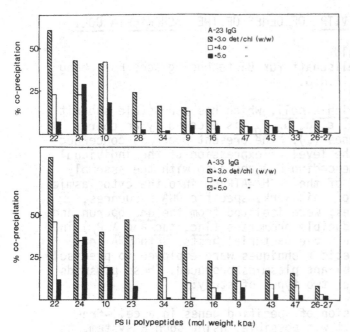

Fig 1. Degree of coprecipitation for PS II polypeptides with the antigenic proteins. The ratio between each polypeptide and the antigenic protein was calculated for the precipitates and divided by the corresponding ratio in the PS II particles.

center were quite depleted. This precipitation pattern was further accentuated at a detergent to chlorophyll ratio of 4. At a ratio of 5 and 10 only the 24 and 10 kDa polypeptides showed significant coprecipitation with the 23 kDa protein, the binding of the 22 kDa polypeptide obviously being more sensitive to the detergents.

When IgG against the 33 kDa protein was used the results resembled those obtained with IgG against the 23 kDa protein. Note also the high coprecipitation of the 23 kDa protein. Mn was found only in low amounts in the precipitates of both IgG preparations.

The identity of the 22, 24 and 10 kDa polypeptides remains unclear. They appear to carry neither chlorophyll, since the thylakoid protein to chlorophyll ratio dramatically increased in the precipitates compared to the starting material, nor Mn. Cytochrome b-559 comigrated with a 9 kDa polypeptide on SDS--PAGE and not with the 10 kDa polypeptide. Due to the close structural association of the 22, 24, and 10 kDa polypeptides with the 23 and 33 kDa proteins we propose that these polypeptides provide the binding site for the 23 and 33 kDa proteins to the thylakoid membrane.

1. Dunahay,T.G., Staehlin, L.A., Seibert, M., Ogilvie, P.D. and Berg, S.P. (1984) Biochim. Biophys. Acta 764, 160-169
2. Åkerlund, H.-E. (1983) in The Oxygen Evolving System of Photosynthesis (Inoue, Y. et al., eds), pp 201-208, Academic Press Japan, Tokyo
3. Berthold, D.A., Babcock, G.T. and Yocum, C. F. (1981) FEBS Lett. 134, 231-234
4. Sundby, C. and Larsson, C. (1984) in Proceedings of the 6th International Congress on Photosynthesis (Sybesma, C., ed) Martinus Nijhoff, The Hague, in press

EXPRESSION IN VIVO AND IN VITRO OF GENES OF THE ESCHERICHIA COLI UNC OPERON.

J.E.G.McCarthy and W. Sebald
Abt. Cytogenetik, GBF-Gesellschaft für Biotechnologische Forschung mbH., D-3300 Braunschweig, F.R.G.

The H^+-ATPase of Escherichia coli, which has a possible subunit stoichiometry of $\alpha_3\beta_3$ γ, δ, ϵ, a, b_2c_{10}, is encoded by an operon (unc) which comprises 9 genes (1). The present paper is concerned with factors relating to the level of expression of the individual genes (particularly those encoding F_0) and also with the assembly of certain of the subunits of the H^+-ATPase into the cytoplasmic membrane. For the purpose of this work, specific DNA sequences, containing one or more genes, were isolated from the unc operon and cloned behind different inducible promoters (lac, tac and λ P_L) in expression vectors using suitable bacterial hosts. Both DNA sequencing and classical genetic techniques were employed to precisely define the series of recombinant plasmids produced. These plasmids were particularly useful for two types of study:

(a) The in vitro expression of specified genes in a cell-free (S30) extract of E. coli. It was possible, using such a system, to tackle the question as to how the ultimate levels of expression of the individual genes of the unc operon is achieved (cf. 2). This question is expecially pertinent to expression of the c gene. Sequences containing the genes b, c (F_0) and δ (F_1) were inserted between the tac promoter and the galactokinase gene on an expression vector (3) such that both the inserted sequence and the galactokinase gene are transcribed as one unit. The level of expression of the galactokinase gene could be used as a reference indicator for the amount of transcription initiated by the tac promoter. The results show that the c gene region of the unc operon possesses a highly potent translation initiation sequence. Manipulation of the DNA sequence in this region has allowed more detailed definition of the nature of this sequence. Further investigations and possible applications of the pre-c gene sequence will be discussed.

(b) The insertion of unc genes behind (at least partially) inducible and relatively strong promoters facilitates further study of the synthesis and assembly of the component subunits of the H^+-ATPase in vivo. Using plasmids bearing for example the c gene cloned behind the tac promoter it has been possible to follow the integration of the c subunit into membranes of a constructed derivative of a lac repressor overproducer strain of E. coli

(JM103) which contains a deletion of most of the <u>unc</u> operon. Furthermore, the effects of overproduction of F_0 subunits in other mutant host strains have been studied.

References

(1) Futai, M. and Kanazawa, H. (1983) Microbiol. Rev. <u>47</u>, 285-312

(2) Brusilow, W.S.A., Klionsky, D.J. and Simoni, R.D. (1982) J. Bacteriol. <u>151</u>, 1363-1371

(3) Russell, D.R. and Bennett, G.N. (1982) Gene <u>20</u>, 231-243

TRANSPORT OF PROTEINS INTO MITOCHONDRIA:
Inhibition of the permeation of aspartate aminotransferase into rat
liver mitochondria by metal-complexing agents.

S. Passarella°, E. Marra°, S. Doonan[+] and E. Quagliariello°
°Istituto di Chimica Biologica and Centro di Studio sui Mitocondri e
Metabolismo Energetico, C.N.R. Bari, Italy
[+]Department of Biochemistry, University College, Cork, Ireland

Since 1976, we have been investigated the uptake of mature
proteins into mitochondria through a model system developed by us.
Selective permeation of aspartate aminotransferase and malate dehydrogenase into isolated rat liver mitochondria has been shown
through different experimental approaches (1). Subsequently, a precursor for aspartate aminotransferase has been reported (2,3).
However, no evidence exists that the pre-piece functions as leader
sequence in directing proteins into or through inner membrane. Thus,
our results are in agreement with the idea that mature proteins can
move through mitochondrial membranes. Our model system could be
valid for the direct investigation of transport features due to its
unique capability in manipulating both enzyme and mitochondria.

Evidence of the existence of a common receptor has been proposed
because of the mutual inhibition shown by both aspartate aminotransferase and malate dehydrogenase and their mercaptoethanol uptake sensitivity.

Evidence is given here of the ability that different metal-
complexing agents such as tiron, neocuproine, phenanthroline, bathophenanthroline and bathocuproine have in inhibiting the uptake of
mitochondrial aspartate aminotransferase into rat liver mitochondria
in vitro (Table I).

REFERENCES

1. Doonan, S., Marra, E., Passarella, S., Saccone, C., and Quagliariello, E. (1984) Int. Rev. Cyt. 91, 141-186.
2. Sakakibara, R., Huyuh, A.K., Nishida, Y., Watanabe, T., and Wada, H.
   (1980) Biochem. Biophys. Res. Commun. 95, 1781-1788
3. Marra, E., Passarella, S., Greco, M., Saccone, C., and Quagliariello, E. (1983) in "Mitochondria 1983" (R.J. Schweyen, K. Wolf,
   F. Kaudewitz eds.) pp.579-592, W. de Gruyter & Co., Berlin

TABLE I

THE EFFECT OF METAL-COMPLEXING AGENTS ON THE PERMEATION OF ASPARTATE
AMINOTRANSFERASE INTO RAT LIVER MITOCHONDRIA IN VITRO.

Mitochondria were incubated at 20°C in 2 ml of a solution of 0.25 M sucrose, 20 mM
Tris-HCl pH 7.25, 1 mM EGTA, containing rotenone (2 ug) and 1 mM sodium arsenite.
After 3 min the additions indicated were successively made at the following concentrations: aspartate (ASP) 12 mM; oxoglutarate (OG) 2.8 mM; mitochondrial aspartate
aminotransferase (mAAT) 3.1 ug. The metal-complexing agents Tiron (TIR), Neocuproine
(NEO), 1,10-phenanthroline (PHEN), Bathophenanthroline (BPHEN), Bathocuproine (BCUP)
were added (25 uM each) where indicated. No effect by these compounds was found on the
enzyme activity. The rate of decrease in fluorescence of the intramitochondrial
NAD(P)H (V) was then recorded and expressed in arbitrary units. V_{M-C} indicates the
rate of decrease in fluorescence in the presence of a metal-complexing agent. The
amount of mitochondrial protein was 2 mg.

	ADDITIONS					V	V_{M-C} (% control)
	t=0	+30"	+30"	+30"	+30"		
MIT	ASP	–	–	–	OG	100	–
MIT	ASP	–	mAAT	–	OG	130	–
MIT	ASP	–	TIR	–	OG	96	100
MIT	ASP	TIR	mAAT	–	OG	91	96
MIT	ASP	–	mAAT	TIR	OG	126	131
MIT	ASP	–	–	NEO	OG	94	100
MIT	ASP	NEO	mAAT	–	OG	94	100
MIT	ASP	–	mAAT	NEO	OG	123	131
MIT	ASP	–	PHEN	–	OG	83	100
MIT	ASP	PHEN	mAAT	–	OG	70	84
MIT	ASP	–	mAAT	PHEN	OG	100	121
MIT	ASP	–	BPHEN	–	OG	87	100
MIT	ASP	BPHEN	mAAT	–	OG	83	95
MIT	ASP	–	mAAT	BPHEN	OG	104	120
MIT	ASP	–	BCUP	–	OG	104	100
MIT	ASP	BCUP	mAAT	–	OG	108	104
MIT	ASP	–	mAAT	BCUP	OG	126	121

Influence of Heme a on the Assembly of Mitochondrial Subunits of
Rat Liver Cytochrome c Oxidase

A. Wielburski, P. Gellerfors and B.D. Nelson
Department of Biochemistry, Arrhenius Laboratory, University of
Stockholm, S-106 91 Stockholm, Sweden

Based on studies on isolated hepatocytes and isolated rat liver
mitochondria (1, 2), we suggested that the assembly of mammalian
cytochrome c oxidase subunits follows a certain temporal sequence.
Although all three mitochondrially translated subunits (I, II, III)
seem to be synthesized concomitantly, there is a lag in the assembly
of subunit I with subunits II and III. This apparent late assembly
of subunit I is of interest since this subunit probably binds heme
in the holoenzyme, raising the possibility that its assembly, and
thus the formation of the biologically active enzyme, could be dependent upon heme binding.

In vitro labeled rat liver mitochondria were used as a test system to relate heme accessibility to assembly of the labeled cytochrome c oxidase subunits. The general protocol was to add freshly
isolated heme a or heme b to in vitro labeled rat liver mitochondria
under swelling conditions. The subunits were then immunoabsorbed
with monospecific antisera. Antiserum against subunit I immunoabsorbed only subunit I from untreated mitochondria, whereas subunits II
and III were co-absorbed from mitochondria pre-treated with heme a
(figure 1). Subunits II and III were not co-absorbed from mito -
chondria pre-tretaed with heme b. Furthermore, heme a in which the
pyridine -NaOH spectra was altered presumably due to changes around
the iron, was ineffective in promoting assembly of subunit I with
subunits II and III. Analogous results were obtained using monospecific antiserum against subunit II (figure 1). Here, subunit I was
co-absorbed only from heme a pre-treated mitochondria. These data
are consistent with the idea that heme a is required for the assembly
of subunit I in rat liver cytochrome c oxidase.

Assembly of radiolabeled cytochrome b was also investigated in
the same system. An antiserum against bc_1 complex was used. This
antiserum contained antibodies against subunits I and II, cytochrome
c_1 and FeS protein, as revealed by immunoblotting. Cytochrome b was
not immunoabsorbed from untreated mitochondria, indicating that it
is not assembled with any of the above subunits after a 30 min labeling period. Furthermore neither heme b or heme a promoted assembly
of cytochrome b in this test system.

Figure 1. Heme-induced assembly of rat liver cytochrome c oxidase subunit I

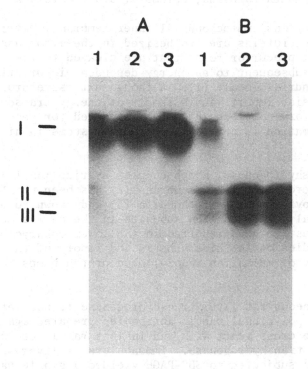

Radiolabeled mitochondria were incubated in vitro in the presence or absence of heme, and then immunoabsorbed with antibodies against subunit I (A) or subunit II (B). Lane 1, heme a added; lane 2, heme b added; lane 3, no heme added. Cytochrome c oxidase subunits I-III are indicated with Roman numerals.

References:

1. A. Wielburski and B.D. Nelson (1983) Biochem. J. 212, 829-834.

2. A. Wielburski (1983) Doctoral thesis, University of Stockholm.

BIOSYNTHESIS OF MITOCHONDRIAL TRANSHYDROGENASE IN RAT HEPATOCYTES.
Licia N. Y. Wu, Ira Lubin and Ronald R. Fisher, Department of Chemistry, University of South Carolina, Columbia, South Carolina 29208 USA

With a few exceptions, mitochondrial inner membrane, matrix, and intramembrane space proteins are synthesized in the cytoplasm as higher molecular weight precursor forms that are cleaved to the mature species during or subsequent to an energy-dependent importation to their intramitochondrial locale (1). Although extensive studies on the site of synthesis, import, and assembly of energy transducing H^+-ATPase and respiratory complexes have been reported for yeast and N. crassa, less information is available on these systems in higher eukaryotes.

NADH$\rightarrow$NADP$^+$ transhydrogenase, which couples hydride ion transfer to H^+ translocation across the inner membrane (2) has been purified to homogeniety from bovine heart mitochondria (3) and shown to exist as a dimer of identical 110 kilodalton subunits in the native membrane (4). The enzyme has not been reported in the lower eukaryotes. We have initiated studies on the biosynthesis and import of this structurally most simple of mammalian redox-linked proton pumps in isolated rat hepatocytes.

Although homogeneous rat liver transhydrogenase is not yet available, monospecific polyclonal rabbit antibodies prepared against the bovine heart enzyme cross-react with and inhibit rat liver submitochondrial particle transhydrogenase. Immunoblots of liver submitochondrial particles subjected to SDS-PAGE yielded a single band corresponding to transhydrogenase having slightly higher (<500 daltons) apparent molecular weight than the purified or membrane-bound heart enzyme. Anti-transhydrogenase specifically immunoprecipitated the enzyme from [^{35}S]-methionine labelled isolated hepatocytes. Labelling of transhydrogenase was prevented totally by cycloheximide, but not by chloramphenicol. This result, coupled with the observation that liver mitochondria do not synthesize the enzyme, demonstrate that it is a nuclear gene product. Rabbit reticulocyte lysates programmed with liver mRNA that was partially purified by sucrose gradient centrifugation synthesized a single immunoprecipitated product. The in vitro product (pTH) was about 1.5 kilodaltons larger than mature transhydrogenase (mTH) derived from labelled cells. Further studies on the processing of pTH by isolated mitochondria and the synthesis of pTH in intact cells will be described.

Supported by NIH Grant GM-22070

References
1. Hay, R., Böhni, and Gasser, S. (1984) Biochim. Biophys. Acta 779, 65-87.
2. Earle, S. R. and Fisher, R. R. (1980) J. Biol. Chem. 255, 827-830.
3. Wu, L. N. Y., Pennington, R. M., Everett, T. D., and Fisher, R.R. (1982) J. Biol. Chem. 257, 4052-4055.
4. Wu, L. N. Y. and Fisher, R. R. (1983) J. Biol. Chem. 258, 7847-7851.

UV-INDUCED CHANGES IN RESTRICTION ENZYME CLEAVAGE PATTERNS OF DNA

H. Martin-Bertram

Abt. Strahlenbiologie, Ges. f. Strahlen- und Umweltforschung mbH,
D-8042 Neuherberg, Federal Republic of Germany

The majority of DNA lesions which are introduced by UV-light of 254 nm are the formation of cyclobutane dimers of adjacent pyrimidines (TT > CT > CC) and photoproducts (nondimers) of pyrimidine nucleoside-cytidine sequences (pyr(6-4)dC, Cyd(5-4)pyo). These lesions are likely to inhibit the action of restriction enzymes of type II, having either CC, CT or TT(C) in the respective recognition site. Besides the normal cleavage pattern beeing observed on agarose gels, there appear new extra bands which contain uncleaved DNA of two and three adjacent fragments. These extra bands can be detected after exposure of DNA-solutions (bacteriophage Lambda DNA) to UV-light from 500 J/m^2 to 8000 J/m^2 with only minor variation in quality and quantity over the whole dose range. Restriction enzymes having no adjacent pyrimidine sequences in their recognition sites produce normal cleavage patterns with DNA beeing exposed even to high fluences of UV-light. Up to 80% of the abnormal cleavage pattern shows bands which contain one uncleaved recognition site, e.g. two adjacent fragments, 20-40% of the extra bands consist of three uncleaved fragments. The variation in the amount of extra bands with two or three uncleaved fragments might be due to the sequence of pyrimidines and the UV-lesions produced.

Uncleaved restriction sites which contain three adjacent pyrimidines (f.e. restriction endonucleases Eco RI, Hind III) become single-stranded at high fluences of UV-light (4000-8000 J/m^2). With nuclease S1 the single-stranded sites were cut and the restriction enzyme pattern appeared then normalized. It is to assume that single-stranded sites in the recognition site arise from bulky structural defects beeing associated with the pyrimidine nucleoside lesions.

592

DIURON-RESISTANCE IN SACCHAROMYCES CEREVISIAE
A.M.Colson[1],A.Fauconnier[2],M.F.Henry[3] and B.Meunier
1. Research Associate F.N.R.S., 2.Boursier I.R.S.I.A.
3. Chargé de Recherche C.N.R.S.
Unité de Génétique, Lab. de Génétique Microbienne,U.C.L.
Place Croix du Sud,4, 1348 Louvain-la-Neuve, Belgium
In yeast, Diuron(3-3,4-dichlorophenyl-1,1-dimethylurea) inhibits
the electron flow of the respiratory chain between the cytochromes
b and c_1. Diuron-resistant mutants(Diu^R) have been selected on a concentration of Diuron which inhibits completely yeast growth.diu^r mutations of mitochondrial heredity were found to belong to the mitochondrial cytochrome b split gene(1,2). Resistance was detected *in vitro*
at the level of the mitochondrial inner membrane due presumably to a
modification of the apocytochrome b leading to a reduced affinity of
the diuron-binding site(s) for the drug.Nuclear Diu^R mutants were also
obtained. Two mutants exhibit their resistance *in vitro* since the electron flow of the respiratory chain remains functional in the presence
of increasing concentrations of Diuron. These mutants could carry a
mutation in a gene coding for a protein of the bc_1 complex other than
apocytochrome b. One nuclear Diu^R mutant did not show *in vitro* resistance at the level of the mitochondrial inner membrane. Its mutation could
be responsible for the overproduction of a diuron-binding protein or it
could affect the transport of Diuron into the cell. This resistance
seems to be specific to Diuron since the mutant remains sensitive to
Antimycin A. The mutation is not centromere linked. Therefore, it cannot belong to the gene controlling multiple drug-resitance in *S.cerevi-
siae* which is centromere linked on chromosome V11(3). Resistance of
the three nuclear Diu^R mutants was not accompanied by a detectable amplification of a DNA segment. This was observed by agarose gel electrophoresis of *EcoR1* and *Hind111* restricted nuclear DNA which had been purified in a CsCl gradient. Genetic mapping of the mutations as well as
molecular cloning of the mutated genes are in progress.

Mutants selected on DL-lactate plus Diuron instead of glycerol were
also isolated. These mutations do behave neither as a mitochondrial nor
as a typical nuclear mutation(4). Total DNA from the parental strain
KL14-4A, a mutant Diu3-743 and a double Diu3-743/329 carrying the
diu3-743 mutation plus the mitochondrial mutation *diu1-329* were analysed by agarose gel electrophoresis after cleavage with *EcoR1* restriction enzyme. Several amplified bands from mitochondrial DNA, 2μDNA
and rDNA(ribosomal DNA) presented their typical *EcoR1* restriction pat-
tern.In addition, four unidentified ampℓified bands were found. Scans
of the corresponding portions of the gel were performed(Fig1).Peaks n°1
and n°5 correspond to two rDNA bands of respectively 2.4Kb and 2.8Kb.
The new bands are of 2.5Kb(peak n°2),2.6Kb(peak n°3),2.7Kb(peak n°4)
and 3.2Kb(peak n°6). Their molecular weights do not correspond to

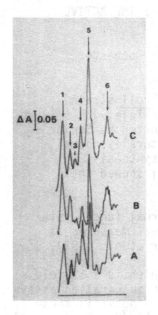

Fig1. EcoR1 restriction pattern of total DNA of low molecular weight (2.4Kb to 3.2Kb) isolated from KL14-4A, Diu3-743 and Diu3-743/329. Gel scans were performed with a Beckmann DU-8 spectrophotometer at 500nm; absorbance varies between 0 and 0.5. Peak n°1:2.4Kb(rDNA), peak n°2:2.5Kb, peak n°3:2.6Kb, peak n°4:2.7Kb, peak n°5:2.8Kb, peak n°6: 3.2Kb. A:KL14-4A, B:Diu3-743, C: Diu3-743/329.
Complete digestions by EcoR1 were performed at 37°C in Tris.HCl pH 7.5 20mM, NaCl 200mM, $MgSO_4$ 20mM, β-mercaptoethanol 20 mM

known bands of mit DNA, 2 μDNA or rDNA. The calculated areas covered by peak n°1(rDNA:2.4 Kb) do not fluctuate among the tested strains On the contrary, the area values obtained for the new bands fluctuate significantly among the strains. The area of peak n°2 is increased in strain Diu3-743(graph B) compared to strain KL14-4A(graph A) and in the double mutant Diu3-743/329(graph C). The area of peak n°3 decreases of 50% in strain Diu3-743 and increases in the double mutant Diu3-743/329. The area of peak n°4 in both mutants show 1/3 area obtained in KL14-4A. The area of peak n°6 increases by a factor of two in mutant Diu3-743 and decreases 3 times in the double mutant. The origin of these amplified DNA bands is presently under investigation using molecular cloning.

1. Colson,A.M.,Luu The Van, Convent, B.,Briquet,M. and Goffeau,A. (1977) Eur.J.Biochem. *74*,521-526
2. Colson,A.M. and Slonimski,P.P.(1979) Molec.gen.Genet.*167*,287-298
3. Saunders,G.W.,Rank,G.H.,Kustermann-Kuhn,B. and Hollenberg,C.P. (1979) Molec.gen.Genet.*175*,45-52
4. Colson,A.M. and Meunier, B. submitted to Current Genetics

A MUTATION IN THE F_1-ATPase β-SUBUNIT WHICH AFFECTS THE ACTIVE
SITE AND CATALYTIC SITE CO-OPERATIVITY.

Thomas M. Duncan and Alan E. Senior
Department of Biochemistry, Box 607, University of Rochester
Medical Center, Rochester, NY 14642, USA.

Properties of the point mutation carried in E. coli strain
AN1543 (uncD484) have been studied. This mutant allele,
which affects the β-subunit of F_1-ATPase, was reported by Senior
et al (1983). It directs the synthesis of a structurally-normal
F_1 which has greatly-impaired ATP hydrolysis and synthesis
rates. Further characterisation of the uncD484 F_1 showed the
following:

1. "Unisite" ATP hydrolysis rate = $0.014s^{-1}$. Normal (unc+) rate
in parallel assays was $0.041s^{-1}$. "Unisite" rate was measured as
described in Wise et al (1984) (see also Grubmeyer et al, 1982).

2. "Multisite" ATP hydrolysis rate = $0.171s^{-1}$, which is a 12-fold
promotion of the unisite rate. Normal (unc+) rate in parallel assays
was $58s^{-1}$, a 1400-fold promotion of unisite rate. (Note: in these
assays formation of bound plus released ^{32}Pi from AT^{32}P was measured).

3a. MgAMPPNP bound to 3 sites with negative cooperativity in the
mutant native enzyme, as it does in normal native F_1 (Cross and
Nalin, 1982; Wise et al, 1983). However, the K_D at the first site
was 9.0μM (normal = 0.3μM) and the K_D at sites 2 and 3 was 110μM
(normal = 20μM).

3b. MgADP bound to 2 sites with equal affinity in the mutant native
enzyme, as it does in normal native enzyme. However, the K_D was
29μM (normal = 3μM).

4. Aurovertin bound to the mutant enzyme to form a fluorescent
complex (Kd = 1.2μM; normal = 3μM). On addition of ADP (up to 100μM)
no enhancement of bound aurovertin fluorescence occurred, in contrast
with normal enzyme where ADP induces a large fluorescence enhancement
with KmADP of 1μM.

5. DCCD inhibited and reacted with mutant enzyme in a fashion similar
to its action on normal enzyme. 10mM Mg^{++} gave considerable protection from DCCD in both enzymes. NBD.Cl inhibited both enzymes
equally.

The uncD484 mutation therefore appears to affect both intrinsic catalytic properties of the catalytic site (Point 1) and cooperativity between catalytic sites (Point 2). The lack of $\alpha \leftrightarrow \beta$ intersubunit conformational interaction, indicated by the aurovertin fluorescence test (Point 4) (see Wise et al, 1981) is consistent with the lack of catalytic site cooperativity (Senior and Wise, 1983). Nucleotide binding to the catalytic sites seemed markedly weaker than normal (Points 3a and 3b). The actual mutated amino acid is not yet known; Point 5 shows it is not the DCCD-reactive E_{192} or the NBD.Cl-reactive residue on β-subunit.

References cited.

Cross, R.L. and Nalin. C.M. (1982) J. Biol. Chem. 257 2874-2881.

Grubmeyer, C., Cross, R.L., and Penefsky, H.S. (1982) J. Biol. Chem. 257 12092-12100.

Senior, A.E., Langman, L.P., Cox, G.B. and Gibson, F. (1983) Biochem. J. 210 395-403.

Senior, A.E. and Wise, J.G. (1983) J. Memb. Biol. 73 105-124.

Wise, J.G., Latchney, L.R. and Senior, A.E. (1981) J. Biol. Chem. 256 10383-10389.

Wise, J.G., Duncan, T.M., Latchney, L.R., Cox, D.N. and Senior, A.E. (1983) Biochem. J. 215 343-350.

Wise, J.G., Latchney, L.R., Ferguson, A.M. and Senior, A.E. (1984) Biochemistry 23 1426-1432.

Supported by NIH grants GM25349 and GM29805 to AES.

PROPERTIES OF PLASMA MEMBRANE ATPase MUTANTS IN YEASTS

A. Goffeau, S. Ulaszewski, J.-P. Dufour
Laboratoire d'Enzymologie, Universite Catholique de Louvain, Place
Croix du Sud, 1, 1348 Louvain-la-Neuve, Belgium

The fission yeast <u>Schizosaccharomyces</u> <u>pombe</u> plasma membrane contains a Mg^{2+}-dependent ATPase (1,2) which operates as an electrogenic proton pump and which is sensitive to vanadate as well as to Dio-9 (ref 3 for a review).

Mutants were selected for growth on solid yeast extract glucose (YEG) medium containing 400 µg/ml of Dio-9. In liquid YEG medium, the minimal inhibitory concentration of Dio-9 was 160 µg/ml for the mutant <u>pma</u>1 compared to 10 µg/ml for the parental strain. The resistance to Dio-9 was due to a mutation in a single nuclear gene. The ATPase activity in the enzyme purified from the Dio-9 resistant strain was sensitive to Dio-9 but resistant to vanadate. The <u>in vivo</u> Dio-9[R] and <u>in vitro</u> vanadate[R] cosegregated in tetrads obtained from crosses with a sensitive strain. The resistances to vanadate and to Dio-9 are thus genetically linked.

The purified enzymes from the mutant and the parental strains were incorporated into phospholipid vesicles. ATP-dependent energization of the reconstituted vesicles was monitored by the fluorescence quenching of 9-amino-6-chloro-2-methoxyacridine, a sensitive pH probe (4). The extent of quenching was similar in the wild type and the mutant although the generation of ΔpH was much slower with the mutant.

These results indicate that the mutant is modified in a gene controlling the vanadate sensitivity of the plasma membrane ATPase. The mutation decreases the rate of ATP hydrolysis activity. The slower proton pumping function of the mutant isolated plasma membrane ATPase is only due to the reduced rate of ATP hydrolysis but not to decreased ATP/H^+ stoichiometry.

Similar <u>pma</u>1 mutants have been obtained from the budding yeast <u>Saccharomyces</u> <u>cerevisiae</u> which show <u>in vivo</u> Dio-9 resistance and <u>in vitro</u> vanadate resistance (5). The growth of <u>pma</u>1 mutants is resistant to several inhibitors: Dio-9, ethidium bromide, N,N'(p-xylylidene)-bis-aminoguanidine (XBAG) and decamethylenediguanidine. These resistances are not expressed in vitro at the level of ATPase activity. However the in vivo XBAG^R and in vitro vanadate^R cosegregate in 16 tetrads.

The pma mutations are linked to the chromosome VII markers leu1, trp5 and ade6. Analysis of meiotic recombination in 164 asci locates pma1 between leu1 and trp5 at 5.2 centimorgan from the centromere. Moreover the pma mutations are closely linked to the multiple drug resistance mutations pdr1-2, pdr1-2 and pdr1-3 (6,7,8). However none of the latter mutants show in vitro vanadate resistance of the ATPase activity.

It is concluded that pma and pdr mutations are probably variants of the same category of pleiotropic drug resistance which have altered uptake properties probably due to modifications of the H$^+$-translocating plasma membrane ATPase activity.

1. Delhez, J., Dufour, J.P., Thines, D., and Goffeau, A. (1977) Eur. J. Biochem. 79, 319-328.

2. Dufour, J.P., and Goffeau, A. (1978) J. Biol. Chem. 253, 7026-7032.

3. Goffeau, A., and Slayman, C. (1981) Biochim. Biophys. Acta 639, 197-223.

4. Dufour, J.P., Goffeau, A., and Tsong, T.Y. (1982) J. Biol. Chem. 25, 9365-9371.

5. Ulaszewski, S., Grenson, M., and Goffeau, A. (1983) Eur. J. Biochem. 130, 235-239.

6. Colson, A.M., Goffeau, A., Briquet, M., Weigel, P., and Mattoon, J. (1974) Molec. Gen. Genet. 135, 309-326.

7. Saunders, G.W., and Rank, G.H. (1982) Can. J. Genet. Cytol. 24, 493-503.

8. Saunders, G.W., Rank, G.H., Kustermann-Kuhn, B.N., and Hollenberg, C.P. (1979) Molec. Gen. Genet. 175, 45-52.

PRODUCTION OF STRAINS OF ESCHERICHIA COLI THAT CAN GROW ON HIGH
CONCENTRATIONS OF UNCOUPLING AGENTS

M R Jones and R B Beechey
Department of Biochemistry and Agricultural Biochemistry, University
College of Wales, Aberystwyth, Dyfed, SY23 3DD, UK

An understanding of the mode of action of classical uncoupling
agents, eg: salicylanilide derivatives [1]; carbonyl cyanide phenyl-
hydrazines [2]; substituted benzimidazoles [3]; is a key step towards
the elucidation of the mechanism of energy conservation. It is particularly significant in the resolution of the current problem of whether
the 'energised intermediate' between electron transport and ATP-
synthesis is delocalised over the entire membrane or more localised
interactions exist. The objective of this work is to develop strains
of E. coli that can grow in the presence of high concentrations of
classical uncoupling agents and then investigate the molecular basis
of the differences between sensitive and insensitive strains.

Our strategy has been to use the Doc S strain of E. coli K12
(lac^i-z$^+$y$^+$a$^+$pro$^-$trp$^-$his$^-$met$^-$) which has a highly permeable outer cell
wall [4]. In this way we can be more confident that any resistance
to uncoupling agents is due to changes in the limiting membrane of
the cell, (where the energy conservation system is located) and not
due to diminishment in the permeability of the outer wall. This
permeability in the outer wall is manifested by a sensitivity to
sodium deoxycholate - a sensitivity not shown by wild type K12
strains. Doc S is further characterised by a constitutive lac operon
and a requirement for proline, tryptophan, histidine and methionine.
These characteristics have been used to confirm that resistant strains
are directly related to the original Doc S strain.

By culturing Doc S in minimal medium under conditions of increasing concentrations (10-100µM) of 4,5,6,7-tetrachloro-2-trifluoro-
methyl benzimidazole (TTFB), we have isolated a strain which will now
grow using succinate as sole carbon source in the presence of 100µM
TTFB - a concentration at which growh of the original Doc S strain is
completely inhibited.

This TTFB resistant strain shows an unchanged sensitivity to
deoxycholate, has retained the constitutive lac operon and shows the
same amino acid requirements as the parent Doc S. Estimates of the
growth yield of the parent and TTFB resistant strains indicate that
the levels of energy conservation are similar.

Studies are currently under way to confirm that the TTFB resistant strain shows a cross resistance to other classical uncoupling agents such as carbonyl cyanide m-chlorophenyl hydrazone (CCCP), 5-chloro-3-t-butyl-2 -chloro-4 -nitro salicylanilde (S-13) and fentrifanil [5]. Studies are also under way to determine whether the uncoupler resistant strain differs significantly from the parent strain with respect to the effect of these uncouplers on the transmembrane electrochemical proton gradient and the rate of substrate oxidation in both whole cells and spheroplasts.

1. Williamson, R.L. & Metcalf, R.L. (1967) Science 158, 1694-1695
2. Heytler, P.G. (1963) Biochemistry, 2, 357-361
3. Beechey, R.B. (1966) Biochem. J., 98, 284-289
4. Ahmed, S. & Booth, I.R. (1983) Biochem. J. 212, 105-112
5. Nizamani, S.M. & Hollingworth, R.M. (1980) Biochem. Biophys. Res. Commun., 96, 704-710

ORGANIZATION OF GENES ON A LINEAR MOLECULE OF MITOCHONDRIAL DNA OF <u>CANDIDA RHAGII</u>

L. Kováč[o], J. Lazowska[+] and P. P. Slonimski[+]
[o]Institute of Animal Physiology, Slovak Academy of Sciences, 900 28 Ivanka pri Dunaji, Czechoslovakia
[+]Centre de génetique moléculaire, Centre nationale de la recherche scientifique, 91 190 Gif-sur-Yvette, France

Mitochondrial DNA forms circular molecules of various length in most eucaryots, including mammals. However, in <u>Tetrahymena</u>, <u>Paramecium</u> and yeast <u>Hansenula mrakii</u> linear molecules of mitochondrial DNA have been described /for review see ref. 1/. This communication reports on the presence of linear mitochondrial DNA in another yeast species, <u>Candida rhagii</u>, and on the organization of some genes in the linear genome.

The yeast was isolated in the Gif laboratory. The cells were cultured on glucose in a semi-synthetic medium and mitochondria were isolated from protoplasts. Mitochondrial DNA was isolated by centrifugation in CsCl gradient containing bis-benzimide. Restriction map was constructed from partial and double digests. Restriction fragments were hybridized with appropriate probes carrying specific genes of <u>S. cerevisiae</u> either cloned in plasmid pBR 322 or located on appropriate mitochondrial "petite" DNA. Sequencing of DNA was done by the chemical method of Maxam and Gilbert.

Mitochondrial DNA of <u>C. rhagii</u> had different buoyant density in CsCl /1.695 g/cm^3/ than that of <u>S. cerevisiae</u> /1.685/ while the buoyant density of nuclear DNA isolated from the two species was identical /1.699/. Restriction mapping showed that the mitochondrial DNA of <u>C. rhagii</u> contained 30.5 kilobases and that its molecule was linear. The linearity of the molecule was borne out by radioactive labelling of 5´ ends of the molecule: only restriction fragments placed at the two terminals of the molecule were substantially labelled. Preliminary sequencing indicated that the two terminals consisted of identical inverted repeats. Even though hybridization of restriction fragments of DNA from <u>C. rhagii</u> with probes prepared from <u>S. cerevisiae</u> was considerably less efficient than homologuous hybridization of the same probes with restriction fragments from S. cerevisiae, the hybridization under conditions of low stringency enabled to locate several mitochondrial genes on the linear mitochondrial genome of C. rhagii:

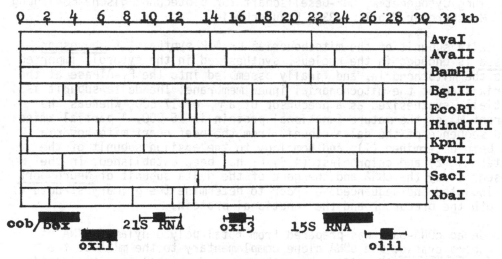

The order of genes in the genome is unique not found so far in other species. Apparently, the placement of genes on the mitochondrial genome may vary without affecting the functionning of mitochondria since the mitochondria of C. rhagii exhibited normal respiration and oxidative phosphorylation.

The linear mitochondrial DNA of C. rhagii may represent a useful system to study the mechanism of replication of linear DNA molecules.

1. Wallace, D.C. /1982/ Microbial. Rev. 46, 208-240

NUCLEOTIDE SEQUENCE OF THE CLONED mRNA AND GENE OF THE DELTA SUBUNIT
OF THE MITOCHONDRIAL ATP-SYNTHASE FROM NEUROSPORA CRASSA

B. Kruse and W. Sebald
Abteilung Cytogenetik, GBF-Gesellschaft für Biotechnologische Forschung
mbH, Mascheroder Weg 1, D-3300 Braunschweig

The delta subunit of the mitochondrial F_1-ATP synthase of Neurospora
crassa is encoded in the nucleus, synthesized in the cytosol, imported
into the mitochondria, and finally assembled into the F_1-ATPase at the
matrix side of the mitochondrial inner membrane. The delta subunit is
initially synthesized as a precursor of a M_r of 17 500, whereas the
aparent M_r of the mature functional protein is 15 000. A partial amino
acid sequence of the delta subunit from the beef heart mitochondria
has been determined (1), and homology to the epsilon subunit of the
bacterial (2) and chloroplast (4,5) F_1 has been established. In the
present study the cDNA and the gene of the delta subunit of Neurospora
was isolated and sequenced in order to determine the primary structure
of both the precursor and the functional protein.

An ordered cDNA-bank was prepared from total polyadenylated mRNA of
Neurospora crassa. One cDNA clone complementary to the mRNA of the
delta subunit was identified by methods based on cell-free translation
of hybridization-selected mRNA and subsequent analysis of the translation products with specific antibodies. Further delta cDNA clones
which occur at a frequency of about 0.03 % were found by colony filter
hybridization. Two of them comprised nearly the complete mRNA including
the poly(A)tail. The gene was cloned and isolated from a partial Sau 3A
digest of total genomic DNA. The DNA sequence of both - the cDNA and
the genomic DNA - was determined.

The gene contains three introns after the 34th amino acid codon (72 bp),
after the 44th codon (120 bp), and within the 96th codon (104 bp). The
cDNA includes an open reading frame of 495 bp which codes for the 165
amino acid residues of the precursor of the delta subunit. This alignment is based on the following information:
The deduced amino acid sequence shows clear homologies to a partial
amino acid sequence of the delta subunit of the bovine heart F_1 (1)
and to the sequence of the epsilon subunit of E. coli (2,3) and chloroplast F_1 (4,5). The start of the preprotein is placed at the first and
only ATG occuring in several hundered bases before the sequence coding
for the mature protein. A sequence of eight bases before the ATG
(ATATCATCATGATG) shows a striking homology to the corresponding region
of other Neurospora mRNA (6). The amino acid composition of the mature
delta subunit is in good agreement with the amino acid sequence if the

protein starts around tyrosine 27 (Fig. 1). . The derived molecular weight of 15 000 was determined by SDS-gel electrophoresis (7).

The established presequence is 26 residues long. It is polar and does not contain a hydrophobic segment like the prepieces of the secretory proteins. Seven basic residues exist (arginines) but no acidic ones. This is remiscent of the presequence of the proteolipid subunit of the mitochondrial ATP synthase , which is also polar and basic (6).

```
        10        20     ↓ 30        40        50        60
MNSLRIARAA LRVRPTAVRA PLQRRGYAEA VADKIKLSLS LPHQAIYKSQ DVVQVNIPAV

        70        80        90       100       110       120
SGEMGVLANH VPSIEQLKPG LVEVIEESGS NKQYFLSGGF AVVQPGSKLS INAVEGYALE

       130       140       150       160
DFSAEAVRAQ IAEAQKIVSG GGSQQDIAEA QVELEVLESL QAVLK
```

Fig. 1: Amino acid sequence of the precursor of the delta subunit of the F_1-ATP synthase.

References:

1. Walker, J.E., Runswick, M.J., and Saraste, M. (1983) FEBS Lett. 146, 393-396

2. Futai, M. and Kanazawa, H. (1983) Microbiol. Rev. 47, 285-312

3. Walker, J.E., Saraste, M., and Gay, N.J. (1984) Biochim. Biophys. Acta, in press

4. Krebbers, E.T., Larinna, I.M., McIntosh, L. and Bogorad, L. (1982) Nucl. Acids Res. 10, 4985-5002

5. Zurawski, G., Bottomly, W. and Whitfeld, P.R. (1982) Proc. Natl. Acad. Sci. USA 79, 6260-6264

6. Sebald, W. and Kruse, B. (1984) in H$^+$-ATP synthases (Papa, S. et al. eds.)ICSU Press (in press)

7. Sebald, W. and Wild, G. (1979) Meth. Enzymol. 55, 344-351

THE FUNCTION OF THE I-PROTEIN OF THE ATP SYNTHASE OPERON OF ESCHERICHIA COLI

Ole Michelsen, Kaspar von Meyenburg, Department of Microbiology, The Technical University of Denmark, DK-2800 Lyngby, Denmark.

The 8 genes for the ATP-synthase of Escherichia coli are organized in an operon at 83.5 min on the E.coli chromosome. The genes are transcribed as a polycistronic messenger-RNA together with a ninth proximal gene, atpI, coding for a 14 kD basic and hydrophobic protein, that has been designated the i-protein. The i-protein is not essential, since a plasmid carrying the 8 structural genes could complement deletions of the atp operon although it did not carry the atpI gene (1). Mutations that only affect the atpI gene have not been described so far.

In order to search for a function of the i-protein, deletions in the atpI gene were constructed. The atpI gene contains a 199 base pairs HindIII fragment; a plasmid carrying the promoter, atpIp, together with the atpB gene but lacking that HindIII fragment could be constructed. This deletion, atpI705, brings the distal part of the atpI gene out of reading frame with the proximal part. An "in-frame-deletion" of 195 base pairs, atpI707, was constructed by opening the plasmid in the HindIII site and filling the ends with Klenow fragment of E.coli DNA-polymerase I. Both deletions were crossed into the chromosome of E.coli. No significant effects on growth rate and yield or expression of the atpE gene were found on different carbon sources and under anaerobic conditions. However, when the pH of the medium was raised above 8.3, the strain carrying the frame shift deletion atpI705 showed a 15% decrease in growth yield compaired to wild type strain, while the strain with the in-frame-deletion atpI707 had normal yield. The decrease in growth yield could be restored by complementation in trans by a plasmid that carried the atpI gene expressed from the atpIp promoter.

The i-protein has a basic aminoterminal end and two hydrophobic domains. The peptides expressed from the deletions have the same basic N-terminal; while the peptide expressed from deletion atpI707 still has one of the hydrophobic domains, the carboxyterminal end of the peptide expressed from deletion atpI705 has become acidic and hydrophilic due to the frameshift and would not be expected to enter the membrane. These results suggest that the i-protein in a membrane protein (2) of which the basic N-terminal domaine is located on the outside of the cytoplasmic membrane and serves to protect the periplasmic proton reservoir from loosing protons under alkaline growth conditions.

1) K. von Meyenburg et al., Mol. Gen. Genet. 188:240-248 (1982).
2) W.S.A.Brusilow et al., J. Bacteriol. 155:1265-1270 (1983÷.

CYANELLE GENOME ORGANIZATION IN CYANOPHORA PARADOXA

H. Mucke[*], W. Löffelhardt[*] and H.-J. Bohnert[+]

[*] Institut für Allgemeine Biochemie der Universität Wien und Ludwig Boltzmann-Forschungsstelle für Biochemie, Währingerstraße 38, A-1090 Wien/Austria

[+] Department of Biochemistry, Biological Sciences West, The University of Arizona, Tucson, Az 85721/USA

Cyanophora paradoxa, a unicellular biflagellate, is an outstanding representative of the so-called endocyanomes, a curious class of photoautotrophic organisms containing cyanelles in hereditary endosymbiosis. Cyanelles serve as chloroplast for their host organism. They exhibit several characteristics of blue-green algae, e.g. a central body, concentrically arranged thylakoids and a lysozyme-sensitive cell wall.

Recently, we were able to show that the cyanelle genome is very similar to chloroplast DNA in its size and general structure (1,2,3). Cyanelle DNA isolated from the commonly used Cyanophora strain LB 555 UTex is a circular molecule 127 kb long and contains two 10,5 kb inverted repeat units dividing the genome in a large (88,5 kb) and a small (17kb) single copy region. Radioactive probes derived from spinach chloroplast protein genes were used to localize the corresponding genes on the cyanelle restriction map (4), and cyanelle tRNAs were used as homologous hybridization probes(5).

As in higher plants, the genes for the α-, β- and ϵ-subunit of the ATPase coupling factor CF_1 (atpA, atpB and atpE) are coded by the photosynthetic compartment itself. No attempt was made to identify the γ- and the δ- subunit, but it seems likely that their genetic information resides in the nuclear genome of the host cell. Both atpB- and atpE - probes hybridize to a 7,6 kb Bgl II restriction fragment, as well as the probes for the large and small subunit genes (rbcL, rbcS) of ribulose-1,5-bisphosphate carboxylase. This gene cluster seems to be extended to the adjacent 3,4 kb Bgl II fragment, which carrys the smaller part of rbcL and in addition the gene for the 32 kd herbicide - binding membrane protein associated with photosystem II (psbA). About 13 kb apart from this cluster the atpA gene was localized on a 6,6 kb Bgl II fragment, whereas the genes coding for two components of the cytochrome b_6/f - complex, b_6 itself (petB) and subunit 4 (petD) were found on an adjacent 5,6 kb Bgl II fragment.

Positive results from Northern hybridizations support the view that no pseudogenes or fortuitous homologies are responsable for these data and that the respective genes are being actively transcribed. No common large transcript could be detected for rbcL and rbcS. Seen in connection with sequencing data (6), these genes are most probably seperated by atpB and atpE.

Preliminary results have also been obtained concerning the localization of ribosomal protein genes using hetero-logous probes from E. coli.

1) Mucke H., Löffelhardt W. and Bohnert HJ., FEBS Lett. 111(2), 347-352 (1980)

2) Löffelhardt W., Mucke H. and Bohnert HJ., in: Endocyto-biology I, pp. 523-530, Editors H. Schenk and W. Schwemm-ler; deGruyter, Berlin and New York 1980

3) Bohnert HJ., Michalowski Chr., Koller B., Delius H., Mucke H. and Löffelhardt W., in: Endocytobiology II, pp. 433-448, Editors H. Schenk and W. Schwemmler; de Gruyter, Berlin and New York 1983

4) Mucke H., Löffelhardt W. and Bohnert HJ., in: Proc. Sixth Int. Congress on Photosynthesis, Brussels 1983; in the press

5) Kuntz M., Crouse EJ., Mubumbila M., Burkard G., Weil JH., Bohnert HJ., Mucke H. and Löffelhardt W. Mol. Gen. Genet., in the press

6) Wasman K. and McIntosh L., private communication

CLONING OF CYTOCHROME OXIDASE AND CYTOCHROME c_1 GENES FROM PARACOCCUS DENITRIFICANS

B. Paetow and B. Ludwig

Institute of Biochemistry, Medical University
D 2400 Lübeck   West Germany

The study of those bacterial respiratory complexes that appear homologous to complexes found in mitochondria has added to our understanding of structure and function of mitochondrial electron transport and energy transduction.

Cytochrome c oxidase from Paracoccus denitrificans, a bacterium known for its close relationship to mitochondria in this respect (1), is much simpler in its polypeptide structure, yet has been shown to fulfill the same functions described for the mitochondrial enzyme, including the ability to act as a proton pump (2,3). While its two subunits do have common determinants with the two largest, mitochondrially coded subunits of the eukaryotic oxidase (2,4), knowledge of their complete sequence would not only allow us a closer comparison, but could also answer whether or not domains of other polypeptides that are individual subunits of the mitochondrial enzyme and have putatively been assigned certain functions (such as subunit III), are found in the sequence of either of the two subunits of the Paracoccus enzyme.

Here we describe attempts to circumvent the problem of a direct protein sequencing analysis by isolating the gene(s) for DNA sequencing.

A Paracoccus gene bank was constructed by inserting sized Mbo I - DNA fragments into the Bam H1 site of the vector pBR 322 . This bank was screened by an immunological technique for possible recombinants expressing the polypeptides of interest. Positives turned up at a frequency of about 1 per 10 000 recombinants.

A set of individual clones was obtained which express the full length cytochrome c_1 polypeptide in E.coli, as judged from SDS-PAGE/immuneblotting.

For cytochrome c oxidase subunit I, some clones express
the full gene with the typical fuzzy migration behaviour of
the polypeptide on SDS gels, while others, possibly lacking
part of the C-terminal portion of the peptide, yield a discrete band of apparently higher molecular weight.

References:

(1) John, P. and Whatley, F.R. (1977) Biochim.Biophys.Acta
    463, 129-153
(2) Ludwig, B. (1980) Biochim.Biophys.Acta 594, 177-189
(3) Solioz, M., Carafoli, E. and Ludwig, B. (1982) J.Biol.
    Chem. 257, 1579-1582
(4) Steffens, G.C.M., Buse, G., Oppliger, W. and Ludwig, B.
    (1983) Biochem.Biophys.Res.Comm. 116, 335-340

Acknowledgement: Supported by DFG Grant Lu 318/1-1

ENERGETICS OF DNA ENTRY IN COMPETENT STREPTOCOCCUS PNEUMONIAE

M.C. Trombe°, D.A. Morrison"
°Centre de Biochimie et Génétique Cellulaires du CNRS et Université Paul Sabatier 31062 Toulouse Cedex, France.
"Department of Biological Sciences University of Illinois at Chicago, Chicago Ill 60680 USA.

DNA transport accros the bacterial cell membrane, is the preliminary step of several genetic exchange systems. In transformable bacteria DNA transfer occurs at a specialized physiological state called competence. In competent S. pneumoniae it is known that double stranded DNA is bound at the surface of the cells, nicked and converted to single strands during entry while oligonucleotides are released in the medium (1). Curently, experimental evidence points to the importance of the protonmotive force in bacterial transformation and it is proposed, as an hypothesis, that the protonmotive force could be the driving force of DNA entry (2). Recently the chemical gradient ΔpH rather than the electrical gradient $\Delta\Psi$ was proposed as the driving force of DNA entry in B. subtilis (3).

We checked the energy requirement of the three steps involved in DNA uptake in S. pneumoniae : Binding, Degradation and Entry. This was achieved first, by using conditions of partial inhibition by the protonophore CCCP. Second by reduction of the surface potential of the bacteria with the cyclophosphazene derivative SOAz (4).

In S. pneumoniae the activity of the protonophore CCCP is highly pH-dependent. At pH 7.95 CCCP inhibited partially DNA binding and DNA degradation (30% to 40% of the control value) with no significant reduction of DNA uptake suggesting that the transmembrane proton gradient was not involved in uptake but was implicated in DNA binding and DNA degradation. Inhibition of DNA degradation by CCCP could result from the inhibition of DNA binding on the surface of the bacteria. However in the same experiment, at pH 7.95, SOAz inhibited DNA binding at the same level as CCCP with no effect on DNA degradation. This suggests that the membrane nuclease activity does require an energized membrane state in competent S. pneumoniae. At pH 6.95 inhibition of DNA binding and degradation by CCCP were enhanced respectively from 38 to 76% and from 30 to 63% when compared to the control (no CCCP). In addition DNA uptake was also reduced by the protonophore.

A possible explanation is that the proton circulation is involved in DNA uptake at that pH. Alternatively it can be suggested that this inhibition of uptake results from the strong inhibition of DNA degradation i.e. 70%. Indeed in these conditions the ratio DNA degradation versus

DNA uptake shifted from 0.7 (in the control experiment) to $\simeq$ 1. This shows that the amount of single stranded DNA resulting from DNA degradation was entirely taken up by the bacteria.

Therefore our results suggest that DNA binding and DNA degradation required an energized membrane as indicated by their sensitivity to the protonophore CCCP. In addition, in S. pneumoniae DNA binding was selectively reduced by SOAz a chemical which reduces the surface potential of model membrane (4) suggesting that the charges on the surface of the bacteria might be involved in DNA binding. On the other hand we would like to suggest that the entry of DNA itself might not require the proton circulation, provided that enough DNA degradation occured. Indeed, it is known that a DNA degradation block can prevent the formation of single stranded DNA, the DNA form which is transported within the cells (5). Consequently the limiting step in the presence of CCCP might be the formation of single stranded DNA as a consequence of the inhibition of DNA degradation. Similar inhibition of uptake is observed when the DNAse activity is inhibited by EDTA or by mutation (6).

References
1. Smith H.O., Danner D.B. and Deich R.A. 1981. Ann. Rev. Biochem. 50 41-68
2. Grinius L. 1980. Febs Let. 113 1-10
3. Van Nienwohen M.H., Hellingwert K.J., Venema G. and Konings W.N. 1982. J. Bacteriol. 151 771-776
4. Trombe M.C., Beaubestre C., Sautereau A.M., Labane J.F., Laneelle G. and Tocanne J.F. 1984. Biochem Pharmacol. in the press.
5. Lacks, S.A. 1962. J. Mol. Biol. 5 119-131
6. Lacks, S.A., Greenberg, B. and Neuberger, B. 1975. J. Bacteriol. 123 222-232

SIZE AND SURFACE CHARGE OF PURPLE AND WHITE MEMBRANE PREPARATIONS
FROM HALOBACTERIUM HALOBIUM.

B. Arrio[+], G. Johannin[+], P. Volfin[+], M. Lefort-Tran[°], L. Packer[*],
A. Robinson[*], S. Wu-Chou[*], E. Hrabeta[*]
[+]Institute de Biochimie, Bat. 432, Universite de Paris XI, 91405
Orsay, France
[°]The Laboratory de Cytophysiologie de la Photosynthesis, Gif Sur
Yvette, France
[*]The Membrane Bioenergetics Group, Applied Sciences Division,
Lawrence Berkeley Laboratory, University of California, Berkeley,
California 94720, U.S.A.

Quasi-elastic light scattering (QELS), negative staining electron
microscopy, and laser doppler velicometry (LDV) were employed to
characterize the size, structure and surface charge of purple
membrane preparations from H. halobium and white membranes from the
mutant strain R_1mW (1-2).

QELS and Electron Microscopic Studies

In suspension, purple membranes exist as stacks of membranes of
about 0.25 uM or larger because membrane-sheets are piled one on top
of the other. The net effect is that the hydrodynamic radius, calculated from the angular dependence of light scattering, is consistent with the particles behaving in suspension as spheres. White
membranes observed by QELS and electron-microscopy were smaller in
size, ca 0.08 uM, and were unstacked.

Upon typsin treatment, purple membrane preparations become highly
aggregated in suspension. Purple membrane preparations after storage
were found to contain protease activity inhibited by benzamidine
indicating typsin-like activity. The stacking may therefore be a
result of low level proteolysis; a finding important in the measurement of protons released during the photocycle of bacteriorhodopsin,
since more stacked preparations should release fewer protons per
bacteriorhodopsin. This suggests an explanation for discrepancies in
M_{412} stoichiometry results for membrane preparations in suspension
and reconstituted into liposomes (3-4).

LDV Studies

Electrophoretic light scattering techniques have been used to make
direct measurements of the surface charge of purple membranes. In
this method, the doppler shift of laser light scattered by particles
moving in an electric field, laser doppler velicometry (LDV) is used
to calculate surface charge density (5). The linear correlation
between electrophoretic mobility and applied electric field. The

rapid and direct measurement of surface charge indicate that surface
electrical responses of purple membranes can be accurately monitored.
Purple membranes and chemically modified membranes were examined.
The results show that purple membranes treated with water soluble
carbodiimides in the presence of nucleophiles to modify carboxyl
residues (6) have a lower surface charge than native purple membranes.
Also native purple membranes have a lower surface charge than white
membrane preparations obtained from the R_1mW mutant. Light induced
changes in surface charge were also measured. These changes correlate
with a slower decay of the M_{412} intermediate of the photocycle such
as is found after using a new light-dependent method for nitration of
tyrosine residues (7). Results of surface charge changes correlate
with spin label (8) and resonance Raman (9) probe methods. Thus, LDV
is an excellent method for directly characterizing surface electrical
properties of purple membranes.

Acknowledgements

Research support from The Centre National de les Recherche
Scientifique (A.T.P. "Transduction de l'energie dans les membranes
biologiques et photosynthesis" n° 3093/79); the Office of Biological
Energy Research, Division of Basic Energy Sciences, U.S. Department
of Energy; and a NATO Grant for International Collaboration in
Research.

References

1 Arrio, B., Johannin, G., Carrette, A., Chevallier and Brethes, D.
(1984) Arch. Biochem. Biophys. 228, 220-229.
2 Arrio, B., Johannin, G., Volfin, P. and Packer, L. (1984) Biophys.
J. 45, 212a.
3 Govindjee, R., Ebrey, T.G. and Crofts, A.R. (1980) Biophys. J. 30,
231-242.
4 Govindjee, R., Ohno, K. and Ebrey, T.G. (1982) Biophys. J. 38,
85-87.
5 Ware, B.R. and Haas, D.D. (1983) in Fast Methods in Physical Biochemistry and Cell Biology (Sha'afi, R.I. and Fernandez, S.M.,
eds.) pp. 174-220, Elsevier, Amsterdam.
6 Packer, L., Tristram, S., Herz, J.M., Russell, C. and Borders, C.L.
(1979) FEBS Lett. 108, 243-248.
7 Lam, E., Seltzer, S., Katsura, T. and Packer, L. (1983) Arch.
Biochem. Biophys. 227, 321-328.
8 Carmeli, C., Quintanilha, A.T. and Packer, L. (1980) Proc. Natl.
Acad. Sci. USA 77, 4707-4711.
9 Ehrenberg, B. (private communication).

BIOENERGETIC INTERACTION OF BACTERIORHODOPSIN AND HALORHODOPSIN IN HALOBACTERIUM HALOBIUM

K. Bickel, B. Traulich and G. Wagner

Botanisches Institut I der Justus-Liebig-Universität
Senckenbergstr. 17-21, D-6300 Giessen, West Germany

Two retinal protein pigments with electrogenic ion transport function are found in Halobacterium halobium: Bacteriorhodopsin (BR) which acts as a light-driven proton pump [1] and Halorhodopsin (HR), a light-driven chloride pump [2,3]. The coexistence of two pigments with bioenergetic function is a puzzling phenomenon. Therefore we used quantitative action spectroscopy to decide on the photoenergetic interaction of BR and HR in intact cells. Photoreceptor mutants of H. halobium are characterized as follows:

R1M1 (BR^+, HR^+); L-33 (BR^-, HR^+); M-18* (BR^+, HR^-).

Fig. 1, right side, shows a diagram of initial rate of light-induced proton uptake in L-33 measured at wavelengths λ = 560 nm and λ = 580 nm, respectively; similarly, initial rate of proton extrusion was followed in M-18. The calculated relative quantum efficiency in L-33 compared to M-18 matches the difference in absorption spectrum of HR relative to BR.

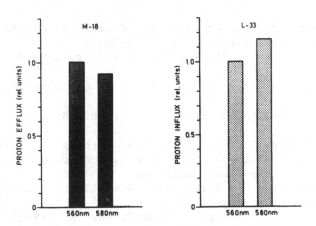

Fig. 1 Rates of light-induced proton translocation in H. halobium mutant strains M-18 and L-33, respectively.
E_o (λ = 560 nm) $\equiv$ E_o (λ = 580 nm) = $2{,}2 \cdot 10^{20}$ Photons$\cdot m^{-2} \cdot s^{-1}$

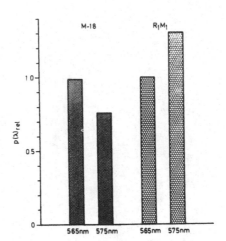

Fig. 2 Relative quantum yield of photophosphorylation in H. halobium mutant strains M-18 and R1M1, respectively.

According to reference data from S-9 [4] which seems similar to R1M1 in terms of BR and HR contents, an amount of 2-5% of HR relative to BR can be estimated for R1M1. Comparing the relative pigment concentrations with relative quantum yield of photophosphorylation, quantum efficiency at wavelength λ = 575 nm appears enhanced by an order of magnitude in R1M1 (Fig. 2). This enhancement seems based on HR interaction with BR as confirmed by the results in M-18. Thus, the presence of BR seems to enhance quantum efficiency of HR in intact cells.

* M-18 was kindly provided by D. Oesterhelt; detailed characterization of M-18 will be given elsewhere in collaboration.

References

1 Oesterhelt, D. and Stoeckenius, W. (1971) Nature New Biol. 233, 149 - 151.
2 Mukohata, Y., Matsuno-Yagi, A. and Kaji, Y. (1980) Saline Environment, 31-37, ed. H. Morishita and M. Masui, Osaka.
3 Schobert, B. and Lanyi, J. K. (1982) J. Biol. Chem. 257, 10309 - 10313.
4 Oesterhelt, D. (1982) Methods in Enzymology, Vol. 88, part I, 10 - 17.

RECONSTITUTION OF PHOTOSYNTHETIC OXYGEN EVOLVING PARTICLES FROM PHORMIDIUM LAMINOSUM

Jane M. Bowes,[o] Antony, W.D. Larkum" and Derek S. Bendall[o]

[o]Department of Biochemistry, University of Cambridge, Tennis Court Road, Cambridge, CB2 1QW, UK.
"School of Biological Sciences, University of Sydney, New South Wales 2006, Australia.

Procedures have been developed to effect the reconstitution of photosystem II (PSII) particles from the blue-green alga Phormidium laminosum into liposomes prepared from lipids extracted from the same organism. The PSII particles, which were highly active in oxygen evolution, were extracted from the thylakoid membranes by detergent fractionation, using lauryldimethylamine N-oxide (LDAO) and dodecyl-β-D-maltoside (1). Diacyl lipids were extracted from Phormidium according to Larkum and Anderson (2). Reconstitution could in principle be achieved by removal of the detergent associated with the particles, in the presence of lipid.

Treatment of a mixture of sonicated lipids and PSII particles with polystyrene beads followed by one freeze/thaw cycle and sonication resulted in association between the lipid and protein. This was indicated by the formation of a single band on a sucrose density gradient. The rate of light-induced oxygen evolution with a lipophilic quinone acceptor was the same as in the original particles. The rate supported by the charged acceptor ferricyanide, was considerably lower than in the particles and more sensitive to DCMU, even at low pH (1). The low rate with ferricyanide may suggest that the majority of centres were incorporated into the vesicles inside out. However, since simple addition of lipid to particles also resulted in a lowering of the rate, the effect may be due to non-specific association of lipid. Moreover, measurements of proton release accompanying oxygen evolution provided no evidence for the formation of transmembrane ion gradients that would indicate a vectorial arrangement of the photosystem in the artificial membrane. The leakiness of the membrane to protons may be due to the presence of residual detergent. The reconstitution has been modified to produce a slower but more effective removal of detergent by dialysis against the polystyrene beads, and the proteoliposomes thus formed generated uncoupler-sensitive proton gradients in the light.

1. Bowes, J.M., Stewart, A.C. and Bendall, D.S. (1983) Biochim. Biophys. Acta, 725, 210-219
2. Larkum, A.W.D. andAnderson, J.M. (1982) Biochim. Biophys. Acta, 679, 410-421

MOLECULAR ORGANISATION OF THE LIGHT-HARVESTING POLYPEPTIDES
FROM RHODOSPIRILLUM RUBRUM

R.A. Brunisholz[o], R.J. Cogdell["], J. Valentine["], and H. Zuber[o]

[o] Institut für Molekularbiologie und Biophysik, ETH-Hönggerberg
   CH-8093 Zürich, Schweiz
["] Department of Botany, University of Glasgow, Glasgow G12 8QQ, U.K.

The B890 antenna complex from Rhodospirillum rubrum contains as
'minimal compositional unit' 2 molecules of BChl a and 1 molecule
of spirilloxanthin non-covalently bound to two apoproteins (α and
β-polypeptide)(1). The α-polypeptides (52 amino acids, B890-α/B870-α)
from Rs. rubrum S1 (wild type, with carotenoids) and Rs. rubrum G-9[+]
(carotenoidless mutant) as well as the β-polypeptides of these
strains (54 amino acids, B890-β/B870-β) have identical amino acid
sequences (2-4). In the case of Rs. rubrum G-9[+] the apoproteins
B870-α and B870-β have their N-termini exposed towards the cytoplasmic surface (5).

In the present study chromatophores (inside-out vesicles) from both
strains (S1 and G-9[+]) have been incubated with proteinase K. In G-9[+]
6 amino acids of B870-α and 16 amino acids of B870-β have been removed from their N-terminal domains. Under the same conditions the
light-harvesting polypeptides of S1 differ in their sensitivity to
proteinase K digestion. This protease removes 5 amino acids of B890-β
from its N-terminal domain whereas B890-α apparently resists to proteolytic digestion. Thus, we conclude that, compared with the situation in G-9[+], different conformations of the N-terminal domains of
the two apoproteins B890-α and B890-β allow binding of spirilloxanthin. Thereby, a more rigid structure is apparently induced which

reduces the sensitivity for proteolytic attack.

Upon proteinase K digestion of the chromatophores from both strains the antenna BChl $\underline{a}$ (λ max of Q_y: 890nm/870nm) retained its spectral in vivo properties indicating that the N-terminal domains of the light-harvesting polypeptides are not involved in BChl $\underline{a}$ binding. These data are discussed on the basis of the transmembrane model of the two apoproteins of the light-harvesting complex from Rs. rubrum (see also abstract H.Zuber).

1) Cogdell, R., Lindsay, J.G., Valentine, J. and Durant, J. (1982) FEBS Lett., 150, 151-154

2) Brunisholz, R.A., Cuendet, P.A., Theiler, R. and Zuber, H. (1981) FEBS Lett., 129, 150-154

3) Gogel, G.E., Parkes, P.S., Loach, P.A., Brunisholz, R.A. and Zuber, H. (1983) Biochim. Biophys. Acta, 746, 32-39

4) Brunisholz, R.A., Suter, F. and Zuber, H. (1984) Hoppe Seyler's Z. Physiol. Chem. (in press, July issue 1984)

5) Brunisholz, R.A., Wiemken, V., Suter, F., Bachofen, R. and Zuber, H. (1984) Hoppe Seyler's Z. Physiol. Chem. (in press, July issue 1984)

RADICAL FORMATION IN CHROMATOPHORES OF RHODOPSEUDOMONAS CAPSULATA
UPON OXIDATION OF THE B880 ANTENNA COMPLEX

F. F. del Campo* and I. Gómez"
*Departamento de Botánica y Fisiología Vegetal, Universidad Autónoma,
Madrid 34
"Instituto de Biología Celular, C.S.I.C.,
Madrid 6, Spain

In the antenna of Rhodospirillaceae two bacteriochlorophyll-protein complex types have been described, named B880 and B800-850 after their approximate absorbance maxima in the near infrared (1).

We have recently reported (2) that oxidation of antenna bacterio-chlorophyll (Bchl) in membrane preparations (chromatophores) of Rhodopseudomonas sphaeroides and Rps. capsulata gives rise to a radical, detectable by electron spin resonance (ESR). The ESR signal shows a g value of 2.0023, similar to that of oxidized reaction center (RC) Bchl, and its first derivative has a peak to peak linewidth (ΔHpp) of about 3.7 gauss, quite distinct from that of oxidized RC Bchl (about 9.5 gauss). The present work was undertaken to elucidate, through the use of chromatophores from antenna mutants lacking either B880 or B800-850, the antenna Bchl complex which is responsible for the above radical in Rps. capsulata.

The strains utilized were SB1003 (wild-type), Y142 (B880$^-$, RC$^-$) and MW442 (B800-850$^-$), kindly provided by Dr. B. L. Marrs. Treatment of both wild type and MW442 chromatophores with ferricyanide produces an ESR signal (g = 2.0023) of variable magnitude and ΔHpp, the larger and narrower signals corresponding to the higher redox potentials (Table I). In contrast, similarly treated Y142 chromatophores hardly exhibit any signal.

Based on the data obtained by mathematical deconvolution of the observed ESR signals, we interpret that the broad 9.2-gauss signal is mainly due to oxidized RC Bchl while the narrower ones result from the addition to that signal of another one of about 3.7 gauss, due to oxidized antenna Bchl.

It seems clear that the Bchl component responsible for the 3.7 gauss signal is B880, since B880-lacking chromatophores do not show it. The different oxidation degree of B880 attained at the various redox potentials would thus affect the magnitude and linewidth of the composite signal both in wild-type and in B800-850$^-$ chromatophores.

The appearance of the 3.7-gauss signal is always accompanied by an absorbance increase at about 1235 nm, different from that corresponding to oxidized RC Bchl, which is located at 1245 nm (3).

TABLE I

Redox potential (mV)	Strain	ESR signal	
		Relative amplitude	ΔHpp (gauss)
440	SB1003	28	9.1
	MW442	18	9.2
	Y142	undetectable	
500	SB1003	77	6.1
	MW442	86	6
	Y142	10	9.2
570	SB1003	326	3.7
	Y142	18	9.2

Bchl concentration, 0.2 mM. Modulation amplitude, 1 gauss.

The herein results resemble those previously obtained with Rds. rubrum chromatophores (4). The generation of a radical with similar spectroscopic properties upon oxidation of B880 Bchl from Rps. capsulata, Rps. sphaeroides and Rds. rubrum suggests a certain structural similarity of the B880 complexes in those Rhodospirillaceae.

References

1 Cogdell, R. J. and Thornber, J. P. (1980) FEBS Lett. 122, 1-8
2 Gómez, I. and del Campo, F. F. (1983) Abstracts of the VIth Internatl. Congress on Photosynthesis, Brussels, p. 193
3 Clayton, R. K. (1962) Photochem. Photobiol. 1, 201-210
4 Gómez, I., Sieiro, C., Ramírez, J. M., Gómez-Amores, S. and del Campo, F. F. (1982) FEBS Lett. 144, 117-120

TOPOGRAPHY AND MOLECULAR ORGANIZATION OF PIGMENT-PROTEIN COMPLEXES IN RHODOPSEUDOMONAS CAPSULATA MEMBRANES.

Gerhart Drews, J. Peters
Institute of Biology 2, Microbiology, Albert-Ludwigs-University, 7800 Freiburg, Fed. Rep. of Germany

The chemical composition, macromolecular organization and topography of the two light-harvesting (LH) complexes B870 and B800-850 and of the photochemical reaction center (RC) have been investigated by chemical cross-linking, immuno- and detergent fractionation and 2D-PAGE (1, 2). Each LH complex contains two bacteriochlorophyll (Bchl) binding polypeptides (Mr5000-8000) (3, 4). A sequence of 20 hydrophobic amino acids, which spans the membrane as -helix, is believed to interact with Bchl via one HIS residue. The B870 polypeptides form oligomeric structures which interact with the H-subunit of RC (5). The three polypeptides of RC (Mr34000-28000; 6) interact with each other but do not form homooligomers. It is proposed that three oligomeric structures of B870 surround one RC. The second antenna complex B800-850 spans also the membrane and forms tetrameric structures of its polypeptides. The B800-850 LH-particles interconnect presumably the RC-B870 particles (7). Size and number of photosynthetic units per membrane area and per cell and the photochemical activity of the membrane can be modified by external factors, i.e.: light intensity and oxygen partial pressure (8).

Supported by the Deutsche Forschungsgemeinschaft

References:

1) Peters, J., Drews, G. (1983). Europ. J. Cell Biol. 29, 115-120.
2) Peters, J., Takemoto, J., Drews, G. (1983). Biochem. 22, 5660-5667.
3) Tadros, M.H., Suter, F., Drews, G., Zuber, H. (1983). Eur. J. Biochem. 129, 533-536.
4) Tadros, M.H., Suter, F., Seydewik, H.H., Witt, I., Zuber, H., Drews, G. (1984). Eur.J. Biochem. 138, 209-212.
5) Takemoto, J.Y., Peters, J., Drews, G. (1982). FEBS Lett. 142, 227-230.
6) Youvan, D.C., Bylina, E.J., Alberti, M., Begusch, H., Hearst, J.E. (1984). Cell, in press.
7) Drews, G., Peters, S., Dierstein, R. (1983). Ann. Microbiol. (Inst. Pasteur) 134 B, 151-158.
8) Reidl, H., Golecki, J.R., Drews, G. (1983). Biochim. Biophys. Acta 725, 455-463.

624

SPIN-LABEL STUDIES OF BACTERIORHODOPSIN

M. Duñach, M. Sabés and E. Padrós
Departament de Biofísica, Facultat de Medicina, Universitat Autònoma
de Barcelona (Bellaterra). Spain

Conformational studies of bacteriorhodopsin (BR) are useful to get
information on the proton pump mechanism. This information can be obtained by the covalent spin-label technique. Recently its application
to carboxyl groups of BR has been reported (1). We now report our results in conformational changes of BR by means of spin-labelled tyrosine and lysine residues.

Tyrosine and lysine residues of BR were covalently spin labelled
with N-(2,2,5,5-tetramethyl-3-carboxylpyrrolidine-1-oxyl)-imidazole
(IMSL) and succinimidyl-2,2,5,5-tetramethyl-3-pirroline-1-oxyl-3-car-
boxylate (SCSL) respectively . Unreacted spin labels were eliminated
by several centrifugations followed by chromatography through a Bio-
Gel A-0.5m column,thus giving delipidated BR*in deoxycholate (DOC) as
described in (2). Tyrosine and Lysine residues from white membrane
(strain JW-5) were labelled according to the same procedure.

In order to determine the amount of bound spin label, the labelled
protein was digested with proteinase K overnight at 37ºC. In such
conditions the EPR spectrum gives a highly mobile signal that was
compared with those originated using standard solutions of known concentration of the free label. For a 100-fold molar excess of IMSL to
total tyrosine content and a 30-fold molar excess of SCSL to total lysine content we found respectively 0.4 moles of modified Tyr and 0.8
moles of modified Lys residues per mole of BR. The low yield of labelled Tyr residues can be due to their
low accessibility and to the hydrolysis of the spin label which would decrease its concentration. The EPR spec
tra of delipidated bacteriorhodopsin
in DOC are indicative of two kinds of
spin-label mobilities. As shown in figure 1 the spectrum of tyrosine labelled BR has more proportion of immobilized component.

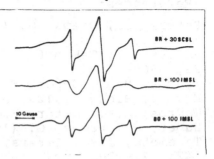

Samples of delipidated BR in DOC
showed a first phase of M412 decay kinetics slower than the corresponding
purple membrane solubilized in DOC (3),
at the same pH. The M412 decay kinetics
of the labelled samples is slower than

*Fig. 1: EPR spectra of de-
lipidated purple membrane
and white membrane in 0.2%
DOC at pH 7.5, labelled in
native conditions.*

that of the non-modified delipidated BR.

EPR spectra of dark and light adapted samples were recorded in order to detect light induced structural changes. No variations in mobility were observed.

We found that the accessibility of the spin labels to Tyr and Lys residues of the white membrane is higher than the one of the purple membrane. Thus, one mole of Tyr and 1.5 moles of Lys residues were labelled per mole of bacterio-opsin. The EPR spectrum of the IMSL is indicative of a higher mobility compared with the case of IMSL-BR (fig.1). These results suggest a less compact structure for bacterio--opsin.

We have also applied the Saturation Transfer-EPR technique to detect changes in the environment of Tyr and Lys residues at the level of the M412 photointermediate. Spin-labelled BR samples in 50% glycerol and 0.12% DOC at pH 7.5, were irradiated with yellow light for 60min at 230K. No changes in mobility could be appreciated. The ST--EPR spectra of the M412 intermediate, corresponding to greatly immo bilized spin labels, were very similar to those of BR at 230K. Previously we have verified the formation of the M412 intermediate using low temperature spectrophotometry (yield>90%) of a delipidated BR sample in the same medium. These are different conditions than those described by other authors to isolate the M412 (4). Taken into account the limit of the method to detect changes slower than 10^{-3} s (5), these results suggest that the conformational changes involving the appearance of the M412 intermediate do not affect the mobility of the spin labels.

We would like to thank Drs. M. Michel-Villaz, J.L. Girardet and J.J. Lawrence for helpful discussions and for use of EPR facilities. Research supported by the CIRIT of the Generalitat de Catalunya and by grant 403/81 from the Comisión Asesora de Investigación Científica y Técnica.

References

1.Herz, J.M., Mehlhorn, R.J. and Packer, L. (1983) J.Biol. Chem. 258,9899-9907
2.Huang, K.S., Bayley, H. and Khorana, H.G. (1980) Proc.Natl. Acad. Sci. USA, 77, 323-327
3.Lam, E. and Packer L. (1982) Arch. Biochem. Biophys. 221, 557-564
4.Becher, B.,Tokunaga, F. and Ebrey, T.G., (1978) Biochemistry, 17, 2293-2300
5.Hyde, J.S. (1978) Methods Enzymol. 49, 480-511

PRIMARY AND SECONDARY ION TRANSPORT IN <u>HALOBACTERIUM</u> <u>HALOBIUM</u> VIA BACTERIORHODOPSIN AND HALORHODOPSIN

A. Duschl, R. Sawatzki and G. Wagner

Botanisches Institut I der Justus Liebig Universität
Senckenbergstr. 17-21, D-6300 Giessen, West Germany

In Halobacterium halobium light-driven ion transport is mediated by two retinal-proteins: Bacteriorhodopsin (BR), acting as an electrogenic proton pump [1] and Halorhodopsin (HR), a light-driven chloride pump [2,3]. Chloride transport was followed by ^{36}Cl uptake in intact cells. Proton and potassium uptake were determined by ion-selective electrodes in cell suspensions. ΔpH was measured by flow-dialysis technique using ^{14}C-methylamine. All experiments were carried out in standard actinic light and under nitrogen conditions. Photoreceptor mutants are characterized as follows: R_1M_1 (BR+,HR+), M-18* (BR+,HR-) and L-33 (BR-,HR+).

HR and BR mediate significant Cl^- import (Fig. 1). The BR-deficient mutant strain L-33 demonstrates action of the primary chloride pump Halorhodopsin, but BR-containing strains show increased Cl^- uptake. This is interpreted as the effect of a secondary chloride transport system, possibly acting via a H^+/Cl^- symport mechanism.

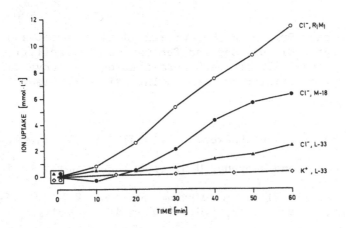

Fig. 1. Uptake of Cl^- and K^+ in light and under nitrogen conditions. Cellular density = $9.3 \cdot 10^9$ cells·ml^{-1}; T = 25 °C; irradiance = 1000 W·m^{-2}, GG 495.

Under comparable experimental conditions, we studied secondary proton movements in membrane vesicles prepared from mutant strain L-33 depending on HR. These vesicles as well as intact cells [4] show a transient net proton uptake up to 15 minutes (Fig. 2). Thus proton uptake mediated by HR cannot compensate the chloride import, which continues linearly over at least 60 minutes.

R_1M_1 shows high rates of potassium uptake in light [5]. In contrast, potassium uptake in L-33 under identical conditions is close to zero (Fig. 1). Here sodium seems to act as counterion for chloride (data not shown).

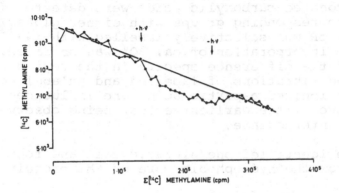

Fig. 2. Flow dialysis of ^{14}C-methylamine in suspended membrane vesicles of mutant strain L-33. Protein concentration = 3.96 mg·ml^{-1}; irradiance = 500 W·m^{-2}, GG 495.

---

*M-18 was kindly provided by D. Oesterhelt; detailed characterization of M-18 will be given elsewhere in collaboration.

References
1. Oesterhelt, D. and Stoeckenius, W. (1971) Nature New Biol. 233, 149-151
2. Mukohata, Y., Matsuno-Yagi, A. and Kaji, Y. (1980) Saline Environment, 31-37, ed. H. Morishita and H. Masui
3. Schobert, B. and Lanyi, J. (1982) J. Biol. Chem. 257, 10309-10313
4. Wagner, G., Oesterhelt, D., Krippahl, G. and Lanyi, J. (1981) FEBS Letters 131, 341-345
5. Wagner, G., Hartmann, R. and Oesterhelt, D. (1978) Eur. J. Biochem. 89, 169-179

MECHANISM OF PROTON TRANSFER IN BACTERIORHODOPSIN (bR)

M. Engelhard[o], B. Hess[o], K. Gerwert", K.D. Kohl[o],
W. Mäntele" and F. Siebert"

[o]Max-Planck-Institut für Ernährungsphysiologie,
Rheinlanddamm 201, 4600 Dortmund 1, FRG
"Institut für Biophysik und Strahlenbiochemie der Univer-
sität Freiburg, Albertstraße 23, 7800 Freiburg, FRG

The molecular events during the photocycle of bacteriorhodopsin (1) have been studied with the method of time-
resolved and static infrared difference spectroscopy. Characteristic spectral changes involving the C=O stretching
vibration of protonated carboxylic acids were detected (2).
To identify the corresponding groups with either glutamic
or aspartic acid, bR was selectively labelled with (4-^{13}C)-
aspartic acid. An incorporation of ca. 70% was obtained.
The comparison of the difference spectra in the region of
the C=O-stretching vibrations of labelled and unlabelled
bR indicates that ionized aspartic acids are influenced
during the photocycle, the earliest effect being observed
already at the K-intermediate.

So far we have identified one glutamic acid and four
aspartic acids from the hydrophobic core of the protein.

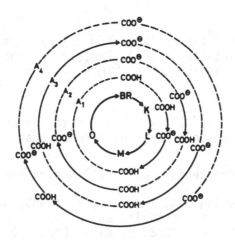

Fig. 1: Photo- and Carboxylic Acid Cycle

Glutamic acid, remaining protonated, is perturbated at the K and M intermediates. Evidence has been provided, that one aspartic acid (A_1) is deprotonated and reprotonated in correlation with the formation of L and M respectively. In the same time-scale a second aspartic acid (A_2) is protonated but it is only deprotonated during the M-L transition. The other two aspartic acids (A_3 and A_4) are protonated later during the photocycle. The protonation of A_3 is correlated with the formation of M whereas A_4 lags behind A_3. Both amino acids are again deprotonated in bR.

Because of the position of the IR-modes, the lack of oxygen-exchange with $H_2^{18}O$ and the insensitivity to the external pH, we concluded that these carboxylic groups do not form hydrogen bonds. Therefore, mechanisms for the proton-pump which are based on a hydrogen-bonded network in a hydrophobic environment can be excluded. In addition, our results do not give evidence for a negative charge near the ß-ionone ring (3).

The observed effects on the carboxylic groups can only be explained if they are located in the hydrophobic core of the bR-molecule in the neighbourhood of the Schiffs base end of the retinal since most of the local charge movements occur in this region. This provides considerable constraints on the models proposed by Trewhella et al. (4). In addition, the results described here, are consistent with a model of the proton-transfer in which two protons are pumped through the membrane.

1. Hess, B., Kuschmitz, D. and Engelhard, M. (1982) in Membranes and Transport (Martonosi, A.N., ed.) vol. 2, pp. 309-318, Plenum Publ. Corp. New York.
2. Siebert, F., Mäntele, W. and Kreutz, W. (1982) FEBS Lett. 141, 82-87.
3. Nakanishi, K., Balogh-Nair, V., Arnaboldi, M., Tsajimoto, K. and Honig, B. (1980) J. Am. Chem. Soc. 102, 7947-7949.
4. Trewhella, J., Anderson, S., Fox, R., Gogol, E., Khan, S., Engelman, D. and Zaccai, G. (1983) Biophys. J. 42, 233-241.

ON THE ROLE OF THE 16, 24 AND 33 kDa POLYPEPTIDES IN PHOTOSYNTHETIC OXYGEN EVOLUTION

Lars-Gunnar Franzén, Lars-Erik Andréasson and Örjan Hansson
Department of Biochemistry and Biophysics, Chalmers Institute of
Technology and University of Göteborg, S-412 96 Göteborg, Sweden

Deoxycholate was used to solubilise the 16 and 24 kDa proteins from spinach thylakoids, resulting in the loss of oxygen evolution. Manganese was retained in the membrane. When the extracted membranes were subjected to a mild heat treatment, the water-soluble 33 kDa protein was selectively released (Table 1). Less than one manganese per reaction center was lost on heating but this loss was not correlated to the solubilisation of protein. Most of the membrane-bound manganese remained EPR-undetectable and could be released by Tris or hydroxylamine. This indicates that the manganese involved in oxygen evolution remains in its native binding site despite the loss of the 33 kDa protein. These results contradict the hypothesis that the 33 kDa protein is responsible for manganese binding at the photosynthetic oxygen-evolving site.

TABLE 1. Solubilisation of manganese and 33 kDa protein from deoxycholate-extracted thylakoids

Method	Extracted Mn/400 chl	Bound Mn/400 chl	Extracted protein, %
0.8 M Tris, pH 8.4	3.2	1.1	82
20 mM NH_2OH	1.4	3.2	0
Heat treatment, 55°C, 3 min	0.9	3.2	72
Tris after heat treatment	2.3	1.1	26
NH_2OH after heat treatment	1.0	2.1	13

To investigate further the role of the 16, 24 and 33 kDa proteins in photosynthetic oxygen evolution, the effect of specific release of these proteins from oxygen-evolving PS II particles was studied. 1 M NaCl selectively liberates the 16 and 24 kDa proteins and inhibits oxygen evolution (Refs. 1,2; Table 2), whereas 1 M $MgCl_2$ solubilises all three proteins (Ref. 3; Table 2). The effect of salts is at or near the water-splitting site as the electron transport from added diphenylcarbazide is not inhibited while the multiline EPR signal from the S_2 state is abolished (Table 2). There is no significant mobilisation of manganese in either case (Refs. 2,3; Table 2),

TABLE 2. Effect of NaCl and MgCl$_2$ on PS II particles

	Activity (%)		Bound protein (%)			Bound Mn /225 chl	Multiline EPR signal (%)	SII$_f$ (%)
	H$_2$O→PPBQ	DPC→DPIP	16	24	33			
Control	100[a]	ND[b]	100	100	100	4.2	100	10
+Tris	0	100	0	0	20	ND	0	100
1 M NaCl	9	100	0	0	100	3.8	5	15
+Tris	0	ND	ND	ND	ND	ND	ND	100
1 M MgCl$_2$	0	100	0	0	0	3.6	5	90
+Tris	0	ND	ND	ND	ND	ND	ND	100

[a]100%=440 μmol O$_2$/mg chl per h   [b]ND=not determined

in contrast to treatment with Tris, which in addition to the three proteins, also releases manganese (2). Tris also prevents rapid reduction of Z$^+$, the secondary PS II donor, which is now observed in EPR as Signal II$_f$ (t$_{\frac{1}{2}}$=0.5 s). After NaCl washing Signal II$_f$ was not observed. This is not the result of a block in the electron transfer between Z and the primary donor, P$_{680}$, since Signal II$_f$ was seen when the NaCl treatment was followed by Tris washing. Thus, while loss of the 16 and 24 kDa proteins prevents cyclic turnover and oxygen evolution, the oxygen-evolving site is still capable of limited rapid electron transfer (<1 ms) to Z$^+$. However, the additional loss of the 33 kDa protein after MgCl$_2$ treatment completely disconnects the water-splitting site with the appearance of Signal II$_f$ as a result (Table 2).

References
1. Åkerlund, H.-E., Jansson, C. and Andersson, B. (1982) Biochim. Biophys. Acta 681, 1-10
2. Kuwabara, T. and Murata, N. (1983) Plant and Cell Physiol. 24, 741-747
3. Ono, T.-A. and Inoue, Y. FEBS Lett. 164, 255-260

LIGHT HARVESTING PIGMENTS AND REACTION CENTER IN CHLOROBIUM LIMICOLA.

P.D. Gerola and J.M. Olson
Institute of Biochemistry, Odense University
Campusvej 55, DK 5230 Odense M., Denmark

In the green bacteria the photosynthetic reaction centers are localized in the cytoplasmic membrane, while the light-harvesting pigment bacteriochlorophyll (Bchl) c is organized in oblong bodies (chlorosomes) present in the cytoplasm close to the inside surface of the membrane. Between the clorosome and the membrane is a baseplate composed of water-soluble Bchl a--proteins to which the chlorosome is probably attached.

By French press treatment and sucrose gradient centrifugation it is possible to separate a fraction enriched in chlorosomes and one enriched in membrane fragments (RCM-I). By treating RCM-I with guanidine HCl it is also possible to solubilize the water-soluble Bchl a--protein (1). We can demonstrate that treatment with high pH or chaotropic agents can solubilize the Bchl a--protein, thereby indicating the involvement of hydrophobic forces and/or H-bonds in binding the protein to the cytoplasmic membrane. The photochemical activity of the membrane fragments depleted of the water-soluble Bchl a--protein (RCM-II) has also been studied.

In the subunit model of the chlorosome rod-element proposed by Olson (2), there are 12-14 Bchl c molecules per subunit. From the circular dichroism (CD) spectra of chlorosome fractions, it is clear that the Bchl c is highly aggregated with very strong exciton interactions giving rise to strong, conservative CD signals. The position of the Q_y-band varies between 730 and 752 nm in various chlorosome fractions, and two types of chlorosomes are defined based on CD spectra. The first type (730 nm $< \lambda(Q_y) <$ 748 nm) was characterized by a positive CD peak on the red side of the Q_y absorption band and a negative peak on the blue side. The second type ($\lambda(Q_y)$ = 752 nm) was characterized by a negative CD peak on the red side, a positive peak on the blue side, and a third (minor) negative peak on the blue side of the positive peak. This second chlorosome CD spectrum is strikingly similar to that of Bchl c dissolved in CH_2Cl_2/hexane (1-2/200) which Smith et al. (3) demonstrated to have an absorption spectrum similar to that for chlorosomes.

Although the CD signals associated with chlorosomes are stronger than those associated with Bchl $\underline{c}$ in CH_2Cl_2/hexane the proposed Bchl $\underline{c}$ oligomers in CH_2Cl_2/hexane are attractive models for the organization of Bchl $\underline{c}$ in chlorosomes.

1. Olson, J.M., Giddings, T.H. and Shaw, E.K. (1976) Biochim. Biophys. Acta 449, 197-208.

2. Olson, J.M. (1980) Biochim. Biophys. Acta 594, 33-51.

3. Smith, K.M., Kehres, L.A. and Fajer, J. (1983) J. Am. Chem. Soc. 105, 1387-1389.

PROPERTIES OF A NADH-PLASTOQUINONE OXIDOREDUCTASE BOUND
TO THE PHOTOSYNTHETIC MEMBRANES OF <u>CHLAMYDOMONAS</u>
<u>REINHARDII</u> CW-15

D. Godde
Lehrstuhl für Biochemie der Pflanzen
Ruhr-Universität Bochum
4630 Bochum, West-Germany

Green algae show a number of photosynthetic reactions
unknown to higher plants. One of these reactions is the
light dependent, DCMU insensitive oxidation of NADH and
NADPH by thylakoid preparations of Chlamydomonas reinhardii. The electrone transport system from NAD(P)H to
photosystem I includes plastoquinone, the cytochrome
b_6/f complex and plastocyanin. It is sensitive against
high concentrations of rotenone, an inhibitor of the
mitochondrial NADH-ubiquinone oxidoreductase (1). This
light dependent oxidation of NAD(P)H via plastoquinone
suggests a NAD(P)H-plastoquinone oxidoreductase bound
to the thylakoid membranes of Chlamydomonas.
Indeed, highly purified thylakoid membranes of Chlamydomonas show high NAD(P)H oxidizing activity with plasto-
quinone-1 as electrone acceptor in the dark. This
activity could be solubilized from the membrane by the
non ionic detergents Triton X-100 (2) and more effectively by lauryl maltoside. The NAD(P)H-plastoquinone oxidoreductase activity was further purified protaminsulfate
fractionation and gradient centrifugation. The spectral
properties of the enriched enzyme indicated the
presence of flavogroups and iron sulfur centers.
As electron acceptors the NAD(P)H-plastoquinone oxidoreductase prefered analogs of plastoquinone-9 like
plastoquinone-1 and DBMIB, known as reducable inhibitor
of the cytochrome b_6/f complex. Ubiquinone-1, a good
acceptor of the mitochondrial NADH-dehydrogenase, is
only reduced with low affinity. The preference of plastoquinone should be expected from an enzyme incorporated
into a photosynthetic membrane. Like the electron
transport from NADH to photosystem I the NADH-plastoquinone oxidoreductase activity is sensitive to relatively high concentrations of rotenone.
To get more information about the molecular structure of
the enzyme photoaffinity labelling was performed.
Using the plastoquinone analog azido-plastoquinone a
polypeptide with a molecular weight of about 70 kD was

intensively labeled. This peptide was also enriched
during the purification method. This might suggest that
this polypeptide is involved in the binding and
reduction of plastoquinone.
The presence of a "respiratory like" NAD(P)H-plasto-
quinone-oxidoreductase in the photosynthetic membranes
of the eukaryotic green alga Chlamydomonas reinhardii
points to similarities with photosynthetic bacteria
and cyanobacteria. In these organisms photosynthetic
and respiratory electron transport comopents are located
in the same membrane.

1. D. Godde, A. Trebst, Arch. Microbiol., 127, 245-252,
   1980

2. D. Godde, Arch. Microbiol., 131, 197-202, 1982

TIME-RESOLVED INCREASE OF THE TRANSMEMBRANE ELECTRIC POTENTIAL IN
HALOBACTERIUM HALOBIUM CELL ENVELOPE VESICLES AND THERMODYNAMIC
MODELING OF THE ASSOCIATED ION FLUXES

S. L. Helgerson[o], M. K. Mathew[o], W. Stoeckenius[o], and E. Heinz"
[o]Department of Biophysics and Biochemistry and Cardiovascular
Research Institute, University of California
San Francisco, CA 94143, U.S.A.
"Department of Physiology, Cornell University Medical College,
New York, NY 10021, U.S.A.

Bacteriorhodopsin (bR) functions as an electrogenic, light-
driven proton pump in H. halobium to generate a protonmotive
force. Previously, the coupling between the bR photocycle and the
transmembrane electric potential ($\Delta\Psi$) component was investigated
in intact cells [1] and cell envelope vesicles [2]. Cell envelope
vesicles were prepared by sonication and then purified by Ficoll-
400/NaCl/CsCl density gradient centrifugation. In agreement with
earlier work, the vesicles developed an intensity-dependent steady-
state $\Delta\Psi$ of 0 to -100mV when continuously illuminated (590±50nm, 0
to 25mW/cm^2). The membrane potential was measured by flow
dialysis of ^{3}H-TPMP$^+$ uptake. Both $\Delta\Psi$ and any actinic illumination-
dependent changes in the bR photocycle reported here can be abolished by the uncoupler carbonylcyanide-m-chlorophenylhydrazone.
In energetically closed vesicles, the decay kinetics of the laser
flash (592.5nm, 7nsec) induced absorbance transients measured at
418nm were inhibited by increased actinic illumination intensities. The decay kinetics were fitted with two exponential decays
having initial amplitudes of M_1 and M_2 and rate constants k_1 and
k_2. Under steady-state actinic illumination, a calibration curve
was obtained establishing the linear dependence of the reciprocal
initial M decay rate, $(M_1k_1+M_2k_2)^{-1}$, on $\Delta\Psi$. Based on this empiri-
cal calibration, the development of $\Delta\Psi$ was resolved with a 2msec
resolution by monitoring M decay rates at varying times after initiating background illumination. The vesicles were loaded with
and suspended in either 3M NaCl or 3M KCl buffered with 50mM HEPES
pH 7.5 and the membrane permeability to protons modified by pretreatment with dicyclohexylcarbodiimide. In each case $\Delta\Psi$ rose
with a halftime of ~75msec. The steady-state $\Delta\Psi$ achieved depended
on the cation present and the H$^+$ permeability, i.e., higher potentials were developed in DCCD-treated vesicles or in NaCl media as
compared to KCl media [3].

Previously, the kinetic equations for the development of a protonmotive force across a membrane containing an electrogenic

proton pump were derived based on the thermodynamics of irreversible processes [4]. This treatment predicted that initiation of proton pumping should give a rapid increase (<100msec) in $\Delta\Psi$ to a value determined by the passive cation permeabilities. A constant driving force (A_{ch}) was assumed such that the rate of the pumping reaction cycle (J_r) could be variable. The basic equations have been modified to account for physiological ion fluxes mediated by the voltage-gated, electrogenic Na^+/H^+ antiporter [5,6] and the DCCD-inhibitable F_O proton-pores of the inactive H^+-ATPase complexes [7] present in the vesicle membrane. The time-resolved membrane potentials have been fitted with these modified equations. Estimates for the linear phenomenological coefficients describing the overall proton pumping cycle ($L_r = 3.5E-11$ $mole^2$/joule·g·sec), passive cation permeabilities ($L_H = 2E-10$, $L_K = 2.2E-10$, $L_{Na} = 1E-11$), and the Na^+/H^+ exchange via the antiporter ($L_{ex} = 1.5E-10$) have been obtained.

(Supported by NIH Program Project Grant GM-27057, NIH Grant GM-26554 and NASA Grant NSG-7151.)

References:

1. Dancshazy, Zs., Helgerson, S.L. and Stoeckenius, W. (1983) Photobiochem. Photobiophys. 5, 347-357
2. Groma, G.I., Helgerson, S.L., Wolber, P.K., Beece, D., Dancshazy, Zs., Keszthelyi, L. and Stoeckenius, W. (1984) Biophys. J. in press
3. Lanyi, J.K., Helgerson, S.L. and Silverman, M.P. (1979) Arch. Biochem. Biophys. 193, 329-339
4. Heinz, E. (1980) In Hydrogen Ion Transport in Epithelia (Schultz, I. et al., eds.) pp. 41-45, Elsevier/North-Holland Biomedical Press, Amsterdam
5. Lanyi, J.K. and Silverman, M.P. (1979) J. Biol. Chem. 254, 4750-4755
6. Cooper, S., Michaeli, I. and Caplan, S.R. (1983) Biochim. Biophys. Acta 736, 11-27
7. Michel, H. and Oesterhelt, D. (1980) Biochemistry 19, 4607-4614

DIELECTRIC SPECTROSCOPY OF THE ROTATIONAL AND TRANSLATIONAL MOTIONS OF MEMBRANE PROTEINS: THEORY AND EXPERIMENT

Douglas B.Kell

Department of Botany & Microbiology, University College of Wales, ABERYSTWYTH, Dyfed SY23 3DA, U.K.

Within the framework of so-called localised theories [1] of energy coupling in electron transport phosphorylation, a knowledge of the dynamic, topological disposition of membrane protein complexes assumes importance [2,3].

In principle [4], dielectric spectroscopy (see e.g. [5]) should provide a powerful tool for assessing the electrical organisation and ion conducting pathways of energy coupling membrane vesicle suspensions. We have therefore developed [6] a computerised, frequency-domain dielectric spectrometer, covering the range 5Hz-13MHz, and have observed, for the first time, dielectric dispersions apparently corresponding to the rotational and translational motions of the protein complexes in Rhodopseudomonas capsulata chromatophores [7,8].

Calculations of the (dielectric) relaxation time for rotational motion of charged membrane proteins of typical size indicate that, for membrane (micro-)viscosities in the range 1-10 poise (0.1-1 Pa s), a dielectric dispersion should indeed be observable with a characteristic frequency in the range 1-10 kHz.

If translational (2-dimensional) diffusion is unrestricted, as in the "fluid mosaic" model (see [9]), the characteristic frequency for relaxation due to translational motion (which scales as the inverse square of the vesicle radius) should lie in the range 0.5-5 Hz for a 'spherical' microbial cell of radius 0.5 μm. In contrast, if lateral diffusion is restricted, say by "long-range" protein-protein interactions, f_c values may be as great as 1-10 MHz, depending on the distance diffused before, on average, a 'barrier' is encountered.

The effects of cross-linking reagents (dimethyl suberimidate and glutaraldehyde) on chromatophores [8] and on cells and protoplasts of P. denitrificans (in preparation) indicate that whilst rotation is limited essentially by frictional (hydrodynamic) forces alone, translational diffusion in these systems is somewhat more restricted. This conclusion contrasts with certain measurements of "lateral electrophoresis" carried out at higher field strengths than those used herein ($\leq$0.3V/cm) [10,11].

Dielectric spectroscopy constitutes a wholly non-invasive approach to such studies. Such measurements might also be used to optimise, and further to understand the mechanisms underlying, electric field-mediated cell fusion [12].

This work was supported by the SERC, U.K.

1) Westerhoff,H.V. et al (1984) FEBS Lett. 165,1-5
2) Kell,D.B. (1984) Trends Biochem.Sci. 9,86-88
3) Kell,D.B. & Westerhoff,H.V. (1984) in Catalytic Facili-
   tation in Organised Multienzyme Systems (Welch,G.R.,ed.),
   in press, Academic, New York.
4) Kell,D.B. & Hitchens,G.D. (1983) in Coherent Excitations
   in Biological Systems (Fröhlich,H. & Kremer,F., eds) pp
   178-198. Springer-Verlag, Heidelberg.
5) Pethig,R. (1979) Dielectric and Electronic Properties of
   Biological Materials. Wiley, Chichester.
6) Harris,C.M. & Kell,D.B. (1983) Bioelectrochem.Bioenerg.
   11,15-28
7) Harris,C.M.,Hitchens,G.D. & Kell,D.B. (1984) in Charge &
   Field Effects in Biosystems (Allen,M.J. & Usherwood, P.N.
   R.,eds). Abacus Press, Tunbridge Wells.
8) Kell,D.B. (1984) Bioelectrochem.Bioenerg., 11, in press.
9) Saffman,P.G. & Delbrück,M. (1975) PNAS 72, 3111-3113
10) Poo,M-m (1981) Ann.Rev.Biophys.Bioeng. 10, 245-276
11) Sowers,A.E. & Hackenbrock,C.R. (1981) PNAS 78,6246-6250
12) Zimmermann,U. (1982) Biochim.Biophys.Acta 694,227-277

SPECIFIC REQUIREMENT OF CATIONS FOR THE PHOTOCYCLE OF BACTERIORHODOPSIN (BR)

K.-D. Kohl, M. Engelhard and B. Hess[o]
[o]Max-Planck-Institut für Ernährungsphysiologie, Rheinlanddamm 201, 4600 Dortmund 1, FRG

The reversible transition from the purple membrane (BR_{trans}^{568}) to the blue membrane (BR_x^{605}) (1,2,3) can be understood as competitive binding of protons and cations at two different sites at the protein. Binding curves obtained by spectroscopy indicate a two step protonation of BR from the purple form BR_{trans}^{568} via an intermediate state BR_x^{568} (λ_{max} = 568 nm; $\varepsilon_{BR}^{568} \approx$ 54.000 l $mol^{-1}cm^{-1}$) to the blue form BR_x^{605} (λ_{max} = 605 nm; $\varepsilon_{BR}^{605} \approx$ 52.000 l $mol^{-1}cm^{-1}$).

An overall binding of $8H^+$ per BR molecule was determined from titration curves. Only one H^+ is required for the bathochromic shift of the absorption maximum. The blue form BR_x^{605} is stable even at pH = 5.4 but is extremely sensitive to cationic contamination confirming the results of (4). We found that BR_x^{605} has its own photocycle that differs from both the native BR and the acid BR^{605} (5). In the BR_x^{605} photocycle, at least, two slower intermediates are found. Low temperature spectroscopy indicates differences in the photochemistry of the primary reaction step.

The regeneration reaction from BR_x^{605} with cations restores the spectrum and the photocycle of the protein in its BR_x^{568} intermediate form. The original state of native BR can be obtained by addition of a small amount of a base (i.e. NaOH). The shape of the absorption spectrum and the kinetics of the photocycle are identical with that of native BR.

In the investigation of the following cations Li^+, K^+, Na^+, Cs^+, Mg^{2+}, Ca^{2+}, Mn^{2+}, Cu^{2+}, Sr^{2+}, Fe^{2+}, Fe^{3+} equilibrium constants were determined for monovalent ($K \approx 3 \times 10^{-3}M$) for divalent ($K \approx 6 \times 10^{-5}$ M) and trivalent ($K \approx 8 \times 10^{-6}M$) cations. A strong influence of ferric ions on the function of BR is observed. No specificity for the binding of cations was detected. Sigmoidal binding curves for Na^+ and all divalent cations might be due to specific and unspecific binding sites at the protein. Because of a single isosbestic point at around 578 nm for all cationic titrations a single reaction for the transition $BR_x^{605} \rightarrow BR_x^{568}$ is suggested. Again only one cation is bound per protein molecule to obtain the configuration of the purple BR; the 1:1 stoichiometry is confirmed by atomic absorption spectroscopy. Thus, we conclude that the purple state of the membrane requires one cation per molecule.

Whereas kinetics of the photocycle are not influenced very strongly by different cations bound to the protein the light/dark adaptation is remarkably affected by the nature of the cations.

The labelling of the cationic binding site with suitable ions is a useful tool to identify the site within the BR-lattice. Furthermore, this site might be involved in the proton pump mechanism, so the restoration of the natural function is an important prerequisite.

Evidence is provided that during the photocycle affinity changes of the cation-binding site occur. This was demonstrated by the light-induced transient Ca^{2+}-release from $Ca^{2+}-BR^{568}$ detected with the Ca^{2+}-specific indicator Arsenazo III.

(1) Oesterhelt, D., Stoeckenius, W. (1971) Nature, 233, 149-152.
(2) Moore, T.A., Edgerton, M.E., Parr, G., Greenwood, C., Perham, R.N. (1978) Biochem. J., 171, 469-476.
(3) Muccio, D.D., Cassim, J.Y. (1979) J. Mol. Biol., 135, 595-609.
(4) Kimura, Y., Ikegami, A., Stoeckenius, W., Photochem. Photobiol. in press.
(5) Mowery, P.C., Lozier, R.H., Chae, Q., Tseng, Y.-W., Stoeckenius, W. (1979) Biochem. 18, 4100-4107.

ORIENTATION AND SOME PROPERTIES OF HYDROGENASES IN PHOTO-
SYNTHETIC BACTERIA

K.L. Kovács and Cs. Bagyinka
Institute of Biophysics, Biological Research Center
Hung. Acad. Sci. Szeged POB 521. Hungary

Hydrogenase (EC class 1.12) catalyses the reversible reaction H_2 $2H^+ + 2e^-$ [1]. Both the forward and backward reaction can be used for assaying enzyme activity. Methyl viologen and benzyl viologen are suitable redox dyes which
have unique membrane permeability properties. It was shown
[2] that unlike the oxidized forms the reduced dyes can
easily penetrate the plasma membrane and this property can
be utilized for the determination of enzyme distribution
and orientation. Various strains each showing H_2 evolution
activity, including *Thiocapsa roseopersicina BBS, Chromatium minutissimum IC, Ectothiorhodospira shaposhnikovii IK,
Rhodospirillum rubrum I, Rhodopseudomonas viridis IV,
Rhodopseudomonas capsulata B-10, Chlorobium limicola forma
thiosulfatophilum L.* were compared.
Protoplast prepared from the cells exhibited unchanged activity. After disrupting the protoplasts by osmotic shock
the activity essentially remained in the membrane fraction
in all strains but *C. limicola* which displayed higher activity in the supernatant fraction. Hydrogen uptake measurements were done in two ways. First, we have measured
the uptake activity of whole cells with electron acceptor
outside the cell. Second, the cells were loaded with electron acceptor utilizing the permeability difference of the
oxidized and reduced forms. Hydrogenase activity was found
in all strains with the electron acceptor outside, but

C. limicola which exhibited activity in both experiments.
We conclude that the hydrogenase of the purple bacteria
Chromatiaceae and *Rhodospirillaceae* is a membrane bound
enzyme with its active center oriented towards the outer
surface of the plasma membrane. Accordingly, electrons are
released to the periplasmic space. The green photosynthetic
bacterium *C. limicola*, on the other hand, seems to have
both membrane-bound and soluble activity.
In light of recent results concerning the proton release
to the cytoplasmic side of the plasma membrane by hydrogenase [3] it is reasonable to assume that the enzyme,
while functioning in uptake direction in vivo, must build
up a concomitant membrane potential. Our membrane potential
measurements which showed a change during hydrogenase action corroborate with this hypothesis. By parallel dialysis
and fluorescence measurements an acidification of the
cytoplasm and an increase in the membrane potential was
observed.

References:

1 Adams,M.W.W., Mortenson,L.E. and Chen, J.S. /1981/
  Biochim. Biophys. Acta 594 105-176
2 Bagyinka,Cs., Kovács, K.L. and Rak, E. /1982/
  Biochemical Journal 202 255-258
3 Vignais,P.M., Henry,M.F., Sim,E. and Kell,D.B. /1981/
  Current Topics in Bioenergetics 12 115-196

IMPORTANCE OF MEMBRANE-INTERFACE PROTONS FOR INITIAL ENERGY TRANSDUC-
TION — FURTHER EVIDENCE FROM ATPase PROTEOLIPOSOMES AND CHLOROPLASTS

R. Kraayenhof, H.S. van Walraven, and J.M.G. Torres-Pereira
Biological Laboratory, Vrije Universiteit, De Boelelaan 1087,
1081 HV Amsterdam, The Netherlands

Previous experiments with different fluorophores, covalently bound to
the thylakoid membrane and the protruding α and β subunits of its ATPase
complex, and with particle electrophoresis demonstrated the energy-linked reorientation of protolytic groups (COO^- and NH_3^+, resp.) and this
strongly suggests the involvement of short-range proton displacements at
the membrane-medium interface (1,2). Here, we present further evidence
for proton and charge rearrangements upon energization by ATP hydrolysis
in (I) proteoliposomes prepared from the ATPase complex and native lipid
of the thermophilic cyanobacterium *Synechococcus* 6716 (3), and (II)
unstacked chloroplast thylakoid membranes.

(I) *ATPase proteoliposomes*. ATP-induced inward proton translocation
was studied with the pH indicators neutral red (NR^+) and cresol red
(CR^-), externally added or trapped inside the vesicles, at different
$MgCl_2$, buffer and ionophore concentrations, and at different pH_{out} and
pH_{in}. The generated electric potential was monitored with the native
carotenoids and added oxonol VI (Ox.VI). Net charge changes at the external interface were followed with a membrane-bound aminoacridine probe
(AA^+). Table I illustrates the experimental system and lists the approximate halftimes of the different pH and $\Delta\Psi$ indicators. Upon addition of
ATP, the first event is the generation of an (intra)membrane potential
($\Delta\Psi_m$). Then alkaliniza-
tion of the outer in-
terfacial layer occurs,
followed by acidifica-
tion of the inner inter-
face. Outer bulk alkali-
nization and inner bulk
acidification occur at
a considerably slower
rate. Mg^{2+} induces NR^+
repulsion and CR^- attraction with respect to
the interfaces, with
concomitant slower and
faster responses, res-
pectively. Valinomycin
enhances the pH changes
but interface and bulk
rates remain different.

Table I. ATP-induced sequential proton dislo-
cations by ATPase proteoliposomes. Reaction
conditions, *cf.*(3); ★, including 20-30 s lag.

reaction system	parameters measured	probes used	ATP-induced response halftimes (s)		
			control	+10mM MgCl$_2$	+25nM Val.
H_{ob}^+		CR^-	180	100	40
H_{oi}^+		NR^+	25	65	12
$\Delta\Psi_m$		Ox.VI Carot.	5	–	–
H_{ii}^+		NR^+	35	80*	20
H_{ib}^+		CR^-	200*	150*	100*

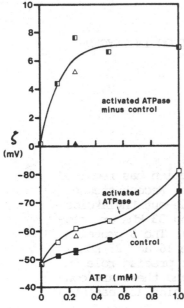

Fig. 1. ATP-dependent increase of electrokinetic potential (ζ) of unstacked thylakoid
membranes. Free-flow electrophoresis (2) in
the dark; 0.25 mM $MgCl_2$ present; $\triangle$,$\blacktriangle$: activated ATPase *plus* 5 and 20 µM DCCD, respec-
tively; no DCCD effects on control.

AA^+ response has a halftime of 5 s and is enhanced by valinomycin (nigericin sensitive).
See (3) and a forthcoming paper for further
details. Thus, even in this relatively slow
energy-generating system interfacial and bulk
gradients remain clearly distinguishable. In
photosynthetic and respiratory electron flow
the kinetic importance of interface protons
for initial energy transduction on the ms
time scale is even more significant.

(II) *Chloroplast thylakoids*. Light-induced
electron flow in broken chloroplasts and subchloroplast vesicles induces electrogenic H^+
displacement, $\Delta\Psi_m$ generation and an increase of electrokinetic potential
(2,4). The latter expresses external net surface charge density (protolytic groups). Fig. 1 shows that in the dark ATP also induces a detectable increase of ζ that is dependent on pre-activation of the ATPase and
DCCD sensitive. These experiments require some Mg^{2+}; moreover, they report on potentials at the hydrodynamic plane of shear, so that the actual surface potentials are higher. Under the same conditions the 'slow'
(ms) carotenoid (or Ox.VI) and AA^+ responses show a similar ATP-dependent generation of $\Delta\Psi_m$ and surface potential as the proteoliposome system. Since the ATP-dependent effects are complimentary to those induced
by electron flow, we consider the observed electric potentials as identical in both forward and backward energy-transducing reactions.

References

(1) Kraayenhof, R. (1977) in Structure and Function of Energy-Transdu-
    cing Membranes (van Dam, K. and van Gelder, B.F., eds.) pp. 223-236,
    Elsevier/North-Holland Biomedical Press, Amsterdam
(2) Schapendonk, A.H.C.M., Hemrika-Wagner, A.M., Theuvenet, A.P.R.,
    Wong Fong Sang, H.W., Vredenberg, W.J. and Kraayenhof, R. (1980)
    Biochemistry 19, 1922-1927
(3) van Walraven, H.S., Lubberding, H.J., Marvin, H.J.P. and Kraayenhof,
    R. (1983) Eur. J. Biochem. 137, 101-106
(4) Schuurmans, J.J., Peters, A.L.J., Leeuwerik, F.J. and Kraayenhof, R
    (1981) in Vectorial Reactions in Electron and Ion Transport in Mito-
    chondria and Bacteria (Palmieri, F. et al., eds.) pp. 359-369,
    Elsevier/North-Holland Biomedical Press, Amsterdam

POLYPEPTIDE RELEASE AND MANGANESE COMPARTMENTATION IN OXYGEN-
EVOLVING PHOTOSYSTEM II PREPARATIONS.

M. Miller and R.P. Cox
Institute of Biochemistry,
Odense University
Campusvej 55, 5230 Odense M
Denmark

The mechanism of photosynthetic oxygen evolution has remained
unresolved despite considerable research effort. Recent studies
on oxygen-evolving PS II preparations and inside-out thylakoids
have revealed that 3 polypeptides of 16, 23, and 33 kDa are closely related to the water-splitting process (1). The 3 proteins
are located at the inner thylakoid surface, the 16 and 23 kDa
apparently being bound electrostatically. Their precise role in
the water-splitting process is still unclear, and there are conflicting reports whether the 16 and 23 kDa proteins are essential
for the process (2-5).

Manganese is known to be essential for photosynthetic oxygen-
evolution and probably functions in the electron transfer process
between water and the PS II reaction center. Under certain circumstances both Mn^{2+} (6), and also Cl^- (7) and H^+ (8), all ions
associated with the water splitting process, can be trapped in
aqueous compartments more restricted than the thylakoid lumen.
From studies on compartmentation in non-vesicular membrane preparations we suggested that the various proteins of the PS II
complex do not form a compact unit, but are arranged around a
central aqueous domain into which Mn^{2+} is liberated (9).

The aim of the present study was to investigate whether the
16, 23, and 33 kDa proteins may be involved in formation of such
a compartment. The experimental material was an oxygen-evolving
PS II preparation obtained by Triton X-100 treatment of pea thylakoids in the presence of $MgCl_2$. The polypeptide composition of
the membrane fragments was measured by SDS-PAGE following inhibition of the oxygen-evolution by a number of different treatments. The inhibitory treatments included both those where endogenous manganese is converted to EPR-detectable Mn^{2+} but retained
within the membrane fragments (e.g. incubation with Zn^{2+}), and
those causing an immediate release of Mn^{2+} into the surrounding
medium (e.g. heat treatment).

The relationship between polypeptide release and the location of chloroplast manganese will be discussed.

REFERENCES:

1. Barber, J. (1984) Trends Biochem. Sci. 99, 79-80.
2. Nakatani, H.Y. (1984) Biochem. Biophys. Res. Commun. 120, 299-304.
3. Miyao, M. and Murata, N. (1984) Febs Lett. 168, 118-120.
4. Ono, T. and Inoue, Y. (1984) Febs Lett. 168, 281-286.
5. Åkerlund, H.-E. et al (1984) Biochim. Biophys. Acta 765, 1-6.
6. Miller, M. and Cox, R.P. (1983) Febs Lett. 155, 331-333.
7. Johnson, J.D. et al (1983) Biochim. Biophys. Acta 723, 256-265.
8. Laszlo, J.A. et al (1984) J. Bioenerg. Biomembranes 16, 37-51.
9. Miller, M. and Cox, R.P. (1984) Biochem. Biophys. Res. Commun. 119, 168-172.

Acknowledgements

M.M. is supported by the Carlsberg Research Foundation.

BIOPHYSICAL CHARACTERISTICS OF PLASTOQUINONE LATERAL MOTION WITHIN
THE THYLAKOID MEMBRANE.

P.A. Millner and J. Barber.
AFRC Photosynthesis Research Group, Department of Biology,
Imperial College, London, SW7 2BB, U.K.

1. Introduction.

The emerging viewpoint on the location of photosynthetic redox-
complexes within the thylakoid membrane is that photosystem 2 (PS2)
is confined to the appressed (granal) membranes whilst the cytochrome
b_6-f complex is probably located at the margins of the grana or in
the stromal membranes (1). This would imply that electrons have to
be transported laterally some tens or hundreds of nm to link the two
complexes, and the hydrophobic redox component plastoquinone (PQ_A)
has been proposed to fill this role (2). The suitability of PQ_A for
this function is considered.

2. Biochemical/Biophysical Properties of the Lipid Matrix.

Lipids of the thylakoid matrix are extremely unsaturated with 80%
of more of the lipid acyl chains possessing three double bonds (4).
This leads to an extremely fluid membrane. In addition, by use of
spin labelled fatty acid probes a transverse fluidity gradient has
been shown (4), with viscosity decreasing to a minimum value at the
bilayer midplane. This value may be as low as 0.01 P (3), ie equivalent to water at 20°C.

3. Location of Plastoquinone within the Lipid Bilayer.

Due to differences in the headgroup structure and tail length of
PQ_A (C_{36}) and thylakoid lipids, of which monogalactosyldiacylgly-
cerol (MGDG, C_{18}) is the predominant species, it is extremely unlikely that PQ_A is able to pack effectively between the thylakoid lipid
acyl chains. Differential scanning calorimetry, monolayer, fluorescence quenching, NMR and ESR studies support this prediction (see 4)
and indicate that it is located in the midplane region of the thylakoid membrane.

4. Rate of Plastoquinone Lateral Motion.

Previous estimates of the lateral diffusion coefficient (D_L) for
PQ_A assume that its motion occurs analogously to that of a phospholipid (2). However due to the properties of the thylakoid membrane
it is likely that PQ_A can "tunnel" along the bilayer midplane with a
microscopic D_L of $10^{-6} cm^2 s^{-1}$. This value would permit PQ_A to diffuse

laterally around 2800nm in the 20ms that represents the half- time
for cytochrome f re-reduction.

5. Distance Traversed by Plastoquinone.

In a typical granum of 250nm radius, the average PS2 complex will
be about 73nm from the nearest cytochrome b_6-f complex at the granal
margin (Fig.1)

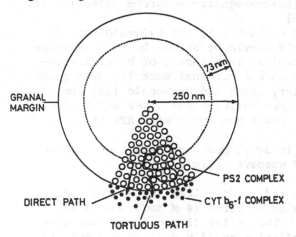

Fig.1 – Representation of a typical granum.

The circle 73nm from the granal circumference divides the granal disc into equal areas and represents the position of an average PS2.

However, this value is the linear distane between PS2 and the b_6-f
complex and the actual distance diffused by a PQ_A molecule may be
much longer due to a tortuous diffusion path being taken between
integral protein complexes (Fig.1). Despite this it is unlidely that
the sub-macroscopic D_L would drop below 10^{-7} cm ^{2}s^{-1}.

6. Conclusion.

The estimated sub-macroscopic D_T value supports the concept that
PQ_A can act as an effective mobile redox component linking PS2 with
the cytochrome b_6-f complex. Although the diffusion time itself may
be rate limiting it is possible also that the exchange reactions
involved in plastoquinol oxidation could control the overall rate of
electron flow from H_2O to NADP.

References.

1. Barber, J. (1983) Plant, Cell and Environ. 6, 311-322.
2. Anderson, J.M. (1980) FEBS Lett. 124, 1-10.
3. Vaz, W.L.C., et al (1982) Biochem. 21, 5608-5612.
4. Millner, P.A. and Barber, J. (1984) FEBS Lett., 169, 1-6.

HALORHODOPSIN-DEPENDENT LIGHT-DRIVEN ATP SYNTHESIS IN HALOBACTERIAL
VESICLES

Y. Mukohata and M. Isoyama
Department of Biology, Faculty of Science, Osaka University,
Toyonaka 560 Japan

Halorhodopsin (1,2) was first described as a retinal protein which
could drive ATP synthesis in *Halobacterium halobium* in light (3,4).
This pigment is distributed widely among halobacterial strains and
under illumination builds up an inside-negative membrane potential (1,
5) by pumping in chloride ions (6).

In bacteriorhodopsin-defective strains (7), the halorhodopsin-dependent photophosphorylation of ADP should be driven by this membrane
potential. The claims (8) of possible involvement of bacteriorhodopsin (9) in the R_1mR strain used in our original work (3) were ruled
out by genetic analysis (7). Later, Clark and MacDonald (10) supported our results with R_1mR envelope vesicles but Weber and Bogomolni
(11) described that halorhodopsin could not synthesize ATP in their
ET-15 cells in light.

Now, a reproducible procedure is developed for preparing envelope
vesicles which can synthesize ATP whenever illuminated.

R_1mR (and L33) cells were cultured as described in (3) for 5 days,
then harvested and washed twice with a buffer (4 M NaCl and 10 mM
PIPES pH 6.8) by centrifugation.   The pellet (2 g protein) was suspended in 100 ml of the "stuffing" solution (3 M NaCl, 0.9 M KCl, 50
mM $MgCl_2$, 10 mM PIPES pH 6.8, 20 mM Pi buffer pH 6.8 and 5 mM ADP),
then sonicated.   The disrupted cells were centrifuged and the pellet
was suspended in the basal stuffing solution (the stuffing solution
without the substrates) and layered on top of 15% Na-tartarate in 4 M
NaCl as in (5).   After centrifugation at 16,000 x g for 45 min at 10°
the cell envelope vesicles were collected from the layer just above
the tartarate layer and washed.   Proteins were determined by the
Lowrey method with BSA as a standard.

Vesicles were suspended in the basal stuffing solution in a glass
vessel kept at 30° and illuminated by yellow light from a 750W slide
projector (10^5 lux). Portions of the vesicle suspension were sampled
at given time intervals and assayed for ATP by a luciferin-luciferase
method.   Membrane potential was estimated with a TPP^+-sensitive electrode (12).

The intravesicular ATP increased as the envelope vesicles were
illuminated in the presence of TPT [triphenyltin, a OH^-/Cl^- exchanger
(13)] which would cancel any pH difference across the vesicle membrane.
In the absence of TPT, ATP was also synthesized in the vesicles which
was strongly buffered at the same pH both sides of the membrane.

Therefore, without any pH gradient, ATP was synthesized in the envelope vesicles only by light energy. The largest amount of ATP was found to be 1,300 pmol/mg protein which corresponds to 0.56 mM. The membrane potential increased from around -50 mV in the dark to more than -150 mV in light. Anaerobiosis in N_2 or addition of 2 mM KCN did not affect the observed ATP level, indicating that the vesicles had lost their respiratory activity.

The membrane potential-driven ATP synthesis was inhibited by DCCD as expected from the earlier results with intact cells (4,14) and vesicles (10). The half-maximum inhibition occurred at around 20 μM. The synthesis was also inhibited completely by SF6847 at 5 μM.

The pH dependence of ATP synthesis in the absence of pH gradient (in the presence of TPT) showed an optimum at 6.5. The activity was lost at pH lower than 5.5 and higher than 7.5, while the membrane potential (steady state) was recorded at -142 ± 5 mV throughout this pH range.

The threshold value for the membrane potential below which no ATP was synthesized, was found to be around -100 mV in the presence of TPT. A little higher value was obtained in the absence of TPT.

This work was supported in part by grants in aid for special project research on bioenergetics #57122006 and 58114006 to Y.M. from the Ministry of Education, Science and Culture of Japan.

1. Mukohata, Y., Matsuno-Yagi, A. and Kaji, Y.(1980)*in* Saline Environment (Morishita and Masui eds.)pp 31-37, BCAS, Japan
2. Mukohata, Y. and Kaji, Y.(1981) Arch.Biochem.Biophys., 206, 72-76
3. Matsuno-Yagi, A. and Mukohata, Y.(1977) Biochem.Biophys.Res.Comm. 78, 237-243   297-303
4. Matsuno-Yagi, A. and Mukohata, Y.(1980) Arch.Biochem.Biophys., 199,
5. Lindley, E.V. and MacDonald, R.E.(1979) Biochem.Biophys.Res.Comm. 88, 491-499
6. Schobert, B. and Lanyi, J.K.(1982) J.Biol.Chem., 257, 10306-10313
7. DasSarma, S., RajBhandary, V. and Khorana, H.D.(1983) Proc.Natl. Acad.Sci.,USA, 80, 2201-2205   587-616
8. Stoeckenius, W. and Bogomolni, R.A.(1982) Ann.Rev.Biochem., 52,
9. Greene, R.V. and Lanyi, J.K.(1979) J.Biol.Chem., 254, 10986-10994
10. Clark, R.D. and MacDonald, R.E.(1981) Biochem.Biophys.Res.Commun. 102, 544-550   -608
11. Weber, J. and Bogomolni, R.A.(1981) Photochem.Photobiol., 33, 601
12. Kamo, N., Muratsugu, M., Hongoh, R. and Kobatake, Y.(1979) J.Membr. Biol., 49, 105-121
13. Selwyn, M.J., Dawson, A.P., Stockdale, M. and Bains, N.(1970) Eur. J. Biochem., 14, 120-126   1234-1238
14. Danon, A. and Stoeckenius, W.(1974) Proc.Natl.Acad.Sci.,USA, 71,

ORGANIZATION OF PHOTOSYSTEM II IN THE ALGA <u>CHLAMYDOMONAS</u> <u>REINHARDTII</u>
AND THE CYANOBACTERIUM <u>APHANOCAPSA</u> 6714

E. Neumann[o], and F. Joset-Espardellier["]

[o]Lehrstuhl Biochemie der Pflanzen, Ruhr-Universität, D-4630 Bochum 1,
 FRG
["]Laboratoire de Photosynthèse, CNRS, F-91190 Gif-sur-Yvette, France

The photosystem II complex of the photosynthetic electron transport
chain in thylakoids contains several proteins. Their number and function is not clear yet. For examination of the acceptor side of photosystem II different methods have been used.

Inhibitors of photosynthetic electron transport have been proven
to be suitable tools for detailed examination of the photosystem II
complex. So far, two different types of inhibitors are known which
both prevent plastoquinone reduction at the acceptor side of photosystem II: "DCMU-type" and phenolic inhibitors (1). For both types of inhibitors, photoaffinity labels are available: azido-atrazine ("DCMU-
type") and azido-dinoseb (phenolic) (2). For our investigations we
have used thylakoids and photosystem II particles (3), which have been
prepared from the cell wall deficient mutant Chlamydomonas reinhardtii
CW15. In thylakoids, azido-atrazine almost exclusively labels a 32-34
kDa protein (D-1 or HBP-protein), whereas azido-dinoseb predominantly
binds to a 51 kDa protein which is part of the photosystem II reaction
center (2). A photosystem II preparation isolated from ^{14}C-acetate
grown algae still contains the D-1 protein, as judged from the high
amount of radioactivity found in the "rapidly turning over" D-1 protein (4). However, electron transport in these photosystem II particles is no longer sensitive toward "DCMU-type" inhibitors, and azido-atrazine does not bind to the D-1 protein (2).

A plastoquinone analogue photoaffinity label (azido-plastoquinone)
(5) in photosystem II particles from Chlamydomonas predominantly binds
to the 51 kDa photosystem II reaction center protein and in addition
to proteins of approximate molecular weights of 47 and 32 kDa (4).

For further characterization of their properties, thylakoids and
photosystem II particles from Chlamydomonas have been subjected to
protease treatment. Changes in gel patterns after protease digestion
and changes in photosynthetic activities in the presence of different
kinds of inhibitors have been correlated.

As already stressed, photosystem II inhibitors can be divided into
"DCMU-type" and phenolic inhibitors. We notice that we also have to
differentiate between compounds of the "DCMU-type" inhibitor class.
This differentiation arose from studies with herbicide resistant
plants and algae. Similarily, from the cyanobacterium Aphanocapsa 6714
two different mutant strains could be isolated. One strain is highly
resistant against DCMU, but still sensitive toward atrazine (6,7).
Contrary, the other strain is resistant against atrazine, but still

sensitive toward DCMU. We have measured photosynthetic electron transport in isolated thylakoids of both mutant strains in the presence of various compounds of the "DCMU-type" and phenolic inhibitor class. Furthermore, we have also studied the binding of various radioactively labeled inhibitors of both classes and of photoaffinity labels as well.

References

1. Oettmeier, W., and Trebst, A. (1983) in "The Oxygen Evolving System of Photosynthesis" (Inoue, Y., et al., eds.), pp. 411-420, Academic Press, Tokyo.
2. Johanningmeier, U., Neumann, E., and Oettmeier, W. (1983) J. Bioenerg. Biomembr. 15, 43-66.
3. Diner, B.A., and Wollman, F.A. (1980) Eur. J. Biochem. 110, 521-526.
4. Neumann, E., Depka, B., and Oettmeier, W. (1984) in "Advances in Photosynthesis Research" (Sybesma, C., ed.), Vol.1, pp. 473-476, Martinus Nijhoff/Dr. W. Junk Publishers, The Hague, Boston, Lancaster.
5. Oettmeier, W., Masson, K., Soll, H.J., Hurt, E., and Hauska, G. (1982) FEBS Lett. 144, 313-317.
6. Astier, C., Vernotte, C., Der-Vartanian, M., and Joset-Espardellier, F. (1979) Plant & Cell Physiol. 20, 1501-1510.
7. Astier, C., and Joset-Espardellier, F. (1981) FEBS Lett. 129, 47-51.

FUNCTIONAL MOLECULAR SIZES OF PHOTOSYNTHETIC COMPLEXES

J H A Nugent[O] and Y E Atkinson"
[O]The Ciba Foundation, 41 Portland Place, London W1N 4BN
"Dept of Botany & Microbiology, University College, London

Radiation inactivation is a proven technique for measuring the functional molecular size of both soluble and membrane bound proteins[1]. The technique allows measurement of membrane proteins in crude preparations, giving the molecular weight of the functional complex rather than that of individual polypeptides.

Information about the functional sizes of photosynthetic reaction centres, oxygen evolution and the cytochrome b_6-f complex would be a valuable addition to the available data on the size of individual polypeptides. This communication will report the preliminary evaluation of the radiation inactivation technique as a method to provide this information.

The method involves selecting a particular activity, e.g. oxygen evolution and using a preparation with a high level of that activity. A series of doses of high energy radiation (16MeV X-rays) are then given to samples and the residual activity measured. A radiation dose versus log activity graph is then constructed which shows the rate of inactivation by the radiation.

A linear response to radiation dose is desirable for molecular size estimations. Assuming a linear response is obtained, the molecular size of the complex can be calculated by calibration using enzymes of known molecular weight.

Experiments are normally performed on freeze-dried samples at room temperature under a partial vacuum. This minimises secondary effects caused by production of free radicals. An alternative method uses frozen samples at cryogenic temperatures. Both methods are needed for photosynthetic complexes as some reactions, e.g. oxygen evolution, are seriously affected by freeze drying.

The initial experiments used an oxygen evolving preparation from the thermophilic cyanobacterium <u>Phormidium laminosum</u>. This preparation has high rates of oxygen evolution ($0.5-1.5$ mmoles O_2mg Chl^{-1}hr^{-1})[2]. Frozen samples were irradiated at 77K. Oxygen evolution was measured using an oxygen electrode and a ferricyanide and 2,6-dimethyl-1,4-benzoquinone mixture as electron acceptors. This activity requires both the oxygen evolving system and the photosystem II reaction centre. A linear dose/activity relationship was obtained. Calibration with yeast alcohol dehydrogenase gave an average functional size of 125Mrad in a range from 99-160kDa. These experiments have been recently reported in [3].

The results so far show that the technique can be applied to photosynthetic complexes. Further experiments are in progress to estimate the sizes of the photosystem I and photosystem II reaction centres. These indicate a functional size below 100 kDa for each reaction centre. Calibration using a wider range of enzymes will reduce the errors on present measurements and allow further discussion concerning which of the already identified polypeptides may be contained in the functional units of photosynthetic complexes.

References

1)   Kempner, E S and Miller, J H (1983) Science 222, 586-589

2)   Stewart, A C and Bendall, D S (1979) FEBS Lett. 107, 308-312

3)   Nugent, J H A and Atkinson, Y E (1984) FEBS Lett. (in press)

THE INHIBITORY ACTION OF ANT 2p* ON PHOTOSYSTEM 2

N.K. Packham and J. Barber
Department of Pure and Applied Biology, Imperial College of Science and Technology, London SW7 2BB, U.K.

In an attempt to identify the mechanism by which ADRY reagents inhibit the oxygen-evolving capacity of chloroplasts, we have studied the effect of ANT 2p on the following photosystem (PS) 2 mediated reactions: photo-reduction of added dichlorophenol-indophenol (DCPIP), photo-oxidation of the high potential form of cytochrome b559, and chlorophyll a fluorescence.

Our results show:

(1) The inhibition of the DCPIP reduction rate by ANT 2p is dependent not only on the concentration of the ADRY reagent but also on the actinic light intensity (Fig.1). The efficacy of the ANT 2p is enhanced at the lowest light intensities.

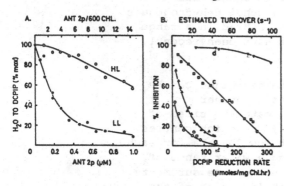

Fig.1A ANT 2p titration either at low (5Wm^{-2}) or at high (55Wm^{-2}) light. Maximal rates of 18 and 114µmoles DCPIP/mg Chl.hr recorded for LL and HL respectively. Fig.1B Intensity dependence of the inhibition caused by (a)90nm, (b)180nm, (c)0.9µM and (d)4.6µM ANT 2p. The potency of ANT 2p was measured at different initial rates.

(2) The inhibition by ANT 2p at low light intensities can be reversed by the addition of diphenylcarbazide, indicating that the oxygen-evolving complex is affected. Diphenylcarbazide is without effect at high ANT 2p concentrations.

(3) Low concentrations of ANT 2p slows the rise kinetics of the flourescence, consistent with a decreased rate of electron flow through PS2. At high concentrations of ANT 2p the fluorescence yield is quenched.

(4) Addition of ANT 2p can cause the photo-oxidation of cyt b559. With low concentrations of ANT 2p, a net oxidation is detected only at low actinic light intensities (Fig.2).

*ANT 2p: 2(3-chloro-4 trifluromethyl) anilino-3,5 dinitrothiophene.

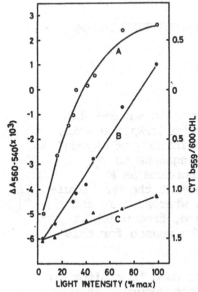

Fig.2 The light-intensity dependence of the ANT 2p-induced redox reactions of cyt b₅₅₉ (a)0.1μM=0.24 ANT 2p/PS2, (b)1μM, (c)10μM.

(4) The dependence of the extent of cyt b_{559} photo-oxidation on the actinic intensity arises from the competition between the oxidation and the re-reduction reactions. The reduction is favoured at high light intensities and in the absence of oxidants of the quinone pool. This redox behaviour of cyt b_{559} matches the intensity dependence of the PS2 inhibition by ANT 2p (1).

(5) The inhibition caused by low ANT 2p at limiting light is enhanced at low cation levels where the amount of excitation energy arriving at PS2 is decreased.

We interpret these results to indicate that ANT 2p has two distinct sites of inhibition on PS2. Low concentrations of ANT 2p appear to affect the oxygen-evolving complex under low PS turnover rates. The inhibition might be due to a cyclic flow around PS2 (2) or, alternatively, to a disruption of the electron donation pathway when the oxygen-evolving complex is in a susceptible oxidation state (1). High concentrations of ANT 2p, however, presumably block the acceptor side of PS2 between the pheophytin and Q_A, which could result in the quenching of the fluorescence yield through the accumulation of the anionic form of pheophytin (3).

Acknowledgements This work was supported by the S.E.R.C. and by the A.F.R.C. We wish to thank Dr. G. Renger for the gift of ANT 2p.

1. Packham, N.K. and Barber, J. (1984) Biochem. J. (in press)
2. Renger, G., Bouges-Bocquet, B. and Delosme, R. (1973) Biochim. Biophys. Acta 292, 796-807
3. Renger, G., Koike, H., Yuasa, M. and Inoue, Y. (1983) FEBS Lett. 163, 89-93

INTRATHYLAKOID RELEASE OF Mn^{++} FROM THE WATER-SPLITTING ENZYME
STUDIED WITH A NEW TYPE OF SPIN LABEL.
Jens Z. Pedersen and Raymond P. Cox
Institute of Biochemistry, Odense University
Campusvej 55, DK-5230 Odense M, Denmark

The manganese associated with the photosynthetic water-splitting enzyme can be removed by various treatments (alkaline Tris,
hydroxylamine, heating) with a concomitant inhibition of oxygen
evolution. (1). It is generally agreed that manganese is released from the inner side of the thylakoid membrane as Mn^{++} .
However, following hydroxylamine or heat treatment the Mn^{++} equilibrates rapidly with the surrounding medium, whereas Tris and
Zn^{++} release Mn^{++} into a confined aqueous space, from which it
diffuses only slowly into the medium (2-5). The reason for this
behaviour is unknown.

The electron spin resonance spectrum can be used to monitor the
release of $Mn(H_2O)_6^{++}$ directly, bound Mn^{++} is not observed at room
temperature due to increased spin lattice relaxation. However, in
the e.s.r. spectrum it is not possible to distinguish between intrathylakoid and external Mn^{++}. We have developed a spin label
method that can be used to measure intrathylakoid Mn^{++} directly.
We use a new type of water-soluble spin labels, the nitronylnitro-
xides, that partition equally between intrathylakoid and external
water phases and have a low membrane solubility. The e.s.r. signal
from the external label can be removed by the membraneimpermeable
spin-broadening agent chromium oxalate (6) and only the signal
from the intrathylakoid label is seen. Release of Mn^{++} causes
spin-spin relaxation between the paramagnetic metal ion and the
free radical, thus the e.s.r. signal is diminished, it will be
restored after diffusion of the Mn^{++} out of the intrathylakoid
space.

Using this method we have confirmed the findings of Miller and
Cox (3-4,7) that externally added Mn^{++} equilibrates rapidly across
the thylakoid membrane while Zn^{++}-released Mn^{++} can not pass the
membrane readily. Results will be presented that explain the modes
of action of the different Mn^{++}-releasing treatments.

References.

1. Amesz, J. (1983) Biochim. Biophys. Acta 726, 1-12.

2. Blankenship, R.E. and Sauer, K. (1974) Biochim. Biophys. Acta 357, 252-266.

3. Miller, M. and Cox, R.P. (1983) FEBS Lett. 155, 331-333.

4. Miller, M. and Cox, R.P. (1984) Biochem. Biophys. Res. Commun. 119, 168-172.

5. Miller, M. and Cox, R.P. (1984) in: Advances in Photosynthesis Research, Vol. I (Sybesma, ed.), pp. 355-358, Nijhoff/Junk, The Hague.

6. Berg, S.P. and Nesbitt, D.M. (1979) Biochim. Biophys. Acta 548, 608-615.

7. Miller, M. and Cox, R.P. (1982) Photobiochem. Photobiophys. 4, 243-248.

STUDIES ON THE KINETICS OF THE FLASH-INDUCED P515 RESPONSE
IN SPINACH CHLOROPLASTS.

R.L.A. Peters and W.J. Vredenberg

Laboratory of Plant Physiological Research, Agricultural University
Wageningen, The Netherlands.

The P515 electrochromic bandshift following a single saturating
light flash in dark-adapted and well preserved chloroplasts shows multi-
-phasic rise and decay kinetics (1). By using double flashes it has been
shown (1) that the single flash response curve can be deconvoluted
into at least two separate responses, called reaction 1 and 2.

Reaction 1, characterized by a fast (ns) rise and a subsequent
single exponential dark decay with a rate constant of approx. $10 \ s^{-1}$
is observed   without interference from other reaction components
in aged chloroplasts, or in broken chloroplasts which have been rethawed
from a frozen preparation. In broken chloroplasts, in all cases where
reaction 1 is the exclusive component, the first order rate constant
of the decay titrates linearly with the concentration of permeant ions
in the absence or presence of ionophores (2). These data confirm earlier
suggestions (3) that reaction 1 reflects the generation and decay of
the delocalized transmembrane electric field induced by the charge
separation in PSI and PSII.

Reaction 2 is characterized by a relative slow increase in absorbance, occurring during the first 100 to 150 ms after the flash, and
a subsequent decay with a first order rate constant of the order of
1 to 2 s^{-1}. The absence of reaction 2 in the presence of DBMIB (3),
its reappearance in the presence of DCMU after the addition of H-donors
to the Fe-S cyt b-f protein complex (4) and its presence in PSI vesicels
(Peters, F.A.L.J. unpublished results), confirm earlier evidence (1) that
it is associated with an energetic process driven by PSI. It has been
suggested by van Kooten (5) that reaction 2 is caused by the lateral
and transversal delocalization of inner-membrane electric fields associated with the liberation of protons in inner-membrane domains near
the Fe-S cyt b-f protein complex. These domains might be connected
via H-conductive channels with other membrane domains which act
as proton sinks (i.e. the ATP synthetase). Models depicting site specific
intramembranal proton processing in the thylakoid have been suggested
(6). In this respect it is of interest to mention that, in conformation
with results of others (7) we have shown that reaction 2 can also be
induced in the dark towards its saturation level by ATP driven proton
translocation (8). However, the stabilization of protons inside the hydro-

phobic membrane is still a matter of discussion.

In contrast to reaction 1, the occurrence of reaction 2 is strongly dependent on the functional integrity of the membrane, i.e. disappears upon ageing, after a temperature shock and, as will be presented at the symposium, is largely reduced in chloroplasts isolated from plants which were grown at low light intensity. It was found that the thylakoids of these plants showed a largely reduced amount of one of the two major membrane lipids, i.e. MGDG, as well as a reduced amount of both cyt b563 and cyt f. It also will be shown that the occurrence of the reaction 2 component of the P515 response is selectively sensitive towards uncouplers of photophosphorylation that act as a proton carrier in the lipophilic phase of the thylakoid membrane. The addition of 10^{-6} mol/l CCCP results in the complete suppression of reaction 2, whereas, at this low concentration hardly any effect could be detected on the kinetics of reaction 1 and on the Hill reaction rate. In contrast, the addition of relatively high amounts (5.10^{-3} mol/l) of the hydrophilic uncoupler NH4Cl did not result in an alteration of the kinetics of the P515 response, neither with respect to reaction 1, nor with respect to reaction 2. At this concentration, the Hill reaction was highly uncoupled.

From these experiments it can be concluded that the reaction 2 component of the P515 response is selectively sensitive towards factors that affect the functional integrity of the thylakoid membrane. These results are in support for our suggestion that reaction 2 is the reflection of an intramembrane electrical event associated with the liberation of protons, presumably near the Fe-S cyt b-f protein complex.

1. Schapendonk, A.H.C.M. (1980) Doct. Thesis Agric. Univ. Wageningen
2. Peters, R.L.A., van Kooten, O. and Vredenberg, W.J. (1983) In Proc. 6th. Int. Congr. Photosynth. (Sybesma, C. ed.) Martinus Nijhoff- Dr. W. Junk Publ. Inc. The Hague, in press
3. Vredenberg, W.J. (1981) Physiol. Plant. 53, 589-602
4. Selak, M.A. and Whitmarsh, J. (1982) FEBS Lett. 150, 286-292
5. Westerhoff, H.V., Helgerson, S.L., Theg, S.M., van Kooten, O., Wikström, M.K.F., Skulachev, V.P., and Dancshazy, Z. (1984) Acta Biochim. Biophys. Acad. Hung. 18, 125-150
6. Dilley, R.A., Tandy, N., Bhatnager, D., Baker, G. and Millner, P. (1981) Proc. 5th Int. Congr. Photosynth. (Akoyunoglou, G. ed.) Vol. 2, Balabon Int. Science Services, Philadelphia, 759-769.
7. Schreiber, U. and Rienits, K.G. (1982) Biochim. Biophys. Acta 682 115-123.
8. Peters, R.L.A., Bossen, M, van Kooten, O. and Vredenberg, W.J. (1983) J. Bioenergetics and Biomembranes. 15, No. 6, 335-346.

CONTAMINANTS OF ANTENNA AND REACTION CENTER PREPARATIONS ISOLATED
FROM RHODOSPIRILLUM RUBRUM S1

J.M. Ramírez and G. Giménez-Gallego
Instituto de Biología Celular, C.S.I.C.,
Madrid 6, Spain

In the photosynthetic membrane, chlorophylls and carotenoids are
noncovalently bound to specific apoproteins. A large fraction of such
pigment-protein complexes belongs to the light-harvesting antenna
that absorbs light and transfers excitation energy to the reaction
center, a different and minor type of pigment-protein complex that
carries out primary photochemistry. Both reaction center and antenna
complexes have been solubilized and purified from several phototrophic bacteria with little alteration of their in situ absorption
spectra, what indicates that the native pigment-protein interaction
has not been significantly disturbed by the preparative procedures.
When analyzed by polyacrylamide gel electrophoresis in the presence
of sodium dodecyl sulfate, followed by Coomassie brilliant blue
staining, the purified reaction center of Rhodospirillum rubrum
contained, in addition to pigments and redox constituents, three
different polypeptides of about 36, 29 and 25 kDa (1). An antenna
complex (B880), isolated recently from the same bacterium (2,3),
showed two different polypeptides of M_r close to 7000 when analyzed
by the same technique.

Serum of rabbits immunized against the isolated B880 antenna
complex was used to detect antigen bands in the polyacrylamide gels
after electrophoretic transfer to nitrocellulose sheets (4). Such
immunochemical method revealed the presence in the B880 preparation
of at least two constituents which had gone undetected by Coomassie
brilliant blue staining. With the aim of checking such finding by an
independent method, the gels were stained with silver (5) and, again,
several bands appeared in addition to those stainable with Coomassie
brilliant blue. Since we observed that chromatography on Sephacryl
allowed to obtain B880 enriched fractions in which the levels of the
constituents nonstainable with Coomassie brilliant blue were highly
reduced, we designed a purification procedure in which a single
chromatographic step on Sephacryl yielded a preparation of the
antenna complex which did not contain the extra constituents and
which showed a near infrared absorption spectrum similar to that of
intact chromatophores. The A_{280}/A_{880} ratio of that preparation was
below 0.4, what indicated a significant improvement of the bacteriochlorophyll to protein ratio over the previous procedure (A_{280}/A_{880} =
0.67 in ref. 3). The stability of those B880 fractions was
considerably higher than that of the previously described preparation

(3), probably because of the presence in the latter of high levels of Triton X-100. Other chromatography fractions were enriched in a carotenoprotein that lacked bacteriochlorophyll. The polypeptides found in those fractions differed from those associated to the B880 complex in their electrophoretic mobilities and/or in their staining properties, what suggested that the carotenoprotein did not result from the B880 complex upon loss of bacteriochlorophyll.

Two of the contaminants that had been detected in the B880 preparations were also present in R. rubrum reaction centers obtained by usual procedures (1,6), as indicated by silver staining of the polyacrylamide gels and by the immunochemical method. The contaminants could be removed from the reaction center preparations without apparent alteration of the absorption spectrum except for a significant reduction of the 280-nm (protein) band. However, the removal of such contaminants, which was achieved by Sephacryl chromatography at high detergent concentrations, went along with some degradation of the heavier subunit (36kDa), due probably to the residual levels of endogenous proteinases which remained in the preparations (6,7).

References

1 Vadeboncoeur, C., Noël, H., Poirier, L., Cloutier, Y. and Gingras, G. (1979) Biochemistry 18, 4301-4308
2 Cogdell, R. J., Lindsay, J. G., Valentine, J. and Durant, I. (1982) FEBS Lett. 150, 151-154
3 Picorel, R., Bélanger, G. and Gingras, G. (1983) Biochemistry 22, 2491-2497
4 Haid, A. and Suissa, M. (1983) Methods Enzymol. 96, 192-205
5 Merril, C. R., Goldman, D. and Van Keuren, M. L. (1982) Electrophoresis 3, 17-23
6 Giménez-Gallego, G., Suanzes, P. and Ramírez, J. M. (1983) FEBS Lett. 162, 91-95
7 Rivas, L. and Giménez-Gallego, G. (1984) Biochim. Biophys. Acta 764, 125-131

MULTIPLE ACTION OF HYDROXYLAMINE ON PHOTOSYSTEM II OF ISOLATED CHLOROPLASTS.

M.K. Raval[o], G.B. Behera" and U.C. Biswal[o]
[o]School of Life Sciences, Sambalpur University, Jyoti Vihar-768 019, Orissa, INDIA.
"Department of Chemistry, Sambalpur University, Jyoti Vihar-768 019, Orissa, INDIA.

Hydroxylamine depending on its concentration has complex action on the chloroplast membranes (1-5). We have studied the kinetics of hydroxylamine action on isolated chloroplasts aging in light or dark and have attempted to locate the specific site(s) on the membrane for the amine action.

Chloroplasts from the leaves of 6-day-old wheat (Triticum aestivum L.CV.Sonalika) seedlings were isolated (6) and incubated at $25\pm2°C$ in continuous white light (70 Wm^{-2}) or in continuous dark. Photosystem II activity was measured in term of DCIP photoreduction (6). Heat inactivation of O_2 evolution was done by incubating the freshly prepared chloroplasts at 50°C for 5 min.

Hydroxylamine at 30 μM neither affects the DCIP photoreduction of isolated chloroplasts nor restores it in heat inactivated samples. Hydroxylamine at 20-50 μM reversibly inhibits O_2 evolution due to specific binding to O_2 evolving complex which is oxidised by 2 flash illumination (1,7). We expose the reaction mixture for 30 sec. to light, so no significant decrease in DCIP photoreduction is observed. The binding of the amine maybe at or on the oxidising side of the heat inactivation site. Hydroxylamine at 1 mM decreases DCIP photoreduction rate by 40% when added to the reaction mixture. After 1 hr. of aging its addition does not decrease the DCIP photoreduction rate (Fig.1A). Upon incubation of isolated chloroplasts with the amine (1 mM or 10 mM) DCIP photoreduction rate decreases immediately by 40% and then progressively decreases to zero by 60 min.both in light and dark (Fig.1B).Hydroxylamine at 1 mM in the reaction mixture does not restore the DCIP photoreduction in heat inactivated chloroplasts. Hydroxylamine at 1 mM releases Mn and inhibits Z P680$^+ \rightarrow$ Z$^+$P680(2-4). Hydroxylamine binds on either side of Z(4). The amine binding between Z and P680 is non-specific and weak and removed by washing (4). The immediate decrease in DCIP photoreduction may be due to hydroxylamine binding between Z and P680 and blocking electron transport to

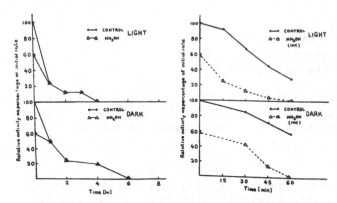

Fig.1. Changes in DCIP photoreduction by chloroplasts(A) with NH_2OH(1mM) in reaction mixture(B) incubated with NH_2OH (1mM or 10mM). 100% equals 66.6 μM DCIP reduced mg^{-1} total chlorophyll hr^{-1}.

A   B

DCIP, which may be reversible in light. Progressive inactivation may be due to release of Mn from its binding site. Binding site between Z and P680 may be sensitive to aging (Fig.1A). Hydroxylamine at 10 mM in the reaction mixture decreases the DCIP photoreduction rate by 40% immediately and restored 33.3% of activity in hydroxylamine or heat inactivated chloroplasts. Hydroxylamine acts ($\geqslant$ 10mM) as a donor to Photosystem II (5) and the donor site may be close to P680.

References:

1. Bennoun,P. and Bouges,B. (1972) in Proc.2nd Int. Photosynthesis Congr.(Forti,G.,Avron,M. and Melandri,A. eds.) Vol.I, pp 568-576,Dr.W.Junk N.V.Publishers, The Hague.
2. DenHaan,G.A., De Vries,H.G. and Duysens,L.N.M. (1976) Biochim.Biophys.Acta 430,265-281.
3. Yocum,F.,Yerkes,C.T.,Blankenship,R.E.,Sharp,R.R. and Babcock,G.T.(1981)Proc.Natl.Acad.Sci.USA 78,4507-4511.
4. Ghanotakis,D.F. and Babcock,G.T.(1983)FEBS Lett.153, 231-234.
5. IZawa,S., Heath,R.L. and Hind,G.(1969)Biochim.Biophys. Acta 180, 389-398.
6. Biswal,U.C. and Mohanty,P.(1976)Plant Cell Physiol.17, 323-331.
7. Bouges-Bocquet,B. (1973) Biochim.Biophys. Acta 292, 772-785.

EPR STUDIES OF PHOTOSYSTEM II

A.W. Rutherford and J.L. Zimmermann
Service de Biophysique, Département de Biologie
CEN Saclay, 91191 Gif-Sur-Yvette cédex, France

Rapid advances in the study of PS II by EPR have taken place over the last few years (for a review see (1)). In this report we summarize recent developments that have taken place in our laboratory.

The acceptor side

1) The primary semiquinone-iron acceptor, Q_A^-Fe, gives rise to two different EPR signals. Firstly a signal with features at g=1.82 and g≈1.67 and secondly a signal with features at g=1.90 and g≈1.64. The type of signal present is determined by the pH, the g=1.90 signal dominating at higher pH values (≤pH 7.5) (2).

2) The secondary semiquinone acceptor also seems to give rise to an EPR signal attributable to a semiquinone-iron complex, Q_B^-Fe (3). In PS II membranes at pH 6.0 the Q_B^-Fe signal is similar to the g=1.90 resonance form of Q_A^-Fe but centred at slightly higher g-value. Q_B^-Fe was generated by low temperature photoreduction of Q_A^-Fe followed by warming to room temperature. The signal was stable at room temperature, was present in the dark in 20 % of the centres and was lost when DCMU was present.

3) The Q_A^-Fe signals are oriented so that the g=1.90, the g=1.82 and the g 1.66 features are maximum when partially dried oriented multilayers of PS II membranes were oriented perpendicular to the measuring magnetic field (4).

4) The split Ph⁻ signal shows an orientation dependent splitting being maximum when the membranes were parallel to the magnetic field and minimum when perpendicular (4).

The donor side

1) The S_2 multiline signal in PS II membranes oscillates with flash number showing amplitude maxima on the 1st and 5th flash (5).

2) 200 K illumination induces formation of the S_2 multiline signal and at the same time a new signal at g=4.1 (Fig. 1). Both signals are absent if illumination is given at 77K or if the membranes are washed in Tris. Only the S_2 multiline signal is formed by flash illumination at 0 to 20°C or by 200 K illumination in the presence of DCMU. When both signals are present at the same time the g=4.1 signal is much less stable to warming than is the S_2 multiline signal. It is concluded that the g=4.1 signal reflects oxidation of an electron carrier close to S_2 and may be a pre S_3 state. Its spectral characteristics are similar to those of

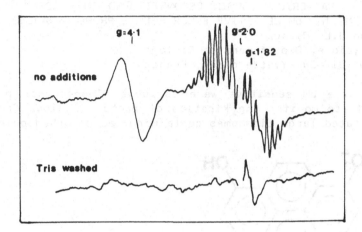

Figure 1 - Light-dark spectra induced by 200 K illumination of PS II membranes. Temp. 8 K, power 8 dB, mod. amp. 2.0 mT.

ferric iron in a rhombic environment but manganese dimers can also give rise to this kind of signal. This signal is probably the same as that reported recently and suggested to be a pre-S_2 state (6). From both sets of data it is suggested that the component may act as a carrier between the S states and D_1, the first donor to P_{680} (5).

3) Neither the S_2 nor the g=4.1 signals show large orientation effects although the former does show some slight effects in the low field wings (4).

4) From orientation studies of the P_{680} triplet it is concluded that the chlorophyll ring is in the plane of the membrane (4).

References
1. RUTHERFORD, A.W. (1983) in The Oxygen evolving system of photosynthesis (Inoue Y. et al, eds) Academic Press, Japan pp 63-69
2. RUTHERFORD A.W., ZIMMERMANN J.L. and MATHIS P. (1984) Proc. 6th Int. Cong. Photosynth., Brussels, in press
3. RUTHERFORD A.W. and ZIMMERMANN J.L. (1984) Biochim. Biophys. Acta, submitted
4. RUTHERFORD A.W. (1984) Biochim. Biophys. Acta, submitted
5. ZIMMERMANN J.L. and RUTHERFORD A.W. (1984) Biochim. Biophys. Acta, submitted
6. CASEY J.L. and SAUER K. (1984) Biophys. J. 45, 217a

PROTON PUMPING BY BACTERIORHODOPSIN RECONSTITUTED IN LARGE LIPO-
SOMES STUDIED USING THE pH SENSITIVE FLUORESCENCE PROBE PYRANINE

M. Seigneuret and J.L. Rigaud
Service de Biophysique, Département de Biologie
CEN Saclay, 91191 Gif-Sur-Yvette cédex, France

We have used the pH sensitive water soluble fluorescence probe pyranine (Fig. 1) to monitor the kinetics of proton gradient forma- tion in reconstituted large liposomes containing bacteriorhodopsin.

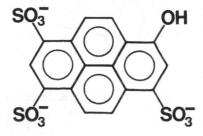

Fig. 1. 8-hydroxy-1,2,6-pyrenetrisulfonate (pyranine)

The distinct excitation spectra of pyranine in its protonated and unprotonated forms allows pH changes to be measured from variations of the emission intensity at 510 nm. Furthermore, the non permeant character of the probe makes it suitable to monitor pH inside liposomes (1,2,3).

In the present study, we have used our newly introduced large (diameter 0.2 μm), unilamellar liposomes of bacterirhodopsin and phospholipids prepared by the reverse phase evaporation technique (4). Our previous results have indicated an inside out orientation of bacteriorhodopsin (4,5). Pyramine was entrapped in the liposomes during reconstitution and externa probe was removed by flotation on a ficoll cushion.

The calibration of the relative emission intensity vs. pH was nearly identical for pyranine free in buffer and pyranine entrapped insde proteoliposomes (apparent shift in pKa 0.05) provided that small amounts of negative charge were present in the lipids (e.g. 10 %) phosphatidic acid 90 % egg phsphatidylcholine). Under these conditions, direct determination of the amount of pyranine bound to the membrane gave values less than 5 %.

Proton pumping by bacteriorhodopsin, occuring upon illumination of light adapted proteoliposomes, could be detected from the fluo- rescence intensity decrease of entrapped pyranine due to internal

acidification. Valinomycin (1 μ M) was necesser y to observe a significant effect.

From the kinetics of fluorescence decay, proton pumping initial
rates, total inward proton fluxes and final stationnary ΔpH values
could be determined. We have studied in details the dependence of
such parameters upon : 1) buffering capacity and initial pH of both
external and internal compartments ; 2) pump/leak ratio of the liposomes (modulated by varying the light intensity, the number of bac-·
teriorhodopsin molecules per liposome and the cholesterol content of
the lipid). Results will be discussed in terms of the dependence of
light induced proton pumping activity of bacteriorhodopsin upon
internal pH and/or ΔpH (see ref. 6).

Δ pH values obtained using pyranine fluorescence were compared to
those derived from external pH potentiometric measurements and
[14]C-methylamine or 9-aminoacridine inside/outside distribution. A
critical comparison of results obtained by the different techniques
will be given.

References
1. Kano K. and Fendler J.H. (1978) Biochim. Biophys. Acta 509, 289-
   299
2. Clement N.R. and Gould J.M. (1981) Biochesmistry 20, 1534-1538
3. Damiano E., Bassilana M., Rigaud, J.L. and Leblanc G. (1984)
   FEBS Lett. 166, 120-124
4. Rigaud J.L., Bluzat A. and Buschlen S.(1983) in Physical chemistry of transmembrrane ion motion (Spach G. ed.) Elsevier pp.457-
   467
6. Westerhoff H.V. and Dancshazy Zs (1984) Trends Biochem. Sci. 10,
   112-117

DIFFERENTIAL INTERACTION OF BAL WITH THE PHOTOSYNTHETIC ELECTRON
TRANSPORT SYSTEM UNDER AEROBIC AND ANAEROBIC CONDITIONS.

Y. Shahak[*], Y. Siderer[†] and E. Padan[o]
[*]Biochemistry Department, The Weizmann Institute of Science, Rehovot,
Israel 76100.
[†]Department of Biological Chemistry and [o]Division of Microbial and
Molecular Ecology, The Institute of Life Sciences, The Hebrew
University, Jerusalem, Israel.

The treatment of mitochondria with BAL (British Anti Lewisite, 2,3-
dimercapto-1-propanol) under air has been shown to inhibit electron
transport between cytochromes b and c (1). It was later reported that
BAL destroyed the mitochondrial Rieske Fe-S center as revealed by ESR
spectrometery (2). We have recently found that the photosynthetic electron transport system is also sensitive to BAL (3). Preincubation of
either cyanobacterial cells (*Oscillatoria limnetica*) or spinach chloroplasts with BAL, resulted in the inhibition of electron transport in
both systems. To obtain a full block, 90 min preincubation was required with 3-10 mM BAL under air and low chlorophyll concentration. The
inhibition was unwashable. The following effects of pretreatment with
BAL were observed: (i) Anoxygenic sulfide dependent hydrogen evolution
by *O. limnetica* was inhibited. The inhibition could be bypassed by
TMPD. (ii) Oxygenic NADP photoreduction by spinach chloroplasts was
blocked. $DCPIPH_2$ or TMPD restored the activity. DCMU insensitive sili-
comolybdate photoreduction was unaffected by BAL treatment. The results
indicate that both PS II (from water to Q) and PS I (from plastocyanin
to NADP in spinach or to the hydrogenase in *O. limnetica*) are not inhibited by BAL treatment. This locates the BAL sensitive site at the
b_6/f complex.

BAL pretreatment of chloroplasts did not prevent cyt f turnover.
Fig. 1 demonstrates that unlike control chloroplasts, in BAL treated
chloroplasts both far red and red light induce the oxidation of cyt f.
The above results and the analogy to mitochondria suggest that the BAL
labile factor is the Rieske protein. However, in chloroplasts which
were fully inhibited by BAL pretreatment, the ESR signal at g=1.89
remained identical to that of control chloroplasts (Fig. 2A). Still,
the photooxidation of the Rieske Fe-S center by PS I was inhibited in
BAL treated chloroplasts. It is therefore suggested that aerobic BAL
treatment specifically blocks electron transfer from the Rieske protein to cyt f without any major destruction of either components.

We have further found that under anaerobic conditions, BAL served
as an electron donor, rather than an inhibitor of photosynthetic electron transport. In *O. limnetica* the rates of sulfide dependent and of
BAL dependent hydrogen evolution were the same. In spinach chloroplast

the rate of BAL → NADP was half that of water → NADP, under optimal conditions. Both reactions were stimulated to the same extent by uncouplers. Electron transport from BAL was insensitive to DCMU, but inhibited by DBMIB, DNP-INT or aerobic BAL pretreatment. Also ESR spectra of spinach chloroplasts which had been frozen during illumination in the presence of DCMU and methylviologen (Fig. 2 B) showed that BAL (added anaerobically) can reduce the photooxidized Rieske center (g=1.89) and plastocyanin (g=2.05). We suggest that under anaerobic conditions BAL donates electrons to the Rieske Fe-S protein either directly or via plastoquinone.

1. Slater, E.C. (1950) Nature 165, 674-676.
2. Slater, E.C. and de Vries, S. (1980) Nature 288, 717-718.
3. Belkin, S., Siderer, Y., Shahak, Y., Arieli, B. and Padan, E. (1984) Biochim. Biophys. Acta, in press.

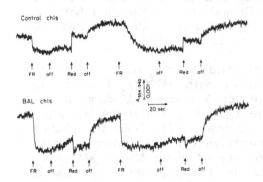

Fig. 1: Cytochrome f redox changes under red and far red illumination in control and BAL treated chloroplasts. Temp = 22°C.

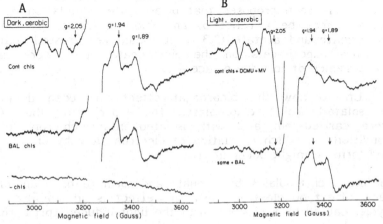

Fig. 2: ESR spectra at 5°K. Chloroplasts were frozen in the dark under air (A) or light under N_2 (B). Microwave power= 10 mW; modulation amplitude = 10G.

REGULATION OF PHOTOSYNTHETIC ELECTRON FLOW BY FORMATE AND BICARBONATE IN ISOLATED INTACT CHLOROPLASTS

Jan F.H. Snel, Dirk Naber and Jack J.S. van Rensen

Laboratory of Plant Physiological Research, Agricultural University Wageningen, The Netherlands

The mechanism of the stimulation of linear electron flow in photosynthesis by bicarbonate (or CO_2) has been investigated and partly elucidated during the past decade. In thoroughly CO_2-depleted broken chloroplasts the Hill reaction is inhibited for more than 95% and a dark-incubation after addition of bicarbonate is sufficient to reactivate the Hill reaction. This stimulation of electron flow by bicarbonate is known as the "bicarbonate-effect". In CO_2-depleted thylakoids electron flow is slowed down between Q_A and Q_B (the primary, resp. secondary quinone-type electron acceptor of PS II) and the equilibration between Q_B^{2-} and the plastoquinone-pool is strongly inhibited. Much of this and earlier work has been reviewed in reference 1.

Recently more attention has been paid to the role of formate in the bicarbonate-effect. Binding studies with $H^{14}CO_3^-$ have shown that formate and bicarbonate compete for a common binding site at the thylakoid membrane (2) and formate appears to inhibit the stimulation of the Hill reaction in a competitive manner (3). Both the binding studies and the Hill reaction measurements yield a dissociation constant (K_d) of the binding site-bicarbonate complex of about 80-100 μM bicarbonate. The bicarbonate concentration in the chloroplast stroma is probably less than 100 μM in vivo and a K_d value of 100 μM bicarbonate implies that linear electron flow would be impaired. This is not observed and it has been proposed that bicarbonate stimulation of electron flow actually may be a relief of an inhibition caused by inhibitory anions as formate and acetate (2,3,4). It should be stressed that most of the previous experiments on the bicarbonate-effect were carried out in the presence of formate of acetate.

Up to now the bicarbonate-effect has been demonstrated only in isolated broken chloroplasts. The experiments that will be shown were carried out also with isolated intact chloroplasts to establish conditions at which regulation of electron flow by bicarbonate (or CO_2) and formate might occur in vivo.

In intact chloroplasts bicarbonate and formate have an additional, indirect, effect on electron flow. This effect is an inhibition of the Calvin cycle, caused by a dissipation of the light-induced ΔpH across the chloroplast envelope (5). The resulting acidification of the stroma can inhibit several enzymes involved in the reduction of CO_2.

Isolated intact chloroplasts were incubated at pH 5.8 in the dark or illuminated in the presence of formate and an artificial electron acceptor at pH 6-7. Based on measurements of linear electron flow (oxygen evolution) or flash-induced P_{515}-absorbance changes it is shown that these treatments induce a "bicarbonate-effect" in isolated intact chloroplasts.

References:

1. Govindjee and Van Rensen, J.J.S. (1978) Biochim. Biophys. Acta 505, 183-213
2. Stemler, A. and Murphy, J. (1983) Photochem. Photobiol. 38, 701-707
3. Snel, J.F.H. and Van Rensen, J.J.S. (1984) Plant Physiol. 75 (in press)
4. Stemler, A. and Jursinic, P. (1983) Arch. Biochem. Biophys. 221, 227-237
5. Enser, U. and Heber, U. (1980) Biochim. Biophys. Acta 592, 577-591

THE TOPOLOGY OF THE PHOTOSYNTHETIC MEMBRANE IN THE BATERIOCHLOROPHYLL B CONTAINING BACTERIUM ECTOTHIORHODOSPIRA HALOCHLORIS

R. Steiner and H. Scheer
Botanisches Institut der Universitat Munchen, Menzingerstr. 67
D - 8000 Munchen 19, FRG

E. halochloris is a halophilic photosynthetic bacterium with an unusual pigment composition and absorption spectrum (Imhoff and Truper, 1977; Steiner et al. 1981; Steiner and Scheer, 1984). Its membrane topology has been studied by proteolytic digestion. Chromatophores (inside-out particles), spheroblasts (rightside-out particles) and isolated antenna complexes were treated with trypsin, thermolysin and proteinase K. Changes in the protein pattern (SDS-PAGE) were correlated with changes in the absorption, absorption difference, circular dichroism and low temperature fluorescence spectra in the visible and near infrared region.

The isolated antenna (Steiner and Scheer, 1984), as well as chromatophores, spheroblasts or whole cells, show a reversible, pH-induced absorption change from a "high-pH" form B800/1020 to a "low-pH" form B800/960. The same spectral changes are seen upon proteolysis, with the important difference that it works with chromatophores only, but not with spheroblasts.

The SDS gels of the photosynthetic membrane preparations show in the high molecular weight range the three typical peptides of the reaction center (RC), and a 34 kDa band tentatively assigned to a peripheral cytochrome c. In the low molecular weight range appear three bands (6.5, 6.0 and 4.5 kDa) assigned to the light-harvesting complex (LHCP). During proteolysis of chromatophores, the "cytochrome" band disappears first, followed by the H and L subunit and later by M. The LHCP is stable until all RC bands have disappeared. After addition of more protease, the 6.5 kDa polypeptide is digested first, followed at last by the two lightest peptides. A plot of the used protease concentration and the 1020 nm absorption, and comparison with the results from SDS-PAGE shows that the 1020 nm absorption is shifted to 960 nm simultaneously with the digestion of the 6.5 kDa peptide. When incubating spheroblasts, only the H subunit of the RC is attacked.

This indicates that only H spans the membrane, whereas the other RC polypeptides and the light-harvesting polypeptides are exposed only to the cytoplasmatic side. A similar result is shown for the membrane of another Bchl b containing organism, e.g. Rp. viridis (Jay, 1983). If we treat the isolated LHCP of E. halochloris with protease, all three polypeptides disappear simultaneously.

The results are discussed in terms of a model, in which the hydrophilic domains of the membrane polypeptides play a role in the membrane organization.

REFERENCES:
Imhoff, J. and Truper, H. (1977) Arch. Microbiol. 114, 115.
Steiner, R., Schafer, W., Wieschhoff, H. and Scheer, H. (1981) Z. Naturforsch. 36c, 417.
Steiner, R. and Scheer, H. (1984) Biochim. Biophys. Acta, submitted
Jay, F., Lambilotte, M. and Muhlethaler, K. (1983) Eur. J. Cell Biol. 30, 1.

THE EFFECT OF CATIONS ON THE MECHANISM OF REGULATION OF EXCITATION
ENERGY DISTRIBUTION BY REVERSIBLE PHOSPHORYLATION OF LHC IN PEA
THYLAKOIDS

A. Telfer, M. Hodges, P.A. Millner and J. Barber
AFRC Photosynthesis Research Group, Dept. Pure and Applied Biology,
Imperial College, London SW7 2BB, U.K.

It has been suggested that the regulation of excitation energy
distribution by reversible phosphorylation of LHC in higher plants is
brought about because of the increase in negative charge on the surface of this complex, due to the addition of phosphate groups (1).
This increase in the surface charge density of a mobile pool of phosphorylated-LHC is thought to lead to intermixing of chlorophyll-
protein complexes which are normally segregated laterally within the
plane of the thylakoid membrane; PSI being restricted to the non-
appressed lamellae and PSII and most of the LHC being restricted to
the appressed regions of the grana (2). In vitro studies have shown
that stacking and heterogeneous distribution of the complexes in non-
phosphorylated (NPhos) membranes are controlled electrostatically (3),
e.g. under poor 'screening' conditions the grana unstack and the complexes randomize in the plane of the membrane.

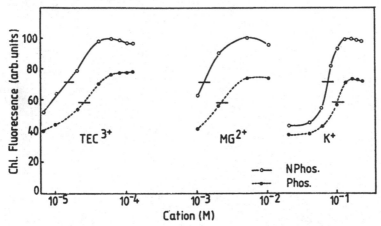

<u>Fig. 1</u> The effect of cations of different valency on chlorophyll
fluorescence from NPhos and Phos pea thylakoids. TEC^{3+}; tris(ethyl-
enediamine) cobaltic cation. Horizontal bars indicate $C_\frac{1}{2}$.

Fig. 1 compares the concentration requirement for the cation-induced
increase in chlorophyll fluorescence in Phos and NPhos membranes. The
order of effectiveness of cations of different valency is typical of
an electrostatic effect for both types of membrane. However the
increase in $C_\frac{1}{2}$ for the fluorescence increase in Phos compared to NPhos

membranes with each cation investigated (previously seen only for Mg^{2+}, refs 4,5) suggests that the surface charge density of Phos membranes is greater than that of NPhos membranes. There is also a cation-independent decrease in fluorescence in Phos membranes which has been attributed to the inability of the phosphorylated-LHC to segregate into the appressed regions of the grana.

We have also measured the effect of LHC-phosphorylation on the cation concentration requirement for stacking both by changes in 90° light scattering and from electron micrographs. Table I shows that the extent of unstacking induced by phosphorylation of LHC is highly dependent on the level of screening cations (in this case, Mg^{2+}) being much more extensive at the lower concentrations. However even at 5mM Mg^{2+} some phosphorylation-induced unstacking (12%) is seen. As this is measured on a linear basis it represents a 23% reduction in area of appressed membrane. This reduction in area qualitatively accounts for the cation-independent decrease in fluorescence which is brought about by migration of a pool of phosphorylated-LHC from the appressed to the non-appressed membranes. We suggest that this dramatic effect of changing the level of screening cation on the extent of unstacking in Phos membranes explains the previously rather varied reports of the amount of unstacking associated with LHC-phorphorylation (see 6).

Table I Effect of Mg^{2+} concentration on % changes in fluorescence (Fl), 90° LS and thylakoid stacking (St) in NPhos and Phos pea thylakoids.

Exp. Conditions		%Δ	%Δ	%Δ
mM Mg^{2+}	Phos	Fl	LS	St
5	−	100	100	100
5	+	64	90	88
2	−	89	96	95
2	+	24	46	39
1	−	47	52	74
1	+	5	4	15
0.5	−	18	29	27
0	−	0	0	6

We conclude that the effect of cations on chlorophyll fluorescence and the structure of NPhos and Phos thylakoids indicates that regulation of excitation energy distribution by protein phosphorylation is mediated by a change in the surface charge characteristics of the membrane surface.

1. Barber, J. (1982) Annu. Rev. Plant Physiol. 33, 261-295
2. Anderson, J.M. (1981) FEBS Lett. 124, 1-10
3. Barber, J. (1980) Biochim. Biophys. Acta 594, 253-280
4. Horton, P. and Black, M.T. (1982) Biochim. Biophys. Acta 680, 22-27
5. Telfer, A., Hodges, M. and Barber, J. (1983) Biochim. Biophys. Acta 724, 167-175
6. Simpson, D.J. (1983) Biochim. Biophys. Acta 725, 113-120

EFFECTS OF BICARBONATE, FORMATE AND HERBICIDES ON PHOTOSYNTHETIC ELECTRON TRANSPORT IN ISOLATED BROKEN CHLOROPLASTS

J.J.S. van Rensen, F. de Koning and J.F.H. Snel

Laboratory of Plant Physiological Research, Agricultural University Wageningen, The Netherlands

Carbon dioxyde is not only required as a substrate for CO_2-reduction, but CO_2 (or bicarbonate) plays also a role in photosynthetic electron transport. CO_2-depleted chloroplasts suspended in a reaction medium containing formate do not show a substantial rate of Hill reaction upon illumination, unless bicarbonate is added. It is not known whether bicarbonate or CO_2 is the active species. For convenience this phenomenon has been named the "bicarbonate-effect". It has been demonstrated that this bicarbonate-effect is caused by an inhibition and reactivation of photosynthetic electron transport between the primary quinone electron acceptor of Photosystem II, Q_A, and the plastoquinone pool (1,2). Herbicides affecting Photosystem II also inhibit between Q_A and plastoquinone (3). We have studied the regulation of photosynthetic electron transport by bicarbonate, formate and herbicides and the mutual interactions of their effects.

Isolated broken chloroplasts are depleted of bicarbonate by incubating the chloroplasts in the dark in a medium containing formate at pH of 6 or lower. The Hill reaction rate in these depleted chloroplasts is then measured with an oxygen electrode after transferring them to a higher pH and using ferricyanide as an electron acceptor. The Hill reaction rate is low, often approaching zero activity. After incubation of these chloroplasts with bicarbonate in the dark the Hill reaction rate is almost completely restored (4). Another method of depleting chloroplasts from bicarbonate is by illumination in the presence of an electron acceptor in a medium at a pH between 6 and 7, and containing formate (5).

There appears to be a Michaelis Menten type of relationship between the uncoupled Hill reaction rate and the bicarbonate concentration in the medium, characterized by a maximal Hill reaction rate (V_{max}) and an apparent reactivation constant (K_r, which is the bicarbonate concentration at half-maximal reactivation). The K_r depends on the concentration of formate and cations in the medium.

In the presence of 100 mM formate the K_r of bicarbonate is about 1 mM, while this K_r could be calculated to be about 80 µM bicarbonate in the absence of formate. Formate appears to be a competitive inhibitor of the bicarbonate stimulation of electron transport. Since formate can be an intermediate in photorespiration, it is suggested that regulation of linear electron flow by formate and bicarbonate may be a mechanism that could link photorespiration to electron flow (6).

In the presence of 50 mM Tris-formate at pH 6.5 the K_r of bicarbonate is about 1 mM in the presence of high cation concentration (100 mM Na^+ or 5 mM Mg^{2+}), while the K_r is about 2 mM at low cation concentration. There is no difference in V_{max}. It is suggested that the binding site of formate and bicarbonate is located below the surface of the thylakoid membrane. This surface is negatively charged. While bicarbonate could diffuse as the uncharged CO_2 and probably binds as bicarbonate, formate moves as a negatively charged anion. The negative charges at the membrane surface affect the movement of the formate anion between its binding site and the bulk phase. By neutralizing the negative charges by cations the exchange of formate for bicarbonate is improved.

The reactivation of the Hill reaction in CO_2-depleted chloroplasts by various concentrations of bicarbonate can be measured in the absence and in the presence of a Photosystem II inhibiting herbicide. It appears that these herbicides decrease the apparent affinity of the thylakoid membrane for bicarbonate (7). Different characteristics of bicarbonate--binding are observed in chloroplasts of triazine-resistant Amaranthus hybridus compared to the triazine-sensitive biotype (8).

In isolated broken chloroplasts bicarbonate and formate have a regulatory effect on electron transport. There is a mutual interaction of bicarbonate with Photosystem II inhibiting herbicides. We are investigating whether these effects are also present in more intact systems like isolated intact chloroplasts or unicellular green algae.

References:

1. Govindjee and Van Rensen, J.J.S. (1978) Biochim. Biophys. Acta 505, 183-213
2. Vermaas, W.F.J. and Govindjee (1981) Proc.Natl.Sci.Acad. B47, 581-605
3. Van Rensen, J.J.S. (1982) Physiol.Plant. 54, 515-521
4. Vermaas, W.F.J., Van Rensen, J.J.S. and Govindjee (1982) Biochim. Biophys. Acta 681, 242-247
5. Snel, J.F.H., Naber, D. and Van Rensen, J.J.S. (1984) Z. Naturforsch. 39c, in press
6. Snel, J.F.H. and Van Rensen, J.J.S. (1984) Plant Physiol. 75, in press
7. Van Rensen, J.J.S. and Vermaas, W.F.J. (1981) Physiol.Plant. 51, 106-110
8. Van Rensen, J.J.S. (1984) in: Weed Science Advances, Vol. I (R.G. Turner, ed.), Butterworths, London, in press

ON THE CHANGE OF THE CHARGES IN THE FOUR PHOTO-INDUCED OXIDATION STEPS OF THE WATER SPLITTING ENZYME SYSTEM S IN PHOTOSYNTHESIS

Ö. Saygin and H.T. Witt
Max-Volmer-Institut für Biophysikalische und Physikalische Chemie, Technische Universität Berlin, Strasse des 17. Juni 135, 1000 Berlin 12, FRG

In the primary act of system II of photosynthesis one electron is transferred from the excited $Chl\text{-}a_{II}$ (P680) (1,2) to the first stable acceptor, a plastoquinone, Q_A-X320 (3,4). The $Chl\text{-}a_{II}^+$ extracts electrons via two electron carriers, D_1 and D_2, from the water splitting enzyme S. Thereby, after four $Chl\text{-}a_{II}$ turnovers, $4\ e^-$ are extracted from S and $2\ H_2O$ are decomposed into $4\ H^+$ and one O_2.

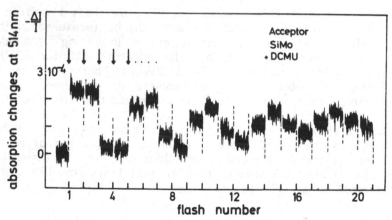

Fig. 1. Absorption changes at 514 nm induced by single laser flashes at isolated O_2-evolving PS II complexes with SiMo as acceptor in the presence of DCMU as a function of the flash number.

After dark adaptation of oxygen evolving PS II complexes and excitation with single flashes, oscillatory absorption changes (stable > 0.5 s) with a periodicity of 4 were detected at 514 nm (see Fig. 1). They have been correlated with the four oxidation states of the water splitting enzyme system S (see Fig. 2). Supposing that the changes are due to electrochromic shifts (5), they might indicate a positive surplus charge in states S_2 and S_3. This means that the electron release pattern 1:1:1:1 is accompanied by a charge formation pattern O:+:+:O and an intrinsic proton release

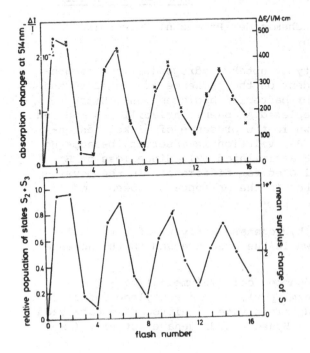

Fig. 2. Top: Absorption changes at 514 nm after the indicated flash, according to Fig. 1. $\Delta\varepsilon$ is related to the concentration of the reaction centers.
Bottom: Relative populations of the calculated S_2 and S_3 states after the indicated flash. On the right, the mean surplus charge of the enzyme S, based on the assumption that S_2 and S_3 are positively charged.

stoichiometry 1:0:1:2 for the transitions $S_0 \rightarrow S_1$, $S_1 \rightarrow S_2$, $S_2 \rightarrow S_3$, $S_3 \rightarrow S_0$. The H^+ release pattern is in accordance with results obtained through pH measurements (6). The charge formation pattern may indicate the valence changes of manganese in the enzyme system S.

References
(1) Döring, G., Stiehl, H.H. and Witt, H.T. (1967) Z. Naturforsch. 22b, 639-644
(2) Döring, G., Renger, G., Vater, J. and Witt, H.T. (1969) Z. Naturforsch. 24b, 1139-1143
(3) Stiehl, H.H. and Witt, H.T. (1968) Z. Naturforsch. 23b, 220-224, (1969) 24b, 1588-1598
(4) Witt, K. (1973) FEBS Lett. 38, 116-118
(5) Witt, H.T. (1979) Biochim. Biophys. Acta 505, 355-427
(6) Junge, W. in Current Topics in Membranes and Transport, Vol. 16, pp. 431-465, Academic Press 1982

BIOENERGETICS OF NITROGENASE IN BLUE-GREEN ALGAE. IV. CONSEQUENCES
OF NICKEL DEPLETION FOR NITROGENASE ACTIVITY IN ANABAENA VARIABILIS

H. Almon and P. Böger
Lehrstuhl für Physiologie und Biochemie der Pflanzen, Universität
Konstanz, D-7750 Konstanz, Germany

As shown recently, the activity of uptake hydrogenase in nitrogen-
fixing blue-green algae is dependent on the presence of nickel in the
culture medium (1,2; comp. [3] for bacterial hydrogenases). Since
hydrogenase activity of nickel-depleted Anabaena variabilis is reduced
in comparison with organisms grown in the presence of nickel, an enhanced rate of net light-induced H_2 evolution is observed, because H_2
produced by nitrogenase cannot be assimilated (2). Since loss of H_2
implies loss of energy, we have looked for differences in the physiology of filaments cultivated either in the presence or absence of
nickel (Table 1).

Table 1: Growth, nitrogenase and hydrogenase activity of batch cultures of Anabaena variabilis grown in the presence and in the absence
of nickel.

Chl = chlorophyll (μg/ml); pcv = packed cell volume (μl/ml); P+A =
phycocyanin + allophycocyanin (mg/ml pcv); C_2H_4 = acetylene reduction
(μmol C_2H_2 reduced/ml pcv x h); H_2 = net rate of nitrogenase-catalyzed
H_2 formation (μmol H_2/ml pcv x h); H_2ase = hydrogenase activity (μmol
H_2/ml pcv x h) [comp. ref. (2)].

| Parameter | Cultivation time | | | | | |
| | 24 h | | 48 h | | 72 h | |
	+Ni	−Ni	+Ni	−Ni	+Ni	−Ni
Chl	3.4	3.4	11.3	9.6	20.2	19.9
pcv	2.1	2.0	5.1	4.7	7.3	7.2
P+A	44.7	43.4	44.8	38.2	29.6	16.6
C_2H_4, light	154.8	182.1	163.7	161.2	145.3	143.4
H_2, light	114.0	133.6	91.1	117.2	47.6	117.8
H_2ase, dark	0.2	0.2	0.7	0.3	4.2	1.8

Growth (measured as increase in chlorophyll content and packed cell
volume) as well as nitrogenase activity (ethylene formation) are not
markedly influenced by nickel depletion. In contrast, the phycobiliprotein (P+A) content of nickel-depleted cells decreases faster than
the control during cultivation time. This is an expression of energy
deficiency, which is not necessarily reflected in the growth parameters used herein. After 48 h, hydrogenase activity increases in both

cultures (though to a different extent). Consequently, differences in the net rate of H_2-evolution become evident only after that time.

As shown in Table 2, nickel concentrations required for hydrogenase activity are very low. Moreover, nickel is accumulated by the filaments regardless of the level of hydrogenase activity (comp. Table 1).

Table 2: Nickel content of cells (filaments) and culture medium at various stages of cultivation as determined by atomic absorption spectroscopy. n.d. = not determined.

Cultivation time (h)	Filaments (μmol Ni/ml pcv)		Culture Medium (μM Ni)	
	+Ni	-Ni	+Ni	-Ni
24	16.2	1.0	2.9	<0.1
48	11.3	0.9	2.7	<0.1
72	5.5	n.d.	1.4	<0.1

We conclude that the main function of uptake hydrogenase in Anabaena is recycling of hydrogen to save energy. During start of cultivation, when the carbohydrate content (glycogen, representing the energetic state) of the filaments is high (A. Ernst et al., this laboratory, paper submitted), the organisms obviously tolerate loss of H_2, whereas during later stages of cultivation, when the carbohydrate content of the filaments decreases, expression of uptake-hydrogenase activity is of energetic advantage (comp. Table 1: content of phycobiliproteins). Data on the glycogen content of +Ni-cultures will be presented in our poster.

An essential role of the "Knallgas reaction" as an O_2-protecting mechanism was not observed, since nitrogenase activity in nickel-depleted cultures is comparable to nitrogenase activity of cultures grown in the presence of nickel.

References

(1) Daday, A. and Smith, G.D. (1983) FEMS Microbiol. Lett. 20, 327-330.
(2) Almon, H. and Böger, P. (1984) Z. Naturforsch. 39c, 90-92.
(3) Thauer, R.K. et al. (1983) Naturwissenschaften 70, 60-64.

HUMAN ANTIMITOCHONDRIAL ANTIBODIES - IS THE ANTIGEN A SIGNAL SEQUENCE?

H. Baum, C. Palmer and P.N. Uzoegwu
Biochemistry Department, Chelsea College (University of London),
Manresa Road, London SW3 6LX, U.K.

Primary biliary cirrhosis (PBC) is a chronic cholestatic liver disease
of unknown aetiology, characterised by high titres of complement-
fixing antimitochondrial antibodies (AMA). AMA are predominantly
directed against a constituent of the inner mitochondrial membrane
(1). Isofixation curves suggest that the antigenic determinant is a
restricted sequence on the antigen, and sensitivity to mercurials
implies that it contains one or more thiol groups (2). The antigen
copurifies with chloroform-released ATPase (3), but is separable from
ATPase enzyme activity (4).

We have applied the Towbin 'blot' procedure (5) to the study of
AMA (6). Samples were dissociated by SDS-PAGE, electroblotted on to
nitrocellulose sheets, treated with serum and then with [^{125}I]-anti
human immunoglobulin. Reactive species were visualised by autoradiography.

The exclusive mitochondrial location of the antigen has been
confirmed. Several antigenic bands are observed in mitochondria
from all species tested, including mammals, protozoa and fungi.
These bands form a pattern characteristic for each species but, (in
the case of mammalian sources), are independent of the organ from
which the mitochondria were isolated. All antigenic species co-
purify with F$_1$ ATPase released from the membrane either by chloroform
(3) or chaotropic ions (7), but represent very minor protein constituents, which do not correspond with the ATPase subunits. The
different antigenic species have been characterised with respect to
MW and sensitivity to various physical and chemical treatments.

Although there is little correspondence between MWs of antigenic
species in different mitochondria, any particular mitochondrial
preparation is apparently able to absorb from PBC serum all antibodies capable of reacting with another mitochondrial preparation,
e.g. beef heart (BH) mitochondria, although exhibiting a completely
different pattern of antigen bands when compared to rat liver (RL)
mitochondria, are able selectively to remove antibodies from PBC
serum, with the result that the adsorbed serum no longer reacts with
BH mitochondria, or RL mitochondria or indeed mitochondria from
hamster brown adipose tissue.

This suggests that the same antigenic determinants exist on proteins
of different MW in different mitochondria. However, evidence from the
study of many PBC sera (105 samples) suggests that the antigenic
determinant may not be identical on the different antigenic species of

any particular type of mitochondria. The pattern of reaction of PBC sera is essentially constant; nevertheless, some sera do not reveal certain of the antigen bands, and the sera may be classified into groups according to the detailed pattern of reaction that they exhibit. The groupings thus identified appear to be independent of the mitochondrial preparation used, confirming the antigenic correspondence between sets of proteins of different MW in different mitochondria. Non-PBC sera (108 samples) did not exhibit these patterns of reactivity.

Experiments with yeast indicate that the antigenic species are cytoplasmically synthesised, (at least in the case of this organism where only 2 antigenic species can be detected, compared with 6 in BH mitochondria), but the expression of the antigen is suppressed under conditions of glucose repression.

The above observations suggest that common epitopes are present on proteins of different MW in different mitochondria. These sets of highly antigenic proteins are present in very small amounts and are of high MW. The evidence is thus compatible with the antigenic sequences being related to 'signal' sequences that have not yet been cleaved from high-MW, unprocessed precursors of cytoplasmically-synthesised mitochondrial constituents. Experiments are under way on yeast mitochondria to test this possibility.

References

1. Berg P.A. & Baum H. (1980) Springer Seminars in Immunopathology 3 355-373
2. Baum H. & Berg P.A. (1981) Seminars in Liver Diseases (Jones E.A., ed.) 1 pp. 309-321, Thieme-Stratton Inc., N.Y.
3. Beechey R.B., Hubbard S.A., Linnett P.E., Mitchell A.D. & Munn E.D. (1975) Biochem. J. 148 533-537
4. Sayers T., Leoutsakos A., Berg P. & Baum H. (1981) J. Bioenerg. Biomemb. 13 255-257
5. Towbin H., Staehelin T. & Gordon J. (1979) Proc. Natl. Acad. Sci. USA 76 4350-4354
6. Baum H. & Palmer C. (1982) EBEC Reports 2 75-78
7. Tzagoloff A., MacLennan D.H. & Byington K.H. (1968) Biochemistry 7 1596-1602

686

FACTORS INVOLVED IN THE INSERTION OF D-β-HYDROXYBUTYRATE APODEHYDROGE-NASE (apoBDH) INTO PHOSPHOLIPID MONOLAYERS OR IN PHOSPHOLIPID VESICLES

J.M. Berrez, F. Pattus°, N. Latruffe and M.S. El Kebbaj
Laboratoire de Biochimie ERA CNRS 1050 "Biologie Moléculaire des Membranes" Université de Franche-Comté 25030 Besançon France
°EMBL 6900 Heidelberg RFG

BDH is an integral protein located in the matrix face of inner mitochondrial membrane (1). Its activity is specifically dependent of PC which allows the binding of the coenzyme (NAD(H)) to the active site i.e. the first step of the catalysis (2). Meanwhile the non reactivating mitochondrial phospholipid (mainly PE and DPG) are able to modify the reactivation process (3).

Progress in the knowledge of BDH-phospholipid interaction have been done some years ago by using a purified soluble apoenzyme in a lipid and detergent free form (4). Informations were obtained on the role of polar head (5) and hydrophobic moiety of phospholipid in the reactivation process.

In this present report, factors involved in the insertion of apoBDH into phospholipid have been demonstrated using the monolayer technique (6).

Injection of the apoenzyme into the subphase is accompanied by an increase in the surface pressure of the phospholipid monolayer. This indicates a penetration of the protein into the film. At an initial surface pressure of 28 mN/m the apoenzyme is no longer able to penetrate mitochondrial PC or PE monolayers while under the same conditions it is able to penetrate DPG film up to a surface pressure of 35 mN/m. This result suggests

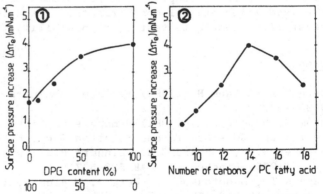

Figure 1 : Influence of different PC/DPG molar ratio on the penetration of apoBDH. Experimental conditions : initial surface pressure : 25 mN/m, subphase buffer 20 mM Tris HCl pH 6.5, 1 mM EDTA, 1 mM DTT, 25°C. ApoBDH (0.32 ug/ml) was injected into the subphase.

Figure 2 : Influence of phospholipid fatty acid chain length on the penetration of apoBDH into PC monolayer. Experimental conditions see legend of figure 1 excepted that initial surface pressure was 21 mN/m.

a strong interaction between apoBDH and DPG molecules in agreement with results obtained using the gel-filtration-centrifugation technique. Indeed we have shown that apoBDH is much more tightly bound to DPG/PE liposomes than pure PC liposomes.

The effect of DPG on the affinity of the apoenzyme for the lipid-water interface was confirmed by the experiments carried out with
mixed PC/DPG films containing increasing amount of DPG (figure 1).
Other negative phospholipid (PA, PG and PS) behave like DPG. There is
a specific decrease in the affinity of the apoenzyme for DPG when the
pH is increased from pH 7.5 to 9.

The influence of fatty acid chain length on the penetration of
the apoenzyme into lecithin monolayer is shown in figure 2. The
increase of surface pressure observed after protein injection increases with the phospholipid chain length up to 14 carbon atoms. The decrease of penetration observed with longer acylchain length could be
explained by the phase transition which occurs between the expanded
state and the condensed state.

The penetration increases with increasing NaCl concentration up
to 0.5 M. This indicates that the charges neutralization enhances the
role of hydrophobic interactions in the BDH penetration process.

Using radioactive $125I$-BDH we observed that a minimum of 75
phospholipid molecules are required to insert one apoBDH molecule.

Conclusions : ApoBDH strongly penetrates into phospholipid films.
The factor involved are :
    - hydrophilic interactions especially between positive charged
aminoacid residues of BDH polypeptide chain and negative charge of
phospholipid polar head, mainly from DPG.
    - hydrophobic interactions between uncharged region of BDH poly-
peptide chain and fatty acid moiety of phospholipid.
    - the fluidity state of phospholipid.
    - the presence of a minimum of phospholipid molecules for BDH
insertion.

References :
1. Gaudemer Y. & Latruffe N. (1975) Febs lett. 54, 30-34.
2. Gazzotti P., Bock H-G.O. & Fleischer S. (1974) Biochem. Biophys.
   Res. Comm. 58, 309-315.
3. Gazzotti P., Bock H-G.O. & Fleischer S. (1975) J. Biol. Chem. 250,
   5782-5790.
4. Bock H-G.O., Fleischer S. (1974) Meth. Enzymol. 32, 374-391.
5. Isaacson Y.A., De Roo P.W., Rosenthal A.F., Bittman R., Mc Intyre
   J.O., Bock H-G.O., Gazzotti P. et Fleischer S. (1979) J. Biol.
   Chem. 254, 117-126.
6. Verger R. et Pattus F. (1982) Chem. Phys. Lipids. 30, 189-227.

ACKNOWLEDGEMENTS

This work was supported by a grant from the CNRS ATP n° 8155.

METALLOCHROMIC PSEUDONUCLEOTIDES

K.-S. Boos and E. Schlimme

Lab.Biol.Chem., University Paderborn, P.O.Box 1621,
D-4790 Paderborn, F.R.G.

Chemically modified nucleotides are successfully employed
as molecular probes for the topochemical and topographical
in situ analysis of enzymatic binding centers or receptor
areas (1).
As far as the mitochondrial adenine nucleotide carrier,
i.e., a chemodynamical receptor is concerned, we could
show that - besides the traditional way of pure chemical
modification (2,3,4) - the strategy of molecule-variation
leads to biochemically interesting pseudosubstrates (5,6).

Starting from the principle of molecular complementarity
and due to a structure-activity study we could find out
the chemo- and biofunctional as well as bioisosteric sub-
structures of these compounds.
Based primarily on these as well as on the findings with
substrate analogs we proposed a new model for carrier-
-mediated adenine nucleotide transport (7).
Moreover, these investigations led to a new concept of
molecule-combination. This concept comprises the chemical
combination of substrate- and inhibitor-relevant sub-
structures to a bifunctional and bioanalog compound, i.e.,
a pseudonucleotide (8).
The inhibitory substructure corresponds to the minimal
structure of metallochromic pseudosubstrates (6). These

diarylazo compounds bear o,o'-positioned hydroxyl groups and are - in analogy to ADP - able to chelate a postulated metal-ion at the cytosolic substrate binding site of the adenine nucleotide carrier protein.

Due to their spectroscopic and chemical properties these pseudonucleotides are of potential usefulness in studying protein-metal-nucleotide interactions.

1) Yount,R.G. (1975) Adv.Enzymol.Relat.Areas Mol.Biol. 43, 1

2) Schlimme,E., Boos,K.-S., Bojanovski,D. & Lüstorff,J. (1977) Angew.Chem. 89, 717

3) Boos,K.-S. & Schlimme,E. (1979) Biochemistry 18, 5304

4) Schlimme,E., Boos,K.-S. & de Groot,E.J. (1980) Biochemistry 19, 5569

5) Boos,K.-S. & Schlimme,E. (1981) FEBS-Lett. 127, 40

6) Boos,K.-S. (1982) Biochim.Biophys.Acta 693, 68

7) Boos,K.-S. & Schlimme, E. (1983) FEBS-Lett. 160, 11

8) Boos,K.-S. (1984) Chemiedozententagung, Konstanz

THE EFFECT OF PROTEIN CONCENTRATION ON THE RATE OF LATERAL DIFFUSION
ON INNER MEMBRANE COMPONENTS OF THE MITOCHONDRIA

B. Chazotte°, E.-S. Wu", and C.R. Hackenbrock°
°Labs for Cell Biology, Dept. of Anatomy, Univ. of North Carolina
Chapel Hill, NC   27514 U.S.A.
"Permanent address: Dept. of Physics, Univ. of Maryland-Baltimore
County, Baltimore, MD, 21228, U.S.A.

Our laboratory has postulated an essential role for lateral diffusion of membrane components in the mechanism for electron transport
in the mitochondrial inner membrane, contending that electron
transport is "diffusion coupled" (1,2). In a diffusion coupled
mechanism, which requires the collision of membrane components, factors that affect the rate of lateral diffusion must in turn affect
component interaction.

The two major parameters in any collision-based mechanism are the
concentration of the reactive components and their diffusion coefficients. Consequently any pertubation of the protein density (concentration) of the inner membrane should affect the rate of electron
transport. Previous work from our laboratory has established that as
the membrane protein density is decreased the rate of electron
transfer between specific redox components also decreases (3).

Fluorescence recovery after photobleaching (FRAP) was employed to
measure the lateral diffusion of the lipid analogues 3,3'-dioctyl-
decylindocarbocyanine (DiI), N-4-nitrobenz-2-oxa-1,3-diazole (NBD-PE)
and selected immunofluorescently labeled redox proteins in intact rat
liver mitochondrial inner membranes. To decrease protein density the
inner membranes were selectively enriched in phospholipid using the
low pH method of Schneider et al. (3). The fusion of the native as
well as the enriched membranes to a diameter suitable for FRAP was
accomplished by the calcium fusion method of Chazotte et al. (4).
The latter technique produces unilamellar, osmotically active
membranes 10-200 microns in diameter. Freeze fracture electron
microscopy confirmed that the fused membranes have randomly distributed proteins with the same density as the phospholipid enriched
membranes from which they are derived.

Lipid lateral diffusion was found to increase in correspondence
with the decrease in membrane protein density ranging from $D=3.9 \times 10^{-9}$ cm^2/sec in the unenriched membrane to $D=1.3 \times 10^{-8}$ cm^2/sec in
the 700% enriched membrane. The temperature dependence, as deter-
mined via an Arrhenius analysis, of lipid diffusion was found to be
greater in the native (protein dense) membranes than in pure lipid

systems and the temperature dependence was found to progressively decrease as the integral membrane protein density decreased.

The overall protein density appears to influence the rate at which membrane lipids as well as catalytically important redox proteins diffuse in the inner membrane. For example, the diffusion of the bc_1 redox protein was faster in 30% enriched membrane than in the native unenriched membrane. These results indicate that membrane protein density can limit the rate of lateral diffusion of mitochondrial inner membrane components.

References

1) Hackenbrock, C.R. (1981) Trends Bioch. Sci. 6, 151-154
2) Gupte, S., Wu, E-S., Höchli, M., Jacobson, K.A., Sowers, A.E., & Hackenbrock, C.R. (1984) Proc. Nat. Acad. Sci. USA, in press.
3) Schneider, H., Lemasters, J.J., Höchli, M., & Hackenbrock, C.R. (1980) J. Biol. Chem. 255, 3748-3756
4) Chazotte, B., Wu, E-S., & Hackenbrock, C.R. (1983) Fed. Proc. 42, 2170

EFFECT OF METHOMYL AND HELMINTHOSPORIUM MAYDIS TOXIN ON THE
BIOENERGETICS OF MAIZE MITOCHONDRIA.
A.Ghazi', A.Bervillé", M.Charbonnier" and J.F.Bonavent".

'Laboratoire des Biomembranes, Université de Paris-XI, 91405 Orsay
"Laboratoire de Mutagénèse, INRA BV 1540, 21034 Dijon, France.

The fungus Helminthosporium maydis severely attacks maize plants
carrying the Texas male sterile cytoplasm (Texas plant), causing the
disease known as Southern corn leaf blight (1).The host specific
pathotoxin responsible for the disease, T-toxin from H.maydis, has
been isolated from culture filtrates and structurally characterized .
Carbamate methomyl, an insecticide whose molecular structure is quite
different from that of T-toxin, has a similar effect when sprayed on
Texas plants (2). Plants carrying one of the other cytoplasms,
Normal(=N), Charrua (=C), or S, are insensitive to T-toxin, and
methomyl. Methomyl and T-toxin have been shown to block oxidative
phosphorylation in mitochondria isolated from Texas male sterile
plants (T mitochondria) but have no action on mitochondria isolated
from Normal cytoplasm (N mitochondria)(3). Nevertheless, the mechanism
of action of these compounds on T mitochondria is unknown. Since
cytoplasmic male sterility is determined by the mitochondrial genome,
and since mutants from T cytoplasm plants, resistant to T-toxin, are
also revertant for male fertility, the understanding of the mechanism
of action of methomyl and T-toxin on mitochondria may shed some light
on the basic biochemical events underlying cytoplasmic male sterility.

We examined the effect of methomyl and T-toxin on the protonmotive
force (PMF) of maize mitochondria. Upon addition of a substrate, N and
T mitochondria maintain a PMF of some 200 mV mainly under the form of
a membrane electrical potential. In the presence of methomyl or
T-toxin, the PMF is absent in T mitochondria, while it is virtually
unchanged in N mitochondria.

A possible effect of methomyl or T-toxin could be a
permeabilization of the internal mitochondrial membrane to ions, and
especially to protons. The mechanism of such an effect could not be
identical to that of the unspecific one of uncouplers, since no effect
is observed in N mitochondria. It is possible to conceive,
nevertheless, that methomyl or T-toxin interact with a membrane
protein, present in T mitochondria and absent (or modified) in N
mitochondria, thus leading to the opening of a channel for protons or
other ions and subsequently to the observed collapse of the PMF.

However, the absence of a PMF in T mitochondria in the presence of methomyl or of T-toxin, which can account for the lack of oxidative phosphorylation, does not explain per se observations concerning the action of these compounds on NADH, and malate sustained respiration: while methomyl and T-toxin induce an increase in the oxidation of NADH, they completely inhibit malate oxidation in T mitochondria. This effect is in striking contrast with the stimulation induced by uncouplers. The action of methomyl and of T-toxin cannot therefore be explained only in terms of uncoupling.

We have shown that malate oxidation can be restored after methomyl or T-toxin treatment by addition of NAD+; this strongly suggests that inhibition of malate oxidation could be due to an efflux of NAD+ out of the mitochondria, induced by methomyl or T-toxin, since, in the absence of NAD+, malate cannot be oxidized to pyruvate or to oxaloacetate. Douce and collaborators (4) have shown that, in potato tuber mitochondria, NAD+ is actively transported and that this transport has a direct influence on malate oxidation. We have shown that NAD+ is also transported in maize mitochondria and that methomyl or T-toxin prevent accumulation of NAD+ or promote efflux of accumulated NAD+ in T mitochondria. This transport is not dependent on the PMF, since it is unaffected by FCCP. Therefore the action of methomyl and T-toxin on NAD+ transport is not a consequence of their action on the PMF.

In conclusion, methomyl and T-toxin have apparently at least two different effects in T mitochondria: i) collapse of PMF which can account for the lack of oxidative phosphorylation ii) release of NAD+ which can account for the inhibition of malate oxidation. Further studies should help to decide whether these two effects correspond to different targets for the toxin or methomyl, or whether they are the consequence of a more general phenomenon.

1- Daly, J.M. and Knoche,H.W. (1982) in "Chemistry and Biology of Pathotoxins exhibiting host selectivity". Academic Press. N.Y.
2- Humaydan, H.S. and Scott, E.W. (1977) Hort.Sci. 12, 312-320.
3- Bednarski, M.A., Izawa, S. and Scheffer, R.P. (1977) Plant Physiol. 59,540-545.
4- Tobin, A., Djerdjour, B., Journet, E., Neuburger, M. and Douce, R. (1980) Plant Physiol. 66:225-229.

HYDROLYSIS OF PURPLE MEMBRANE LIPIDS BY PHOSPHOLIPASE D

Félix M. Goñi and Arturo Muga
Department of Biochemistry,Faculty of Science,University of
the Basque Country,P.O. Box 644,Bilbao,Spain.

Halobacterium purple membrane is a very interesting system because it contains a single protein,bacteriorhodopsin,
excepcionally well characterized (1).The phospholipid composition of the purple membrane is very peculiar,and phospholipase A_2 is unable to act because there are no ester
bonds.No phospholipase C or D activity has been reported
either.In this paper we describe the effect of phospholipase D on purple membrane and derived liposomes.

A series of experiments have been carried out on the
major phospholipid of the membrane,an ether analogue of a
phosphatidylglycerophosphate (PGP). This lipid was isolated
(2), as a barium and sodium salt;liposomes were readily
formed upon hydration.Treating the liposome suspension
with phospholipase D,turbidity increases (providing a con-
venient assay procedure) and L-α-glycerophosphate is relea-
sed.The enzyme effect on PGP (sodium salt) liposomes is
shown in Fig.1. No major difference has been detected between the behaviour of Na^+ and Ba^{++} salts.Enzyme activity
is shown as a change in turbidity (ΔA_{500}) or as phosphate
released in the form of glycerophosphate and assayed in the
form of P_i after suitable digestion.No inorganic phosphate
is released as such by the enzyme.Phospholipase D activity
shows a pH optimum at 5.3.The Arrhenius plot is non-linear
with a discontinuity near 19°C.The enzyme requires Ca^{++}
(optimum concentration 45 mM) and its activity is suppre-
sed by EGTA or bovine seroalbumin.Both,sodium cholate and
glycerol enhance the enzyme activity as shown in Fig. 2.

Experiments have also been carried out using purple membrane as substrate.Phospholipase D is also active on the
intact purple membrane,and L-α-glycerophosphate is also a
product.The effect of enzyme activators and inhibitors
appears to be the same as in the case of PGP liposomes. It
will be interesting to see the extent to which lipid modification alters bacteriorhodopsin properties.

1. Stoeckenius, W.,Lozier, R.H.,and Bogomolni, R.A. (1979)
   BBA 505,215-278
2. Kates, M.,Yengoyan,L.S. (1965) BBA 98,252-268

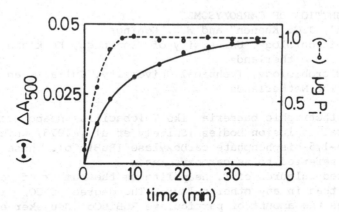

Fig. 1. The effect of phospholipase D on purple membrane de
rived liposomes (PGP, see text) as a function of
time. (●) increase in turbidity; (O) phosphate
liberation (in the form of glycerophosphate, see text).
Assays were carried out at 30°C.

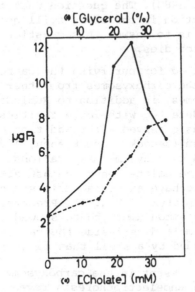

Fig. 2. Activation of phospholipase D activity by (●) sodium
cholate and (O) glycerol. Enzyme activity was measured
at 30°C as phosphate liberation.

STRUCTURE AND FUNCTION OF CARBOXYSOMES
Y.A. Holthuijzen', J.G. Kuenen" and W.N. Konings'
'Department of Microbiology, University of Groningen, Kerklaan 30,
9751 NN Haren, The Netherlands
"Department of Microbiology, Technical University, Julianalaan 67A,
2628 BC Delft, The Netherlands

Many chemolithotrophic bacteria like Thiobacillus neapolitanus
contain polyhedral inclusion bodies (Shively et al., 1973) which
contain ribulose-1,5-bisphosphate carboxylase (RuBisCo). These
microbodies are referred to as carboxysomes.

In CO_2 limited cultures of T. neapolitanus the number of carboxysomes is higher than in any other culture. The degree of CO_2 limitation determines the amount of particulate RuBisCo (Beudeker et al.,
1981). Beside RuBisCo, many other enzymes were reported to be present
inside the carboxysomes: all the enzymes of the Calvin-cycle and a
set of enzymes that made it possible for malate to act as reductant
inside the microbodies (malate shuttle, Beudeker and Kuenen, 1981).
These observations, however , could not be confirmed by other investi-
gators (Cannon and Shively, 1983). The question what the role of the
carboxysomes is in CO_2 fixation is therefore still not answered.
The aim of our studies is to answer this question and to analyse
the structure of these microbodies.

A new method was developed for purifying the carboxysomes resulting in full seperation of the carboxysomes from other cell-material.
In these purified carboxysomes, in addition to RubisCo, only phospho-
glycerate kinase could be detected with low activities. Stimulation
of CO_2 fixation in vitro was observed after addition of ATP.
The malate shuttle mechanism (Beudeker and Kuenen, 1981) could not
be detected in our purified carboxysome preparations. Malate did not
stimulate CO_2 fixation and no malate dehydrogenase activity was found.
The carboxysomes seem to have an icosahedric structure in Pt-
shaded electron microscope pictures (Fig. 1). Preparations treated
with 1% uranylacetate or 3% ammoniummolybdate showed an orderly
arrangement of the RuBisCo molecules inside the carboxysomes. The
10 nm molecules are surrounded by a shell that maintains its icosahedric shape after breakage.

The number of proteins present in the carboxysomes is not completely elucidated. Crossed immunoelectrophoresis revealed the presence
of at least 4 to 5 proteins and iso electric focusing and SDS-PAA gelelectrophoresis indicate that even more proteins are present.
Also the nature of the shell of the carboxysomes is still unknown.

It has been reported that carboxysomes of Nitrobacter winogradskyi
and N. agilis contain DNA (Westphal et al, 1979). No evidence for the
presence of DNA in carboxysomes of T. neapolitanus could be obtained.

However, large amounts of DNA were attached to and perhaps penetrating the carboxysomes. This DNA appears to be of chromosomal origin.

Incubation of our purified carboxysomes with ribulose diphosphate and CO_2 results in the formation of five sofar unknown metabolites indicating the involvement of more than one enzyme in CO_2 fixation in the carboxysomes.

We are currently investigating the nature of these metabolites and the enzymes that are responsible for their production.

Fig. 1. A carboxysome of Thiobacillus neapolitanus (16 x 20.000), Pt-shaded, angle 15°.

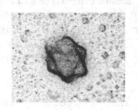

References

Beudeker, R.F., G.A. Codd and J.G. Kuenen (1981) Arch. Microbiol. 129: 361-367
Beudeker, R.F. and J.G. Kuenen (1981) Febs Lett. 131: 269-274
Cannon, G.C. and J.M. Shively (1983) Arch. Microbiol. 134: 52-59
Shively, J.M., F. Ball, D.H. Brown and R.E. Saunders (1973) Science 182: 584-586
Westphal, K., E.Bock, G. Cannon and J.M. Shively (1979) J. Bact. 140: 285-288

THE LACTOSE PERMEASE OF E. COLI: EVIDENCE IN FAVOR OF A DIMER.
C. Houssin', M. le Maire'' and E. Shechter'

'Laboratoire des Biomembranes, Université de Paris–Sud, 91405 Orsay
'' Centre de Génétique Moléculaire, C.N.R.S., 91190 Gif (France).

The lactose permease of E. coli is an integral membrane protein
that catalyses a sugar/H+ symport. The protein has been purified to
homogeneity (1); the molecular weight of a single polypeptide chain is
46,500. Although several hypothesis have been proposed (2,3), the
state of association of the protein in the membrane is still unknown;
its knowledge is important in understanding the molecular mechanism of
transport. We report the characterization, by gel filtration and
analytical ultracentrifugation, of the lactose permease in an
homogeneous non–denaturing detergent, dodecyl octaethylene glycol
monoether (C(12)E(8)).

Exchange of octyl glucoside for C(12)E(8). The protein solubilised
in octyl glucoside and purified as described in (1) has a strong
tendency to precipitate. Therefore, we exchanged octyl glucoside for
C(12)E(8) in which the protein is more stable. The exchange is carried
by binding the protein reversibly to a carboxymethyl cellulose (CM)
column at low ionic strength and washing with buffer containing
C(12)E(8); in addition, the washing eliminates the excess
phospholipids. The protein was then eluted from the column by
increasing the ionic strength.

Chemical characterization of the protein/detergent/phospholipid
complex. Bound C(12)E(8) was determined using radioactive detergent;
bound phospholipids was determined by extracting the lipids and
analyzing their fatty acids by gas chromatography. 0.2 g of detergent
and 0.15 g of phospholipids are bound per g of protein. This
corresponds to a total of 27 alkyl chains per polypeptide chain of
molecular weight 46,500.

Gel filtration: determination of the Stokes radius. The Stokes
radius of the complex is determined by gel filtration on a Sephacryl S
300 column. The protein elutes as a single peak which does not
coincide with the void volume of the column. This excludes the
presence of large aggregates. However, the peak is asymmetric: the
maximum is reached very steeply but declines more slowly suggesting
the presence of minor component(s) of smaller sizes. The Stokes radius
of the major component associated with the maximum of the peak is 5.3
nm.

Analytical ultracentrifugation. The sedimentation coefficient of the complex was determined by sedimentation velocity of the fraction eluted from the CM cellulose column. The calculated s (20 C, water) is 3.79 S.

We have also performed sedimentation equilibrium experiments. The representation: ln (protein concentration) as a function of the square of the radial distance, was not a straight line indicating the presence of more than one molecular species, 25 % being in the form of monomer and 20 % in the form of higher oligomers. In addition, calculation of recovery (4) indicated that more than 50 % of the complex has precipitated, probably as a result of the high concentration reached at the bottom of the cell. Therefore, in this case, sedimentation equilibrium was not suitable to study the state of association of the lactose permease.

Molecular weight of the complex. The average molecular weight of the protein/detergent/phospholipid complex calculated from the combination of the sedimentation coefficient and the Stokes radius is 105,000. Allowing for the bound detergent and bound phospholipids, this corresponds to protein molecular weight of 90,000, i.e., to the size of a dimer. The presence of smaller molecular weight species was inferred from these techniques and confirmed by sedimentation equilibrium measurements. These minor components represent at most 25 % of the total material.

We suggest that the functional state of the lactose permease in the membrane is a dimer. Furthermore, the detergent seems to bind as a monolayer on the protein (54 alkyl chains per dimer for an aggregation number of 120) as already reported for polyoxyethylene detergents (5).

1 – Newman, M.J., Foster, D.L., Wilson, T.H. and Kaback, H.R. (1981) J. Biol.Chem. 256, 11804–11808.
2 – Goldkorn, T., Rimon, G., Kempner, E.S. and Kaback, H.R. (1984) Proc. Natl. Acad. Sci. (US) 81, 1021–1025.
3 – Wright, J.K., Weigel, U., Lustig, A., Bocklage, H., Mieschendahl, M., Muller-Hill, B. and Overath, P. (1983) FEBS Lett. 162, 11–15.
4 – le Maire, M., Lind, K.E., Jorgensen, K.E., Roigaard, H. and Moller, J.V.(1978) J. Biol. Chem. 253, 7051–7060.
5 – le Maire, M., Kwee, S., Andersen, J.P. and Moller, J.V. (1983) Eur. J. Biochem. 129, 525–532.

SUBFRACTIONATION OF THE RAT LIVER MITOCHONDRIAL OUTER MEMBRANE

V. Jancsik[1], B.D. Nelson, M. Lindén and L. Ernster, Department of
Biochemistry, University of Stockholm, Stockholm, Sweden

[1]permanent address: Institute for Enzymology, Academy of Sciences,
Budapest, Hungary

Recent studies (1) provide evidence for points of attachment
between the inner and outer mitochondrial membranes which are regulated by the energetic state of mitochondria or by free fatty acids.
Attachment points between the two membranes have also been proposed
to explain import of cytoplasmic precursor peptides into the mitochondrial matrix or inner membrane (2). In this report, we present
the results from our initial attempts to locate such attachment
points by graded digitonin fractionation of the outer membrane.

The rational for these experiments was provided by the studies
of Schnaitman and Greenawalt (3) on the digitonin fractionation of
rat liver mitochondria. These authors showed that adenylate kinase
(AK) was released from the intermembrane space at lower concentrations of digitonin than required for release of monoamine oxidase
(MAO) from the outer membrane. We have repeated this result using
the conditions of Schnaitman and Greenawalt in which fat free BSA is
present during digitonin treatment. We further found, however, that
the order of release of MAO and AK was reversed in the absence of
BSA, i.e., MAO is released at low concentrations of digitonin whereas
AK remains associated with the mitochondrial particles.This finding
suggests that the association of MAO with the inner membrane is somehow promoted by BSA. BSA need not be present to produce its effect
on the release of MAO. It is sufficient to wash mitochondria once
with 0.5 mgBSA/ml (pre-treated mitochondria) prior to digitonin
treatment (Fig. 1).

To exclude possible artifacts due to the use of digitonin, breakage of the outer membrane and release of AK was also induced by hypotonic shock (4). BSA pre-treated RLM diluted in 0.125M or 0.025M
sucrose exhibited a time-dependent release of AK and MAO. Release
was more rapid at lower osmolarities. However, release of AK was always more rapid than the release of MAO, a result which is consistent
with those obtained with digitonin as well as with the idea that MAO
remains associated with mitochondria under conditions in which AK is
released from the intermembrane space.

Further analysis of the proteins released during digitonin titration of BSA-treated RLM was done by SDS-PAGE and by immunoblotting
using specific antiserum against rat liver porin. Groups of peptides
were detected by coomaisse blue staining which were released conco-

mitantly with either AK or MAO. Porin was released together with AK (Fig. 1). Thus, a certain domain of the outer membrane (porin containing) is released together with intermembrane space components (AK) under conditions in which a second outer membrane domain (containing MAO) remains attached to mitochondrial particles.The role of BSA in promoting attachment of MAO is reminiscent of its role in inducing contact points between the two membranes (1). More detailed studies are underway to determine if MAO is located at such a contact point.

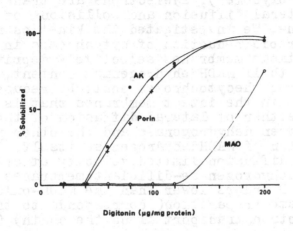

Fig. 1. Release of specific proteins from BSA pre-treated rat liver mitochondria. Adenylate kinase (AK) and monoamine oxidase (MAO) were measured enzymatically. Porin was measured by immunoblotting.

References.

1) Klug, G.A., Krause, J., Östlund, A.K., Knoll, G. and Brdiczka, D. (1984) Biochim. Biophys. Acta 764 272-282.

2) Daum, G., Gasser, S., and Schatz, G. (1982) J. Biol. Chem. 257, 13075-13080.

3) Schnaitman, C. and Greenawalt, J.W. (1968) J. Cell. Biol. 38, 158-175.

4) Pfaff, E. and Schwalbach, K. (1967) in Mitochondrial Structure and Compartmentation. (Quagliarello et al., eds.) Adriatica Editrice, Bari, pp. 346-350.

DYNAMIC ASSEMBLY MODEL FOR RESPIRATORY CHAIN ORGANIZATION IN BACTERIAL MEMBRANE. THE ROLE OF PROTEIN-LIPID INTERACTIONS IN THE ASSEMBLY STABILIZATION

A.S.Kaprelyants, D.N.Ostrovsky
A.N.Bach Institute of Biochemistry, USSR Academy of
Sciences, 117071 Moscow, USSR

There are at least two possibilities for electron transport chain (ETC) functioning in the membrane: I) the electrons are transported by a multienzyme assembly of carriers ("oxysome"), 2)electrons are transported on account of lateral diffusion and collisions of carriers in the membrane. We investigated the kinetics and the level of anaerobic reduction of cytochromes in Micrococcus lysodeikticus membranes selectively deprived of nine tenth of their NADHdehydrogenase content. Under these conditions thecytochromes could be reduced to the same degree as in the intact membranes that is indicative of existance either of lateral diffusion of the ETC component(s) between dehydrogenase and the other part of ETC or diffusion of NADHdehydrogenase itself. At the same time, initial diffusion-limited velocity of cytochromes reduction in dehydrogenase-dificiant membranes by NADH was almost by 2 orders lower than the NADH-oxidase activity of the same preparation(corresponds to the velocity of the electron transport along the chain) (I).

After crosslinking of proteins in the isolated M. lysodeicticus membranes with glutaric dialdehyde the membrane retained 50-60% their capacity to oxidize NADH in comparison with the untreated ones. The size of the NADH-oxidase target in intact and crosslinked membranes calculated from the radiation inactivation experiments corresponds to 47 and 150 kDa. The target size for NADH-dehydrogenase activity was equal(47 kDa) in both preparations (2). We conclude that: I)a rapid exchange of individual components(or complexes), that is lateral diffusion, exists between different chains, 2)long-living multienzyme assembly of ETC exists only under artificial conditions (crosslinking), 3)ETC is organised in the form of a dynamic assembly having finit life time, 4)effectiveness of electron transport through transient complex is much higher than that by lateral diffusion and collisions of carriers.

In oder to establish a possible correlation between

the expression of the boundary lipid (B.L.) and the NADH-
oxidase activity, the temperature dependences of the
membranes of bacteria grown at 14 and 38° were investigated. The state of B.L. in M.lysodeikticus membranes is
assessed by comparing the excimerization parameters of
fluorescent probe pyrene in the free bilayer and in the
vicinity of proteins using energy transfer from the proteins tryptophanyls onto pyrene(3). B.L. begins to dissa-
per when a difinite temperature (T_{melt}) is reached. The
T_{melt} for B.L. correlated with growth temperature. A similar temperature dependence was observed with NADH-oxidase activity, i.e. inhibition of activity at $T > T_{melt}$.
Incorporation of cis-unsaturated fatty acid (UFA) into
membranes markedly decreased the structural heterogeneity of lipids and caused a simultaneous inhibition of
NADH-oxidase activity. NADHdehydrogenase activity was
shown to be much more resistant to the UFA action. Saturated fatty acid failed to alter either lipid structural
state or enzymatic activity (4). We proposed that the
equlibrium "respiratory assembly $\rightleftharpoons$ mobile carriers" is
governed by the state of nearest to proteins lipid
layer(s). Homogenization of the boundary and free lipid
by UFA and high temperature is expected to trigger the
assembly dissociations, when elrctron transport is diffu-
sion-controlled, and procceds on account of carriers
collisions, this being less effective than the intra-
complex transport.

1.Kaprelyants A.S., Ostrovsky D.N., 1975, Biokhimiya
   40:1210-1215.
2.Zinoveva M.E., Kaprelyants A.S., Ostrovsky D.N.,
   1983, Molec.cell.biochem. 55:141-144.
3.Dergunov A.D., Kabishev A.A., Kaprelyants A.S.,
   Ostrovsky D.N., 1981, FEBS Letters, 131:181-185.
4.Kaprelyants A.S., Dergunov A.D., Ostrovsky D.N.,
   1983, Biokhimiya, 48:2049-2055.

NATURAL MITOCHONDRIA WELL CONSERVE THEIR STATE IN ORGANISM

M.N.Kondrashova and E.V.Grigorenko
Institute of Biological Physics, Academy of Sciences of the USSR,
142292 Pushchino, Moscow Region, USSR

The complex of conditions is elaborated for physiological investigations of isolated mitochondria (M) providing good conservation of
the state of M in organism (1). Some details of standard procedures to
investigate M were found to induce artificial acceleration of respiration rate. This masks an increase of respiration arising under excitement of organism in isolated M. An essential condition for revealing
physiological changes of respiration in isolated M is providing the
existence of aggregations (A) in suspension of M, which distinguishes
it from usual preparations of "intact" M represented by separate granules. Aggregated state of isolated M is a manifestation of their natural property to form associations in cell.

A of M in isolated preparations are observed under the following
conditions.
(1) Using concentrated suspension of M under storage (60–80 mg prote-
in/ml, MP, for rat liver M) and high concentration of M during incubation (4–5 MP). A high concentration under storage is more significant
than that under incubation.
(2) Exclusion of washing M.
(3) Exclusion of any procedure destroying A of M: homogenization and
pipetting stock suspension in the course of its preparation.
(4) Sampling M for incubation by reliably cooled pipette.
(5) Using M for experiments within 40–60 min after isolation.
(6) Providing quiet state of intact (control) animals by avoiding any
additional excitement.

Size of A differs in diluted suspensions (2–4 MP) obtained from
two stock suspensions 75 and 40 MP. Diluted M originating from 75 MP
suspension are represented mainly by large A 18–25 μ^2 (71%), while
those originating from 40 MP suspension are represented mainly by
small A 6–12 μ^2 (78%). Fig. 1 shows that respiration of M originating
from less concentrated suspension is somewhat higher (I,II), which we
explain by dissipation of A. It is of importance that this increase
in respiration is connected with a decrease of the difference between
M taken from quiet and excited animals, in this case those under 24
hour stress (I,II). Washing M results in a greater acceleration of respiration and in a complete loss of the difference between M from quiet and stressed animals (III). Physiological difference in respiration of M diminishes during storage in cold due to a decrease of respiration of excited animal M and to an increase of respiration of quiet

animal M (Fig. 2). Meantime, a dissociation of large A into smaller is observed. The data described suggest that the presence of A in isolated M may be considered as an evidence for conservation of their natural properties.

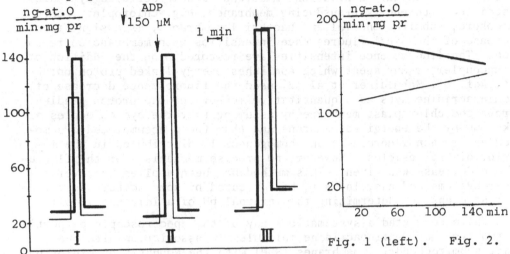

Fig. 1 (left).   Fig. 2.

Fig. 1. The dependence of physiological difference in respiration of M upon their concentration and washing. Thin line, quiet animal; thick line, 24 hour immobilization stress. MP under storage and incubation correspondingly: I, 75 and 4.5, unwashed; II, 40 and 2.2, unwashed; III, 75 and 4.5, washed. Incubation media 0.3 M sucrose, 5 mM tris-HCl, 50 mM KCl, 3 mM KH_2PO_4, 3 mM $MgCl_2$.

Fig. 2. A time dependence of physiological changes in phosphorylating respiration of M conditions as for I in Fig. 1.

References:
(1) M.N.Kondrashova, E.V.Grigorenko et al. (1981) Biofizika (USSR) 26, 687-691.

DO THE ENERGY-LINKED FLUORESCENCE DECREASES OF AMINOACRIDINES QUANTI-
TATIVELY REPORT THE TRANSMEMBRANE PROTON GRADIENT OF SUBMITOCHONDRIAL
MEMBRANES? C. P. Lee, Department of Biochemistry, Wayne State Univer-
sity School of Medicine, Detroit, Michigan 48201, U.S.A.

One of the unique observations made when 9-aminoacridine dyes are
associated with energy transducing membranes, e.g. chloroplast, chromatophore, submitochondrial and bacterial membranes, is a significant
decrease of the dyes' fluorescence intensities upon membrane energization. The fluorescence intensities are restored upon the addition of
an uncoupler or any agent which abolishes energy-linked proton uptake
(cf. Ref. 1). Schuldiner et al (2) used the fluorescence decrease of
9-aminoacridine dyes as a quantitative measure of the proton gradient
across the chloroplast membrane by assuming that the dye molecules are
taken up by the energized membrane and that these accumulated dye molecules are non-fluorescent and homogeneously distributed in the interior of the vesicles. However, no precise mechanism for the fluorescence decrease was given. This method has been applied to several
other systems and has, in fact, been treated by some workers as a con-
venient means for determining the internal pH of acidic vesicles (3).

We have conducted a systematic study of the spectroscopic properties
of two types of 9-aminoacridine molecules in association with beef
heart submitochondrial membranes (SMP) with the emphasis on exploring
their similarities and differences. The dyes we have studied are:
9-aminoacridine (9AA), 9-amino-3-chloro-7-methoxy-acridine (9ACMA) and
quinacrine (QA). Evidence derived from the absorption and the fluorescence excitation and emission spectra, and the quantum yield of QA,
either alone or associated with energized (E) and non-energized (NE)
SMP, revealed that the energy-linked fluorescence decrease results
from protonation of the monoprotonated species of QA ($QA \cdot H^+$) to form
the diprotonated species ($QA \cdot H_2^{++}$), and the $QA \cdot H_2^{++}$ formed are tightly
bound to the membrane. The extent of protonation depends on the concentration of H^+ in the membrane generated through energization and on
the concentration of $QA \cdot H^+$ determined by the pH of the assay medium(4).
The fluorescence intensity of 9AA (pKa = 9.75) is independent of pH
over the range from 7.5 to 9.0, since the protonated species predominates (>82%). That the fluorescence intensity of 9ACMA (pKa = 8.80)
at pH 7.5 is greater than that at pH 9.0 indicates that the protonated
form is more fluorescent than the neutral one. The quantum yields of
9AA and 9ACMA are virtually constant over the pH range from 7.0 to 9.0
in contrast to that of QA which at pH 9.0 is approx. 3 times that at
pH 7.0. The energy-linked fluorescence decrease of 9AA and 9ACMA must
therefore result from mechanisms other than protonation of the neutral
species, presumably through the formation of non-fluorescent complexes
with membrane components possessing conjugated π-bonding systems (5).
Specific binding of QA, 9AA and 9ACMA to E-SMP is further supported by

the inhibition of the energy-linked fluorescence decrease and the concomitant decrease in binding of these dyes by local anesthetics (5,6).

The degree of polarization (P) of QA, nearly zero in solution, increases significantly when associated with NE-SMP and further increases, by more than 2 fold, upon energization. A distinct break at $15^{o}C$
in the plot of P associated with E- (but not NE-) SMP vs 1/T was seen:
a steady increase in P with a decrease in temperature from 27 to $15^{o}C$;
further decrease in temperature has virtually no effect on P. The P
of 9AA and 9ACMA associated with either E- or NE-SMP is, however, virtually identical to that of either dye alone in the medium, approx.
zero. This suggests that P reflects only the fluorescence of those
dye molecules that are "free" in the medium and that those bound to
the membrane are non-fluorescent. These data indicate that the molecular rotation of the emitting species of QA associated with E-SMP is
strongly hindered which may result from specific interactions between
the membrane and the QA molecules via the long side chain substitution
at the 9-amino group of the acridine nucleus (5).

Identification of the specific binding sites of QA and 9ACMA with
SMP has been pursued with photoaffinity analogs in which the 3-chloro
substitution of QA and 9ACMA was replaced with an azido group (7,8).
Analyses of the labelling patterns, enzyme activities and fluorescence
excitation and emission spectra provided evidence which indicate that
there are specific binding sites of QA and 9ACMA with E-SMP located
in the hydrophobic regions of the membrane.

In conclusion, QA probes the H^{+} content, while 9AA and 9ACMA probe
charged components associated with conjugated π-bonding systems,
generated in energized SMP. Both events occur in the membrane phase
and result in apparently similar fluorescence decreases. Estimation
of the proton gradient across submitochondrial membranes induced upon
energization based on the method of Schuldiner et al (2) would be
severely overestimated.

1. Kraayenhof, R., Brockelhurst, J.R. and Lee, C.P. (1976) in Con-
   cepts in Biochemical Fluorescence (Chen, R.F. and Edelhoch, H.,
   eds.), pp. 767-809, Marcel Dekker, New York.
2. Schuldiner, S., Rottenberg, H. and Avron, M. (1972) Eur. J.
   Biochem. 25, 64-70.
3. Rottenberg, H. (1979) Methods Enzymol. 55, 547-569.
4. Huang, C.S., Kopacz, S.J. and Lee, C.P. (1977) Biochim. Biophys.
   Acta 259, 241-249.
5. Huang, C.S., Kopacz, S.J. and Lee, C.P. (1983) Biochim. Biophys.
   Acta 722, 107-115.
6. Mueller, D.M. and Lee, C.P. (1982) FEBS Lett. 137, 45-48.
7. Mueller, D.M., Hudson, R.A. and Lee, C.P.(1982) Biochem.21,1445-53.
8. Kopacz, S.J. and Lee, C.P. (1983) Fed. Proc. 42, 1943.

THE COUPLING BETWEEN TRANSPORT AND PHOSPHORYLATION OF FRUCTOSE IN INSIDE/OUT VESICLES OF <u>RHODOPSEUDOMONAS SPHAEROIDES</u>

J. Lolkema, G. Robillard
Department of Physical Chemistry, University of Groningen, 9747 AG Groningen, The Netherlands

The bacterial phosphotransferase systems are believed to catalyse the concomitant transport and phosphorylation of hexoses and hexitols. The transport is from the outside to the inside of the cell. An absolute coupling between transport and phosphorylation has however been questioned in the literature (for review see ref. 1). We have tested the coupling by analyzing the kinetics of fructose phosphorylation by inside/out vesicles of <u>Rhodopseudomonas sphaeroides</u>. In such vesicles the fructose binding site on the carrier enzyme is inside the vesicle. The kinetics of the phosphorylation shows two populations of carriers, one with a K_M of 8 µM, the other with an apparently much lower affinity. The low affinity carriers can be converted into high affinity carriers by pretreating the membranes with a concentration of deoxycholate that can be shown to make the membranes permeable to small molecules. The characteristics of the kinetics of the low affinity carriers is consistent with an enzymatic process that is preceded by a slow diffusion step. At this point we conclude that our membrane preparation is a mixture of open and closed vesicles. All substrates have free access to their appropriate binding sites on the carrier in an open membrane structure, leading to the real affinity constant of 8 µM. In case of the closed membrane vesicles a much lower affinity is measured because fructose first has to enter the vesicle before it can be phosphorylated. The diffusion is slow relative to the phosphorylation, making the internal concentration much lower than the external concentration under steady state conditions. Measurement of the internal concentration under phosphorylating conditions confirms this conclusion.

Quantitative analysis of the rate of diffusion and studies in which the activity of the phosphotransferase system is altered by changing the phosphoryl donor concentration show that the diffusion is a process which is also catalysed by the phosphotransferase system.

We conclude that phosphorylation is strictly coupled to transport. The sugar can not be phosphorylated from the cytoplasmic side of the membrane. The phosphotransferase system, however, can transport the sugar without phosphorylation. The latter activity also requires the presence of the phosphoryl donating PTS components and PEP.

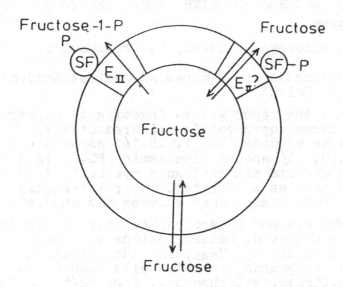

Figure 1. Fluxes of fructose through the membrane of an inside/out
vesicle when the phosphotransferase system is turning over.
The passive diffusion at the bottom of the figure is kinetically irrelevant.
E_{II}: carrier enzyme, SF: PTS phosphoryl donor,
P: phosphoryl group.

Reference
Hays, J.B. (1978) in: Bacterial Transport (Rosen, B.P., ed.),
Chapter 2, p. 43-102, Marcel Dekker, Inc., New York.

INTERACTION OF ASCORBATES WITH THREE ENZYMES CONTAINING COPPER

Z. Machoy, P. Wieczorek, G. Gałka, R. Królikowska,
U. Bilska
Department of Biochemistry, Pomeranian Medical Academy,
Szczecin, 70-111, Poland

The purpose of the paper was to investigate substrate specificity of three cuproprotein oxidoreductases, namely: cytochrome oxidase /EC. 1.9.3.1/, ascorbate oxidase /EC. 1.10.3.3/ and ceruloplasmin /EC. 1.12.3.1/. L-ascorbic acid and its six analogues synthetized in our laboratory were used as substrates. Their interaction with the three above - mentioned enzymes was studied.

Spectrophotometric and polarographic /oxygen electrode/ methods were employed. Determinations that were made covered the values K_m, V_{max}, and the influence exerted by other compounds, intentionally chosen and added, such as citrate, cytochrome c, ions Cu^{2+} and Fe^{2+}.

Ascorbate oxidase oxidizes ascorbates depending on their structure /on number of carbons and spatial configuration/, the best substrate being L-ascorbate and analogues containing six carbons. Ceruloplasmin also displays ascorbate oxidase properties towards the used ascorbates in the following order: L-ascorbate > D-isoascorbate. Cytochrome oxidase /cytochrome aa3/ in the absence of cytochrome c forms with ascorbates the oxido-reducing system /1/. The oxidizing of ascorbates is directly accompanied by cytochrome a reduction in the range from 44 to 82 %. Effectiveness of cytochrome a reduction was decreasing in the following sequence: galactoascorbate, glucoascorbate, xyloascorbate, araboascorbate, ramnoascorbate, ascorbate and isoascorbate. Differences and dependences were found in the influence of compounds, added to the reacting medium, on the behaviour of three oxidoreductases investigated /Table 1/.

Table 1. Direct ascorbate oxidation by three investigated enzymes and added compounds

Parameters and added compounds	ascorbate oxidase	ceruloplasmin	cytochrome oxidase	autooxidation of ascorbate
K_m /mM/ at 265 nm	0.13		0.026	
K_m /mM/ O_2-uptake	0.23	4.7 [2]	/0.0/	
V_{max} at 265	2200		2257	
V_{max} O_2-uptake	3800	1200	/0.0/	
Opt. pH	5.5-7.0 [4]	5.5 [2]	5.07	f/pH/
Cyt. c^{2+} O_2-uptake	0.0	0.0	371	
μM O_2/μM Cu^{2+}				528 [5]
μM O_2/μM Fe^{2+}				161 [5]
Citrate /% of inhibition/	0.0	94 [3]	59	

References:
1. Gałka, G., Machoy, Z. and Ogoński, T. /1981/ Pol. J.Chem. 55, 1783-1793
2. Curzon, G. and Young, S.N. /1972/ Biochim.Biophys. Acta 268, 41-48
3. Osaki, S., McDermott, J.A. and Frieden, E. /1964/ J.Biol.Chem. 239, 3570-3575
4. Nakamura, T., Makino, N. and Ogura, Y. /1968/ J.Biochem. 64, 189-195
5. Gałka, G. /1978/ Ann.Acad.Med.Stet. 24, 217-238

INTERACTION OF DOXORUBICIN(ADRIAMYCIN) WITH THE PHOSPHATE
TRANSPORT PROTEIN (PTP) AND CARDIOLIPIN OF THE INNER
MITOCHONDRIAL MEMBRANE(IMM)

M.Müller ,D.Cheneval,R.Moser,R.Toni and E.Carafoli
Laboratorium für Biochemie III,ETH-Zentrum,CH-8092 Zürich
*Medizinisch-Chemisches Institut,Bühlstr. 28,CH-3012 Bern

Doxorubicin forms specific complexes with negatively charged phospholipids(1). The association constant for the do-
xorubicin·cardiolipin complex has been calculated to be1.6
x 10^6 M^{-1}. Cardiolipin is one of the three major phospholipid components of the IMM and possesses two negative
charges at physiological pH.
Purification and activity of the PTP has been demonstrated
to be cardiolipin-dependent(2). Doxorubicin inhibited the
transport activity of the purified PTP reconstituted in
asolectin vesicles, possibly by complexing essential cardiolipin molecules(3).
Since the interaction between cardiolipin and doxorubicin
is of ionic type, we have found that increasing the ionic
strenght of the incubation medium with KCl prevents completely the inhibition of the phosphate transport by doxorubicin at about 70 mM. This suggests that the PTP·cardio-
lipin complex is accessible from the aqueous phase, since
K^+ ions apparently reach the doxorubicin binding site on
the PTP·cardiolipin complex.
To further investigate the interplay between PTP, cardiolipin and doxorubicin the PTP was modified by the hydrophilic reagent succinic anhydride. Succinylation of the
PTP was performed under SH-groups-protecting conditions
by adding mersalyl to mitochondria prior to their solubilization at pH 8 and 9. The modification of the PTP at pH
8 resulted in a preparation identical to the untreated
protein but doxorubicin could not inhibit the transport
of the treated PTP reconstituted in liposomes. When succinylation was performed at pH 9 complete and irreversible
inactivation of the PTP followed(4).
The finding that the accessibility of cardiolipin to the
PTP was not modified by succinylation at pH 8, whereas the
reactivity to doxorubicin was lost, is suggestive of the
presence of two distinct binding sites, one for cardiolipin and one for doxorubicin on the PTP.

An hypothetical model to interpret the results is given in Fig. 1. In the PTP, modification of positive charge 1 at pH 8 would limit the accessibility of doxorubicin(A) to cardiolipin(CL). Succinylation at pH 9 would result in the modification of both positive charges 1 and 2, resulting in the irreversible inactivation of the PTP.

Doxorubicin was also used to determine the transversal distribution of cardiolipin across the IMM. It was found that about 56% of the total cardiolipin was located on the cytoplasmic side of the IMM. In this case the distribution of cardiolipin is considered as slightly asymmetric and not completely asymmetric as previously reported by others This implies that cardiolipin should have a function in the cytoplasmic layer of the IMM. It was extensively repor ted that cardiolipin was found in several isolated comple xes of the oxidative phosphorylation. We have evidences that cardiolipin is the binding receptor for the mitochon drial creatinekinase.

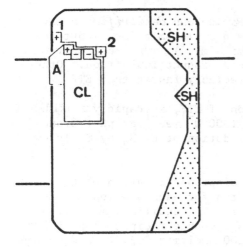

Figure 1

Hypothetical model of the PTP and its binding sites for cardiolipin(CL), doxorubicin(A) and SH-reagents.

1. Goormaghtigh,E.,Chatelain,P.,Caspers,J. and Ruysschaert J.M. (1980) Biochim. Biophys. Acta 597, 1-14
2. Mende,P.,Hütter,F.J. and Kadenbach,B. (1983) FEBS Lett. 158, 331-334
3. Cheneval,D.,Müller,M. and Carafoli,E. (1983) FEBS Lett. 159, 123-126
4. Müller,M.,Cheneval,D. and Carafoli,E. (1984) Eur. J. Biochem., in press

REACTIONS OF ETF AND ETF-CoQ OXIDOREDUCTASE

R.R. Ramsay, D.J. Steenkamp and M. Husain.
Dept. Biochemistry and Biophysics, University of California, San Francisco, and Molecular Biology Division, Veterans Administration, San Francisco, California 94121, USA

ETF-Q oxidoreductase catalyzes the reoxidation of reduced ETF with Q_1 as the electron acceptor (1). It is essential for β-oxidation of fatty acids, for catabolism of some amino acids and for the operation of the mitochondrial one carbon cycle. A kinetic assay was designed based on the natural electron donor and acceptor:

glutaryl-CoA $\rightleftharpoons$ glutaryl-CoA $\longrightarrow$ ETF $\longrightarrow$ ETF-Q $\longrightarrow Q_1$
crotonyl-CoA $\rightleftharpoons$ dehydrogenase $\qquad\qquad$ oxidoreductase

Both ETF and Q can form anionic semiquinones (2) but only $ETF^{\cdot-}$ is stable under anaerobic conditions. Its interaction with the other electron carriers can readily be investigated by monitoring changes in the visible spectrum. In particular, the relative rates of the two redox steps, $ETF_{ox} \rightleftharpoons ETF^{\cdot-} \rightleftharpoons ETFH_2$ have been studied.

Glutaryl-CoA dehydrogenase (GDH) from *Paraccocus denitrificans* catalyzes the rapid reduction of ETF_{ox} (V_{app} = 6.2 $\mu mol.min^{-1}mg^{-1}$) but reduces $ETF^{\cdot-}$ more slowly (V_{app} = 1.3 $\mu mol.min^{-1}mg^{-1}$). Progress curves for the reduction of $ETF^{\cdot-}$ showed pronounced product inhibition. Other dehydrogenases similarly reduce ETF_{ox} appreciably faster than $ETF^{\cdot-}$.

$ETF^{\cdot-}$, generated by dithionite titration of ETF, is rapidly reoxidized by Q_1 and a second order rate constant of 1300 $M^{-1}sec^{-1}$ at pH 7.8 and $25^\circ C$ was determined. In contrast the reoxidation of $ETFH_2$ by Q_1 is extremely slow.

ETF-Q oxidoreductase catalyzes a rapid disproportionation of $ETF^{\cdot-}$ to a mixture of $ETF^{\cdot-}$, ETF_{ox} and $ETFH_2$. Spectra for each species and for the equilibrium mixture are shown in Fig. 1. The equilibrium constant is 1 at pH 7.8. The inset of Fig. 1 shows a double reciprocal plot from which was calculated K_{ETF} = 8 μM and V = 200 $\mu mol.min^{-1}mg^{-1}$. To determine whether this unusual reaction was a general feature of the enzymes interacting with ETF, GDH (0.3 μM) was added to $ETF^{\cdot-}$ (9.5 μM) but disproportionation proceeded at a rate of only 0.032 $\mu mol.min^{-1}mg^{-1}$.

In the presence of Q_1, ETF-Q oxidoreductase catalyzes the oxidation of $ETFH_2$, with V_{app} = 50 $\mu mol.min^{-1}mg^{-1}$ at 171 μM Q_1. The kinetic constants for the overall reaction shown above are V = 46 $\mu mol.min^{-1}mg^{-1}$ K_{ETF} = 0.23 μM, K_Q = 6 μM with α = 6.3. These results suggest that

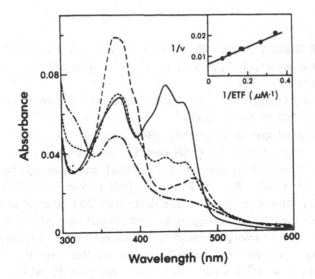

Fig. 1. Disproportionation of ETF·⁻ by ETF-Q oxidoreductase. ETF (13.9 μM in 25 mM HEPES (pH 7.8), containing 10% ethylene glycol, 30 μM glucose, glucose oxidase (1 unit), catalase (24 units), was reduced to ETF⁻ by titration with dithionite. Disproportionation was initiated by adding 98 ng ETF-Q oxidoreductase. The spectra are: ——— ETF ; — — — ETF·⁻ ; — · — · ETFH₂; ·····the equilibrium mixture after disproportionation (E₃₇₂ = 7300). Inset: Double reciprocal plot for the rate of disproportionation (μmol.min⁻¹ mg⁻¹).

reoxidation of ETFH₂ could be fast enough to account for the rate of the overall reaction. In addition, the relative magnitudes of spectral changes at several wavelengths in stopped flow experiments indicated that disproportionation of ETF·⁻ took place even in the presence of Q, rather than direct reoxidation of the semiquinone.

The data indicate that ETF·⁻, the product of reduction by GDH, is rapidly removed by disproportionation to ETF + ETFH₂ catalyzed by ETF-Q oxidoreductase. Reoxidation of ETFH₂ by Q₁ in the presence of ETF-Q oxidoreductase is adequate to account for turnover. Thus disproportionation is likely to contribute to the kinetic mechanism of ETF-Q oxidoreductase.

References

1. Ruzicka, F.J. and Beinert, H. (1977) J. Biol. Chem. 252, 8440-8445.
2. Husain, M. and Steenkamp, D.J. (1983) Biochem. J. 209, 541-545.

PROPERTIES OF MAJOR MYELIN PROTEINS PURIFIED FROM BOVINE BRAIN.

P.Riccio°, J.P.Rosenbusch", F.Pattus", A.Bobba°, A.Lustig[+], A.Tsugita"
and E.Quagliariello°.
°Istituto di Chimica Biologica, Facoltà di Scienze, Università di Bari
and C.S.M.M.E., CNR, 70126 Bari Italy.
"European Molecular Biology Laboratory, 6900 Heidelberg, FRG.
[+]Biozentrum, University of Basel, 4056 Basel, Switzerland.

The most representative proteins of central nervous system myelin
are the "encephalitogenic" Basic Protein (MBP) and the Proteolipid
Protein (PLP) (1-2). These proteins account for 30% and 50% of total
myelin protein respectively. Structure and function of these major
proteins are not yet known, probably because of the denaturing
conditions presently used for the purification of MBP and PLP (3-5).

We have developed a mild procedure for the purification of both
MBP and PLP from bovine brain myelin using the nonionic detergent
Octylpolyoxyethylene for solubilization (6).

Contrary to the conventionally prepared MBP, our purified MBP is
not water soluble. In fact, the requirement of detergents for its
solubilization from myelin is absolute (7). MBP is isolated in
monomeric state as a protein-lipid-detergent complex. Purified MBP
shows the novel property to bind most myelin lipids. As detected by
thin layer chromatography, the lipids present in purified MBP are:
phosphatidylethanolamine, phosphatidylserine, cholesterol, cerebro-
sides and sulfatides. Three other lipids were observed but not
identified with certainty. The content of phospholipid phosphorus is
1.04 mg Pi/mg protein. In agreement with this binding, for the first
time MBP shows an ordered structure. In fact, as detected by
circular dichroism analysis, it contains 45-50% of β-sheet and 6-10%
of α-elix structures.

On the other hand the aminoacid composition of our lipid-binding
MBP is very similar to that described for lipid-free Basic Protein
prepared in the traditional way (8). The N-terminus is blocked. The
C-terminus is Ala-Arg, which differs from the expected Ala-Arg-Arg.
The identity is confirmed by the immunoblotting technique using
lipid-free Basic Protein antibodies. As expected, the isoelectric
point of MBP is above pH 10.

Instead, no lipids can be detected in the Proteolipid Protein
purified in Octylpolyoxyethylene. The content of phospholipid phosphorus is very low: 0.023 mg/mg protein. On the basis of sedimen-

tation equilibrium experiments and on an apparent subunit molecular weight of 25 KDa, it can be deduced that purified PLP is present in the form of lipid-free trimers. Recently the presence of lipid-free hexamers have been found in PLP preparations in either deoxycholate or Triton X-100 (9).

In order to find some functional characteristics of purified PLP and MBP, the conductance properties of both proteins after reassembly in planar bilayers has been studied. Asymmetric planar bilayers with both PLP and MBP on different layer sites can be formed from Asolectin phospholipids-cholesterol vesicles containing the reconstituted proteins. Following application of an electrical potential across the bilayer, initiation of discrete conductance steps is observed. The possible induction of channels for cations ($P_{K^+} > P_{Na^+}$) suggests the participation of both proteins in ion transport and justify further studies in this direction.

REFERENCES

1. Gregson,N.A. (1983) in Multiple Sclerosis, Pathology, Diagnosis and Management (Hallpike,J.F., Adamy,C.W.M. and Tourtellotte,W., eds.) pp. 1-27, Chapman and Hall, London.
2. Agrawal,H.C. and Hartman,B.K. (1980) in Protein of the Nervous System, 2nd edition (Bradshaw,R.A. and Schneider,D.M., eds.) pp. 145-169, Raven Press, New York.
3. Deibler,R.G., Martenson,R.E. and Kies,M.W. (1972) Prep. Biochem. 2, 139-165.
4. Tenenbaum,D. and Folch-Pi,J. (1966) Biochim. Biophys. Acta, 115, 141-147.
5. Gonzales-Sastre,F. (1970) J. Neurochem. 17, 1049-1056.
6. Quagliariello,E. and Riccio,P. (1984) 16th FEBS Meeting, Moskow, Abstr. n. 292.
7. Riccio,P., Simone,S.M., Cibelli,G., De Santis,A., Bobba,A., Livrea, P. and Quagliariello,E. (1983) in: Structure and Function of Membrane Proteins (Quagliariello,E. and Palmieri,F., eds.) Elsevier Science Publishers B.V.
8. Eylar,E.H., Brostoff,S., Hashim,G., Caccam,J. and Burnett,P. (1971) J. Biol. Chem. 246, 5770-5784.
9. Smith,R., Cook,J. and Dickens,P.A. (1984) J. Neurochem. 42, 306-313.

CHARACTERIZATION AND FUNCTIONAL STUDIES OF PLASMA MEMBRANES ISOLATED FROM RABBIT SKELETAL MUSCLE.

Edmond ROCK, Christian NAPIAS and Jean CHEVALLIER.
I.B.C.N. du C.N.R.S., 1 rue Camille Saint-Saëns 33077 Bordeaux Cedex FRANCE.

An improved procedure was developed for the isolation of skeletal muscle plasma membranes (1). This method included a DNase treatment of the homogenate prior to the isolation of membranes by differential and sucrose density centrifugation techniques. We obtained two fractions, F1 and F2, banding at the 11-26 % and 26-30 % sucrose layers interfaces respectively, which showed enrichment in biochemical (acetylcholinesterase, 5´-nucleotidase and adenylate cyclase activities, ouabain and insulin binding) and chemical (sialic acid content, cholesterol/phospholipid molar ratio) plasma membrane markers.

The sidedness analysis (2), using acetylcholinesterase and the ouabain binding activities in the presence or the absence of deoxycholate, indicated that these two fractions were composed of 55 % of inside-out, 20 % right-side-out, both sealed vesicles and 25 % of "leaky" structures.

The ability of F1 and F2 to hydrolyse ATP in the presence of specific cations were tested :

- the $(Na^+-K^+-Mg^{++})$-ATPase activities were 0.87 +/- 0.28 and 0.67 +/- 0.28 umol Pi/min/mg protein for F1 and F2 respectively. Optimal activity was found for a Na^+/K^+ ratio of 3 at pH 7.5 and at 37°C.

- the $(Ca^{++}-Mg^{++})$-ATPase activities were 1.02 +/- 0.14 and 1.79 +/- 0.06 umol Pi/min/mg protein for F1 and F2 respectively. Optimal activity was obtained with 5 uM calcium in the medium. In addition this activity was stimulated by calmodulin (half maximum activation Ka of 30 nM) and totally inhibited by 50 uM of trifluoperazin. On the other hand we observed a 16 % stimulation of the calcium ATPase by 50 uM of the adenylate kinase specific inhibitor Ap5A (p1,p5-Di(adenosin-5´)pentaphosphate).

Taken together, all these results led us to consider that these fractions were very suitable for the study of ionic fluxes involved at the muscle plasma membrane level.

BIBLIOGRAPHY :
1- ROCK, E., LEFAUCHEUR, L. and CHEVALLIER, J., (1984), Biochem. Biophys. Res. Commun., (in press).
2- SEILER, S. and FLEISHER, S., (1982), J. Biol. Chem.,257, 13862- 13871.

SUBUNIT INTERACTIONS OF THE MANNITOL CARRIER PROTEIN FROM THE E. COLI PHOSPHOENOLPYRUVATE DEPENDENT PHOSPHOTRANSFERASE SYSTEM; THE PROTEIN CAN BE EXTRACTED AS A DIMER FROM THE MEMBRANE.

F.F. Roossien and G.T. Robillard, Department of Physical Chemistry, University of Groningen, 9747 AG Groningen, The Netherlands

The bacterial PEP dependent phosphotrasferase system (PTS) catalyzes the phosphorylation and transport of a number of hexoses and hexitols across the membrane via the following reactions (1):

$$\text{PEP} + \text{HPr} \xrightarrow{\text{EI}} \text{P-HPr} + \text{Pyruvate}$$

$$\text{Sugar}_{out} + \text{P-HPr} \xrightarrow{\text{EII}} \text{sugar-P}_{in} + \text{HPr}$$

This report deals with studies on both the purified and membrane bound mannitol specific EII(EII^{mtl}).

Fig. 1 shows an experiment in which EI, HPr and purified EII^{mtl} in Lubrol detergent micelles were incubated with [^{32}P]-PEP (2), and subsequently electrophoresed in the presence of SDS. The autoradiogram (Fig. 1) and protein pattern (not shown) of the gel completely coincide, indicating that all three PTS proteins form phosphorylated intermediates in agreement with kinetic measurements (1, 3).

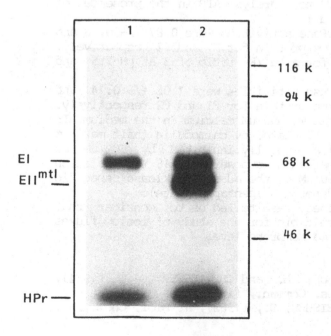

Fig. 1. Autoradiograms obtained after SDS-gel-electrophoresis of a control sample containing EI, HPr and [^{32}P]-PEP (lane 1) and an identical sample in which purified EII^{mtl} was included (lane 2).

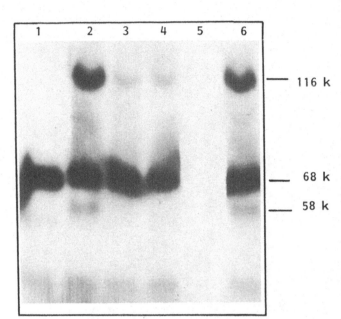

Fig. 2. Autoradiogram after SDS-gelelectrophoresis of membranes incubated with EI, HPr and [^{32}P]-PEP. Lane 1 contains a control sample showing the band of phosphorylated EI. Membranes were phosphorylated by EI, HPr and [^{32}P]-PEP and subsequently incubated for 2.5 min (lane 2) or 9.5 min (lane 6). Mannitol was added at t = 3 min to a portion of the reaction mixture and samples were taken at 3.3 min (lane 3), 3.6 min (lane 4) and 9 min (lane 5).

Membrane fragments obtained from mannitol-grown E.coli ML 308-225 catalyze the PEP dependent phosphorylation of both glucose and mannitol, meaning that they contain both the glucose and mannitol specific Enzymes II. When those fragments are incubated with EI, HPr and [^{32}P]-PEP and subsequently electrophoresed in the presence of SDS, the autoradiogram of the gel shows two bands at 58 and 116 k Daltons, in addition to the bands of P-EI and P-HPr (Fig. 2, lanes 2 & 6). Treatment of the phosphorylated membranes with mannitol for 0.3 and 0.6 min (lanes 3 & 4 respectively) results in a substantial decrease in the intensity of the 58 and 116 k Dalton bands. All proteins are dephosphorylated after 6 minutes incubation with mannitol (lane 5) because all the ^{32}P is transferred to mannitol-P which electrophoreses in the salt front. A similar treatment with glucose had no direct effect on the band pattern. We conclude, therefore, that the 58 and 116 k Dalton bands originate from EII^{mtl} monomers and dimers respectively. The interaction between the subunits of the dimer are not abolished by the addition of up to 5% SDS. However, the non-ionic detergent lubrol, which is present during the purification of EII^{mtl}, is capable of transforming the EII^{mtl} dimers into monomers.

References
(1) Roossien, F.F., Blaauw, M. and Robillard, G.T. (1984) Biochemistry, in press.
(2) Roossien, F.F., Brink, J. and Robillard, G.T. (1983) Biochem. Biophys. Acta 760, 185-187
(3) Misset, O., Blaauw, M., Postma, P.W. and Robillard, G.T. (1983) Biochemistry 22, 6163-6169.

References

ISOLATION AND CHARACTERIZATION OF "WARM" MITOCHONDRIA IN DIFFERENT
CONDITIONS AND SPECIES

A. Sánchez, C. Moreno, and M. Gosálvez
Department of Experimental Biochemistry. Clínica Puerta de Hierro,
National Center of Medical Research, Madrid - 35, Spain

Progress on the isolation methodology and biophysical-biochemical
characterization of rat liver and rabbit kidney mitochondria, isolated
in twenty minutes at room temperature with minimal mechanical stress
("warm" mitochondria), has been made in our laboratory from the moment
of the first description of this new preparation, which may be useful
for different fields of biochemistry and bioenergetics (1). Several
physiological and pharmacological conditions influence the quality of
the preparations and differential biochemical-biophysical characteristics with respect to the corresponding standard (cold isolated) mitochondria. Immunization of the animal with the pooled human serum seems
to exert a good effect, increasing the quantity of mitochondria bound
to polymeric tubulin, as seen by the isotopic colchicine-binding assay. Other influential, but not yet fully asserted factors seem to
include the use of vitamin B and several hormones and pharmacological
drugs. The best preparations of "warm" mitochondria display several
characteristics with respect to respiration, oxidative phosphorylation, respiratory control ratio and ion transport, but the quantitative definition of this method is not yet possible due to variability
from preparation to preparation, nor has there yet been defined a
standardization of the animal treatments which yield the best preparation. Future lines of development of the "warm" mitochondria preparation, which may reflect "in vitro" particular conditions of the mitochondria "in vivo" in certain tissues, will be discussed, as well as
the physiological conditions, given that morphological studies have
described an association of mitochondria with microtubules and micro-
filaments (2-5).

1.- L. Pezzi et al. Federation Proceedings. Vol. 41 (1982).
2.- A.P. Aguas. Ultrastructure Res. 74, 175 (1981).
3.- E. Ball, S.J. Singer. Proc. Natl. Acad. Sci. USA 79, 123 (1982).
4.- P. Mose-Larsen, R. Bravo, S. Fey, V. Small, and J.E. Celis.
    Cell 31, 681 (1982).
5.- B. Baccett, E. Bigliardi, A.G. Burrini and V. Pallini. Gamete
    Research 3, 203 (1980).

A MORPHOMETRIC APPROACH TO THE MOLECULAR ARCHITECTURE OF THE INNER MEM-
BRANE OF RAT LIVER MITOCHONDRIA
K. Schwerzmann and E.R. Weibel
Department of Anatomy, University of Berne, 3000 Bern 9, Switzerland

The mitochondrial inner membrane contains most of the enzymes of the
energy transducing system, notably the proteins of the respiratory chain
and the ATP-synthesizing machinery. The molecular architecture of this
membrane, i.e. the distribution and the density of proteins as well as
their spatial arrangement, is of great interest in the elucidation of
the mechanism of energy transduction in mitochondria.

In this study, we present some structural and functional parameters
relevant to the molecular architecture of the inner membrane of rat
liver mitochondria. Since biochemical measurements in general do not
provide structural information, we have chosen a combined approach of
stereology and biochemistry (1). Morphometry, a stereological method,
applied on electron microscopic pictures of mitochondria yielded precise estimations of mitochondrial volume and membrane surface areas.
Of the same preparation of isolated mitochondria, biochemical measurements were performed (rates of O_2-consumption, concentration of the
respiratory complexes) and related to the structural data obtained by
morphometry. This combined procedure gave then data of the molecular
architecture of the inner membrane of mitochondria, such as the density of respiratory complexes in the membrane and estimates of the
distances between the redox partners, as well as the electron flow
along the inner membrane (Table).

The results confirm the notion of the inner mitochondrial membrane
being densely packed with proteins (2). The avarage nearest distance
(a probability value, assuming random non-interacting distribution)
between any two functional complexes of the inner membrane can be expected to be less than 60 Å, or 70 Å to 140 Å between reaction partners
of the electron transfer chain. The actual distances are still considerably smaller because of the dimensions of the various protein complexes themselves.

Thus, the short spacing between the proteins in the membrane and
their lateral mobility (3) make lateral diffusion itself as a limiting
factor for the rates of electron transfer (4) very unlikely. Our results rather support the view that there is a high degree of interaction between the proteins in the inner mitochondrial membrane due to
long-range intramolecular forces (electrostatic forces, van der Waal

forces) which may cause the proteins to form transient (functional) complexes and determine the rates of electron transfer.

TABLE. Structural and Functional Characteristics of the Inner Membrane of Rat Liver Mitochondria

Area Inner Membrane/mg Protein	$560.2 \ (52.0)^a$
Maximal Electron Flow (e^-/sec . μm^2)	35.9×10^3
- Pyruvate + Malate	
- Succinate	125.4×10^3
- cyt c + Ascorbate /TMPD	249.5×10^3
Density in the Inner Membrane of Components of Oxidative Phosphorylation (Number/um^2)	
- complex I	428^b
- complex II	856^b
- bc_1-complex	846 (23)
- cyt c	2559 (131)
- cytochrome oxidase	$2420 \ (284)$
- $F_1 \cdot F_0$-ATPase	2420^b
Mean Nearest Distance Between Redox Partners ($\overset{\circ}{A}$)^c	
- complex I complex II	140
- complex II bc_1-complex	122
- bc_1-complex cyt c	86
- cyt c cytochrome oxidase	71

[a] Standard Error of the Mean

[b] Estimates Based on 1 complex I, 2 complex II and 6 $F_1 \cdot F_0$ per 6 aa_3 cytochromes

[c] Calculated as $0.5/\left(d_1 + d_2\right)^{1/2}$ (d's are densities)

References

1) Weibel, E.R. (1978) Stereological Methods, Vol. I, Academic Press New York
2) Capaldi, R.A. (1982) Biochim. Biophys. Acta 694, 291 - 306
3) Gupte, S.A., Wu, E.S., Hoechli, L., Hoechli, M., Jacobson, K., Sowers, A.E., and Hackenbrock, C.R. (1983) Proc. Natal. Acad. Sci. 81, 2606-2610
4) Hackenbrock, C.R. (1981) Trends Biol. Sci. 6, 151 - 154

THE INTERNAL VOLUME OF SPINACH THYLAKOIDS

U. Siggel, Max-Volmer-Institut, Technische Universität
Berlin, D-1000 Berlin 12, FRG

The internal volume of thylakoids has been reinvestigated
because it is of special interest for two reasons. Firstly
its magnitude and variation with parameters of the solution
are relevant to the question of chemiosmosis and give
information about the charges within the thylakoids, whichare
related to the electrical potentials. Secondly its value
has to be known for the calculation of the internal pH
which in turn is essential for the regulation of photo-
sythetic electron transport and phosphorylation.
The investigation presented here has two objectives :
critical examination of the method and determination of
the internal volume in the dark state as a function of
electrically effective parameters. In order to guarantee
exactly reproducible conditions as to the solute concentrations spinach thylakoids from only two batches have
been used. This advantage was accompanied by the disad-
vantage that thylakoids had to be used which were stored
in liquid nitrogen and rethawed before use. Their quali-
ties may not always be identical with those of freshly
prepared thylakoids.

Method. The method commonly used consists in the
separation of the thylakoids from the medium by centrifugation through a layer of silicon oil, as introduced by
Klingenberg et al. (1) for mitochondria. Later the method
was also applied to thylakoids (2)   and intact chloroplasts (3) . The volume is measured by the water content
of the probe. Internal and externally adherent water are
discriminated by radioactive double labelling. ^3H_2O
measures the total water and an impermeant ^{14}C-labelled
solute the external water. Sorbitol(4) ,   sucrose (3),
inulin(2,4)   and dextran(1,3) have been used as impermeant compounds. Inulin turned out to mark only part of
the external space (4). Thus sorbitol was favoured in the
past.
The reinvestigation shows that the apparent internal
volume is time-dependent. It decreases with a halftime
of some 15 min (at pH=7 and 5 mM KCl). The interpretation
is that at least our rethawed thylakoids are permeable to
sorbitol. Before information is collected the value at
10 min incubation is taken as a measure of the inner volume.

Now sucrose was chosen as a larger molecule and hoped to be more suitable. But it turned out to pemetrate in the internal space as well. Additionally it is probably adsorbed in the membrane space. Negative values of the apparent volume are calculated for a incubation longer than 20 min. Thus sucrose seems to be less suitable than sorbitol.

Results on the osmotic behaviour. The calculated apparent value for the internal volume, determined with sorbitol, decreases with increasing concentration of electrolyte. A limiting value of some 50 l/Mol Chl (at pH=8) is attained at low concentration (5 mM KCl or 0.5 mM MgCl$_2$). The internal volume is strongly dependent on the pH õf the medium. There is a minimum around pH = 5. When changing the pH to 8, the volume increases by a factor of 7 and 3 for 5 mM and 50 mM KCl respectively. The qualitative interpretation of the results is possible within two different theories.
Within the framework of Donnan theory charged Donnan groups lead to an osmotic pressure difference in case of constant volume and to an increase of the internal volume if the pressure difference is eliminated by transport of solvent. The reason for the dependence of the volume on a permeant electrolyte is the variation of charge compensation with the concentration of electrolyte. At low concentration the charge of the Donnan groups is mainly compensated by a surplus of counter-ions. At higher concentration the deficit of co-ions becomes more important. Within the surface charge theory the charges at the inner membrane surface lead to repulsive forces between facing membrane sections. The equilibrium distance is dependent on the number of charges, given by the internal pH, and the screening of charges by the counter-ions of the electrolyte.
A quantitative description is only possible if all the charged groups and their dissociation constants are known. An approximate description is already possible with a very simple Donnan theory, in which the Donnan groups themselves do not contribute to the osmotic pressure. The charges are described by the minimum of two proton equilibria.

References
(1) M. Klingenberg and E. Pfaff(1967) in Methods Enzym.X,680
(2) Gaensslen a. R.McCarty(1972) Analyt.Biochem.48,504
(3) H.W. Heldt et al.(1973) Biochem.Biophys.Acta 314,224
(4) H.Rottenberg,T.Grunwald a.M.Avron(1972) Eur.J.Biochem.25,54

SUBSTRATE SPECIFICITY OF ASCORBATE OXIDASE AND CERULOPLASMIN

P. Wieczorek, Z. Machoy
Department of Biochemistry, Pomeranian Medical Academy
Szczecin, 70-111, Poland

The purpose of the paper was to investigate the substrate specificity of ascorbate oxidase and ceruloplasmin, as well as to determine the influence of structures of the substrates on their affinity with regard to both of the mentioned enzymes

Ceruloplasmin was obtained from "Biomed" firm and it was used without additional purification. Ascorbate oxidase was isolated from green zucchini by the method /1/ of Avigliano et al. The activities of enzymes were determined by using polarographic method /Clark's electrode//2/ while for ascorbate oxidase the spectrophotometric method /disappearance of absorption at 265 nm/ was additionally employed /3/. For investigating ascorbate oxidase 0.05 M phosphate buffer at pH 7.0 was used, as was

0.01 M acetate buffer at pH 5.5 for ceruloplasmin. The following chemical compounds were exploited as substrates of enzymes: L-ascorbate, D-xyloascorbate, D-isoascorbate, D-glucoascorbate, D-galactoascorbate, p-phenylenediamine /PPD/, N,N-dimethyl-p-phenylene-diamine /DPD/ and o-dianisidine. Maximal velocity Vmax was measured and Michaelis constants /K_m/ were determined. The results established by polarographic method for both enzymes are presented in Table 1 while the results obtained by spectrophotometric method for ascorbate oxidase are depicted in Table 2.

It has been concluded that
1. The structure of substrates exerts an effect on their affinity to ascorbate oxidase. The structural analogue D-iso claims greater affinity than natural substrate of L-ascorbate.
2. As far as the investigated substrates are concerned ceruloplasmin reveals more limited substrate specificity than ascorbate oxidase.

Table 1.   Polarographic measurements

		l-asc	d-izo	PPD	DPD	o-diani-sidin
Cerulo-plasmin	K_m /mM/	4.0	8.1	0.2	0.14	0.6
Cerulo-plasmin	$V_{max} / \dfrac{mM\ O_2}{mg \cdot min} /$	1.2	0.85	8.0	3.0	9.6
Ascorbate oxidase	K_m/mM/	0.23	0.027	0.0	0.0	0.0
Ascorbate oxidase	$V_{max} / \dfrac{mM\ O_2}{mg \cdot min} /$	3.8	0.18	0.0	0.0	0.0

Table 2.   Spectrophotometric measurements

		l-asc	d-ksylo	d-izo	d-gluko	d-galacto
Ascorbate oxidase	K_m/mM/	0.13	0.17	0.02	0.83	0.05
Ascorbate oxidase	$V_{max} / \dfrac{mM}{mg \cdot min} /$	2.2	0.17	0.14	0.41	0.09

References:
1. Avigliano, L., Gerosa, P., Rotilio, G., Finazzi Agro,
   A., Calabrese, L., Mondovi, B. /1972/ Ital.J.Biochem.
   21, 248-255
2. Capietti, G.P., Majorino, G.F., Zucchetti, M. and
   Marchesini, A. /1977/ Anal.Biochem. 83, 394-400
3. Gerwin, B., Burstein, S.R., Westley, J. /1974/
   J.Biol.Chem. 249, 2005-2008

INTERACTION OF DDT WITH INSECT MITOCHONDRIAL ADENOSINE TRIPHOSPHATASE.
II. COMPARISON BETWEEN THE INHIBITION BY DDT, OLIGOMYCIN AND DCCD.

H. M. Younis, M. E. Abdel-Ghany and N. M. Bakry
Department of Plant Protection, Faculty of Agriculture, University of
Alexandria, Alexandria, Egypt.

In a previous work (1) we have demonstrated that DDT inhibits insect
mitochondrial ATPase in a way similar to that of oligomycin and DCCD.
The three inhibitors strongly inhibit membrane bound insect ATPase
activity at concentrations that do not affect the activity of the water
soluble F_1-ATPase (1). Moreover, extraction of an active soluble ATPase
from DDT-pretreated mitochondria was possible (1). Thus the general
features of the inhibition by DDT resemble those of oligomycin and DCCD,
which indicate, by analogy, that the site of action of DDT must reside
in the membrane sector of the ATPase. Having these similarities, we have
tried in the following experiments to investigate whether DDT shares
DCCD or oligomycin in a common site or whether it does have its own
different site.

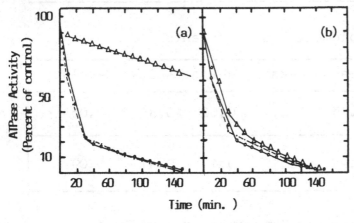

FIG. (1): Inhibition of the ATPase activity of bovine heart mitochondria in (a) & that of cockroach coxal muscle in (b) by 2 μM DDT,△——△, 2 μg oligomycin,●——●,and 2 μM DCCD,●-----●,.

In another piece of work (2) we found that DDT exhibits a degree of
selectivity towards the inhibition of the enzyme from insects. Therefore,
we compared the effects of the three inhibitors on the enzyme activity
of insect and bovine heart mitochondria. The data in Figure 1 a&b show
that the three inhibitors were equally effective against the insect
enzyme, however, DDT was much less potent against the enzyme from bovine
heart compared to the other two inhibitors. Furthermore, the effect of
DDT was characterized by a negative response to temperature, a phenomenon
which is similar to what happens during poisoning of insects by DDT,
whreas, the inhibition by DCCD or oligomycin was not affected by temperature under these conditions (Fig. 2).

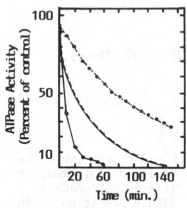

FIG. (2): Comparison between the inhibition of the insect enzyme activity by p,p'-DDT, 2 uM, at 25°C, •----•, and at 5°C, •———•, oligomycin (2 ug) and DCCD (2 uM) both at 5°C and 25°C, ══════,.

We, therefore, conclude that insect mitochondria must contain a selective site of action for DDT which is different from that of oligomycin or DCCD. Further studies have indicated that this site is an intrinsic protein component of insect mitochondrial membranes with an apparent molecular weight of 20 K dalton (3).

(Supported by U.S. National Science Foundation under the grant No: INT 7801467).

References:

1. Younis, H. M., Telford, J. N. and Koch, R. B. (1978) J. Pestic. Biochem. Physiol. 8, 271-277.
2. Younis, H. M., Abdel-Ghany, M. E. and Ibrahim, H. Z. (1983). Proc. 5th Arab Pesticide Conf. Tanta University, Egypt. Vol. II: 279-286.
3. Younis, H. M., S. A. Abou-Sada, A. H. El-Sebae and M. E. Abdel-Ghany. (1982). Reports 2nd, EBEC. L. B. T. M. CNRS editeur, Billeurbanne. pp. 53-54.

EVIDENCE FOR THE DIRECT MECHANISM BY ATP SYNTHESIS OF H^+-ATPases.

Edmund Bäuerlein
Abteilung Membranbiochemie, Max-Planck-Institut für Biochemie D-8033 Martinsried bei München, FRG.

I. Introduction

The chemical hypothesis of ATP synthesis by H^+- ATPases could be considered the less probable, because a phosphorylated intermediate appears to be excluded by two experiments using different approaches: 1) if (β,γ-^{18}O; γ -^{18}O) ATPγ S is applied in the ATPase reaction with $H_2^{17}O$, inversion of the configuration is found in the analysis of chiral inorganic (^{16}O, ^{17}O, ^{18}O) thiophosphate, indicating direct attack of water on ATP [1], 2) if tritiated propylhydroxylamine, a trapping reagent for acyl phosphates, is added to TF_1.Fo liposomes in the ATP synthase reaction of proton jump experiments, modification of the proteolipid, but not of the catalytic β-subunit, is observed. This indicates a superprotonized carboxyl group HO-$\overset{+}{C}$=OH as part of the proton transfer 2 . The chemi-osmotic and conformational hypothesis therefore remain to be discussed here.

II. What is the minimum chemistry of ATP synthesis?

At least two molecules react to form a covalence and water, ADP + Pi $\longrightarrow$ ATP + H_2O; ionisation of phosphate anions is not shown in this equation, because this implies the mechanism of ATP synthesis, as you will see below. It must be stressed, that one proton is consumed in the water formation per turnover.

III. Single turnover ATP synthesis by soluble H^+-ATPases; consequenses for the mechanism.

In single turnover synthesis of ATP, reported first on CF_1 [3] and confirmed with MF_1 and TF_1 in the presence of dimethylsulfoxide, one ADP bound per mole CF_1 is phosphorylated to ATP in the presence of high concentrations of medium phosphate at pH6.0-7.2. The crucial questions are, 1) why ATP is not released by binding of ADP and Pi at the next site, as expected by the conformational hypothesis and 2) what energy drives the synthesis of this one ATP. Energy apparently is not stored in the protein conformation or introduced by the expected binding of the substrates, because otherwise synthesis of ATP would continue at each site, ressembling a perpetuum mobile. Thus "energization" of one site, capable of ATP synthesis, may be ascribed to the presence of a catalytic proton, which is, for example, well defined by an essential tyrosyl function [4]. In contrast to mechanisms, proposed by W.S. Allison or J.H. Wang, this proton is consumed in water formation and has to be restored for neutralisation of the phenolate-ion. This may explain why single turnover ATP synthesis stops. The tightly bound ADP and medium P_i are bound at the site where the catalytic proton of the tyrosyl group is present, whereas both substrates can only enter the next site when a proton has arrived at the catalytic center.

IV. The two possibilities to restore the catalytic proton.

1) The easiest way to restore the catalytic proton appears to be binding of $H_2 PO_4^-$ to the catalytic center providing both, one proton for water formation and one for neutralization of the phenolate ion. Soluble ATPases do not seem to use the latter mechanism, otherwise single turnover ATP synthesis would not stop, but continue. 2) The catalytic proton is restored by proton tanslocation as postulated by the chemi-osmotic hypothesis. This mechamism explains why single turnover ATP synthesis stops. At the site where ADP is tightly bound, the catalytic proton is present. It is used for ATP synthesis, if phosphate is added to the medium at pH 6,0-7,2. The next site can bind ADP only if the catalytic proton is restored, thus promoting ATP release from the first site. Soluble F_1-ATPases, therefore, need a device for the directed transport of protons to the related active site. This machinery is probably the corresponding membrane part F_0. This detailed direct mechanism includes the conformational changes of the triple site mechanism, which may be called direct mechanism with rotating site.

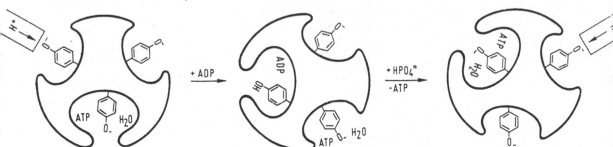

The evidence for it should be supported by experiments which confirm functional and structural linkage between proteolipid and β-subunit as well as by those experiments which may show only one undissociated phenolic OH-group of these three essential tyrosines per F_1-ATPase.

1. Webb, M.R., Grubmeyer, Ch., Penefsky, H.S. and Trentham, D.R. (1980), J. Biol. Chem. 255, 11637-11639
2. Skrzipczyk, M.J., Küchler, B. and Bäuerlein, E. (1983) FEBS Lett. 157, 343-346
3. Feldmann, I.R. and Sigman, D.S. (1982), J. Biol. Chem. 257, 1676-1683
4. Esch, F.S. and Allison, W.S. (1978) J. Biol. Chem. 253, 6100-6106
5. Wang, I.H. (1983), Ann Rev. Biophys. Bioenerg. 12, 21-34

Author Index